energy science

Constants, Symbols, and Conversion Factors

kilo (k) = 10^3; mega (M) = 10^6; giga (G) = 10^9;
tera (T) = 10^{12}; peta (P) = 10^{15}; exa (E) = 10^{18};
femto (f) = 10^{-15}; pico (p) = 10^{-12}; nano (n) = 10^{-9};
micro (μ) = 10^{-6}; milli (m) = 10^{-3}; centi (c) = 10^{-2};
deci (d) = 10^{-1}; 1 billion = 1000 million

1 kilowatt-hour (kWh) = 3.6 MJ

1 British thermal unit (Btu) $\approx$ 1.055 kJ

1 therm = 10^5 Btu $\approx$ 105.5 MJ

1 quad = 10^{15} Btu $\approx$ 1.055 EJ

1 EJ $\approx$ 278 TWh

1 barrel of oil equivalent (boe) $\approx$ 6 GJ

1 tonne of oil equivalent (toe) = 41.868 GJ
    = 11.63 MWh

1 calorie (cal) $\approx$ 4.18 J

1 Calorie (food) $\approx$ 4.18 kJ

1 megaelectron volt (MeV) = 10^6 eV
    $\approx 1.602 \times 10^{-13}$ J

1 horsepower (hp) $\approx$ 0.746 kilowatts (kW)

1 TWh per year $\approx$ 114 MW

1 metric tonne (t) = 1000 kg $\approx$ 2205 lb

1 US ton = 2000 lb

1 imperial ton = 2240 lb

1 mile $\approx$ 1.609 km

1 mph $\approx$ 0.447 m s^{-1}

0 °C = 32 °F $\approx$ 273 K

100 °C = 212 °F $\approx$ 373 K

1 barn (b) = 10^{-28} m^2

1 hectare (ha) = 10^4 m^2 $\approx$ 2.47 acres

1 acre $\approx$ 0.405 ha

1 square kilometre = 10^6 m^2 = 100 ha
    $\approx$ 0.386 square miles

1 US ton per acre $\approx$ 2.24 t ha^{-1}

1 MW per square kilometre = 10 kW ha^{-1}

1 kW m^{-2} = 100 mW cm^{-2}

1 m^3 = 1000 litres = 1000 dm^3 $\approx$ 35.3 ft^3

1 US gallon $\approx$ 3.79 litres

1 imperial gallon $\approx$ 1.20 US gallons $\approx$ 4.55 litres

1 barrel of oil = 42 US gallons of oil

1 mile per gallon (US) $\approx$ 0.425 km litre^{-1}

1 pascal (Pa) = 1 N m^{-2}

1 bar = 10^5 Pa

1 atmosphere (atm) $\approx$ 1.013 bar

1 US dollar (\$) $\approx$ 0.8 UK pound (£) $\approx$ 0.9 euro (€) (2019)

Densities:

air $\approx$ 1.2 kg m^{-3} (20 °C); water $\approx$ 1000 kg m^{-3};

natural gas $\approx$ 0.8 kg m^{-3}

petrol (gasoline) $\approx$ 730 kg m^{-3};

ethanol $\approx$ 790 kg m^{-3}; diesel $\approx$ 840 kg m^{-3};

biodiesel $\approx$ 880 kg m^{-3}

Energy densities (LHV):

carbohydrates $\approx$ 15 MJ kg^{-1}; ethanol $\approx$ 27 MJ kg^{-1};

biodiesel $\approx$ 38 MJ kg^{-1}; diesel $\approx$ 43 MJ kg^{-1};

petrol (gasoline) $\approx$ 43.5 MJ kg^{-1};

coal (bituminous) $\approx$ 27 MJ kg^{-1};

natural gas $\approx$ 34.6 MJ m^{-3}

Avogadro number $N_A \approx 6.022 \times 10^{23}$ mol^{-1}

Gas constant $R \approx 8.314$ J mol^{-1} K^{-1}

Boltzmann constant $k \approx 1.381 \times 10^{-23}$ J K^{-1}

Stefan–Boltzmann constant
    $\sigma \approx 5.67 \times 10^{-8}$ J K^{-4} m^{-2} s^{-1}

speed of light $c \approx 2.998 \times 10^8$ m s^{-1}

Planck's constant $h \approx 6.626 \times 10^{-34}$ J s

$hc \approx 1240$ eV nm

magnitude of electron charge $e \approx 1.602 \times 10^{-19}$ C

$e^2/4\pi\varepsilon_0 \approx 1.44$ MeV fm

Faraday constant $\approx$ 96 485 C mol^{-1}

gravitational constant $G \approx 6.674 \times 10^{-11}$ m^3 kg^{-1}s^{-2}

acceleration due to gravity $g \approx 9.81$ m s^{-2}

radius of Earth $\approx$ 6378 km

Solar constant $\approx$ 1.37 kW m^{-2}

mass of electron $m_e \approx 9.109 \times 10^{-31}$ kg

mass of proton $m_p \approx 1.673 \times 10^{-27}$ kg

mass of neutron $m_n \approx 1.675 \times 10^{-27}$ kg

permittivity of vacuum $\varepsilon_0 \approx 8.854 \times 10^{-12}$ C V^{-1} m^{-1}

permeability of vacuum $\mu_0 = 4\pi \times 10^{-7}$ V s^2 C^{-1} m^{-1}

energy science

principles, technologies, and impacts

fourth edition

John Andrews and Nick Jelley

OXFORD
UNIVERSITY PRESS

Great Clarendon Street, Oxford, OX2 6DP,
United Kingdom

Oxford University Press is a department of the University of Oxford.
It furthers the University's objective of excellence in research, scholarship,
and education by publishing worldwide. Oxford is a registered trade mark of
Oxford University Press in the UK and in certain other countries

First edition 2007
Second edition 2013
Third edition 2017

Impression: 3

Published in the United States of America by Oxford University Press
198 Madison Avenue, New York, NY 10016, United States of America

British Library Cataloguing in Publication Data
Data available

Library of Congress Control Number: 2021937402

ISBN 978–0–19–885440–1

Printed in Great Britain by
Bell & Bain Ltd., Glasgow

Preface

Harnessing the Earth's energy resources has been a source of technical inspiration since ancient times. Energy devices have transformed civilization beyond the wildest imagination of our predecessors. But our reliance on fossil fuels is now threatening dangerous climate change through the cumulative emissions of carbon dioxide from their combustion, and putting many millions of lives at risk. Fortunately, energy generated from solar, wind, and hydropower power does not have these damaging consequences. And in many parts of the world, solar and wind generators are becoming the cheapest source of power. But can we handle the variability of these supplies, and can renewables satisfy the global demand for energy? Can we make the transition away from fossil fuels quickly enough? Will nuclear power or capturing carbon dioxide from the atmosphere play a significant role? What are the options?

These questions have to be answered. What our present civilization does in the next few decades will have a profound effect on the lives of future generations and the state of the planet. One approach is to assume that market forces and governments will sort it all out, but the energy field is not short of uninformed or politically motivated opinions and commercial interests. This book is for those who prefer to make up their own minds, through an understanding of the science and issues involved, and the impact of the various technologies on society and the environment.

The idea for writing this book originated from undergraduate lecture courses given by the authors in Bristol and Oxford. The main focus of the book is to explain the physical principles underlying each technology and to discuss each of the technologies and their environmental, economic, and social impacts. It describes all the key areas of energy science, covering fossil fuels, renewables, and nuclear energy, and the effect of the dramatic fall in cost of solar power and wind power. It describes what climate change is being caused by global warming, why immediate action is required, and the challenges involved.

Energy is a broad subject that crosses the boundaries between the traditional scientific and engineering disciplines. It is not essential to have a background in any particular discipline in order to use this book, apart from a general knowledge of science and mathematics to about high school standard. The aim is to enable students, professionals, and lay-readers to make quantitative estimates and form sound judgements, and to be aware of the vital importance of decarbonizing our energy supply by 2050. The material is presented in such a way that it can be understood on different levels. All the important results are described qualitatively using, where necessary, straightforward mathematical methods with numerical **Examples**. More difficult mathematical **Derivations** and items of supplementary information are contained in **Boxes**. These boxes and derivations can be bypassed by those who do not wish to consider such detail. Students are encouraged to work through the **Exercises** at the end of each chapter. The exercises are designed to be informative and thought-provoking; and for those who want to gain a deeper understanding of some of the more difficult points, some starred exercises have been added in each chapter.

Chapter 1 presents a brief history of energy technology, together with a review of the present global energy demand, the evidence for global warming, the need for decarbonization, and the key characteristics of energy sources. Chapter 2 describes the essentials of thermal physics and its application to the greenhouse effect, the efficiency of thermal engines, and in understanding chemical equilibria, and also gives a brief account of the fluid mechanics used in explaining the exploitation of fluid-based devices. Chapter 3 looks at the global use of fossil fuels, their combustion in thermal power plants, and the outlook for carbon capture and storage of emissions from fossil-fuel powered processes. Chapter 4 describes the use of biomass in both the developing and developed world, and Chapter 5 discusses solar thermal energy, heat pumps, and the extraction of geothermal power.

Chapter 6 looks at hydropower, tidal power, and wave power, and Chapter 7 wind power, both onshore and offshore. Chapter 8 describes photovoltaic devices and the impact of their falling cost. Chapter 9 looks at nuclear energy, in particular fission reactors, and discusses the potential of fusion. Chapter 10 describes the principles of electricity generation and transmission, the handling of the increasing amount of variable renewable generation, and electrical energy storage in batteries or in stored gases (e.g. power-to-hydrogen). Chapter 11 looks at the energy demand in buildings, industry, and transport. Finally, Chapter 12 highlights the central issues concerning the impact of energy on society, with particular attention to the importance of transitioning from fossil fuels to renewables by 2050 to avoid risking dangerous climate change.

All the chapters are self-contained and can be read in any order, and Chapter 2 can be used just as a reference.

New to this edition

A significant amount of new material has been added on our understanding of climate change, the need to limit the global cumulative emissions of CO_2, the challenges in reducing our reliance on fossil fuels, and the global actions needed to decarbonize our energy supply. Other additions include: the use of heat pumps and of power-to-hydrogen for decarbonizing heating, the limitations of biomass, the importance of developments in offshore wind, solar PV, and battery technology, and the impact of distributed generation and demand response on integrating renewables. The fourth edition contains more **Case studies** of energy projects taken from around the world of general interest, and includes more **Exercises** for students to test their understanding of the subject.

Online resources

 The book is accompanied by online resources at www.oup.com/he/andrews_jelley4e, which feature additional materials for both students and lecturers.

For students:

- **Multiple-choice** questions to check your understanding as you progress through the text.
- Weblinks for each chapter pointing you towards useful external sites.

For lecturers:

- **Figures** and **tables** from the book in electronic format.
- Full **solutions** to end-of-chapter exercises.

Acknowledgements

The authors would like to express their appreciation to their families, to their editors, Jonathan Crowe, Dewi Jackson, Philippa Hendry, Kathryn Rylance, Sophie Ladden, and Lucy Wells, and to their friends and colleagues for numerous ideas, criticisms, and helpful suggestions, in particular to David Andrews, Katherine Blundell, Niel Bowerman, Gerard van Bussel, David Cherns, Peter Cook, Steve Cowley, Conyers Davis, Leon di Marco, George Doucas, Nigel Dowrick, Mahieddine Emziane, Nick Eyre, Kieran Finan, Godfrey Gardner, Chris Goodall, Wina Graus, Nick Green, John Hannay, Julian Hasler, David Howey, Oliver Inderwildi, William Ingram, Tegid Jones, Bruce Levell, Chris Llewellyn-Smith, James Loach, Mike Mason, Malcolm McCulloch, Helen O'Keeffe, Robert Orzanna, Robert Paynter, Bruce Pilsworth, Samuel Poncé, John Pye, Rachel Quarrel, Moritz Riede, Alex Schekochihin, Margaret Stevens, Robert Taylor, Andrew Tindal, Marina Topouzi, Peter Wakefield, Justin Wark, Mike Watson.

Contents

1 An Introduction to *Energy Science*

✔ **List of Topics**

- ☐ History of energy technology
- ☐ Global energy usage
- ☐ Global warming and climate change
- ☐ Why did the ice ages occur?
- ☐ Implications for society

- ☐ Decarbonization
- ☐ Sustainable development
- ☐ Characteristics of fossil fuel and renewable energy sources
- ☐ Energy units and concepts

➔ **Introduction**

The Earth's ecosystem is ultimately powered by sunlight and for millions of years animals and plants coexisted in a continuous cycle of energy exchanges. However, when primitive humankind discovered how to make fire, this process began to shift irreversibly. With fire, it was possible to cook meat, deter predators, and fashion metals into tools and lethal weapons. In the last two centuries, humankind has discovered how to convert heat into electricity, the most versatile and convenient form of energy. Electricity has enabled astonishing advances to be made in science and engineering, transforming civilization and making life far more comfortable than that of our predecessors. However, in the process it has created a consumer society that treats electricity and other forms of energy as commodities that should be available on demand.

In the relatively short period since the start of the Industrial Revolution (in the second half of the eighteenth century), a sizeable fraction of the fossil reserves of the planet that took hundreds of millions of years to evolve have been significantly depleted—in every year we consume what took about a million years to lay down. The emissions of carbon dioxide (CO_2) and other products of combustion are now having a noticeable impact on the global climate.

The threat to life in the next few decades due to these emissions could be dire unless humankind can rise to the greatest challenge it has faced since its emergence as the dominant species on Earth. An effective strategy to combat global warming needs a drastic reduction in our consumption of fossil fuels, and the rapid development of alternative technologies to replace the fossil fuel industry.

Energy conversion is a disparate subject, but it is possible to obtain a good understanding of the essentials by applying basic physical principles. At the same time, it is important to appreciate the economic, social, and environmental implications involved in the various forms of energy conversion. Energy issues tend to be open-ended and controversial. Addressing them with an independent mind is a rewarding and intellectually stimulating exercise. It is important to be objective, to pay attention to fact rather than opinion, to challenge assumptions, and always to look for constructive solutions.

1.1 **A Brief History of Energy Technology**

Throughout history the harnessing of energy in its various forms has presented great intellectual challenges and has stimulated much scientific discovery. The energy technologies of today are the result of advances in scientific understanding, technical inspiration, and gradual improvements in engineering design over many centuries. By the time of the Roman Empire, water engineering was already a well-established technology. Thousands of years earlier, irrigation systems had greatly enlarged the area of farming land around the River Nile and increased the prosperity of ancient Egypt. An illustration of the ingenuity of the early engineers is demonstrated by **Archimedes' screw** (Fig. 1.1), a device used to extract water from rivers, empty grain from the holds of ships, and clear water from flooded mines (e.g. Rio Tinto in Spain). Existing records show that the device was used to irrigate the Hanging Gardens of Babylon (before the birth of Archimedes!), and it is still widely used today by the water and chemical industries.

In order to raise water, Archimedes' screw is encased in a hollow cylinder. One end is then immersed in water, and the water trapped in the hollows of the screw is transported upwards by rotating the device about its axis (see Exercises 1.6 and 1.35).

Waterwheels (see Chapter 6 Section 6.1) existed in the ancient world and by 1000 AD were common throughout Western Europe. Early waterwheels were very inefficient but designs gradually improved over the centuries (see Fig. 6.1). A technological breakthrough was made in 1827 by the French engineer Benoît Fourneyron with the invention of a turbine, which used fixed guide vanes to direct water between the blades of a rotating runner. The design of the vanes and the blades enabled most of the kinetic energy of the incident flow to be captured. The **Fourneyron turbine** (see Section 6.1) drove the development of modern turbines, leading to the emergence of hydropower as one of the major providers of electricity today.

Windpower is another old technology. Windmills existed in Persia in the 10th century (see Chapter 7 Section 7.1), and sailing vessels existed on the Nile 5000 years ago. Early sailing

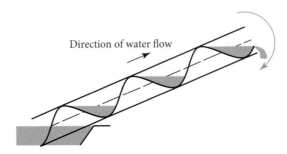

Direction of water flow

Fig. 1.1 Archimedes' screw.

Fig. 1.2 Lateen sails (© iStock.com/duncan1890).

ships had a single mast with a square sail, and were propelled along by their sail when there was a following wind. But these ships could also travel obliquely into the wind by angling their sail to just catch the wind. The sail curved, and the wind created a pressure difference across it that pushed the sail in a direction perpendicular to that of the wind—just as the air flowing over an aircraft wing gives lift. Early Mediterranean galleys, such as the Roman trireme, combined oars and sails, as did the Viking ships; but most merchant vessels were sailing ships. Triangular sails, called lateen sails that may have evolved as a simpler and cheaper alternative to square rigged sails, became well established by the fifth century, and could give ships greater manoeuvrability and enable them to sail more closely into the wind. (Fig. 1.2).

The earliest recorded **steam engine** was a toy device, invented by **Hero of Alexandria**, in the first century AD (Fig. 1.3). It essentially comprised a hollow metal sphere filled with steam, supported by two pivots. The steam is allowed to escape through two bent spouts, and the momentum of the steam jets produces a reaction in the opposite direction on the spouts, which causes the sphere to rotate. The Industrial Revolution was made possible by the emergence of steam power. However, before the first commercial steam engines appeared on the scene there were serious misunderstandings about the nature of vacuum and air pressure that needed to be resolved. In 1644, **Torricelli** (a follower of Galileo) invented the mercury barometer. He proved that the rise of the column of mercury was due to the difference in pressure

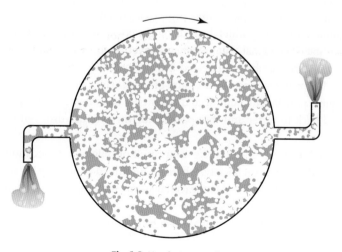

Fig. 1.3 Hero's steam engine.

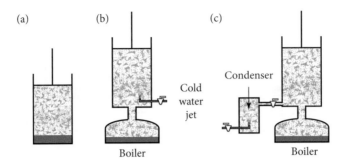

Fig. 1.4 Evolution of design of early steam engines: (a) Papin; (b) Newcomen; (c) Watt.

between the atmosphere and the partial vacuum above the mercury inside the column. The next major breakthrough came about through the invention of the piston air pump, by **von Guericke** in 1650. In one spectacular demonstration, he took two identical metal hemispheres (known as the 'Magdeburg hemispheres', after the town where von Guericke was burgomaster and a military engineer during the Thirty Years' War) and placed them together to form a complete sphere. The hemispheres were in touching contact along a flange, but there was no method of fixing. Von Guericke pumped out the air inside the sphere and invited two teams of eight horses to pull the hemispheres apart; they were unsuccessful, because the force exerted by the air was too great (see Exercise 1.4).

In 1666 the Académie des Sciences was established in Paris, and the Dutch scientist, **Christiaan Huygens**, was one of the founding members. Huygens appointed two young assistants, **Gottfried Leibniz** and **Denis Papin**, who modified a von Guericke air pump to conduct experiments with gunpowder. A small charge of gunpowder was exploded inside a cylindrical chamber containing a tightly fitting piston. The air was expelled through two leather valves, thereby creating a partial vacuum. The piston then collapsed because of the imbalance in air pressure across it, and useful work was done in the process. Huygens realized the enormous potential of this discovery, from new forms of transport to powerful engines that could revolutionize industry.

Not surprisingly, gunpowder explosions proved to be too difficult to control! Papin, with help from Leibniz, tried using water instead of gunpowder, exploiting the fact that water expands to 160 times its original volume when converted to steam. Papin placed a quantity of water inside a piston chamber and heated it from the outside by a flame. The pressure of the steam raised the piston against the pressure of the air (Fig. 1.4 (a)). In order to return the piston to its original position, he poured cold water over the outside of the cylinder so that the steam condensed back to water and the space inside the cylinder dropped to below atmospheric pressure.

Papin was forced to flee France in 1685 as a result of religious persecution. He settled in London and worked at the Royal Society, where he continued to develop his **steam engine**. In 1690 he applied for a patent to build commercial steam engines, but the patent was awarded instead to **Thomas Savery**, a military engineer, who had been strongly influenced by Papin's ideas. Papin's friendship with Leibniz meant that he did not get the support of **Isaac Newton**, who was locked in a bitter dispute with Leibniz over which of them was the inventor of calculus. Papin died a pauper in 1712.

Savery's steam engines consumed huge amounts of coal and proved to be uneconomic. The first successful steam engine was built in 1712 by **Thomas Newcomen**, a blacksmith. Newcomen had discovered by accident that the steam inside a Savery steam engine condensed suddenly after some cold water had leaked into the steam chamber. Newcomen exploited this effect by installing a pipe to squirt a jet of cold water directly into the steam chamber (Fig. 1.4 (b)). The Newcomen steam engine was a large structure with a long horizontal beam that rocked to and fro, with the rise and fall of the piston. Although it could only perform about five or six strokes a minute, it was capable of lifting large volumes of water from flooded mines. In the early Newcomen engines, a boy attendant was employed to open and shut two taps, one that allowed steam into the cylinder and the other that turned on a jet of cold water to condense the steam. This was a very monotonous job but, one day, a young lad called Humphrey Potter had a bright idea. He wanted to play with his friends and decided to connect a cord from each of the taps to the beam, so that the taps opened and closed at just the right moments in the cycle. It worked, and his invention was incorporated into all Newcomen steam engines.

A much more efficient steam engine was patented by **James Watt** in 1769. Watt was an instrument-maker and was working at Glasgow University as an assistant to Professor Black, who had discovered the latent heat of steam (see Section 2.1). Watt took a keen interest in steam engines and the properties of steam. When asked one day to repair a malfunctioning model of a Newcomen steam engine, he calculated that about 80% of the heat was lost in heating the walls of the steam cylinder. Watt deduced that it would be much better to condense the steam in a separate chamber (known as the **condenser**), so that the temperature of the walls in the piston chamber could be maintained and thereby conserve heat (Fig. 1.4 (c)).

Watt needed money to exploit his idea and formed a partnership with a wealthy iron foundry owner, **Matthew Boulton**. This partnership gave Watt the finances needed to develop a commercial steam engine, but he soon ran into a major technical hitch: the cylinder castings were distorted and allowed too much steam to escape. Fortunately for Watt, a breakthrough in the manufacture of cannons provided him with the solution he needed. Cannons were constructed as thick-walled cylindrical tubes, but irregularities in the casting process meant that cannonballs often missed their target. In 1775 John 'Iron Mad' Wilkinson produced a solid cast iron block, from which he bored a smooth cylindrical hole of exactly the right shape and size. This improved the ballistics of cannonballs and enabled Watt to build leak-tight steam engines. The first Boulton and Watt steam engines were sold in 1776, and by 1824 they had produced 1164 machines.

Surprisingly, perhaps, it was not until the middle of the nineteenth century that the concept that heat is a form of energy was finally accepted. Heat was originally thought to be a fluid, known as **caloric**, that was deemed to flow from hot bodies to cold bodies and could neither be created nor destroyed. The caloric theory was a remnant of the science of the ancient Greeks, who believed that all matter consisted of four basic substances: air, fire, earth, and water. The caloric theory was shown to be erroneous by **Benjamin Rumford**, an American scientist who had worked as a spy for the British during the American War of Independence. Rumford fled to Europe, where he married the widow of Antoine Lavoisier, one of the joint discovers of oxygen (and who was guillotined during the French Revolution in 1794). Rumford found occupation in Bavaria, where he improved the manufacture of cannons and was made a Count of the Holy Roman Empire. Cannons were bored under water, and the boring process made the water boil. Rumford observed that the water boiled for only as long as the boring process was continued. He deduced that caloric was apparently being produced by friction, in contradiction to the belief that it was uncreatable. Later, in the 1840s, an amateur scientist, **James Joule**, showed that heat and mechanical energy are equivalent, and that **energy is conserved** in the process.

In 1824, a young French scientist, **Nicholas Carnot**, proved that the maximum possible efficiency of an ideal heat engine depends only on the values of the hot and cold temperatures between which it operates. Carnot proved that the result was independent of the nature of the working fluid, but his explanation assumed the validity of the caloric theory: 'The motive power of heat is independent of the agents employed to realize it; its quantity is fixed solely by the temperatures of the bodies between which is effected, finally, the transfer of caloric' (see Exercise 1.7).

Steam power continued to advance through the nineteenth century with the development of steam trains and ships, powered by reciprocating steam engines. In 1884, **Charles Parsons** invented the rotary steam turbine. His great innovation was to realize that the power in a high-pressure jet of steam could be exploited more efficiently if the pressure was dropped in stages across a set of turbine blades, rather than in a single step, resulting in a very compact and powerful engine. In order to demonstrate the superiority of his invention to a sceptical British Admiralty, he fitted his vessel *Turbinia* with three rotary steam turbines. He had the audacity to appear with *Turbinia*, uninvited, at the display of Her Majesty's fleet for Queen Victoria's Diamond Jubilee at Spithead. He was chased by the Navy's fastest patrol boats, but they were no match for *Turbinia*. Parsons had made his point, and the rotary steam turbine was accepted.

At the same time as the development of steam power, great advances were being made in understanding the nature of electricity and magnetism, which resulted in the invention of a profusion of devices which made the use of electricity a practical reality and enables us to perform tasks today that would have been unimaginable in earlier centuries. One of the first of these was the **electric battery**, invented by **Alessandro Volta** in 1799. This was an electrochemical cell consisting of two electrodes—one copper and the other zinc—separated by an electrolyte containing sulphuric acid, and an electric current flowed when the circuit was completed by a wire. The original purpose of the device was to disprove a theory advocated by **Luigi Galvani**, who had discovered that a dead frog's leg could be made to twitch when the two ends were touched by different metals. Galvani claimed that this was evidence of 'animal electricity' and thought it proved that the secret of life was electricity. Although Volta proved that the effect was actually due to the conduction of electric current through the frog's leg, Galvani's idea had excited the popular imagination, even inspiring an 18-year-old Mary Shelley to write a gothic novel about the Frankenstein monster, a hapless creature brought to life by passing a large electric current through a dead body!

In the nineteenth century, **Michael Faraday**, a self-taught scientist, made great discoveries in electromagnetism. Faraday had begun his career at the age of 12 as a bookbinder. This gave him the opportunity to read learned books on science, and he was eventually appointed by **Sir Humphry Davy** as an Assistant at the Royal Institution, in London. During the 1820s, Faraday was intrigued by two recent discoveries: one by **Oersted**, that the needle of a compass

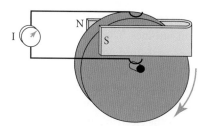

Fig. 1.5 Faraday's rotating disc.

is deflected at right angles to the direction of flow of an electric current in a wire, and the other by **Ampère**, that two current-carrying wires exert a force on each other. In 1831, Faraday published his **laws of electromagnetic induction**, based on his own discoveries that a current is set up in a closed circuit by a changing magnetic field and also in a loop of wire when moved through a stationary magnetic field.

Faraday also showed that a steady current is induced across a rotating copper disc between the poles of a strong magnet (Fig. 1.5). This result led to the invention of the **electric motor** by **Moritz von Jacobi** and of the **dynamo**, which helped the introduction of electric lighting. The early dynamos produced very spiky outputs; the first device to produce a smooth current was the **Gramme** dynamo, using a continuous loop of wire wrapped around a rotating iron ring. Electric telegraphy and electroplating were two of the first useful applications of electricity, followed by arc-lighting for public service. The first patent for an incandescent lamp was awarded to the American inventor and entrepreneur, **Thomas Edison**, in 1879. He used a loop of carbonized cotton thread that glowed in a vacuum for over 40 hours. Earlier, in 1860, **Joseph Swan** in England had patented the world's first **light bulb**, but it had a short lifetime and was inefficient. During the 1870s, Swan improved the vacuum inside his bulbs and formed the Swan Electric Light Company in 1881. The Edison and Swan companies were rivals in the development of the incandescent lamp, and eventually merged in England as the Edi-Swan Company.

In order to capitalize on his invention of the incandescent lamp, Edison patented his electric distribution system in 1880. In 1881, he built the world's first **power station** at Holborn Viaduct in London, which produced 160 kW of power for lighting and electric motors. In the following year he built a similar power station in New York City, to provide electric lighting for Wall Street and the banking community. There was intense rivalry between Thomas Edison's direct current system and a system using alternating current, pioneered by **George Westinghouse** and by **Nikola Tesla** who invented the AC induction motor that electrified power generation in industry. Edison staged public events to highlight the dangers of alternating current, with live electrocutions of dogs, cats, and even an elephant! However, the alternating current system became the one to be generally adopted worldwide.

The latter half of the nineteenth century also saw the invention of the first practical 4-stroke spark ignition **internal combustion engine** by **Nikolaus Otto** in 1876, followed by the first modern car built by **Karl Benz** in 1886, and the first controlled powered flight in 1903 by the **Wright brothers**. The period from about 1870 to 1914 has been called the **second industrial revolution**, which saw the expansion of electricity, automobiles, steel, the telegraph, and the introduction of mass production.

The first large-scale **hydroelectric power station** was built in 1895 on the US side of Niagara Falls, using Fourneyron turbines. The first half of the twentieth century witnessed a massive construction program of coal-fired power stations and hydroelectric plants, and the construction of large-scale grids to transmit power from power plants to population centres. The importance of electric power was recognized by twentieth-century politicians as essential to national prosperity and economic growth (in the old Soviet Union, Lenin even defined communism as 'Soviet power plus the electrification of the country'), and by the early 1990s, over 60% of all homes around the world were receiving electric power.

The second half of the twentieth century saw the development of **nuclear power**. The idea of producing electricity from nuclear fission reactors was an afterthought of the Manhattan Project, a secret military enterprise by the Allied powers in the Second World War to build an atomic bomb. The early fission reactors were used for producing materials for nuclear weapons. Reactors solely for electricity generation did not appear until the latter half of the 1950s. The early years of nuclear power were heralded as a new era of cheap, inexhaustible, and safe electricity. In the 1970s, the developed world became more dependent on nuclear power after a series of large jumps in oil prices following the Arab–Israeli War in 1973. The most striking example was in France, where the government decided in the interests of national energy security to commit the country to nuclear power as the principal means of generating electricity; ~75% of French electricity is currently generated from nuclear power.

A number of high-profile accidents at various nuclear plants around the world, and growing public concern about the disposal of nuclear waste, eroded public confidence in nuclear power during the late 1970s and 1980s. The worst incidents were a partial meltdown of a pressurized water reactor at **Three Mile Island** (USA) in 1979, and a complete meltdown of an RBMK reactor at **Chernobyl** (Ukraine) in 1986. Human error and design faults were found to be significant factors in both cases. As a result there was a general improvement in nuclear safety standards worldwide and greater international support for the effective regulation of civil nuclear installations. However, the impact of these accidents considerably slowed the building of new reactors. The more recent accident in 2011 in **Fukushima**, Japan, which was the result of a series of tsunamis following a massive undersea earthquake that measured 9.0 on the Moment scale, set back the deployment of nuclear power in some OECD countries, but in other countries, e.g. China, it is still expected to play a significant role in combating global warming.

Alternative energy was a neglected area until the **oil price shocks of the 1970s**. Western governments then began to sponsor research programmes into various alternative energy technologies with the aim of reducing their dependence on oil. Funding was, however, on a much smaller scale than that for nuclear power. **Wind power** was the first alternative energy technology to become commercially viable, benefiting from low capital costs and tax breaks, and from knowledge of blade design in the aircraft industry. Small wind turbines were already widely established in the USA by the 1930s, for pumping water for irrigation and generating electricity, but the development of large wind power turbines in the 1970s and 1980s waxed and waned with the price of oil. Since ~2000, however, wind power has grown rapidly, and by 2019 it was generating ~5% of the world's electricity.

Wave power also attracted a lot of interest in the 1970s, but it was soon recognized that the capital costs were high and that most devices were unable to withstand severe storms at sea. Nonetheless, a few devices are still under development, particularly ones which are submerged and tethered to the sea floor. There was also renewed interest in **tidal power** and **geothermal power**.

The **photovoltaic effect** was discovered in 1837 by Edmund Becquerel, but it was not until 1883 that a photocell was produced and then only with a very low efficiency. Efficient cells followed the accidental discovery of a silicon photovoltaic device by Russell Ohl in 1940, and in 1954 a 6% efficient **silicon photocell** was produced in the Bell Laboratories. However, their high cost meant that they were restricted to niche applications; for instance, in computing and the space programme. Over the last few decades, mass-production methods have reduced costs by an enormous amount, and now solar energy farms are commercially viable and expected to make a major contribution globally. See Table 1.1 for relative power scales.

On a different front, the growth of the **biomass** industry towards the end of the twentieth century—the generation of electricity and production of biofuels from organic matter—has slowed in the last decade with concern over carbon dioxide emissions and over conflicts with food production. But it is an important source of energy in the developing world, and its attraction is that it is potentially low carbon, in that the carbon dioxide produced by burning the material is reabsorbed by new crops, in a continuous cycle.

At the same time as renewable generation has been developing, a number of energy related devices have emerged which are already impacting on the energy scene in the twenty-first century. One of these is the **rechargeable lithium-ion battery**, developed in the 1980s and 1990s, and used in portable electronic devices such as laptops and mobile phones. For their pioneering work on lithium batteries, Stanley Whittingham, John Goodenough, and Akira Yoshino were awarded the 2019 Nobel Prize in Chemistry.

Table 1.1 Power scales

Device	Power (kW)
Treadwheel (AD 0)	0.2
Tour de France cyclist (uphill)	0.5
Strong horse	0.7
Newcomen steam engine (1712)	4
Fourneyron water turbine (1832)	37
Parsons' steam turbine (1900)	10^3
Smith–Putnam wind turbine (1942)	1.3×10^3
Sizewell B nuclear power station (1992)	1.2×10^6
Drax coal power station (1986)	3.9×10^6
Large solar PV farm (2019)	0.5×10^6
Large wind farm (2019)	0.5×10^6

The lithium-ion battery also has a massive potential for the **electric vehicles** of the future, and could provide energy storage for **smart grid systems**. Other devices are the **insulated-gate bipolar transistor**, which is fast becoming the high-power semiconductor device of choice for switching in modern appliances, electric cars and trains, and grid systems, and **LEDs**, which have transformed lighting globally by being very efficient and long-lasting. The use of **hydrogen**, produced by electrolysis using renewably generated electricity, for providing clean heat and powering fuel cells for heavy duty vehicles and ships, as well as for energy storage, is also taking off.

To complete this brief historical overview, Table 1.1 compares the power scales involved in a small selection of energy-related devices, from antiquity to the present day. It is a tribute to the achievement of humankind in applying scientific knowledge for the benefit of society, but it also demonstrates that modern civilization has become dependent on consuming vast amounts of energy in order to maintain a comfortable lifestyle.

1.2 Global Energy Usage

We use energy all the time in our daily life: when we make a call on our mobile, boil a kettle, or drive a car. Energy is vital for a good quality of life: in providing warmth (and increasingly air-conditioning), in producing food, and in powering technology; and in the last two hundred years, we have increasingly relied on fossil fuels, as Fig.1.6 shows.

During this time, the world's population has also grown enormously: 1805 1 billion; 1928 2 billion; 1975 4 billion; 1999 6 billion; 2019 7.7 billion. The average person's wealth, as measured by the world's total gross domestic product (GDP) per capita, has also increased, approximately tripling in the last 50 years. This represents a significant improvement in the standard of living, which can be measured using the **Human Development Index (HDI)**, a quantity that combines indicators of educational attainment, life expectancy, and income.

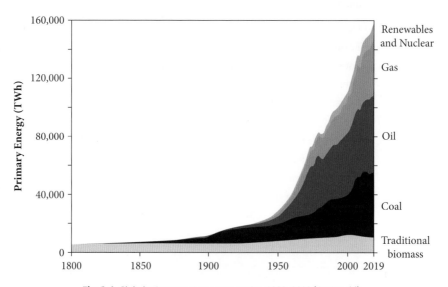

Fig. 1.6 Global primary energy consumption 1800–2019 (OurWorld).

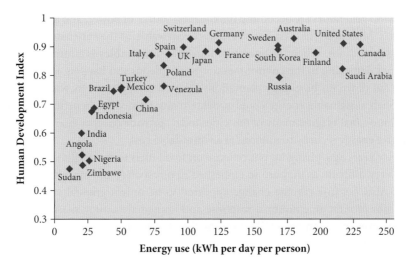

Fig. 1.7 The dependence of the Human Development Index (HDI) on the primary energy use in 2013 (UNDP, WB). The energy use is the average power consumption in a country, which is mainly in buildings, industry, and transport, divided by its population.

While it does not include any measure of inequality or of natural capital, it does emphasize that development is not just about economic growth.

A good correlation is found for low values of the HDI of a country with its energy use, as shown in Fig. 1.7. However, a large spread in energy consumption per person is seen between different highly developed countries. Although some of this reflects differences in climate, considerable scope exists for reducing consumption by improvements in efficiency and changes in lifestyle.

Less developed countries will seek to improve their standard of living by increasing their average energy use. In particular, just under one billion people (13% of the global population) were still without electricity in 2018. Even a small amount of electricity per person can enable access to mobile phones, computers, the web, lighting, TV, and refrigeration. The cost of energy is an important factor in the economy of a society, and for many decades, until recently, this meant that fossil fuels were often the preferred choice.

However, burning coal, oil, and natural gas to supply energy, pumps huge amounts of carbon dioxide (CO_2) into the atmosphere, and also produces harmful pollutants that damage our health and the environment. It would be easy to carry on as we are, and there are enough deposits of fossil fuels to last for several hundred years. But the level of carbon dioxide in the atmosphere, which has already started to seriously disrupt our climate, would cause dangerous climate change before the end of this century because of global warming, and put many millions of lives at risk. And right now, the air pollution is already causing seven million premature deaths every year.

1.2.1 Clean Energy: Renewables

Fortunately, some of the energy we use does not have these damaging consequences, in particular energy generated by solar, wind, and hydro power. And in many parts of the world, solar and wind generators are becoming the cheapest source of power, and a viable alternative to fossil fuels. Moreover, these energy sources are renewable, as they are naturally replenished within days to decades. When the energy produced by **renewables** is affordable, and its generation is not damaging to the environment or to people (as can happen, for example, when forests are cut down for bioenergy plantations), then the supply of renewable energy is **sustainable**. Such sustainable energy could provide many immediate benefits. It could reduce the air pollution that plagues many of our cities today, provide cheaper energy and many new jobs, and give energy security to millions at an affordable cost.

Only a few years ago, giving up our reliance on fossil fuels to tackle global warming would have been very difficult, as they are so enmeshed in our society and any alternative was very expensive. Also, because climate change appeared to be only a gradual and distant threat, which did not prompt an emotional response and immediate action, many individuals and governments were reluctant to act. But now the threat is much closer, and moving away from fossil fuels to renewables has become essential.

1.2.2 **Present Energy Demand**

The annual global demand for energy is an enormous number of kWh, and can be more easily grasped in terms of terawatt hours (TWh), where a TWh equals a thousand million (a billion) kWh. The global demand in 1800 was about 6000 TWh for around 1 billion people; and was nearly all provided by traditional biomass. In 2018, it was 25 times more (157,000 TWh) for 7.6 billion people. Figure 1.6 shows the contribution to the global total **primary energy consumption** provided by the main energy sources, about 86% of which was from fossil fuels in 2018. Renewables include hydropower, wind, solar, biofuels, and geothermal power, with the fastest growth in the last decade seen in wind and solar farms. Traditional biomass refers to the biomass used in the developing world to provide heat.

About a third of the total energy consumption is used mainly in the conversion of fossil fuels to electricity and refined fuels and products. The rest is called the **final energy demand** and is the energy consumed by users: about 100,000 TWh per year as energy to provide services, and about 10,000 TWh as products (called non-energy), such as plastics and bitumen. Of the energy, about 10% is as heat from traditional biomass in the developing world, 22% as electricity, 38% as heat predominantly from fossil fuels, and 30% in transport. Both the heat and electricity are mostly used in industry and buildings. Petrol (gasoline) and diesel provide nearly all of the fuel used for transport. The approximate breakdown of the final energy demand in 2018 is shown in Fig. 1.8, estimated by reducing the fossil fuels' primary contributions by 35%.

We see that providing heat is just as important as electricity. Both can be measured in kWh, but while electricity can be fully changed into heat, as in an electric oven, only a fraction of the energy in the form of heat can be converted into electrical energy, as some is necessarily lost to the surroundings. In a thermal steam power plant, the chemical energy bound up in a fossil fuel is transformed as the fuel is burnt to thermal energy. This heats water to generate steam, which expands across turbine blades that turn an electrical generator. Only part of the heat is transformed to electricity; the rest is transferred to the environment as waste heat when the

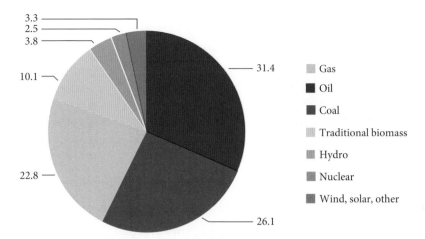

Fig. 1.8 Approximate percentage of global final energy demand of ~110,000 TWh by source in 2018 (OurWorld).

steam is condensed, completing the cycle. The fraction converted is increased by raising the temperature of the high-pressure steam, but this is limited by the strength of the boiler tubes at high temperature. A typical percentage of heat converted to electricity in a modern thermal steam power station is about 40%. For a combined cycle gas turbine (CCGT) plant, which operates at a higher temperature, the percentage can be as large as 60%.

Similarly, only a fraction of the heat generated in an internal combustion engine can be transformed into the energy of motion (kinetic energy) of a vehicle; a typical average efficiency is 25% for a petrol and 35% for a diesel fuelled car, while for diesel trucks and buses the efficiency can be about 45%. Electric motors, on the other hand, have efficiencies of around 90%, so electrifying transport would reduce the consumption of energy significantly. This is an example of the synergy between improved efficiency and renewable energy, which will help in providing the energy that the world needs.

The share of the generation of electricity by fuel type in 2018 is illustrated in Fig. 1.9. The percentage from renewables other than hydro was 9.8%, up from 2.8% in 2008, with wind going from 1.1%, to 4.8% and solar from 0.06% to 2.1%. Nuclear power's percentage has fallen though from 13.6% to 10.1% while hydropower's has remained about the same. The fossil fuel contribution has only decreased slightly from 67% to 64%.

China's economy has been growing rapidly over the last two decades, and its combustion of coal has increased by a large amount; the growth in its energy demand is expected to decrease over the next couple of decades, balanced partially by an increase in consumption in India and other developing countries. The polluting effect and high carbon intensity of coal is causing coal to be displaced by gas and renewables. Shale oil is making the US self-sufficient in oil, while Asia is increasingly reliant on oil imports. Shale gas is expected to grow, as is nuclear and hydro generation, mainly in Asia. But the largest growth is expected in renewables throughout the world.

The rise in the global emissions of CO_2 associated with the combustion of fossil fuels has slowed in the last decade and was approximately constant for 2014-2016, but rose 2.7% by

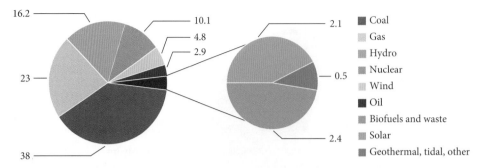

Fig. 1.9 Electricity generation (%) by fuel in 2018: total 26,730 TWh (IEAelec).

2018 and another 0.6% to 36.8 $GtCO_2$ per year in 2019.This indicates that the world is only slowly starting to decouple emissions and economic growth. This is coming about through fuel switching to sources with lower carbon dioxide emissions per unit of energy (gas is about half that of coal, see Table 1.2), the continued increase in renewables, particularly in wind and solar PV, and through energy savings. However, continuing such CO_2 emissions would put the world at risk of very significant climate change owing to the associated global warming, and the world needs to decrease fossil fuel combustion quickly.

1.3 Global Warming and Climate Change

Since the Industrial Revolution there has been a sharp increase in the burning of fossil fuels. This is important for the global climate because CO_2 is what is called a **greenhouse gas**. The main greenhouse gases in the atmosphere are water vapour, CO_2, and methane. These gases absorb infrared radiation, and this has a significant effect on the temperature of the Earth's surface. We can see why this happens by first considering the temperature of the Earth with an atmosphere completely transparent to all radiation. The Sun's radiation would then impinge directly on the Earth's surface, with about 10% reflected and the rest absorbed. This would heat the surface until it was at a temperature of around −1 °C, at which temperature it would radiate energy back out into space at the same rate at which it was received from the Sun.

Now consider the effect of adding an atmosphere that is largely transparent to the Sun's radiation, which is mainly in the visible, but absorbs significantly in the infrared. About 30% of the Sun's radiation is reflected and the rest is absorbed by the Earth's surface. Radiation emitted by a surface that is near room temperature is mainly in the infrared, so the atmosphere near the ground absorbs the Earth's radiation and heats up. As a result, the atmosphere radiates infrared radiation both upward and back to the Earth's surface, which now receives more radiant energy than with a completely transparent atmosphere, and becomes hotter. The warmed air near the ground rises because of its lower density, and expands and cools, since the air pressure falls with altitude. Other air falls and this convective heat transfer maintains an approximately constant fall in temperature with height up to about 10 km. The warming of the atmosphere and ground continues until the temperature at the height where radiation can escape into space is about −28 °C, at which temperature the atmosphere emits energy at the same rate as the Earth receives it. The ground temperature is then around 15 °C.

This is a simplification of what happens in our atmosphere, but it illustrates the key processes. The trapping of infrared radiation and the consequent temperature rise is called the **greenhouse effect**. It is actually only a small part of the temperature rise in greenhouses, which is mainly from reduced convection, but the name has persisted. In 1827 the great French mathematician Jean Fourier was the first scientist to realize that the atmosphere acts like a greenhouse. However, it was not until 1856 in America that **Eunice Foote** first identified the heat trapping properties of CO_2 and she speculated that increasing its atmospheric concentration would raise global temperatures. This was three year before the Irish scientist **John Tyndall** identified water vapour, carbon dioxide, and methane as the main gases in the atmosphere that absorb infrared radiation. In particular, he realized that without the greenhouse effect 'the Sun would rise upon an island held fast in the grip of frost'. The principal greenhouse gas is carbon dioxide, and its amount, prior to the burning of fossil fuels, was maintained at an approximately constant level by natural processes, which include photosynthesis that removes CO_2, and respiration and decomposition that generate CO_2.

1.3.1 Global Warming and Cumulative CO₂ Emissions

The greenhouse effect causes a temperature rise of about 15 °C, so it is very important in determining and maintaining our global temperature. However, increasing the concentration of greenhouse gases by burning fossil fuels increases the effective emission height (radiating level) at which the air is thin enough for the infrared radiation to escape. The temperature there is lower, so the radiation emitted is less than that received by the Earth (the deficit is called **radiative forcing**), and the temperature of the surface and atmosphere rises until the temperature at the effective emission height is again about −28°C to restore equilibrium. Or to put it another way: the thickness of the absorbing lower atmosphere (the insulating blanket) increases, and as a result the surface temperature of the Earth rises. [We discuss the greenhouse effect in a simple radiative convective model of the atmosphere in Chapter 2 (Box 2.1)].

A very important greenhouse gas is water vapour (and water droplets) and is the reason why land cools much quicker at night when there is no cloud cover. But it does not control the Earth's temperature. Rather, its amount is essentially determined by the Earth's temperature and increases by about 7% per 1 °C warming of the atmosphere. The effect of adding carbon dioxide, methane, CFCs, and other greenhouse gases to the atmosphere is to raise the global temperature, which increases the amount of water vapour in the atmosphere and that in turn further raises the temperature. Doubling the amount of CO_2 alone would cause a warming of approximately 1.2°C, but the corresponding increase in water vapour approximately doubles this rise to about 3°C; so, the water vapour feedback effect is very significant.

CO_2 concentrations have risen from about 280 parts per million by volume (ppmv) in 1750, just before the Industrial Revolution, to ~407 ppmv in 2018. The characteristic timescale for an excess of water vapour in the atmosphere to disappear is a few days but, for other greenhouse gases and for the response of the interactions between the oceans and atmosphere, the characteristic timescales are typically 10–1000 years. Aerosols, such as those arising from the burning of fossil fuels (e.g. SO_2 aerosols which produce acid rain), tend to have a cooling effect on the climate and, at current levels, approximately cancel out the warming effect of the non-CO_2 greenhouse gases such as methane. The net result is that the level of CO_2 equivalent gases (CO_2eq) is close to that of CO_2 alone.

The lifetime of carbon dioxide added to the atmosphere is about 200 years before most of it is absorbed. The emissions of other non-CO_2 greenhouse gases, of which methane is the largest, make up about a quarter of the total, However, methane has a much shorter lifetime of about 10 years, which means that, although it is a much more powerful greenhouse gas than carbon dioxide, it is the emissions of carbon dioxide that largely determine the global mean surface warming. To a good approximation, it is not the rate but the **cumulative emissions of CO_2** that determine the global mean surface temperature rise. The diminishing effectiveness of more atmospheric CO_2 in causing warming, as the radiative forcing is dependent on the logarithm of its concentration, is compensated for by the decreasing effectiveness of the oceans to absorb both carbon dioxide and the additional radiant heating from the added CO_2 (see Chapter 2 Section 2.3.3).

1.3.2 Rise in Global Surface Temperatures

The change in the average global surface temperature since the mid-twentieth century is shown in Fig. 1.10, relative to the period 1850–1879 when carbon dioxide concentrations were

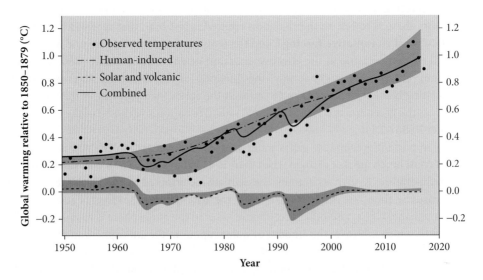

Fig. 1.10 Observed rise in global surface temperatures (HadCRUT4 data) compared with the combined effect of solar and volcanic and human-induced warming. The bands are 5-95% confidence ranges. (GWI).

close to pre-industrial levels. The significant rise in the average global surface temperature that has occurred since pre-industrial times (~1750) and, in particular, over the last 50 years, is called **global warming**. It has been explained by physical models of the Earth's temperature that take into account the emission of greenhouse gases (GHGs), in particular CO_2, due to human activity, and the Intergovernmental Panel on Climate Change (IPCC) has concluded that it is extremely likely that this has been caused predominantly by human-induced emission of GHGs. Including these **anthropogenic** emissions gives good agreement with observations over the period of the study (1950–2017); the temperature would have been expected to have remained about the same as in the 1970s if only natural causes (solar and volcanic) were included in the period 1970–2017. The relatively short-lived dips are caused by aerosols from volcanic eruptions, and the approximate 0.2 °C rise seen by 1950 reflects the cumulative global carbon dioxide emissions by then. A very slight cooling was observed in the 1960s and 1970s and was associated with an increase in aerosols from fossil fuel burning as well as a slight dip in natural forcing.

The measured temperatures for the period 2003–2013 were roughly constant, but over the longer period 1990–2017 the observations are consistent with a steady rise in temperature as predicted by the global warming arising from an increasing level of GHGs. Fluctuations caused by natural causes, such as an El Niño or a La Niña, which is a warming or cooling of the surface of the sea in the Pacific that occurs every few years, will cause deviations that can mask a steady rise in temperature. Studies of the changes in global temperatures in the past millennia have provided valuable information about the variability arising from natural causes, and how the Earth's climate is particularly sensitive to small changes in temperature that in the past were caused by natural events, but are now being caused by human activity [see Box 1.1].

1.3.3 Climate Change

The emission of greenhouse gases, in particular CO_2, mainly from the combustion of fossil fuels, but also from land use changes (such as replacing forests with crops), causes not only global warming but other changes. The melting of glaciers and of the Greenland and Antarctic ice sheets contribute to sea-level rises, (as does the thermal expansion of the sea as it warms) and the loss of ice is affecting animal habitats. At the same time, acidification of the oceans, through more dissolved CO_2, threatens shellfish and other sea creatures, and is causing large amounts of coral to die worldwide. Many species have changed their geographic locations due to changes in climate arising from global warming. A small rise in temperature can heighten the risk of droughts and can affect crop yields: even a 1 °C warming reduces the average wheat yield by about 6%. Since the major greenhouse gases mix throughout the atmosphere, climate change is a global problem, as it affects shared natural resources.

Global warming is generally larger over land than over sea, but in the Arctic, ice melting exposes more water, which is less reflective than ice, so more sunshine is absorbed which amplifies the warming there. The region is now over 2 °C hotter than in the 1970s and significant areas of the Arctic are ice-free as a result. This polar amplification increases the chance of blocking patterns of high pressure associated with extreme heat waves, such as the one experienced in Europe in the summer of 2018. It also brings an increased risk of changes in the

Box 1.1 Why did the ice ages occur?

The occurrence of an ice age was proposed by Louis Agassiz in the first half of the nineteenth century. It explained the formation of moraines far south of existing glaciers. [It is now understood that there have been several glacial cycles in the last five hundred thousand years]. A couple of decades later James Croll showed how the gravitational forces of the Sun, Moon, and plants would slightly affect the Earth's motion and orientation periodically. This could change the distribution of sunlight over the Earth (the solar forcing), and reduce the amount of sunlight during winter in the northern hemisphere. He thought the heavier accumulation of snow could alter winds and the Gulf stream and induce an ice age, but his arguments were not widely accepted.

Milankovitch refined the calculations in the 1920s and 1930s, and the slight periodic changes in the Earth's motion are named after him. There are three main components: the longest is due to the change in the eccentricity of the Earth's orbit, with a period of ~100,000 years, the next longest (~40,000 years) is caused by the variation in the angle (currently ~23.5°) that the Earths' axis is tilted relative to the plane of its orbit, and the shortest (~26,000 years) arises from the precession of the Earth's axis.

Milankovitch suggested that if the sunlight became weak enough then snow and ice would not melt in the summer but continue to build up and reflect more sunlight, which would amplify the cooling. But it appeared that the effect from a variation in solar forcing was too small to account for glaciation. It was not until the mid-1980s that the mechanism was found, when it was discovered from the analysis of ice cores that the level of CO_2 was much lower in the cold period of a glacial cycle than in the warm inter-glacial times. As temperatures fell, carbon dioxide was absorbed by the oceans which increased the cooling, as its greenhouse effect was reduced, giving a positive feedback, see Fig. 1.11, and vice-versa. Methane levels were also found to rise and fall. This greenhouse gas

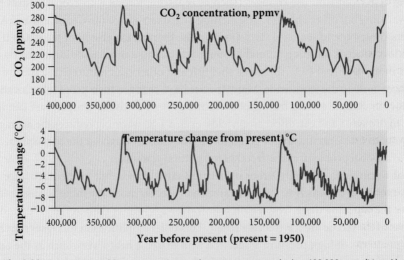

Fig. 1.11 Fluctuations in CO_2 concentrations and temperature over the last 400,000 years (Vostok).

feedback would also explain why a variation in solar forcing in the northern hemisphere gave a world-wide effect. As an instance, the end of the last ice age is thought to have been due to a rise in temperature associated with a Milankovitch cycle in which the huge ice sheets covering Canada and Europe melted. The resulting flow of fresh water into the North Atlantic disrupted the heat flow to northern latitudes and to southern latitudes, which caused the Antarctic to warm and release CO_2 that accelerated the warming.

Climate models can now reproduce the glacial cycles and the temperature variations brought about by CO_2 and CH_4 variations; in particular, a doubling in the concentration of CO_2 causing about a 3 °C rise. This climate sensitivity is in the same range as predicted for our present climate. We can see that variations of around ±5 °C have caused massive climate change in the past. [It is important to note that the existence of a correlation between global temperature and atmospheric carbon dioxide concentration does not indicate which one caused the other, nor indeed whether both were the result of some other effect.] While past variations in temperature were triggered by natural variations in solar forcing, the current unprecedented rise in human-induced greenhouse emissions is driving our current global warming. And while these are delaying the next ice age, the warming could cause catastrophic climate change long before unless we drastically reduce emissions quickly.

jet stream, which can, for instance, cause cold Arctic air to move and bring very cold weather further south. Now that the atmosphere is generally hotter because of global warming, clouds can hold more water vapour. This means that when severe storms occur, rainfalls can be much heavier and severe flooding is more likely. Snowfalls too can be much deeper.

Ocean currents can also be affected. The Earth receives more radiation per square metre on a horizontal surface at the Equator than at the poles and as a result energy flows poleward via winds and ocean currents. A major current is the **Gulf stream** which brings warm water from the Gulf of Mexico up past the East coast of the USA and across to Europe, where it warms western European countries. It is part of the ocean conveyor belt that transports heat around the world. The evaporation and freezing of the surface water in the Gulf Stream in the North Atlantic increases its salinity, and hence its density, and the water sinks and then flows south at depth. This **thermohaline** circulation has been weakening, and would slow even more with a decrease in the salinity of northern waters from greater ice melting. A slower Gulf stream raises the sea level along the USA eastern seaboard, increasing the risk of flooding, and also causes higher temperatures there.

So even one degree of global warming can seriously disrupt our climate and significantly increase the frequency of extreme weather events and of wild (forest) fires. The effects of global warming of greater than 2 °C could be catastrophic for many parts of the world. It could accelerate the melting of glaciers in the Andes and western China, threatening the water supply of millions of people as the seasonal stores that glaciers provide (which hold water in the winter and release it in the summer) are lost. The rise in sea levels would flood low-lying islands, such as the Maldives, and inundate coastlines, e.g. Bangladesh and parts of the East coast of the US, leading to massive movements of populations. There would also be an increased threat to human health from higher temperatures, particularly in tropical and subtropical regions.

(a)   (b)

Fig. 1.12 (a) Polar bear habitat (© iStock/Howard Perry), istock. (b) Flooding in Asia (Editorial credit: 1000 Words/ Shutterstock.com).

Biodiversity is also likely to be irreversibly altered, with the possible extinction of a significant percentage of plant and animal species (see Fig. 1.12). Ironically, the poorest nations, who have contributed the least to climate change, will be the most affected by it. The livelihoods of indigenous people and of many millions working the land are already affected by extreme droughts, flash flooding, and erratic seasons. And such extreme weather is precipitating migration.

1.4 Implications for Society

If the world continues using energy produced mainly from fossil fuels, the danger is considerable that the temperature rise by 2100 will be too high to avoid devastating effects. In its fifth assessment, the **Intergovernmental Panel on Climate Change (IPCC)** found that by the end of the twenty-first century, if the world continued with energy produced mainly from fossil fuels, a business-as-usual (BAU) scenario, the temperature rise by 2100 relative to the average over 1861–1880 would be likely to be about 4 °C. As a result, there is now an overwhelming consensus in the scientific community, and increasingly among politicians, that decisive action must be taken now to reduce our carbon emissions. In the **Paris Agreement** of 2015, nations agreed to limit global warming to 2 °C and, if possible, move towards a limit of 1.5 °C. Although President Trump announced in 2017 that he intended to withdraw the US from the Agreement, a decision reversed by President Biden in 2021, many States in the US, notably California, remain committed to lowering emissions. In addition, President Biden has committed to reduce emissions by at least 50% from 2005 levels by 2030.

A further report in 2018 by the IPCC found that the risks were significantly lower for a global warming of 1.5 °C rather than 2 °C. Limiting the rise to 1.5 °C could spare hundreds of millions of people from climate-related disasters by 2050 and reduce the loss of biodiversity significantly. But while action is clearly vital and urgent for the long-term health of the world, as the warming is already ~1.1 °C, limiting emissions has many immediate benefits for the quality of life of communities, and it is these, in particular, that are motivating people to move away from fossil fuels to clean forms of energy, notably wind and solar power.

As explained in Section 1.3.1, the lifetime of CO_2 in the atmosphere is around 200 years, and it is the **cumulative emissions of CO_2** that principally determine the amount of global warming. Since pre-industrial times, the amount emitted up to 2017 was about 2200 thousand million (giga) tonnes of carbon dioxide, which has caused a temperature rise of approximately 1.1 °C. Allowing for non-CO_2 effects, we must limit further emissions of CO_2 to 580 giga-tonnes to restrict global warming to about 1.5 °C. This means rapidly reducing our dependence on fossil fuels—if we carry on as we are (emitting about 37 gigatonnes of CO_2 a year from fossil fuels), we will have exceeded that temperature by 2035, only a short time away: **We need to reach net-zero by around 2050**.

It is important to note that it is not only energy-related emissions that matter: CO_2 given off as a result of land-use change (e.g. deforestation) is also a major concern. The percentage contributions from different sources of CO_2 and from other gases to global anthropogenic greenhouse gas emissions in CO_2eq in 2010 are shown in Fig. 1.13 (a) and from different sectors in Fig. 1.13 (b), where the contribution from electricity production (plus a relatively small amount

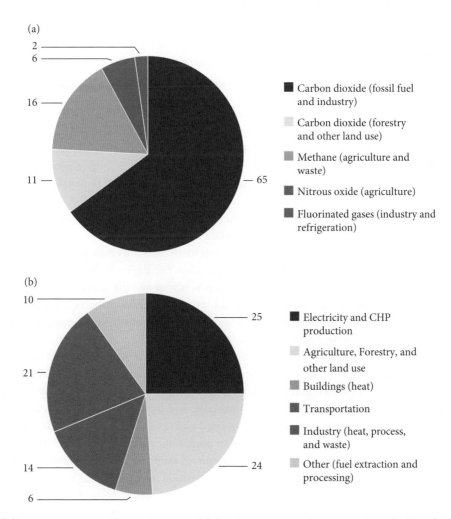

Fig. 1.13 The percentage contributions in 2010 to global anthropogenic greenhouse gas emissions in CO_2eq from (a) different sources of CO_2 and other gases; (b) different sectors.
Source: Adapted from Box 3.2, Figure 1 from the Fifth Assessment Report of the Intergovernmental Panel on Climate Change.

of combined heat and power, CHP) is shown separately. [The percentage of primary energy consumption from fossil fuels was the same in 2019 as in 2010, at 86% (Ourworld).]

We can see from Fig. 1.13 (b) that power production and agriculture, forestry, and land-use changes each account for a quarter of the total emissions, with heat production in buildings and combustion of petroleum derived fuels (petrol and diesel) accounting for another fifth. Emissions associated with heat production and processes in industry give rise to a further fifth. From this we see that decarbonizing heat is just as important as decarbonizing electricity, as their associated emissions are about the same, followed closely by decarbonizing transport. Electrification of transport and of heating buildings gives improved efficiency and helps lower energy demand. Ways to reduce consumption and the demand for energy are essential, since they decrease the rate required for decarbonizing the power supply. Emissions from land clearances must also cease, and significant reforestation must occur.

1.4.1 **Decarbonization**

The challenge facing the world is to provide the energy societies need to raise their standard of living but at the same time to drastically reduce our dependence on fossil fuels. We will need to increase our supply from renewables very significantly, and in particular the supply of electricity; and the cost of renewable energy must be lower than energy from fossil fuels, which it is already for solar, wind, and hydropower in many parts of the world.

Decarbonizing heat, for which there is about an equal demand as for electricity, is particularly difficult. Biomass can provide such heat, but currently about 2.7 billion people jeopardize their health by cooking with traditional biomass on stoves that give off both gaseous (CO) and particulate pollutants, a problem that is discussed in Chapter 4 Section 4.2.2. Renewably produced electricity can provide clean heat, either directly, or through heat pumps, or through power-to-gas in which electricity is used to make a combustible fuel.

Since ~80% of our energy presently comes from fossil fuels, this means decarbonizing our energy supply quickly (since the resulting temperature rise depends only on the amount of carbon emitted, and not when, the rate of reduction is lessened the sooner it is begun). This process would initially mean **fuel switching** to sources with lower carbon emissions per unit of energy, e.g. from coal to gas or renewables, and aiming to satisfy our energy requirements from just renewable, nuclear, and carbon capture plants by 2050.

To help achieve this goal will also require reducing the global demand for energy but without denying the benefits to a society that are associated with energy (see Fig. 1.7). A significant reduction that meets such a condition can be made by improving **efficiency**; e.g. by changes in building and industrial technology, and in transport. Buildings and industry each account for about 28% and 39%, respectively, of energy-related global CO_2 emissions, and these could be reduced significantly through improvements in insulation, reducing losses in driven devices, and by the use of heat pumps and LEDs. Transport accounts for another approximately 22% of energy-related emissions, and significant savings would be made through the introduction of electric cars and fuel-cell powered trucks and trains. Another important measure is **energy conservation**, whereby energy savings are made through changes in behaviour; e.g. lowering thermostats, turning lights off, and using public transport rather than cars.

It is important that the emissions (**carbon footprint**) associated with all activities are reduced. The carbon footprint of an activity is the amount of carbon dioxide (CO_2) or CO_2eq emissions associated with an activity, and, if ongoing, per year. For example, the average carbon footprint of a UK household has been estimated as 26 t CO_2eq per year. Of this, about one third is direct emissions, e.g. space heating, driving, and hot water, and two thirds indirect or embedded emissions, e.g. those arising in the production and shipping of household goods. (Some estimates of carbon footprints omit embedded emissions.) The reductions possible through efficiency improvements, energy conservation and other aspects of energy savings are discussed in more detail in Chapters 11 and 12.

The scale of the problem requires a broad international commitment to stand any chance of success. The Paris Agreement in 2015 was encouraging, but the proposed intended nationally determined contributions (INDCs), if all implemented, have been estimated to limit global warming to only 3.2 °C by 2100. There is much to be done, and the drive to provide sufficient affordable low-carbon energy and to reduce the demand for energy must continue. As the world adjusts to the effects of the **COVID-19** pandemic, governments must ensure that there is sufficient investment and support for renewables, which can not only provide the clean energy all countries need, but also supply employment and energy security. But these sources must be sustainable ones.

1.4.2 Sustainable Development

The choice of energy source or power plant is strongly dependent on its cost and on the benefits that accrue, and should be in line with **sustainable development**. This aim requires policies or actions that provide for the present generation but do not compromise the needs of future generations. An example would be the construction of a solar photovoltaic (PV) farm, rather than a fossil-fuelled power station, to generate electricity in regions of good sunshine where the **levelized cost of energy (LCOE)** is economically competitive. This action would avoid the emission of CO_2, which would exacerbate global warming and generally worsen future climatic conditions, and could provide energy at a competitive cost for the present community.

The LCOE is one key component, but there are associated costs when the source of supply is variable, and transmission costs. There are also **externalities**, which are costs arising from the effect of the generation on the environment; for example, pollution from fossil fuel combustion, and deforestation from biofuel plantations. These are considered in an **environmental impact assessment** (EIA). A **life-cycle analysis** (LCA) will give the carbon dioxide emissions and these can be costed through a carbon price. The benefits can be cleaner air that reduce health care bills; also, jobs can be created that can reduce welfare costs.

There are also considerations of **energy security**. This is the ability of a country or region to meet its own energy demand, even when its supply is threatened, from resources within that area or under its control; threats include severe weather, plant failures, industrial disputes, or terrorism. Generally, a country with good natural energy resources, such as strong winds or high solar insolation, will have better energy security than one which is dependent on imports, such as gas through a transcontinental pipeline. Then there are the challenges in meeting the three requirements of energy security, affordable energy for all, and environmental sustainability: the **energy trilemma**.

The issues arising through interrelated demands for water, food, and energy, all of which are vital to society, also need to be addressed. For instance: a dam may provide hydropower and water storage for irrigation but could affect agriculture downstream and displace people;

bioenergy crops may help in producing electricity and provide employment but reduce food production. The analysis of such issues can help obtain better trade-offs between competing demands and using limited resources more sustainably.

Evaluating the costs and benefits, a cost–benefit analysis (CBA), is the basic method used when trying to decide whether to embark on a given project, and when the project also involves an assessment of the risk, the analysis is called risk–cost–benefit analysis (RCBA). The benefits and risks of a project may well affect different groups of people. In the siting of a nuclear power station, for instance, the local population is at the highest risk from an accident at the site, while a much larger group benefits from the power produced. There needs to be a balance between a widespread consultation and a consequently slow decision process, and a small review panel which could produce a much quicker decision. The latter helps government planning on, for example, national energy security, as well as on the international issue of tackling global warming, while the former is more likely to gain public confidence.

The public perception of risk is often different from that of experts, who tend to focus on the probability of an accident. Risks are perceived to be greater when they are involuntary, uncontrollable, and potentially catastrophic. These concerns contribute to what has been called the **dread factor**. Conversely, risks are more acceptable if they are voluntary, controllable, and limited. Nuclear power is an example of an industry with a high dread factor; the actual number of deaths per year due to nuclear power is low compared with much more familiar incidents such as fatal car crashes.

Public involvement and trust in the utility companies and government agencies involved are key factors that can help speed the approval of a project. It is important that the public are aware of the benefits that the project will bring, both to them and to others. It is also important not to assume that the general public cannot understand technological arguments—if you believe a project to be safe then you should be able to explain why you think so. This raises the question of how you make a system reliable and safe. One method is by **adding redundancy**, so that if one component fails another independent one takes over. An ideal system would be **fail-safe** in the event of a fault; for example, a heat sensitive switch that would turn a machine off in the case of overheating without requiring any operator intervention. Improving safety involves costs, and one way of dealing with possible conflicts between safety and profitability is to appoint a **regulator**: an independent authority that can enforce safety requirements, maintenance schedules, and inspections.

We now look at the characteristics of different energy sources.

1.5 **Characteristics of Fossil Fuel and Renewable Energy Sources**

Table 1.2 gives the characteristics of some important sources of fossil fuel and renewable power generation. For fossil-fuel power plants, typical capacities are ~1 GWe (where e refers to electrical output), while large wind and solar PV farms are now typically ~500 MWe. Of particular importance are the amounts of greenhouse gases that are emitted, the analysis of which is called a life-cycle analysis (LCA). Other significant factors are their capacity factors, and the amount of land, or sea, required, which is determined by the power density (MW km^{-2}), and their levelized cost of energy (LCOE).

Table 1.2 Capacity factor, area required, LCA, and LCOE (USA) of some fossil and low-carbon sources (OpenEI, Lazard2019)

Source	Capacity factor (%)	Area for 1 TWh per year (Square km)	Life-cycle analysis (gCO$_2$ per kWh)	Levelized cost ($ per MWh)
Coal	~60	~0.2	979	66–152
Gas (CCGT)	~50	~0.2	477	44–68
Nuclear	~90	~0.2	12	118–192
Solar PV	~10–25	~7.5–20	44	32–44
Wind onshore	~35	~50–90	11	28–54
Wind offshore	~50	~30–50	11	~89

1.5.1 Life-cycle Analysis (LCA) and External Costs (Externalities)

Accounting for *all* the emissions involved in producing energy from a particular source is called a **life-cycle analysis**. It calculates the amount of CO_2 (and other gaseous emissions) per kWh of energy produced. For fossil fuels, the amounts are ~980 g per kWh from coal and ~480 g per kWh from a natural gas combined cycle gas turbine (CCGT) power plant, whilst the burning of agricultural wastes in a thermal power plant gives less than 30 g per kWh. An LCA should include all stages of the process; e.g. for biomass, the emissions associated with any land change, the use of fertilizers, tractors, and processing need to be included. In the combustion of municipal solid waste (MSW) in a power plant (also referred to as **energy from waste**, EfW), although the fuel is mainly organic, the combustion is not carbon-neutral because some of the material is derived from fossil fuels (typically 20–40%), and gives ~360 g per kWh. Typical values for several sources are shown in Table 1.2. These amounts are called the **carbon intensity** of the technology or fuel.

The CO_2 emissions per kWh of electricity generated for nuclear, wind, and solar are more than an order of magnitude less than for coal- and gas-fired power stations. There are technologies under development that would reduce the emissions from these fossil fuel plants significantly: these are called **carbon capture**. In carbon capture, the carbon dioxide emitted in the combustion of the fuels is captured chemically and subsequently released, either to be stored in carbon capture and storage (CCS), or utilized to make a fuel or other chemical in carbon capture and utilization (CCU). While the technology is already available, the capital costs are very high, and there has been concern about the viability of long-term capture. The increase in the capital requirement has increased the cost of electricity significantly, and the lack of any effective carbon price or long-term guarantees on revenue, and the uncertainty over its long-term future, has meant that little significant progress has been made to date (as of 2019). We describe CCS and CCU in Chapter 3 Section 3.7.

Renewable energy has substantial environmental benefits, and these can be given a monetary value from the amount of CO_2 saved and other emissions such as SO_2 that causes acid rain. The external cost (mainly environmental) of coal-fired generation has been estimated to be 4 ¢ (€ cents) per kWh compared to 0.2 ¢ per kWh for wind energy. If this external cost were included, then the price of electricity in Europe from coal would roughly double.

A related concept to the LCA is the energy returned on energy invested ratio (ERoEI), where the energy inputs and outputs are the equivalent fossil fuel energy values. For a PV installation in Southern Europe a ERoEI of ~30 is typical for a $T = 30$-year lifetime, so the energy payback time (EPBT), which is given by EPBT = T/ERoEI, is ~1 yr.

1.5.2 **Capacity Factor**

When comparing the outputs of power plants using different sources of energy, it is important to note that the maximum continuous or rated power output, called **capacity**, is usually quoted for a plant. For typical fossil-fuel plants, the average annual power output as a fraction of the rated power, called the **capacity factor** or **load factor** of the plant, is ~55%, and for a nuclear power plant it is ~90%. For wind and solar farms, the capacity factor is less: ~35–50% for wind and ~12–25% for solar. The capacity factor of nuclear power is high, since plants tend to operate most of the time as baseload generators, because varying their output is less straightforward than with other types of plant. The values for coal and gas reflect the balance of supply and demand and its effect on prices, while the values for wind and solar also reflect the variable nature of the supply.

It is also important to realize that the efficiency of conversion, as, for example, in the conversion of the energy in sunlight to electrical energy in a photovoltaic cell (the efficiency of PV), refers to the amount of the incident light energy that is converted to electrical energy, and does not take into account the amount of time the Sun was shining. This time is taken into account by the capacity or load factor. The annual output in MWh of electricity of a power plant is its rated power in MW times its capacity factor times the number of hours in a year (8760).

1.5.3 **Area Requirement**

The power density of wind is such that ~50-90 square kilometres are required on land to produce 1 TWh of electricity in a year. Much of this area is available as farmland, but in some countries with high population densities, e.g. in the UK, a wind farm's visual impact can make it difficult to obtain planning permission, and has led in the UK to increased development offshore. Offshore capacity factors are higher and the area of sea required for 1 TWh per year is ~30-50 square kilometres. At 70 km^2 per TWh, the fraction of the UK land area needed to produce the generation in 2019 of 324 TWh would only take up 9% of the UK land area; while the UK's coastline (using a 50 km ruler) is ~3400 km, so at 40 km^2 per TWh only 8% of the area of sea within 50 km of the shoreline would be required. The total final energy demand of the UK is close to 1700 TWh, so a variety of low-carbon sources will be needed.

The mix required depends on location. In India, with its high solar intensity, photovoltaics could supply a significant fraction of final demand, and the space required to generate 1 TWh per year is about 15 square kilometres. From Fig. 1.7 a final energy demand per capita (currently ~2/3 of the primary) needed for a good HDI is ~60 kWh per day per person. With a population of 1.35 billion and an area of 3.3 million km^2, about 14% of the land would be needed to meet the total demand; and in Africa, with 1.22 billion and 30 million km^2, the percentage would be only 2%.

1.5.4 **Variable Sources**

However, solar and wind, two major forms of renewable generation in the future, are both variable in output and not always available. While wind and solar power can complement each other, with wind speeds typically higher in the winter than in the summer, and vice versa, there can still be significant gaps in availability. As the penetration of renewables (i.e. the fraction of electricity generated) increases, this variability becomes more acute. Having energy storage would help alleviate this problem, as would managing the demand for energy by, for example, moving electricity demand from the evening to the daytime, and by using interconnectors to make available generators located over a large region or number of countries. At the moment there is some curtailment of renewable generation at times, but experience has shown that high percentages of electricity generated by renewables can be handled. In Chapter 10 Section 10.6, we will discuss how the variability of renewables is being addressed.

1.5.5 **Resource Potential**

There is ample area in the world to easily provide all our power requirements using renewables, and in the developing world solar and wind can provide local, i.e. distributed, generation of electricity, avoiding the need for electrical grids, and empower hundreds of millions of people. How much electricity (or heat) a particular technology could supply is called its *potential*, with several different potentials defined.

The **technical potential** is the fraction of the gross potential after unsuitable areas, e.g. those covered with ice, have been excluded. In addition, environmental and social limitations further reduce the available area. Two further estimates of energy supply are often quoted, but it should be noted that definitions of potential vary. One is the **economic potential**, which is that fraction of the technical potential that is economically competitive. The other is the **accessible potential**, which is the amount of the technical potential that can be utilized by a particular time. For renewable and nuclear electricity generation, the accessible potential can be given in terms of the electricity generation in TWh per year by that time.

The accessible potential will depend on the cost of alternative supplies, and be affected by policies such as a carbon tax, the effects of competing land use, planning permission, grid limitations, the rate of construction of new generators, and in particular the levelized cost of the energy (LCOE) generated.

1.5.6 **Levelized Cost of Energy**

The **levelized cost of energy** (LCOE) is the cost to produce energy, generally electricity, in cents kWh^{-1}, taking into account the interest rate, R, on the capital $C_{capital}$ needed to build a plant, the number, N, of years of operation, the cost of operations and maintenance (O&M), and the cost of any fuel. The revenue that a generator must bring in over the lifetime of the plant in order to break even must be sufficient to:

- pay off the capital loan plus all the interest that the capital would have earned during the period of the operation of the plant;

- pay the annual maintenance and operations of the plant and the cost of any fuel.

The levelized electricity cost (LCOE) is given by:

$$LCOE = (CRF \times C_{capital} + O \& M_f)/(8760 \times CF) + O \& M_v + Fuel \tag{1.1}$$

where CF is the capacity factor (e.g. 0.5 for a wind farm), 8760 is the number of hours in a year, and $C_{capital}$ is the capital cost in dollars per kW. The fixed operations and maintenance costs ($O\&M_f$) are taken as a fraction of the capital cost, while the fuel costs (*Fuel*) and the variable operations and maintenance costs (O&Mv) are dependent on the amount of electricity. The capital recovery factor (CRF) is the fraction of the capital cost per year that gives the annual revenue required to pay off the capital loan plus all the interest, and is given by

$$CRF = R/[1 - (1 + R)^{-N}] \tag{1.2}$$

(see Derivation 1.1 at the end of this chapter).

When the interest rate is negligible, CRF equals one over the number, N, of years of operation; so, for instance, for a power plant with a lifetime of 30 years, then 1/30 of the capital would have to be paid off each year from the annual revenue (neglecting O&M and fuel costs). But for a typical interest rate (sometimes called **discount rate**) of 6%, CRF is over two times larger.

For renewables that have no fuel costs and only a fixed O&M of 4% each year of the capital cost, the LCOE is given by:

$$LCOE = 1.29 C_{capital} (k\$ \text{ per kW})/CF \quad \$\text{cents per kWh} \tag{1.3}$$

and we can use this to give a rough estimate of the price of electricity from solar PV and wind. For example, in India, for solar PV where capacity factors with the high sunshine levels can be 0.25 and the cost per kW of capacity ~$800 (2019), LCOE would be estimated as 4.2 $cents per kWh and would undercut fossil fuel production (see Table 1.2).

However, with a discount rate of 10%, the price would 5.3 $cents per kWh, which is a significant increase. The discount rate or what is called the **cost of capital** is therefore particularly important in determining the LCOE. This rate depends on many factors, in particular the confidence in the project being built on time and to cost, as well as on the market interest rates.

EXAMPLE 1.1

A 200 MW wind farm cost 300 million dollars. The interest rate and period of operation are 8% and 30 years, its capacity factor is 35%, and the operations and maintenance charge each year, is 4% of the capital cost. Calculate the LCOE.

The cost of capital is $1500 per kW, and from eqn (1.2), the CRF is 0.089. So from eqn (1.1):

$LCOE = (CRF + 0.04) C_{capital}/(8760 \times 0.35) = 0.129 \times 1500/3066 = \$0.063$ per kWh equivalent to $63 per MWh or 6.3 cents per kWh.

1.5.7 **Learning curve estimation**

It is common experience from a broad range of technologies that costs fall as production increases. In particular, costs decrease roughly linearly with cumulative production when plotted on a log/log scale. The **learning rate** for a technology is the percentage reduction in costs for a doubling in cumulative production; typically, the learning rate is between 10% and 30% for industrial products and tends to be larger in the developing stages. It is the *global* cumulative production that matters, not the production in a particular country. Although it appears rather simplistic, learning curves have been shown to provide good future cost estimates—the technique was first noted in the construction of aircraft in 1930s. The tendency for the slope of a learning curve to decrease with time as a technology becomes more mature needs to be allowed for when extrapolating into the future. To calculate the time when the cost will have fallen to a particular value requires assumptions about the rate of production. If that is described by a compound annual growth rate (CAGR) of g per cent and the learning rate is r per cent, then the number of years $n(x)$ before the cost is $1/x$ of what it is today is given by:

$$n(x) \approx 6930 \ln(x)/rg \qquad (1.4)$$

Therefore, if the learning rate and CAGR were both 20% then it would take about 19 years for the cost to halve.

Learning is seen in **economies of scale**, which is the reduction in costs of manufacturing when the number or size of the goods being produced is increased. A larger output will reduce costs when there is a fixed cost, such as for setting up a machine, irrespective of the number produced. Increasing the size of a product can also lead to economies when assembly costs remain the same. Larger-scale production may improve operational efficiencies.

An example of a learning curve is that for the levelized cost of energy from onshore wind, shown in Fig. 1.14. The learning rate between 1985 and 2014 was 19%. There was a rise in the price of turbines in 2004, mainly due to an increased cost of materials and a strong demand, which reflected a lack of manufacturing capability. Over-supply can also distort a learning

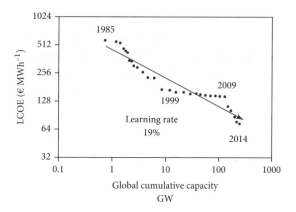

Fig. 1.14 Learning curve for global onshore wind. The solid line shows a learning rate of 19% (*Source*: Bloomberg).

curve, temporarily causing the price to drop. Another example of a learning curve is given for photovoltaic panel production in Chapter 8 (see Fig. 8.19).

The sharp fall in PV over the last decade has meant that PV has achieved grid parity in many regions of the world. As has been discussed (Goodall 2016), this situation may mark a very significant switch from fossil to PV. The installed PV capacity at the end of 2015 was 242 GW, and taking an annual growth rate that falls linearly from 25% to just over 3% by 2050 would give a global installed capacity of ~25,000 GW, a capacity capable of supplying 40,000 TWh towards the global electricity demand in 2050, assuming a capacity factor of ~0.2.

The learning rate for PV panels was ~19% for the period 1980–2015 and ~15% for the LCOE from utility scale PV over the last six years. If we take 6 cents kWh^{-1} as the cost in 2015, then with a learning rate of 15% the cost in 2050 would be only 2 cents kWh^{-1}. [A global capacity of 25,000 GW is $2^{6.7}$ times greater than in 2015 so the price would be $(0.85)^{6.7} = 0.33$ times smaller.] Innovations in technology such as combining a perovskite layer onto a silicon solar cell and obtaining ~30% efficiency, as well as innovations in manufacturing, would help maintain the learning rate.

1.6 Energy Concepts and Units

Energy comes in many different forms: the **kinetic energy** from the speed of an object as in the energy of the wind and a rotating shaft, the **thermal energy** of a heated fluid, the **potential energy** of a raised mass of water as in a dam, the **chemical energy** in the chemical bonds of a fuel such as petrol or natural gas, and the **nuclear energy** from the nucleus of an atom. We use technology to transform energy from one form to another more useful form, as when we burn fossil fuels in a power station and change the chemical energy in the fuel to electricity, which can be transmitted to buildings and industry. Not all of the heat generated from fossil fuels can be transformed to electricity, and the heat rejected makes up a significant part of the difference between the primary and final energy demand. Generally, when energy is converted some of it is lost to less useful forms: energy cannot be destroyed or created. We now define energy and the most important units relating to energy and power.

Energy is the ability to do work. The work done by a constant force F in moving a body a distance s in the direction of the force is Fs. The work done by a constant torque τ in rotating a shaft through an angle θ is $\tau\theta$ (τ and θ are both about the same axis), and power is the rate of work done. For a constant force the power is Fu, where $u = ds/dt$ is the velocity and F is in the direction of u; and for a constant torque τ, the power is $\tau\omega$, where ω is the angular velocity of the shaft. Electrical power is given by the product of the voltage V that drives a current, and the current I, i.e. VI.

In order to appreciate the magnitudes of physical quantities and to be able to make comparisons and estimates, it is essential to have a sound grasp of physical units. The SI system of units is used throughout this book, but many different units are also used for energy and power. A comprehensive list of units can be found at the front of the book, with conversions to both the SI and the imperial system of units. Table 1.3 defines the key physical quantities related to energy and power, and provides some useful conversions.

Table 1.3 Energy-related units and conversions

Quantity	Unit	Definition
Force	newton (N)	Force required to accelerate 1 kg by 1 m s^{-2}
	pound-force (lbf)	Weight of 1 pound ≈ 4.45 N
Energy	joule (J)	Work done by a force of 1N in moving 1 kg by 1 m
	kilowatt-hour (kWh)	$10^3 \times 60 \times 60 = 3.6 \times 10^6$ joules $= 3.6$ MJ ≈ 3412 Btu
	EJ	10^{18} joules ≈ 278 TWh ≈ 23.9 Mtoe
	calorie	Energy to heat 1 g of water by 1 °C ≈ 4.2 J
	Btu	Energy to heat 1 lb of water by 1 °F ≈ 1.055 kJ
Power	watt (W $=$ J s^{-1})	1 joule per second $=$ 1 volt-ampere
	horsepower	550 ft lb per second ≈ 0.746 kW
Fuel equivalence	tonne oil equivalent (toe)	41.868 GJ $=$ 11.63 MWh ≈ 1.5 tonne hard coal
	barrel oil equivalent (boe)	42 US gallon ≈ 159 litres of oil ≈ 6 GJ

We often consider the change in the value of a quantity, and it is important that the sign of the change is consistent. For example, when high-pressure steam expands through a turbine it transfers energy to the turbine, enabling the turbine to do work by rotating an electric generator. In this process a quantity called the enthalpy H of the steam decreases. We define the change in the value of a quantity as the **final value minus the initial value**. So, in this example, the change in enthalpy ΔH is negative.

Work and thermal energy are both forms of energy and arise in the transfer of energy. Heat is thermal energy in transfer, and work is the transfer of other types of energy, such as potential to kinetic. Heat is also called heat transfer, and it occurs from a hotter to a colder material. It is important to realize, as noted earlier, that it is not possible to convert thermal energy entirely into work—the percentage converted depends on the temperatures at which heat is absorbed and rejected. Heat and thermal energy are often used interchangeably, but the difference should be noted.

While heat is important, electricity is more useful as it can be easily transported and converted to other useful forms. For example, it can be used to power electric cars, and if the electricity supply were decarbonized then the emissions from transport as well as the energy used could be significantly decreased, since electric motors are much more efficient that internal combustion engines (ICEs). And, as we will see in Chapter 5 Section 5.2, when we consider heat pumps, an amount of electrical energy E can be used to move several times E as heat either from the environment to inside a building when heating, or vice versa when cooling.

It is important to be able to understand the principles behind energy science and to be able to quantify, for example, the amounts of energy that a technology could produce, and for this, an understanding of some mathematical methods is useful. A very general method is that of **dimensional analysis**, which enables physical expressions to be derived using simple algebra from a sound knowledge of units, and this is described in the following section. For other methods we have included details in derivations in or at the end of chapters.

1.7 **Dimensional Analysis**

Dimensional analysis is a very useful tool for (i) checking whether physical expressions are algebraically consistent, and (ii) deriving the algebraic form of physical formulas. It exploits the fact that the fundamental units on each side of an equation must be identical. Each variable is expressed in terms of the fundamental units of mass M, length L, time T, ..., raised to some arbitrary power α, β, γ ... The unknown powers α, β, γ ... are then determined by equating the powers of like units and using simple algebra. Care is needed in choosing the relevant physical variables, and it is sometimes possible to obtain more than one solution to a particular problem. It is therefore desirable to have an independent means of checking the result.

As an example, verify that the units of pressure p and the expression $\rho g h$ are the same. Expressing the individual symbols in terms of their fundamental physical units, we have:

$$[p] = [\text{Force/area}] = MLT^{-2}/L^2 = ML^{-1}T^{-2},$$
$$[\rho g h] = (ML^{-3})(LT^{-2})(L) = ML^{-1}T^{-2}$$

EXAMPLE 1.2

Derive an algebraic formula for the potential energy per unit wavelength per unit width of wave-front V of a surface water wave in terms of the density ρ, amplitude a, and acceleration due to gravity g.

Assume an algebraic expression of the form $V = k\rho^\alpha a^\beta g^\gamma$, where k is a dimensionless constant. Then:

$$\frac{[\text{potential energy}]}{[\text{wavelength}] \times [\text{width}]} = \frac{[\text{force}] \times [\text{distance}]}{[\text{length}]^2} = \frac{ML^2T^{-2}}{L^2} = MT^{-2}$$

So:

$$MT^{-2} = (ML^{-3})^\alpha L^\beta (LT^{-2})^\gamma = M^\alpha L^{-3\alpha+\beta+\gamma} T^{-2\gamma}$$

Equating powers, we have $\alpha = 1$, $-3\alpha + \beta + \gamma = 0$, $-2\gamma = -2$, yielding $\alpha = 1$, $\beta = 2$, $\gamma = 1$. Hence, $V = k\rho a^2 g$. To derive the form of k it is necessary to have some knowledge of fluid mechanics (see Chapter 2). So, tripling the wave amplitude gives nine times the potential energy in the wave-front.

EXAMPLE 1.3

Derive an algebraic formula for the power P in a wind in terms of the wind speed u, the area swept out by a wind turbine's blades A, and the density of air ρ.

Assuming an algebraic expression of the form $P = ku^\alpha A^\beta \rho^\gamma$ where k is a dimensionless constant, then:

$$[\text{power}] = \frac{[\text{energy}]}{[\text{time}]} = \frac{ML^2T^{-2}}{T} = ML^2T^{-3} = (LT^{-1})^\alpha (L^2)^\beta (ML^{-3})^\gamma = L^{\alpha+2\beta-3\gamma} T^{-\alpha} M^\gamma$$

Equating powers on both sides, we have $\alpha + 2\beta - 3\gamma = 2$; $-\alpha = -3$; $\gamma = 1$, so $\alpha = 3$; $\beta = 1$; $\gamma = 1$. Hence:

$$P = kA\rho u^3$$

This shows the importance of finding locations with a good wind speed, since the power depends on u^3.

SUMMARY

- The greenhouse effect is a natural phenomenon caused by the absorption of solar radiation by the atmosphere, raising the temperature on the surface of Earth by about 15 °C.

- The main greenhouse gases are water vapour, carbon dioxide, and methane, with the amount of water vapour determined by the Earth's temperature.

- The characteristic timescale for an excess of water vapour in the atmosphere to disappear is a few days, but, for other greenhouse gases and for the response of the interactions between the oceans and atmosphere, the characteristic timescales are typically 10–1000 years.

- Carbon dioxide concentrations have risen from about 280 parts per million by volume in 1750 to about 407 parts per million by volume in 2018.

- Net carbon dioxide emissions need to fall to zero by ~2050 in order to restrict the temperature rise, i.e. global warming, (compared to pre-industrial times) to 1.5 °C.

- Continuing to rely predominantly on fossil fuels for our energy—the business-as-usual scenario—could cause a temperature rise of ~4 °C by 2100 and put the world at risk of very dangerous climate change.

- Decarbonizing our electricity and energy (which includes heat) supply, coupled with energy savings, is essential in order to combat climate change.

- The cost, area required, and availability of low-carbon sources are particularly important, and the handling of variable supplies through energy storage, interconnectors, and demand management will become increasingly necessary.

- Dimensional analysis is a useful technique for deriving and checking algebraic relationships between physical quantities.

FURTHER READING

Bloomberg New Energy Finance. (2017). *Learning curve for onshore wind* (Bloomberg).

Botkin, D.B. and Keller, E.A. (2014). *Environmental science*. Wiley, New York. Good discussion of issues about energy and the environment.

Coopersmith, J. (2010). *Energy: the subtle concept*. Oxford University Press, Oxford. Excellent history of how the concept of energy evolved.

Goodall, C. (2016). *The switch*. Profile Books, London. Stimulating argument for the case that photovoltaic cells are now cheap enough for the world to switch from fossil fuels to solar PV.

Jelley, N. (2017). *A dictionary of energy science.* Oxford University Press, Oxford. On Oxford Reference: *www.oxfordreference.com.* Provides descriptions of important terms used in all aspects of energy science.

Jelley, N. (2020). *Renewable Energy: A Very Short Introduction.* Oxford University Press, Oxford. Describes the main renewable sources, emphasizing their importance in tackling climate change.

MacKay, D.J.C. (2009). *Sustainable energy: without the hot air.* UIT, Cambridge. Very readable introduction to alternative energy and energy policy (but dated on solar PV and wind power).

Maslin, M.A. *Climate Change—a very short introduction*, 3rd edn (Oxford University Press, 2014) Very useful overview.

Richie, H. and Roser, M. Our World in Data, *ourworldindata.org/energy* (OurWorld).

Uglow, J. (2002). *The lunar men.* Faber, London. Interesting account of the friends who launched the Industrial Revolution.

data.worldbank.org/indicator/ World Bank (WB).

en.openei.org/apps/LCA/ Life-cycle analysis data (OpenEI).

hdr.undp.org/en/data Human development reports (UNDP).

history.aip.org/climate/index.htm History of the understanding of global warming.

www.iea.org/reports/electricity-information-overview (IEAelec).

www.ipcc.ch/report/ar5/ IPCC fifth assessment report (IPCC 2014).

www.lazard.com Levelized cost of energy (Lazard).

www.nature.com/nature/journal/v399/n6735/full/399429a0.html Petit, J.R. et al. (1999). Climate and atmospheric history of the past 420,000 years from the Vostok ice core in Antarctica. *Nature* 399, 429 (Vostok).

www.nature.com/scientificreports A real-time global warming index (GWI).

www.tsp-data-portal.org Energy and climate data (TSP).

www.wikipedia.org Online encyclopaedia, articles on energy-related topics.

www.worldenergy.org World Energy Council, neutral overview of global energy scene (WEC).

www.wri.org World Resources Institute, environmental think-tank.

EXERCISES

1.1 Can global warming be combated by energy conservation alone?

1.2 Assuming the volumes of the Greenland and Antarctic ice caps are $2.85 \times 10^6 \, \text{km}^3$ and $25.7 \times 10^6 \, \text{km}^3$, respectively, the radius of the Earth is 6378 km, and 70% of the Earth's surface is covered by sea, estimate the rise in sea level if both ice caps were to melt completely.

1.3 Describe the development of the steam engine.

1.4 Estimate the atmospheric force exerted on von Guericke's hemispheres. Diameter 51 cm; assume air pressure $= 10^5 \, \text{Nm}^{-2}$.

1.5 Is the quality of life of a society better or worse without electricity?

1.6 Construct a model of an Archimedes' screw using scrap materials.

1.7 Rewrite in modern terminology Carnot's statement: 'The motive power of heat is independent of the agents employed to realize it; its quantity is fixed solely by the temperatures of the bodies between which is effected, finally, the transfer of caloric.'

1.8 During the second phase of the Papin steam engine cycle the outer surface of the cylinder is cooled to condense the steam so that the piston returns to its original position. Using dimensional analysis, derive a relationship for the timescale for the inner surface of the cylinder to be significantly cooled, in terms of the thermal conductivity k, density ρ, specific heat capacity c, and wall thickness d. Why is an engine of this design very inefficient?

1.9 The number of children under 15 was about 2 billion in 2012, and is expected to be the same in 2072. The global population was 7 billion in 2012. Estimate the global population in 2072.

1.10 Explain the difference between primary energy consumption and final energy demand and why moving to renewables will help in reducing the demand for energy.

1.11 Discuss the evidence for an enhanced greenhouse effect occurring over the last 50 years.

1.12 In what way has the study of the cause of the ice ages helped with understanding the present climate change?

1.13*A region has a temperature variation that can be described by a Gaussian distribution with mean 25 °C and standard deviation 5 °C. The chance of the temperature being above 40.5 °C is 1 in 1000. Prolonged exposure to temperatures above 40 °C can be dangerous to health. If the mean temperature in this region rises by (a) 2 °C and (b) 4 °C and the standard deviation remains the same, calculate the chance that temperatures are above 40.5 °C. Discuss the relevance of your result to global warming.

1.14 The global mean temperature rise is proportional to the cumulative CO_2 emissions with a rise of ~0.5 °C per 1000 $GtCO_2$. Estimate what the temperature rise would be if we were to maintain our emissions at (a) the current level until 2100, or (b) decrease them steadily to zero by 2100.

1.15 Discuss what factors are important when considering the choice of a power plant.

1.16 What percentage of the land would be needed to provide 60 kWh per day per person from photovoltaics in (a) South America and (b) Italy?

1.17 Why are capacity factors for PV farms typically less than ~0.3, even in very sunny regions?

1.18 How does improving the efficiency of conversion of a wind turbine or solar cell affect (a) the capacity factor, (b) the power density, and (c) the levelized cost of energy?

1.19 Estimate what the capital cost per kW would need to be to make PV generation cost-competitive with gas (CCGT) generation in a region where the capacity factor is 0.15.

1.20*Show that the capital sum $C_{capital}$ needs to increase by $F = (1 + R)[(1 + R)^N - 1]/NR$, where R is the discount rate, when the construction time takes N years.

1.21 A 1 GW_e nuclear power plant cost $2500 million to build. The plant has a capacity factor of 85% and a lifetime of 30 years. Assume a discount rate of 10% and annual cost of fuel and of $O\&M_v$ of $5 per MWh and $10 per MWh, respectively. Calculate the cost of electricity when the construction time is (a) five years, (b) ten years.

1.22 Decommissioning the plant described in Exercise 1.21 will cost $1000 million. Work out what annual payment would be required to have a value of $1000 million after 40 years. Hence calculate the cost of electricity allowing for the decommissioning costs.

1.23 Explain why the effect of inflation at $I\%$ is to reduce the discount rate from $R\%$ to $(R-I)\%$.

1.24 Calculate the effect on the LCOE for solar PV power from a farm with a life of 30 years, capital cost of $800/kW, discount rate 8%, capacity factor 0.3, and O&M$_f$ of $10 kWy^{-1} of changing a) the discount rate from 8% to 6%, b) the cost per kW from $800 to $600, c) the capacity factor from 0.3 to 0.375, and (d) the life to 37.5 years. Comment on the relative importance of these changes.

1.25 Explain what actions are required to provide energy for everyone to have a good standard of living, but without putting the world at risk of dangerous climate change.

1.26* The cost $C(P)$ of an article decreases with increasing global production P and is given by $C(P) = aP^b$. Show that b is related to the fractional fall r in costs when production doubles by $b = \ln(1-r)/\ln 2$. If the production after n years is related to the initial cumulative production P_0 by $P_n = (1+g)^n P_0$, where g is the compound annual growth rate, show that the number of years $n(x)$ for the initial cost C_0 to drop by a factor x is given by $n(x) = -\ln 2 \ln x/[\ln(1-r)\ln(1+g)]$. For small r and g, show that $n(x) \approx 0.693\ln(x)/rg$.

1.27 Using eqn (1.4) and Figs 1.14 and 7.21, estimate when the LCOE from onshore wind will be: (a) one half and (b) one third of its value today.

1.28 Show that the physical dimensions of $\frac{1}{2}mu^2$ are consistent with the physical dimensions of work.

1.29 Convert (a) 1 MJ into Btu, (b) 800 kg of oil equivalent per year into kW, (c) EJ to TWh.

1.30 Using a dimensional argument, show that the aerodynamic drag force F_D on an aeroplane travelling at a speed u is proportional to ρAu^2, where ρ is the density of air and A is the frontal area of the aeroplane. How does the drag force depend on the altitude of the aeroplane?

1.31 A car weighing 1500 kg travels at 108 km h^{-1} and experiences a drag force of $\frac{1}{2}C_D\rho Av^2$ (where C_D=0.4, $\rho = 1.2$ kg m^{-3} is the air density, A= 2.5 m^2 is the frontal area, and v is the speed of the car in m s^{-1}), and a rolling resistance of 1% of its weight. Estimate the energy required in joules for a 100 km journey when (a) the journey is on the flat, (b) 10% of the journey is up and down 5% gradient hills, and (c) the car stops and starts five times during the journey. Comment on your result.

1.32 Derive an expression for the characteristic time taken for viscous effects in a fluid to dissipate in terms of the distance x (L), density ρ (ML^{-3}), and the coefficient of viscosity μ (ML^{-1}T^{-1}).

1.33 Derive a dimensionless parameter based on specific heat c (L^{2}T$^{-2}\theta^{-1}$), coefficient of viscosity μ (ML^{-1}T^{-1}), and thermal conductivity k (MLT$^{-3}\theta^{-1}$), where θ represents the dimensions of absolute temperature.

1.34 Table 1.4 shows the rise in the average global near-surface temperature and the atmospheric concentration of carbon dioxide for each year over the period 1979–2005. Analyse whether a statistical correlation exists between the two sets of data.

1.35 Consider an Archimedes' screw consisting of a helical tube of wavelength λ wound around a cylinder of radius a. Prove that the angle of elevation of the axis of the cylinder, θ, for the device to be able to raise water is such that $\tan\theta \leq \dfrac{2\pi a}{\lambda}$.

Table 1.4 Carbon dioxide[1] concentrations and global temperature differences[2] ΔT (°C) for the period 1979–2005

	1979	1980	1981	1982	1983	1984	1985	1986	1987
CO_2	336.53	338.34	339.96	341.09	342.07	344.04	345.10	346.85	347.75
ΔT	0.06	0.10	0.13	0.12	0.19	−0.01	−0.02	0.02	0.17
	1988	1989	1990	1991	1992	1993	1994	1995	1996
CO_2	350.68	352.84	354.22	355.51	356.39	356.98	358.19	359.82	361.82
ΔT	0.16	0.10	0.25	0.20	0.06	0.11	0.17	0.27	0.13
	1997	1998	1999	2000	2001	2002	2003	2004	2005
CO_2	362.98	364.90	367.87	369.22	370.44	372.31	374.75	376.95	378.55
ΔT	0.36	0.52	0.27	0.24	0.40	0.45	0.45	0.44	0.47

[1]*Source*: Hadley Centre for Climate Prediction and Research (Crown copyright).
[2]*Source*: Mauna Loa Observatory, NOAA (temperature difference with respect to the average temperature over the period 1961–90).

Derivation 1.1 Levelized cost of energy

An amount of Q pounds invested today would be worth $Q(1 + R)^n$ pounds after n years if the interest (discount) rate is R. So if A pounds are received in n years time, its present value is $A/(1 + R)^n$. For example, £100 invested today at 5% interest would be worth £110.25 after two years; conversely, the value of £110.25 of revenue two years from now would have a present value of £100.

The **present value** V_P of a series of annual payments A made over N years is therefore:

$$V_P = A/(1+R) + A/(1+R)^2 + A/(1+R)^3 + \cdots + A/(1+R)^N$$

V_P is the sum of a finite geometric series with a common ratio equal to $1/(1 + R)$ and is given by:

$$V_P = A[1 - (1+R)^{-N}]/R$$

We are now in a position to calculate the cost of producing energy by a power plant and the rate of return on the capital invested to build it. If we subtract the capital cost from the present value of the revenue, we get the **net present value** V_{NP}:

$$V_{NP} = V_P - C_{capital} \tag{1.5}$$

The **rate of return** R_{return} is given by finding the discount rate R_{return} such that $V_{\text{NP}} = 0$, i.e. when:

$$C_{\text{capital}} = A[1-(1+R_{\text{return}})^{-N}]/R_{\text{return}} \qquad (1.6)$$

Generally, a requirement for investment is that the R_{return} is greater than the discount rate.

We can calculate the **levelized cost of energy** LCOE by finding what annual revenue A_{cost} at the given discount rate R would make V_{NP} zero. This revenue A_{cost} is given by:

$$C_{\text{capital}} = A_{\text{cost}}[1-(1+R)^{-N}]/R \qquad (1.7)$$

and the fraction of capital recovered each year equals:

$$CRF = A_{\text{cost}}/C_{\text{capital}} = R/[1-(1+R)^{-N}]$$

When there are no operations and maintenance and fuel costs, the levelized cost of energy LCOE is then obtained by dividing A_{cost} by the energy produced per year E, i.e.

$$LCOE = A_{\text{cost}}/E \equiv (CRF \times C_{\text{capital}})/(8760 \times CF) \qquad (1.1)$$

where CF is the capacity factor, 8760 is the number of hours in a year, and C_{capital} is the capital cost in dollars per kW.

 For further information and resources visit the online resources
www.oup.com/he/andrews_jelley4e

2 Essentials of Thermal, Chemical, and Fluid Energy

→ **Introduction**

Thermal energy plays a vital role in modern civilization both at a domestic level and in industrial processes, notably in the conversion of fossil fuel and other sources of heat, such as nuclear reactors, biomass converters, solar collectors, and geothermal energy, into electrical energy, as well as directly in industrial processes and for heating.

In this chapter we explain the basics of thermal physics, starting from the concepts of thermal energy and heat transfer, and show how they can be applied to heating systems and the greenhouse effect, vital for maintaining the diversity of life on the planet, and to global warming. Next, we explain how the laws of thermodynamics relate to thermal power cycles and impose a fundamental limit on their efficiency for conversion of heat into work, as in a power station or in a car. We also describe various thermodynamic quantities which are useful in explaining how particular energy devices function and in describing energy changes in chemical reactions; in particular, the combustion of fuels and the storage of energy in batteries.

We then give a brief summary of the basic physical properties of fluids, as these are used in many energy devices; for example, in transferring thermal energy in heat engines and kinetic energy in wind and water turbines. We derive the conservation laws of mass and energy for a fluid in which viscous effects are ignored (known as an ideal or inviscid fluid), and show how they can be applied to situations of practical interest to derive useful information about the flow. Also, we describe the effect of viscosity on the motion of a fluid around an aerofoil, and show how the flow determines the forces acting on the aerofoil, which is fundamental to the operation of wind, and certain types of water turbines.

2.1 **Heat and Temperature**

Temperature is a characteristic of the thermal energy of a body due to the internal motion of molecules. Two bodies in mutual thermal contact are said to be in thermal equilibrium if they are both at the same temperature. Temperature was originally defined in terms of the freezing point and boiling point of water, but the modern definition of temperature is based on the efficiency of an ideal fluid working in a Carnot cycle (Section 2.4), and is independent of the properties of any particular material.

In general, apart from when a material changes phase (e.g. from solid to liquid), the temperature of any material increases as it absorbs heat. The **heat** ΔQ required to raise the temperature of unit mass of a material by an amount ΔT is given by:

$$\Delta Q = c\Delta T \tag{2.1}$$

The coefficient c is roughly independent of temperature and is called the **specific heat** of the material. The original unit of thermal energy was the **calorie**, defined as the energy needed to increase the temperature of one gramme of liquid water by one degree Celsius, but thermal energy is now usually measured in joules. The energy equivalence of the two units is given by:

$$1\,\text{calorie} \approx 4.18\,\text{joules} \tag{2.2}$$

During a change of phase of a material, heat is absorbed and the temperature remains constant. The heat ΔQ required to change the phase of unit mass of material is called the **latent heat** L. Thus:

$$\Delta Q = L \tag{2.3}$$

The **latent heat of evaporation** (from liquid to gas) is typically about an order of magnitude larger than the **latent heat of fusion** (from solid to liquid) (e.g. 2260 c.f. 334 J g^{-1} for water).

2.2 **Heat Transfer**

There are three basic forms of heat transfer: **conduction**, **convection**, and **radiation**. These are very important in describing heat exchangers, which are central to thermal power plants, and also in understanding the Earth's energy balance with solar irradiation and how the greenhouse effect arises.

2.2.1 **Conduction**

Conduction is the transfer of thermal energy within a body due to the random motion of molecules. The average energy of the molecules is proportional to the temperature. Consider a bar of length d and cross-sectional area A, with one end at a fixed temperature T_1 and the other at a fixed temperature T_2, where $T_1 > T_2$. The more energetic molecules at the hot end transfer kinetic energy to the less energetic molecules at the cold end. In the **steady state**, the rate of flow of heat $\dot{Q}$ is constant along the length of the bar and is given by **Fourier's law of heat conduction**:

$$\dot{Q} = kA \frac{(T_1 - T_2)}{d} \tag{2.4}$$

where k is called the **thermal conductivity**.

It should be noted that eqn (2.4) applies only in the steady state. In practice, it takes time for a solid body to establish a steady-state temperature distribution. For **unsteady heat conduction**, the timescale to establish a steady state is determined by the characteristic time t for an isotherm to diffuse a distance x, and for a material with **thermal diffusivity**, $\kappa = k/\rho c (\mathrm{m^2\ s^{-1}})$, where k is the thermal conductivity, c the specific heat, and ρ the density, is given by:

$$t = \frac{x^2}{\kappa} \tag{2.5}$$

The algebraic form of eqn (2.5) is easily derived by dimensional analysis (see Exercise 2.1), and how heat diffuses along a bar is explored in Exercise 2.2.

2.2.2 Convection

Many processes in thermal power plants involve the transfer of heat between two fluids, by conduction through the wall separating the fluids and by heat transfer through the fluids. Heat transfer in fluids involves both conduction through the fluid and the transport of heat through the bulk motion of the fluid, which is generally dominant, and together this is called **convective heat transfer**.

When a stationary cold fluid is in contact and above a hot surface, a layer of fluid adjacent to the wall is heated by thermal conduction. Hot fluid in this layer will then rise through buoyancy forces, as it is less dense than the surrounding cold fluid, and there is an upward heat transport (**natural convection**).

In **forced convection**, a cold fluid flows over a hot surface, and the rate of heat transfer from the surface to the fluid is greater than that for a stationary fluid. Initially, a layer of fluid adjacent to the wall is heated by thermal conduction, and this hot fluid is then transported into the body of the fluid away from the surface. As a result, the net heat transfer into the body of the fluid is greater than that in natural convection, and is generally much larger than by heat conduction alone.

In both **natural and forced convection**, the rate of heat transfer per unit area is usually expressed in the form:

$$\frac{\dot{Q}}{A} = Nu \frac{k(T_s - T_f)}{D} = h\Delta T \tag{2.6}$$

where T_s is the temperature of the hot surface, T_f is the temperature in the body of the cold fluid, $\Delta T = T_s - T_f$, Nu is a dimensionless parameter known as the **Nusselt number** (the ratio of the convective to the conductive heat flux), and $h = Nuk/D$ is the **heat transfer coefficient**, where D is a characteristic length called the **hydraulic diameter**. The choice of D depends on the geometrical set-up; e.g. for heat transfer from a pipe, it is appropriate to take D to be the diameter of the pipe. For non-cylindrical shapes $D = 4A_d/P_d$, where A_d is the cross-sectional area of the duct and P_d is the wetted perimeter.

Dimensional analysis shows that the Nusselt number for forced convection can be expressed as a function of two other non-dimensional parameters: the **Prandtl number**, $Pr = c\mu/k = v/\kappa$, and the **Reynolds number**, $Re = \rho uD/\mu = uD/v$, where v is the kinematic viscosity ($v = \mu/\rho$, where μ is the dynamic viscosity, see Section 2.11), also called momentum diffusivity, κ is the thermal diffusivity ($\kappa = k/c\rho$), and u is the velocity of the fluid. Pr depends only on the properties of the material and is a measure of the relative magnitude of the thermal and viscous boundary layers.

The numerical value of the Nusselt number in forced convection for a given Pr and Re is usually obtained from empirical correlations. These are often expressed in the following form $Nu \propto Re^a Pr^b$; the magnitude of the indices, a and b, are indicative of the relative importance of the different driving forces (see Example 2.1).

EXAMPLE 2.1

Water is heated in the economizer (see Section 3.10) of the boiler in a power station from 35 °C to 165 °C. The tube wall is at 360 °C and its diameter D is 50 mm.

Calculate the length L of tube required when the flow speed is 1 m s^{-1}, given the correlation:

$$Nu = \frac{0.04 Re^{3/4} Pr}{1 + 2Re^{-1/8}(Pr - 1)}$$

Mean temperature is 100 °C, where $\rho = 961$ kg m^{-3}, $v = 2.93 \times 10^{-7}$ m^2 s^{-1}, $Pr = 1.74$, $k = 0.68$ W m^{-1} °C^{-1}, $c = 4216$ J kg^{-1} °C^{-1}.

$$Re = 1 \times 0.05/(2.93 \times 10^{-7}) = 1.71 \times 10^5, \text{ so the flow is turbulent}$$

$$Nu = 441, \quad \text{so} \quad h = Nuk/D = 5998 \text{ W m}^{-2} °C^{-1}$$

The rate of heat input, $h \times$ (surface area) $\times \Delta T$, must equal the rate of increase of sensible heat in the water, $u \times \rho \times$ cross-sectional area $\times c \times$ increase in temperature, i.e.

$$h \times \pi D \times L \times (360 - 100) = u \times \rho \times (\pi D^2/4) \times c \times (165 - 35), \text{ so}$$

$$L = 1 \times 961 \times 0.05 \times 4216 \times 130/(4 \times 5998 \times 260) = 4.22 \text{ m}$$

In **heat exchangers**, which are widely used for thermal energy transfer, such as in boilers and in heat recovery units, a hot fluid transfers heat to a cold fluid via a thin wall of thickness d and thermal conductivity k that separates the fluids. The overall heat transfer coefficient, h, depends on that for convection from the hot fluid to the wall, h_i, conduction through the wall, k/d, and convection into the cold fluid, h_o. If the temperatures of the hot and cold fluids and hot and cold wall surfaces are T_i, T_o, T_h, and T_c, respectively, then from eqns 2.4 and 2.6 (since the heat flux $F = \dot{Q}/A$ is constant from the hot to cold fluid) we have:

$$h_i(T_i - T_h) = (k/d)(T_h - T_c) = h_o(T_c - T_o) = h(T_i - T_o) = F$$
$$T_i - T_h = F/h_i; \quad T_h - T_c = F/(k/d); \quad T_c - T_o = F/h_o; \quad T_i - T_o = F/h$$
$$F/h_i + F/(k/d) + F/h_o = T_i - T_o = F/h$$

$$\text{So} \quad h = \left(\frac{1}{h_i} + \frac{1}{h_o} + \frac{d}{k} \right)^{-1} \tag{2.7}$$

We will use this expression in Exercise 2.6, where we look at the performance of a counterflow heat exchanger.

The heat transfer through a surface into a liquid is enhanced when the surface temperature passes the boiling point as a large amount of heat (latent heat) is absorbed in the **phase change** forming vapour bubbles. The rate of bubble formation also increases sharply when the surface temperature is a little above the boiling point (the superheat). This process, called nucleate boiling, gives a very high heat transfer for a constant small temperature difference. Likewise, large heat transfers are obtained on condensing a vapour. These conditions are ideal for heat engines, and also enable very high effective thermal conductivities for heating and cooling devices by using heat pipes.

A common design of a **heat pipe** is a tube lined with a porous material (the wick) in which a small amount of liquid is sealed. One end is attached to the heat source and the other to the heat sink. At the hot end the liquid absorbs heat and vaporizes. The vapour travels quickly along the tube to the cold end where it condenses and ejects its latent heat to the sink. The resultant liquid is transported by capillary action back along the wick to the hot end to continue the heat transfer. Effective thermal conductivities more than an order of magnitude higher than with conduction can be achieved (see Exercise 2.8*). Heat pipes are used, for example, in cooling electronic components and in solar thermal units (see Chapter 5 Section 5.1).

2.2.3 Radiative Heat Transfer

Radiative heat transfer is the transport of energy by electromagnetic waves. Unlike conduction and convection, heat can be transferred by radiation in a vacuum. Opaque materials absorb, reflect, and emit radiation, while transparent materials, such as glass, also transmit. A black surface absorbs visible light, and one that absorbs all incident radiation of any wavelength is called a **black body**.

A good practical approximation to a black body is a cavity with a small pinhole that connects it with the outside environment. Since nearly all of the radiation entering the pinhole is absorbed by the surface inside the cavity before it can escape back out of the pinhole, the absorptivity α is very close to unity and the energy distribution of the radiation emitted by the pinhole is effectively determined by the temperature of the surface inside the cavity.

The amount of energy radiated per unit area per second (i.e. the power per unit area) from a surface at a temperature T is given by the **Stefan–Boltzmann law**:

$$\mathcal{P}_e = \varepsilon \sigma T^4 \tag{2.8}$$

where $\sigma \approx 5.67 \times 10^{-8}$ W m^{-2} K^{-4} is the Stefan–Boltzmann constant, and ε is the **emissivity** of the surface, a dimensionless number that varies between 0 to 1 depending on the nature of the surface, with a black body having $\varepsilon = \alpha = 1$.

The wavelength distribution of the emitted light from a black body depends only on the absolute temperature, and is given by the Planck function (see Exercise 2.12). Fig. 2.1 shows the black-body spectrum for three different temperatures: 6000 K, 288 K (15 °C), and 473 K (200 °C). The first two approximate the spectra emitted by the Sun and the Earth, respectively, the third is from a surface at 200 °C, and all are normalized to have the same intensity. The wavelength where the intensity is a maximum is inversely proportional to the absolute temperature (Wien's law) and lies in the visible for $T = 6000$ K and in the infrared for both $T = 288$ K and 473 K (see Exercise 2.12).

The emissivity and absorptivity of a surface (or substance) depend on the spectral emissivity ε_λ and absorptivity α_λ over the range of wavelengths emitted by a black body at the same temperature. ε_λ and α_λ are generally a function of temperature and wavelength and satisfy the condition:

$$\alpha_\lambda = \varepsilon_\lambda$$

i.e. the fraction of incident radiation of wavelength λ absorbed by a surface equals the fraction of black-body radiation emitted with wavelength λ. For example, the surface of white paint at room temperature has $\alpha_\lambda (=\varepsilon_\lambda)$ quite close to zero in the visible but quite close to unity in the infrared. Conversely, a surface designed to maximize the absorption of solar radiation will have $\alpha_\lambda (=\varepsilon_\lambda)$ close to unity in the visible (i.e. it will appear black), and will have $\alpha_\lambda (=\varepsilon_\lambda)$ close to zero in the infrared to minimize radiation loss, as the peak of the emitted radiation from a surface heated up to 200 °C (see Fig. 2.1) lies in the infrared (see Exercise 2.10).

Glass is a selective material since it transmits visible light well but absorbs infrared radiation. This property is what increases the solar gain in buildings and in greenhouses, as the glass absorbs the infrared radiation emitted by the internal surfaces that have been heated by the Sun's rays (the solar gain). The temperature of the glass is usually quite close to ambient, so the re-radiated radiation, which is emitted both inward and outward, has a low intensity as it depends on T^4. Glass windows also reduce convective losses, which otherwise would generally dominate.

An opaque surface at a temperature T absorbs radiation from the environment, which is at a temperature T_0, as well as emitting radiation. The emissivity ε_0 of the surface when at T_0 equals its absorptivity for black-body radiation at T_0. Assuming its absorptivity at T is the same as at T_0, i.e. ε_0, and approximating the radiation from the environment to be that of a black body, the *net* rate of emission per unit area per second is given by:

$$\mathcal{P} = \mathcal{P}_e - \mathcal{P}_a = \varepsilon \sigma T^4 - \varepsilon_0 \sigma T_0^4 \tag{2.9}$$

where ε is the emissivity of the surface at T.

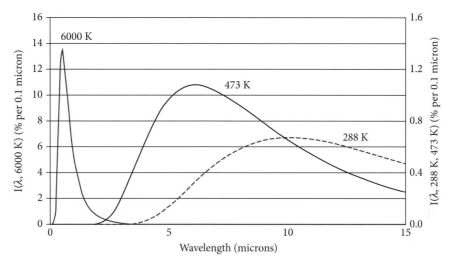

Fig. 2.1 Black-body spectra for $T = 6000$ K, 473 K, and 288 K.

The temperature of the outer surface of the Sun determines the flux of radiation incident on the upper atmosphere of the Earth (see Example 2.2 and Chapter 8 Section 8.1). Also, radiation is the dominant mode of heat transfer in the furnace of a fossil-fuel power plant (see Chapter 3 Section 3.6).

EXAMPLE 2.2

Estimate the solar power per square metre incident at the Equator. Assume that the surface temperature of the outer surface of the Sun is 5830 K, the emissivity is unity, the radius of the Sun is 6.96×10^8 m, and the distance from the Sun to the Earth is 1.52×10^8 km.

The total power emitted by the Sun is the power per unit area per second, eqn (2.8), multiplied by its surface area, i.e.

$$P_s = \sigma T^4 \times 4\pi r_s^2 \approx (5.67 \times 10^{-8}) \times (5.83 \times 10^3)^4 \times 4 \times \pi \times (6.96 \times 10^8)^2 \approx 3.99 \times 10^{26}\,\text{W}$$

The fraction of solar power incident on 1 m^2 at the equator is the solid angle subtended from the Sun,

$$\Omega = \frac{1}{4\pi d^2} \approx \frac{1}{4 \times \pi \times (1.52 \times 10^{11})^2} \approx 3.44 \times 10^{-24}$$

where d is the distance from the Earth to the Sun. Hence the incident solar power per unit area at the Equator is $P_s\Omega \approx (3.99 \times 10^{26}) \times (3.44 \times 10^{-24}) \approx 1.37\,\text{kW m}^{-2}$.

2.3 The Greenhouse Effect

The effect of the atmosphere on the transmission of radiation is very important in determining the temperature at the surface of the Earth. As discussed qualitatively in Chapter 1 Section

1.3, the absorption of infrared radiation by the atmosphere gives rise to the **greenhouse effect**, which raises the Earth's surface temperature by about 15 °C.

As estimated in Example 2.2, the solar radiation incident on the Earth's upper atmosphere has an intensity of $S = 1.37$ kW m^{-2}; this number is known as the **solar constant**. A fraction of this radiation, called the **albedo**, is reflected by clouds in the atmosphere and by the surface of the Earth back into space. The albedo A is close to 30%. The remaining radiation is absorbed by an area $(1 - A)\pi R^2$ of the Earth's surface (i.e. $(1 - A)$ times the cross section facing the Sun), where R is the radius of the Earth. The effect of the albedo is as if the solar intensity is $(1 - A)S$ over the whole cross section of the Earth (πR^2) facing the Sun. As a result, the surface of the Earth heats up until it emits as much radiation as it receives. The radiation emitted from a surface at room temperature is in the infrared. We ignore the geothermal heat flux at the surface of the Earth, which is much smaller than the incident solar flux.

We first consider what would happen if the Earth's atmosphere did not absorb any of the incident solar radiation (which is mainly in the visible part of the spectrum), nor any infrared radiation from the Earth's surface. Let A_0 be the value of the albedo under these conditions and S be the incident solar intensity, and assume the Earth's surface has an emissivity of 1 (i.e. it acts like a black body). In equilibrium, when the Earth's surface is at a temperature T, we can see that:

$$\left(1 - A_0\right) S\pi R^2 = 4\pi R^2 \sigma T^4$$

noting that radiation is emitted by the whole of the Earth's surface (area $= 4\pi R^2$); with no reflection off clouds in the atmosphere, the remaining reflection from the Earth gives $A_0 \approx 0.1$. Putting $S = 1.37$ kW m^{-2} yields $T = 272$ K $= -1$ °C; i.e. some 15 °C colder than what we observe. (If A_0 is taken as ~0.3 the temperature would be ~255 K.)

We now consider a more realistic description, where we neglect the small absorption by the atmosphere of the incident solar radiation but include the absorption of most of the infrared radiation emitted by the Earth, with a fraction t transmitted. The surface of the Earth loses energy by radiation, but also by the evaporation of water and its subsequent release of latent heat in the atmosphere and by thermals. More of the convective and surface infrared radiation is absorbed in the lower atmosphere than in the upper atmosphere, so the temperature is higher nearer the surface. This means that the down radiation is higher than the up radiation (see Fig. 2.2).

We will assume that (a) the surface loss is cS_E where S_E is the infrared radiation emitted by the Earth, with c greater than one to account for the convective losses, and (b) the flux radiated downward by the atmosphere is gS_a, with g greater than one to account for the lower atmosphere being hotter, where S_a is the flux emitted upward by the atmosphere. Under these conditions, the atmosphere would heat up to some temperature T_a such that the energy radiated into space plus the fraction transmitted of infrared radiation radiated by the Earth and the fraction of incident solar radiation reflected was equal to what it received; i.e.

$$S_a 4\pi R^2 + tS_E 4\pi R^2 + AS\pi R^2 = S\pi R^2 \tag{2.10}$$

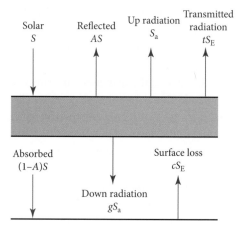

Fig. 2.2 Greenhouse effect.

The energy loss by the Earth equals the solar energy incident on the Earth plus the infrared radiation radiated downward by the atmosphere, i.e.

$$cS_E 4\pi R^2 = (1-A)S\pi R^2 + gS_a 4\pi R^2 \tag{2.11}$$

Multiplying the first equation by g and adding gives:

$$4\left(gt+c\right)S_E = (1+g)(1-A)S$$

Assuming the emissivity of the Earth is 1.0, then $S_E = \sigma T_E^4$. Substituting yields:

$$T_E^4 = \frac{(1+g)(1-A)S}{4(gt+c)\sigma} \tag{2.12}$$

Approximate values for the parameters t, g, and c are: $t = 0.1$, $g = 1.33$, and $c = 1.3$. As $t < c$, we can see from this expression that as the downward infrared radiation emitted by the atmosphere increases, i.e. g increases as CO_2 levels rise, the Earth's surface temperature T_E increases. Taking $A = 0.3$, the mean temperature of the surface of the Earth is 288 K.

The upward radiation from the atmosphere is given by:

$$S_a = \frac{(c-t)(1-A)S}{4(c+gt)} \tag{2.13}$$

From this expression, increasing g, corresponding to rising CO_2, causes S_a and hence T_a to decrease. The increase in the surface of the Earth's temperature T_E means that more infrared radiation is transmitted through the atmospheric window. As a result, the temperature at the stratopause must drop so that the radiation at that level can be in balance. This small fall in the temperature of the stratopause has been observed and is consistent with more CO_2 rather than a larger incident solar radiation, which would cause both T_E and T_a to increase. [The effect of more CO_2 on the ozone in the stratosphere also causes a lowering in T_a of a similar magnitude.]

The rise in surface temperature due to the increased absorption of infrared radiation from the anthropogenic emissions of additional greenhouse gases is called **global warming**. While this model is a simplification, it shows that increasing the concentration of greenhouse gases will raise the surface temperature of the Earth.

The lower layer of the atmosphere, the troposphere, is the part of the Earth's atmosphere close enough to the Earth's surface to be strongly thermally coupled to it, so that extra heating (or cooling) of either warms (or cools) both together. The troposphere contains ~80% of the mass of the atmosphere and is typically about 20 km deep near the Equator and 10 km near the poles. Within the troposphere the transfer of heat is by convection and radiation and the temperature falls approximately linearly with height. This **lapse rate** is determined by convection (see Box 2.1 on radiation-convective equilibrium).

Box 2.1 Radiative–convective equilibrium

The Earth's surface radiates infrared radiation that is absorbed to a greater or lesser extent by the atmosphere dependent on its wavelength. We can approximate this absorption by taking an average depth of atmosphere above which the density of air is sufficiently low that all of the infrared radiation escapes to space. Assuming no absorption of the incoming solar radiation then in this model the temperature at this **effective radiating level** (ERL), which is ~5–10 kilometres high, will be ~255 K. At this temperature the incoming and outgoing radiation flux balance when in equilibrium, taking the Earth's albedo as 0.3.

Below this height, convection and radiation dominate over conduction as the method of heat transfer within the atmosphere, with convection determining the temperature profile. A small 'parcel' of air in contact with the ground that is heated by conduction and radiation will rise, because its density is lower than that of the surrounding air, and so will a parcel of air higher in the atmosphere that is heated by radiation. As the parcel of air rises it expands (since the pressure falls) and it does work pushing against the atmosphere. There is no significant heat loss by conduction or by radiation, as heat transfer by convection is much faster. Hence, the motion is almost adiabatic, and the internal energy and therefore the temperature of the air both fall. For a dry atmosphere the rate of change of temperature with height, the **lapse rate**, is given by $dT/dz = -g/c_p$, where c_p is the specific heat at constant pressure and g is the Earth's gravitational acceleration (see Exercise 2.16*). Substituting the value of c_p for air at STP (standard temperature and pressure, which is 0 °C and 1 bar) gives a lapse rate of ~10 °C km^{-1}. The atmosphere is generally moist and a typical lapse rate is ~5 °C km^{-1}; the smaller rate is caused by the latent heat given out as the moisture condenses when the parcel of air cools on rising.

The effect of increasing the level of greenhouse gases (GHGs) in the atmosphere is to increase the ERL of the atmosphere. The consequence of this is to raise the surface temperature, as illustrated in Fig. 2.3, as the lapse rate stays the same.

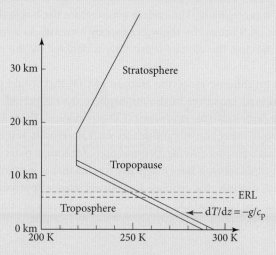

Fig. 2.3 A schematic plot of the variation in temperature of the atmosphere with height, illustrating the effect of a change in the height of the effective radiating level (ERL) of the atmosphere caused by a change in the concentration of greenhouse gases.

Increasing the concentration of greenhouse gasses in the atmosphere raises the height above which the absorption is sufficiently low that effectively all the infrared radiation escapes to space. As the temperature of this effective radiating level (ERL) is fixed (i.e. to balance the incoming and outgoing radiant energy), the temperature of the surface of the Earth rises, since the lapse rate stays the same.

2.3.1 Radiative Forcing

The radiative forcing that causes global warming is defined as the difference in the absorbed and radiated energy per unit area in Wm^{-2} at the top of the troposphere (tropopause) caused by changes in greenhouse gas concentrations, solar radiation, or other climatic influences such as aerosols. The IPCC estimates the radiative forcing relative to pre-industrial conditions (1750). The radiative forcing for 2011 from the increase in CO_2 concentration is estimated to be in the range 1.4–2.2 Wm^{-2}, and from the effect of aerosols 0 to −2 Wm^{-2}; the total is in the range 1.1–3.3 Wm^{-2}.

Roughly speaking, of the outgoing long wavelength radiation (OLR), about half comes from water vapour and half originates from the surface, clouds, and other emitters. An increase in CO_2 reduces the OLR initially and the surface temperature increases until the height of the ERL has risen sufficiently that its temperature and emission is the same as it was before. But the rise in temperature causes more water to evaporate increasing the absorption above the ERL so the surface temperature has to rise further. To a good approximation the relative humidity remains constant in the atmosphere, which means that the absorption by the mass of water vapour above the ERL is only a function of the temperature of the ERL, and therefore increases to what it was originally. The surface temperature rise has to double to compensate for this reduction in OLR caused by the water vapour increase. As a result, the temperature rise per amount of forcing, the climate sensitivity, is doubled by the positive water vapour feedback.

For the different greenhouse gases, the amount of CO_2 that would give the same change in the absorbed infrared radiation in the atmosphere over a period of time (usually 100 years) is termed the *carbon dioxide equivalent* (CO_2eq) amount. It can be applied not only to greenhouse gases but to other climatic influences such as aerosols and land-use changes. For example, methane has a positive forcing 28 times that of the same mass of CO_2 (global warming potential of 28), while aerosols have a negative forcing that is accounted for by subtracting the CO_2eq amount. Currently, the estimated negative forcings approximately cancel out the positive forcings, making CO_2 and CO_2eq concentrations similar at 395 ppmv and 430 ppmv in 2012 (IPCC), respectively, where ppmv stands for parts per million by volume.

2.3.2 Global Mean Surface Temperature

The resultant mean surface temperature of the Earth as a result of a radiative forcing F is determined by the energy balance $F = N + R$, where R is the additional outward radiation to space caused by the global warming of the Earth's surface temperature by ΔT, and is linearly related to ΔT by $R = \alpha \Delta T$, where α is the climate feedback parameter. N is the global mean net downward radiation, which equals the rate of change of the heat content of the climate system and is zero when equilibrium has been reached. To a good approximation, N is the heat flow into the oceans since this takes up 90% of the downward flux. We will approximate the radiative forcing as that due to an increase in CO_2, and will consider first the surface warming when long-term equilibrium has been reached; i.e. $N = 0$.

When $N = 0$, the radiative forcing $F = R = \alpha \Delta T$ can be used to estimate the equilibrium change in the global surface temperature ΔT_E through the relation $\Delta T_E = F/\alpha = \lambda F$, where λ ($=1/\alpha$) is called the climate sensitivity parameter, and is ~0.8 K W^{-1} m^2. But this warming is only fully realized after centuries because of the long oceanic time constants. Additional carbon dioxide mixes quickly in the troposphere in a matter of weeks, but the time for both carbon dioxide and temperatures in the oceans to come into equilibrium is much longer: surface layers over decades, and deep ocean over centuries. The surface layers of the oceans are well mixed by winds, but there is only a slow exchange of surface waters and waters deep below, and it takes many centuries to thoroughly mix. What matters for the environment, climate, and for energy policy is what happens to surface warming over the next hundred years.

2.3.3 Surface Warming Proportional to Cumulative Emissions of CO_2

The transient climate response to emissions (*TCRE*) is the ratio of surface warming to cumulative emissions over a time scale that allows the shallow mixed layer of the ocean to approach equilibrium with the increased forcing but not with the deep ocean; this is a period of about 100 years. Climate models find that *TCRE* is a constant to a good approximation for a wide range of emission scenarios. That global warming is proportional to the cumulative emissions is now being used to set targets for global emissions. And it is the *cumulative emissions of carbon dioxide* that largely determine the global warming, because of their size and CO_2's long lifetime of around 200 years.

This result arises from a compensation between three main effects governing the temperature rise. One is how the temperature is determined by the amount of CO_2 in the atmosphere, the second is how this amount depends on the cumulative emissions, and the third is how the ocean heat uptake varies. The change in amount of CO_2 sets the radiative forcing and this in turn causes a rise in surface temperature given by:

$$F = R + N = \alpha \, \Delta T + N \tag{2.14}$$

The net downward heat flow, N, will fall with time as the surface waters heat up and stratify, so the rise ΔT per degree of forcing will increase with time. This can be modelled by letting $N = k \, \Delta T$, with k decreasing with increasing emissions. The forcing is dependent on the factor a by which the CO_2 in the atmosphere has increased through:

$$F = s \ln a \equiv s \ln \left(1 + \varphi \, \Sigma C / C_a\right) \tag{2.15}$$

(see exercise 2.17) where φ is the air fraction of the cumulative carbon dioxide emissions ΣC, and C_a is the initial amount of carbon dioxide in the atmosphere. The forcing $F \approx \varphi \, s \Sigma C / C_a$, for small ΣC, and can be accurately represented by letting s decrease with increasing ΣC. Then *TCRE*, which is the surface temperature warming divided by the cumulative emissions ΣC, is given by:

$$TCRE = (\varphi \, s / C_a)/(\alpha + k) \tag{2.16}$$

While emissions are sustained and CO_2 levels are rising, *TCRE* is approximately constant. This arises because the declining value of s is compensated for by the falling value of k as ΣC increases, and by the air fraction φ increasing as stratification slows the ocean uptake of carbon dioxide. Moreover, this compensation holds for a variety of different emission scenarios. The effects of absorption by biomass on land, as well as by the oceans, and of other GHGs must be taken into account, but the long-term warming is largely determined by the cumulative emissions of CO_2.

Simulations of what happens if emissions are halted show that the warming remains about constant for many decades to centuries as the slow absorption of CO_2 by the oceans, which gives a cooling, is offset by the warming caused by the continuing decrease in heat uptake by the oceans. Only after many thousands of years, when equilibrium is restored through the slow mixing of surface and deep ocean waters, will a significant fraction of the emitted CO_2 be in the ocean.

The amount of CO_2 taken up by the ocean is more than the aqueous fraction because of the formation of bicarbonate ions through $CO_2(aq) + H_2O \leftrightarrow H_2CO_3 \leftrightarrow H^+ + HCO_3^-$. Some of the hydrogen ions interact with the carbonate ions in the ocean, $H^+ + CO_3^= \leftrightarrow HCO_3^-$. The net result is a slightly increase in acidity and decrease in carbonate ions, and these changes, called **ocean acidification**, adversely affect marine life by, for instance, affecting shell formation. [This can be offset by increasing the alkalinity by the addition of bicarbonate ions, which will sequester carbon in the oceans.]

There is also an increase in the dissolved inorganic carbon (DIC $= CO_2(aq) + CO_3^= + HCO_3^-$). The ratio of the percentage increase in dissolved carbon dioxide (which equals that in atmospheric CO_2) to that in DIC is known as the **Revelle factor** R ($\approx DIC/CO_3^=$), and is about 10. Larger R implies less CO_2 that can be taken up by the ocean.

About half of the emitted CO_2 since the start of the Industrial Revolution has been absorbed by the oceans, with currently about a quarter of the annual CO_2 emissions absorbed there, another quarter by land plants and trees, with the remaining half added to the atmosphere.

2.4 Laws of Thermodynamics and the Efficiency of a Carnot Cycle

The maximum possible efficiency of a thermal power plant can be obtained from the laws of thermodynamics without needing to consider the details of the fluid flow and heat transfer processes involved in the various stages of the plant.

The **first law of thermodynamics** is a statement of energy conservation taking thermal energy into account. Consider the system enclosed by the control volume V shown in Fig. 2.4. By energy conservation, the sum of the heat input to the system Q and the work done on the system W is equal to the change in the internal energy ΔU of the system, i.e.

$$Q + W = \Delta U \qquad (2.17)$$

In order to convert heat into useful work in a **steam power plant**, the working fluid undergoes a change of phase at different stages in a **closed cycle** (i.e. the working fluid is reused), from liquid water, to a two-phase mixture of water and steam, to dry steam, and back to liquid water (Fig. 2.5). The key stages in the cycle are:

1. compressor	(also known as the **boiler feed pump**) work W_{com} done on the system to compress cold water from sub-atmospheric pressure to high pressure;
2. boiler	heat Q_1 added to the system to convert cold water into steam;
3. turbine	work $-W_t$ done by the system (i.e. by steam) on the turbine blades;
4. condenser	heat $-Q_2$ lost from the system to the environment in converting steam back to cold water.

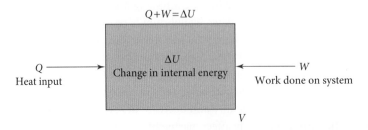

Fig. 2.4 First law of thermodynamics (note the convention that work done on the system and heat input are positive quantities).

After each complete cycle, the working fluid has the same internal energy U, so the net change in internal energy is zero, or $\Delta U = 0$. By the first law of thermodynamics (eqn (2.17)) we have:

$$(Q_1 + Q_2) + (W_{\text{com}} + W_t) = 0$$

Hence the efficiency of the process is given by:

$$\eta = \frac{[\text{net work output}]}{[\text{heat input}]} = \frac{-W_t - W_{com}}{Q_1} = \frac{Q_1 + Q_2}{Q_1} = 1 - \frac{(-Q_2)}{Q_1} = 1 - \frac{|Q_2|}{|Q_1|} \tag{2.18}$$

Thus, the efficiency is unity minus the ratio of the heat output $|Q_2|$ in the condenser and the heat input $|Q_1|$ in the boiler. For a perfect system there would be no heat loss in the condenser and all the heat supplied in the boiler would be used to do useful work. However, the supplied heat increases the dispersal of energy or disorder (i.e. **entropy**, see Section 2.8) of the steam, which is at best unchanged when the steam does work when passing through the turbine. As a result, some heat must be lost by the working fluid in the condenser to the environment, thereby reducing the molecular disorder of the fluid (i.e. entropy) back to its original value, the amount depending on the temperature of the condenser. Hence $|Q_2| > 0$ and $\eta < 1$. It follows that there is an upper limit to the efficiency of a thermal power plant, and the thermal energy ejected heats the external environment.

From a thermodynamic point of view, the fact that the system is less than 100% efficient is a consequence of the **second law of thermodynamics**: that no system operating in a closed cycle can convert all the heat absorbed from a heat reservoir into the same amount of work. Carnot proved that the maximum possible efficiency of a heat engine operating in a closed cycle between two heat reservoirs depends only on the ratio of the absolute temperatures of the reservoirs, i.e.

$$\eta_C = 1 - \frac{T_2}{T_1}, \text{ since } \frac{-Q_2}{Q_1} = \frac{|Q_2|}{|Q_1|} = \frac{T_2}{T_1} \text{ in a Carnot cycle} \tag{2.19}$$

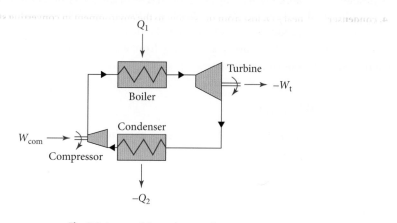

Fig. 2.5 Layout of thermal power plant.

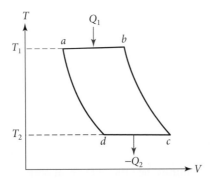

Fig. 2.6 Carnot cycle for perfect gas.

where T_1 and T_2 are the absolute temperatures of the upper and lower reservoirs, respectively, measured in kelvin.

This result follows (See Exercise 2.20) from considering an ideal gas with an equation of state $pV = nRT$, where n is the number of moles of the gas, operating in the closed cycle *abcda* shown in Fig. 2.6. Sections *ab* and *cd* are isotherms at T_1 and T_2, respectively; *bc* and *da* are reversible adiabatics (i.e. no heat transfer with the surroundings). This closed cycle is called a **Carnot cycle**.

2.5 Useful Thermodynamic Quantities

There are six key quantities that are useful in describing the thermodynamics of a thermal power plant: temperature T, pressure p, specific volume v (volume per unit mass, i.e. the reciprocal of density), specific internal energy u, specific enthalpy h, and specific entropy s. In general, only *two* thermodynamic quantities are needed to completely specify the thermal state of a system.

Note: Unless otherwise stated, we assume that u, h, and s are all per unit mass and use lowercase symbols, and drop the prefix *specific* in specific internal energy, etc.

In addition, the Gibbs energy G (sometimes called Gibbs free energy) is very important for determining the conditions under which a chemical reaction takes place spontaneously and the maximum amount of work available, e.g. from a battery. For chemical changes, the Gibbs energy per mole is normally used.

2.5.1 Internal Energy

The concept of **internal energy**, U, has already been introduced from the first law of thermodynamics, eqn (2.17),

$$\Delta U = Q + W$$

which states that the increase or change in internal energy of a system equals the heat flow into the system plus the work done on the system. It is a statement of the conservation of energy. The internal energy of an ideal gas is only a function of its temperature alone.

2.5.2 Enthalpy

Enthalpy is defined as:

$$h = u + pv \qquad (2.20)$$

It is useful (see Derivation 2.1) for describing:

(i) heat transfer at constant pressure (e.g. in boilers and condensers) or the heat absorbed in a chemical reaction at constant pressure from its surroundings, where the change in enthalpy $(h_2 - h_1)$ is equal to the heat input Q:

$$h_2 - h_1 = Q$$

For example, the combustion of carbon is an exothermic reaction, and the amount of heat given out equals the decrease in enthalpy, $-\Delta H$.

$$C(\text{graphite}) + O_2(g) \rightarrow CO_2(g) \quad \Delta H = -393 \text{ kJ mol}^{-1}$$

(ii) adiabatic $(Q = 0)$ compression or expansion (e.g. in compressors and turbines) in ideal fluid flow, where the net work done on the system is equal to the change in enthalpy,

$$W = h_2 - h_1$$

when changes in kinetic and potential energy of the fluid can be neglected.

(iii) passage of a fluid through an expansion valve where no work is done so the enthalpy is constant:

$$h_1 = h_2$$

In a chemical reaction, the enthalpy change is particularly useful because it gives the heat absorbed at constant pressure (a typical practical condition), as it takes account of both the internal energy change plus the change in energy that necessarily occurs from the work associated with the change in volume at constant pressure (called expansion work).

Derivation 2.1

(a) Absorption of heat by a gas at constant pressure

Consider a unit mass of gas contained in a piston tube at constant external pressure p (Fig. 2.7). Heat is added to the gas (by heating the outside of the tube) or is generated internally through a chemical reaction. The absorbed or generated heat changes the internal energy of the gas and performs expansion work.

Suppose that an elemental amount of heat $đQ$ is required to expand the volume from v to $v + dv$. The work done on the gas is $đW = -pdv$. (The same work is done on an uncontained gas expanding against a constant pressure.) By the first law of thermodynamics (eqn (2.17)), the change in internal energy is given by $du = đQ - pdv$. Since p is constant, we can put $pdv = d(pv)$ and rewrite $đQ$ in the form:

$$đQ = du + d(pv) = dh$$

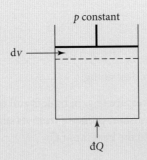

Fig. 2.7 Absorption of heat by a gas in a cylinder causing a piston to move.

Integrating both sides shows that at constant pressure the total heat absorbed equals the change in enthalpy:

$$Q = h_2 - h_1$$

(b) Adiabatic compression or expansion of a flowing fluid

Fig. 2.8 (a) shows a process in which fluid moves from the inlet A to the outlet B and work is done on the fluid by rotating a turbine shaft immersed in the fluid. (We ignore any changes in the kinetic and potential energy of the fluid.) Suppose the pressure is p_1 at A and p_2 at B. The work done in moving a unit mass of fluid through a volume v_1 at A is $p_1 v_1$ and through a volume v_2 at B is $-p_2 v_2$. Thus the net work done on the fluid is $(p_1 v_1 - p_2 v_2)$. If the amount of work added to the fluid by the turbine compressing it is W_t, then the total work done on the system is:

$$W = W_t + (p_1 v_1 - p_2 v_2)$$

From the first law of thermodynamics (eqn (2.17)), we have $Q = -W + \Delta u$, where Q is the heat flow into the system. For adiabatic processes, $Q = 0$, so $W_t = (p_2 v_2 - p_1 v_1) + u_2 - u_1 = \Delta h$, or:

$$W_t = h_2 - h_1$$

(a)   (b)

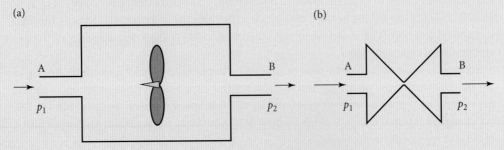

Fig. 2.8 (a) Flow of fluid past a turbine. (b) Flow through an expansion valve.

When compressing water adiabatically ($\Delta s = 0$), its volume hardly changes and its temperature and internal energy only change a very small amount, so to a good approximation $v_1 = v_2$ and $u_1 = u_2$. The compressive work is then given by:

$$W_t = (p_2 - p_1)v_2.$$

If an expansion valve is substituted for a turbine, see Fig. 2.8 (b), there will be a drop in pressure but no work done and ideally no heat flow. From the above, $\Delta h = 0$, so for this adiabatic expansion:

$$h_1 = h_2$$

2.5.3 Entropy

The concept of **entropy** arises from the second law of thermodynamics. It is a measure of how well dispersed the energy is in a system, and determines how a system will evolve. Essentially, there are two types of process whereby a system can change from one state to another: **reversible processes** and **irreversible processes**. In a reversible process, both the system and the surroundings can recover their original states, and this can be achieved by changing the system so slowly that it remains in quasi-static thermal equilibrium throughout the process. In an irreversible process, however, the system and the surroundings are changed in such a way that they are unable to return to their original states (e.g. a scrambled egg). When heat flows into a system at a temperature T, the system's energy increases and becomes more dispersed. The change in entropy is given by:

$$\Delta S = \frac{\Delta Q_{rev}}{T} \tag{2.21}$$

where ΔQ_{rev} is the heat supplied reversibly to a system at an absolute temperature T. There is therefore no change in entropy in a reversible adiabatic process.

In a reversible process the total change in entropy of a system and its surroundings is zero, whereas in an irreversible process there is a net increase in entropy. The second law can be expressed as:

$$\Delta S \geq 0$$

There is no change in entropy over a *complete* Carnot cycle, so the absorption of heat at T_1 causes an increase in entropy of Q_1/T_1 that is equal to the decrease in entropy $-Q_2/T_2$ (NB Q_2 is negative) from the ejection of heat, so:

$$-Q_2/Q_1 = |Q_2|/|Q_1| = T_2/T_1$$

from which, using eqn (2.18), we obtain the **Carnot efficiency** as: $\eta_C = 1 - |Q_2|/|Q_1| = 1 - T_2/T_1$. It should be realized that the concept of reversibility is an idealization that is unachievable in practice.

On a microscopic level, the molecules of a gas in a container occupy quantum mechanical states (microstates) with discrete energy levels. The entropy of the gas is determined by the number of ways W the energy of the particles can be allocated to the different microstates by the Boltzmann relation (where k is Boltzmann's constant-sometimes denoted k_B)

$$S = k \ln W \qquad (2.22)$$

The internal energy of the system $U = \sum_i n_i \varepsilon_i$ where n_i are the number of particles in the state with energy ε_i. A change in U is given by:

$$dU = \sum_i \varepsilon_i dn_i + \sum_i n_i d\varepsilon_i$$

where the first term corresponds to $dQ = TdS$ and the second to dW. The entropy increases when heat is added as the n_i change, but is constant when work is added as the n_i are constant.

When methane is burnt in air, the energy released in the reaction is transferred to the carbon dioxide and water molecules as heat, so the entropy increases. Other examples of entropy increasing are ice melting or a chemical dissolving in water. We will see that combining the changes in entropy and enthalpy through the Gibbs function (see Section 2.7) is very useful in describing chemical reactions. But first we will consider the maximum amount of work that is available from a process, and how this is related to the generation of entropy.

2.6 Maximum Available Work: Exergy

In any reversible isothermal process, the temperature of the fluid and of the heat source or sink are equal. In practice this condition is only approximated when the heat flow (and hence the power output) is very small. Practically useful heat flows require significant temperature differences, and these temperature differences lead to a reduction in the work that can be performed. The maximum amount of work available is called the **exergy**, X, which is a property of both the system and its surroundings.

In the transfer of heat Q across a temperature difference $(T_1 - T_2)$, the initial exergy is given by the Carnot efficiency as:

$$X = Q(1 - T_a / T_1) \qquad (2.23)$$

when the ambient temperature is T_a, as that is the maximum amount of work that could be performed. Likewise the final exergy is $Q(1 - T_a/T_2)$. The reduction in exergy in this heat transfer is $T_a(Q/T_2 - Q/T_1)$. The entropy generation in this process is given by $\Delta S = (Q/T_2 - Q/T_1)$, so the reduction equals $T_a\Delta S$.

In general, a reduction in exergy is associated with the generation of entropy. The ratio of the actual work to the maximum possible work is the exergy (or exergetic) efficiency. The analysis of the exergy efficiency of the processes that make up a system and its surroundings identifies where thermodynamic losses are greatest. As an example, consider an electrical resistance space heater.

The input exergy equals the electrical energy dissipated in the heater E_h, since electrical energy can be converted (in principle) entirely into work. The output is air heated to about 30° (T_{heated}) above the ambient temperature (T_a). While the energy efficiency is close to 100%, as nearly all the electrical energy has been converted to the thermal energy in the heated air, the exergetic efficiency is much less.

As heat is extracted from the warm air to perform work, the temperature of the air falls. When the air has a temperature T the maximum work from a small amount of heat δq is given by the Carnot efficiency as $\delta q(1 - T_a/T)$. As shown in Derivation 2.2, the total work that can be obtained after E_h has been extracted is $E_h(1 - T_a/T_{ln})$, where T_{ln} is the logarithmic mean temperature, and equals $\sim 0.05 E_h$, so the exergetic efficiency is only $\sim 5\%$.

Derivation 2.2 Exergy of heated air

The total amount of work that can be extracted from air initially at T_i and finally at the ambient temperature T_a is given by the integral of $\delta q(1 - T_a/T)$ from T_i to T_a. Noting that the extracted heat $đq = -c\,dT$ and $E_h = c(T_i - T_a)$, where c is the specific heat of air (assumed constant), then:

$$W = -\int_{T_i}^{T_a} c\left(1 - \frac{T_a}{T}\right)dT = c(T_i - T_a) + cT_a \ln\left(\frac{T_a}{T_i}\right)$$

Hence $\qquad W = E_h\left(1 - \frac{T_a}{T_{ln}}\right)$, where $T_{ln} = (T_i - T_a)/(\ln T_i - \ln T_a)$

For $T_i = 318$ K (45 °C) and $T_a = 288$ K (15 °C) then $T_{ln} = 302.8$ K and $W = 0.049 E_h$.

Thermodynamically, space heating can be achieved with 100% exergetic efficiency for an ideal process, and this would be with an ideal heat pump (see Chapter 5 Section 5.2). In practice, a heat pump can have a performance approaching 50% that of an ideal pump for a 30-degree temperature difference; i.e. an exergetic efficiency of 50%. We see that the exergetic efficiency can be a much more useful measure than the energy efficiency.

As another example, consider a thermal steam-power station operating between 850 K and 300 K with an ambient temperature of 288 K. The overall energy efficiency of the plant might be $\sim 33\%$, with most of the lost thermal energy going to the environment in the 300 K condenser. Although this is a significant fraction of the total fuel thermal energy input, it only has a small exergy value, since the heat is transferred at close to the ambient temperature. The largest loss of exergy occurs in the heat transfer into the boiler and in the combustion of the fuel, which for fossil fuels occurs at temperatures around 1300 K.

While exergetic efficiency is important, cost also has to be taken into account in practical situations. For example, in thermal power stations, temperature differences are needed for sensible amounts of heat flow, so to improve the exergetic efficiency, while maintaining the same output power, would require a larger heat transfer area, i.e. a larger and more expensive heat exchanger. Hence, there is a compromise on the size of the heat exchanger that depends on the relative cost of the fuel and the heat exchanger. When fuel costs are comparatively small, the thermal power plant will tend to be optimized for maximum power (see Chapter 3 Section 3.14).

2.7 Chemical Reactions: Gibbs Energy and Spontaneous Reactions

The statement of the second law of thermodynamics—that all practical processes proceed in a way that increases the total entropy—means that in chemical processes a reaction will take place if the change in total entropy, ΔS_t, which equals the change in entropy of the reacting system, ΔS, plus that of the surroundings, ΔS_s, increases; i.e.

$$\Delta S_t = \Delta S + \Delta S_s > 0 \tag{2.24}$$

The heat absorbed by the reacting system from its surroundings at constant pressure is the change in enthalpy ΔH in the reaction. This flow of heat causes the entropy of the surroundings to change by $\Delta S_s = -\Delta H/T$, where T is the temperature of the system and surroundings. The condition expressed by eqn (2.24) is therefore equivalent to:

$$T\Delta S - \Delta H > 0 \tag{2.25}$$

i.e. the heat generated in the system by the change in entropy of the reacting system ($T\Delta S$) less the heat absorbed from the surroundings must be positive.

The **Gibbs energy**, G, of a system is defined as:

$$G = H - TS \tag{2.26}$$

Hence, under conditions of constant pressure and temperature, the condition for a reaction to occur spontaneously is given by (since T the absolute temperature is positive):

$$\Delta G = \Delta H - T\Delta S < 0 \tag{2.27}$$

When a reaction is exothermic ($\Delta H < 0$) and the entropy of the products is higher than the reactants ($\Delta S > 0$) then the reaction will always be spontaneous, but will not be spontaneous when the reaction is endothermic and accompanied with a decrease in entropy. When ΔH and ΔS are either both positive, or both negative, then whether the reaction will proceed or not depends on the relative magnitude of the heat absorbed from the surroundings and the heat generated by the change in entropy in the reaction.

For example, consider the change in phase of water from a liquid to a gas at a pressure of 1 bar and a temperature of 298 K. It takes energy to break apart the bonds that hold the water molecules together in the liquid phase, and this latent heat of vaporization is given by the change in enthalpy: at 1 bar and 298 K, $\Delta H° = 44.1$ kJ mol^{-1}. The entropy change is positive, as water is more disordered in the gas phase than in the liquid phase, and $\Delta S° = 0.119$ kJ K^{-1} mol^{-1}, but not sufficiently so that $T\Delta S°$ outweighs $\Delta H°$ and the change in the Gibbs energy $\Delta G°$ is 8.64 kJ mol^{-1}. Hence, at 1 bar and 298 K water does not spontaneously change from its liquid to its gas phase. (The standard state symbol ° refers to standard conditions, i.e. 1 bar pressure, but does not imply anything about temperature or phase, which have to be specified.)

2.7.1 Reactions in Equilibrium and Le Chatelier's Principle

We can estimate the temperature at 1 bar when the forward and backward reaction rates in the water (vapour)–gas shift reaction,

$$CO + H_2O \leftrightarrow CO_2 + H_2 \tag{2.28}$$

are equal, i.e. the reaction is in equilibrium, for equal quantities of reactants and products by neglecting the change with temperature of $\Delta H°$ and $\Delta S°$ and finding T such that $\Delta G = \Delta H°$ $- T\Delta S° = 0$. Since $\Delta H° = -41.1$ kJ mol^{-1} and $\Delta S° = -0.042$ kJ mol^{-1} at 1 bar and 298 K, then this gives $T = 980$ K, close to the observed temperature of ~1080 K. Above this temperature, the reaction will produce an increasing amount of CO and H_2O with a corresponding decrease in the amount of H_2 and CO_2. This causes ΔG to decrease until the change in concentration of the gases corresponds to $\Delta G = 0$ and equilibrium is reached.

An increase in temperature at constant pressure of an exothermic reaction in equilibrium therefore favours the reactants. This is an example of an important rule (Le Chatelier's principle), which follows from the above discussion on the change in the Gibbs energy given by the second law of thermodynamics, and states: 'When a reaction in equilibrium is perturbed, the equilibrium is altered in the direction that reduces the perturbation.'

2.7.2 Maximum Amount of Work in a Chemical Reaction

The decrease in the Gibbs energy $-\Delta G$ in a chemical reaction also gives the maximum amount of non-expansion work that can be extracted from the reaction; see Derivation 2.3. In a hydrogen fuel cell (see Section 11.4.3) the overall reaction is given by:

$$H_2 + \frac{1}{2}O_2 \rightarrow H_2O, \quad \Delta H° = -285.8 \text{ kJ mol}^{-1} \tag{2.29}$$

The entropy of the gases decreases in this process, since the number of moles is reduced and the product is a liquid. In a reversible process the entropy of the entire system, reactants and surroundings, remains constant. As a result, at 1 bar pressure an amount of heat equal to $-T\Delta S°$, where $\Delta S°$ is the change in the specific entropy of the gases, is transferred to the surroundings; i.e. the entropy of the surroundings increases by the same amount that the entropy of the system decreases. The amount of energy available as electrical energy is the decrease in the Gibbs energy $-\Delta G°$, where:

$$\Delta G° = \Delta H° - T\Delta S° = (-258.8 + 48.7) = -237.1 \text{ kJ mol}^{-1} \tag{2.30}$$

at a temperature of 298 K. The Gibbs energy is therefore a very important quantity, as it determines both the maximum amount of work available in a chemical reaction and the conditions under which a reaction occurs spontaneously.

Derivation 2.3 The maximum amount of non-expansion work in a reaction

We will consider a chemical reaction occurring at constant temperature and pressure; e.g. as in a battery discharging. The change in the Gibbs energy dG is given by:

$$dG = dH - d(TS)$$
$$= dU + d(pV) - d(TS)$$
$$= dQ + dW + pdV - TdS$$

as p and T are constant. Since the reaction is reversible, $dQ = TdS$ and $dW = -pdV + dW_{add}$, where dW consists of expansion work and additional work, such as electrical work. Substituting for dQ and dW gives:

$$dG = dW_{add}$$

Hence the additional work done on the system is the increase or change in the Gibbs energy, which is equivalent to the statement that the maximum amount of work that can be extracted from a reaction at constant p and T is the decrease in Gibbs energy $-\Delta G$.

2.7.3 Dependence of the Gibbs Energy on Concentration

For compounds in a solution of volume V, the partial pressures p_i of the different molecules are related to their concentrations c_i by $p_i = (n_i/V)RT = c_iRT$. In a reaction the concentrations and hence partial pressures change. We can find an expression for the change in the Gibbs energy by considering the expansion of an ideal gas, for which $pV = nRT$. When n moles of an ideal gas change pressure from p_1 to p_2 at a constant temperature, the change in the Gibb's function is given by:

$$\Delta G = \int_{p_1}^{p_2} \frac{nRT}{p} dp = nRT \ln\left(\frac{p_2}{p_1}\right)$$

This follows from $dG = Vdp - SdT$ for a gas. If we take p_1 to be 1 bar as our standard pressure p° then:

$$G_2 = G^\circ + nRT \ln(p_2/p^\circ)$$

Since the partial pressures p_i of the different molecules in a solution are proportional to their concentrations c_i, the Gibbs energy of each compound can be written by analogy as:

$$G_i = G_i^o + nRT\ln(c_i/c^o)$$

where the concentration of the standard state c^o is often taken to be 1 molar. The Gibbs energy per mole is the chemical potential μ, so the chemical potentials of the compounds in solution are given by:

$$\mu_i = \mu_i^o + RT\ln(c_i/c^o) \tag{2.31}$$

(NB This only holds at sufficiently low concentrations that the interactions between the molecules in solution can be ignored, as in an ideal gas; when this does not hold an effective concentration, called the activity, is used.)

Consider the chemical reaction:

$$A + 2B \leftrightarrow 2C + D \tag{2.32}$$

This can be written in the form $\sum_i v_i N_i = 0$, where $v_A = -1$, $v_B = -2$, $v_C = 2$, and $v_D = 1$. When 1 mole of A and 2 moles of B react, the resulting change ΔG in the Gibbs free energy per mole of reactants is given by:

$$\Delta G = \sum_i v_i \mu_i \tag{2.33}$$

If the initial concentrations c_i are c_A, c_B, c_C and c_D, then using eqn (2.31) the change ΔG is given by:

$$\Delta G = \sum_i v_i[\mu_i^o + RT\ln(c_i/c^o)] = \sum_i [v_i\mu_i^o + RT\ln(c_i/c^o)^{v_i}]. \tag{2.34}$$

which can be re-expressed as:

$$\Delta G = \Delta G^o + RT\ln Q \tag{2.35}$$

where the reaction quotient Q is given for eqn (2.32) by $Q = \dfrac{a_C^2 a_D}{a_A a_B^2}$, where the activity $a_x = c_x/c^o$ for sufficiently low concentrations, and ΔG^o by

$$\Delta G^o = -RT\ln K \tag{2.36}$$

where K, the equilibrium constant, is the value of the reaction quotient Q when the reaction is in equilibrium.

2.7.4 Rate of Chemical Reactions and Catalysts

A chemical reaction will occur spontaneously, under conditions of constant pressure and temperature, if the change in Gibbs energy is negative. But the rate of reaction will depend on the temperature and pressure of the reactants and on the reaction mechanism. An example of a reaction with a simple mechanism is:

$$NO_2 + CO \leftrightarrow NO + CO_2$$

For the forward reaction, the probability that NO_2 and CO molecules collide is proportional to the product of their concentrations $[NO_2][CO]$. As the NO_2 and CO molecules come closer, the interaction between them is at first repulsive and then attractive, i.e. there is a potential barrier that has to be overcome for the reaction to occur. The height of this barrier is called the *forward activation energy*, E_{fA}. An estimate for the probability that the molecules interact is the probability that they have sufficient kinetic energy to overcome the barrier, which depends on the temperature of the gas as $\exp(-E_{fA}/k_BT)$, where k_B is the Boltzmann constant. The overall forward reaction rate dR_f/dt is then given by:

$$dR_f/dt = k[NO_2][CO]$$

where k is the **rate constant** and is given in this model by $k = A\exp(-E_{fA}/k_BT)$, with A determined by the diameter of the molecules and their relative speed, proportional to $T^{1/2}$. In the transition-state model, the reaction proceeds by the two molecules forming a transient state at the activation energy E_{fA} (i.e. at the top of the potential barrier). The activation energy for the reverse reaction E_{rA} of $NO + CO_2$ forming $NO_2 + CO$ is higher, and the difference equals the change in enthalpy of the reaction, $E_{fA} - E_{rA} = \Delta H$.

A reaction can be accelerated by using a catalyst to bring together the reactants and to provide a route with lower activation energy. An example is the use of a metal, such as copper, to facilitate the formation of methanol via the reaction:

$$3H_2 + CO_2 \rightarrow CH_3OH + H_2O$$

The interaction of hydrogen molecules with the metal atoms on the surface of the catalyst gives rise to hydrogen atoms bonded to metal atoms, $H_2 + 2Cu \rightarrow 2Cu\text{-}H$. These reactive hydrogen atoms then combine with carbon dioxide molecules that bind to adjacent metal atoms on the surface to form (via a series of intermediary reactions) methanol.

The activation energy in the reaction of hydrogen and carbon dioxide when both are bonded to the catalyst is lower than when isolated hydrogen and carbon dioxide molecules interact. The catalyst provides an alternative route for the reaction to occur with lower activation energy. The catalyst is neither consumed nor chemically changed in the process.

2.7.5 Combustion

The key chemical reaction in the case of fossil fuels is that of **combustion**. This is the reaction of a substance with oxygen, and provides the energy for many thermal engines, as in internal combustion engines, gas turbines, and fossil fuel or biofuel fired power plants. An example is the combustion of methane, the principle component of natural gas:

$$CH_4 + 2O_2 \rightarrow CO_2 + 2H_2O(\ell) \quad \Delta H° = -889\,kJ\,mol^{-1} \tag{2.37}$$

The reaction is *exothermic*, and under standard conditions (1 bar, 298 K) H_2O is a liquid. The heat of combustion is $-\Delta H°$ and is called the *higher heating value* (HHV). In many practical applications the gaseous products of a combustion are above 100 °C and H_2O is a vapour, and under these conditions the energy released is less by essentially the latent heat of vaporization of water (since the sensible heat released is small). This lower value is called the *lower heating value* (LHV) and is related to HHV by:

$$HHV = LHV + n \times 44 \text{ kJ mol}^{-1} \qquad (2.38)$$

where n is the number of moles of water condensed in the combustion of one mole of fuel.

In the combustion of methane, eqn (2.37), the bonding of the atoms has changed from four C—H bonds in methane and two double O=O bonds in oxygen to two double C=O bonds in carbon dioxide and four O—H bonds in water. The number of bonds (counting a double bond as two bonds) is the same before and after combustion, and the energy released (the heat of combustion) comes about from the difference in these bond energies.

The bond energy of O=O is less than that of C=O by 306 kJ mol^{-1} and that of two C—H bonds is less than that of two O—H bonds in H_2O by about 110 kJ mol^{-1}. This would give an estimate of about 832 kJ mol^{-1} for methane, quite close to the actual value of 889 kJ mol^{-1}, and shows that the main contribution to the heat of combustion comes about from the relatively weak double bond in oxygen, with the rest from the stronger O—H bonds in water compared to the C—H bonds in the fuel.

This model can be generalized to predict the heat of combustion of a general organic compound $C_cH_hO_o$. The formula for the HHV is:

$$HHV = -418(c + 0.3h - 0.5o)\text{kJ mol}^{-1} \qquad (2.39)$$

which gives good agreement for many organic compounds (standard deviation 3.1%). The factor in the brackets multiplying 418 is approximately the number of O_2 molecules combusted per molecule of fuel.

A similar analysis shows that this formula is also a good approximation when applied to compounds containing nitrogen as well as carbon, hydrogen, and oxygen. For example, in the combustion of ammonia,

$$4NH_3 + 3O_2 \rightarrow 2N_2 + 6H_2O \quad \Delta H° = -1531 \text{ kJ} \qquad (2.40)$$

applying eqn (2.39) predicts $\Delta H° = -1505$ kJ (note that there are 4 NH_3 molecules on the left-hand side of the equation) within 2% of the experimental value.

EXAMPLE 2.3

Estimate the LHV of octane, the principal component of petrol (gasoline), and the mass of carbon dioxide produced when 1 kg of octane burns.

The combustion of octane is represented by the reaction

$$C_8H_{18} + 12.5O_2 \rightarrow 8CO_2 + 9H_2O$$

From eqn (2.41) the HHV $= -418(8 + 18 \times 0.3) = 5601$ kJ mol^{-1}. The LHV is given by eqn (2.40) as HHV $-44 \times 9 = 5205$ kJ mol^{-1} The molecular weight of octane is $(8 \times 12 + 18) = 114$ g. Therefore, the LHV $= 45.7$ MJ kg^{-1} compared with the measured value of 44.4 MJ kg^{-1}; a 3% error.

From the equation describing the combustion, 114 kg of octane produces $(8 \times 44) = 352$ kg of CO_2, so 1 kg produces 3.09 kg of CO_2, or 0.84 kg of carbon. (NB The 3.67 factor between the amount of emissions as C or as CO_2.)

2.8 Basic Physical Properties of Fluids

In Sections 2.4, 2.5, and 2.6 we showed that it is possible to describe the energy transfer processes in boilers, condensers, and turbines using basic thermodynamic principles *without* a detailed knowledge of the fluid flow processes involved in each device. However, in order to understand energy conversion processes in hydropower, wave power, and wind power, a basic knowledge of fluid mechanics is essential.

2.8.1 The Bulk Physical Properties of a Fluid

Density (ρ). Mass per unit volume of a fluid. Unless otherwise stated, it is assumed throughout the book that the density of a fluid is constant (called **incompressible flow**); the variations in pressure arising from fluid motion (see Example 2.6) are small in comparison with atmospheric pressure. The unit of density is kg m^{-3}. ($\rho_{water} \approx 10^3$ kg m^{-3} and $\rho_{air} \approx 1.2$ kg m^{-3} at $T = 20$ °C and $p = 1$ atm.)

Pressure (p). Force per unit area in a fluid. Pressure acts in the direction normal to the surface of a body immersed in a fluid. The unit of pressure is the pascal (Pa; 1 Pa $= 1$ N m^{-2}; 1 atm ≈ 1 bar $= 10^5$ Pa).

Viscosity. Force per unit area due to internal friction in a fluid arising from the relative motion between neighbouring elements in the fluid. Viscous forces act in the direction tangential to the surface of a body immersed in a flow (see Section 2.11).

2.9 Streamlines and Stream-tubes

A useful concept for visualizing a velocity field is to imagine a set of **streamlines** parallel to the direction of motion at all points in the fluid. Any element of mass in the fluid flows along a notional **stream-tube** bounded by neighbouring streamlines (Fig. 2.9). In practice, streamlines can be visualized by injecting small particles into the fluid. For example, smoke can be used in **wind tunnels** to investigate the flow over wings, turbine blades, cars, buildings, etc.

2.9.1 Mass Continuity

One of the fundamental laws of fluid mechanics is **conservation of mass** (also known as **mass continuity**). Consider the flow along a stream-tube in a steady velocity field. Suppose that

Fig. 2.9 Streamtube (ρ is constant).

the speed of the fluid and the cross-sectional area of the stream-tube at any point are u and A, respectively. By definition, the direction of flow is parallel to the boundaries of the stream-tube, so the fluid is confined to the stream-tube and the mass flow per second is constant along the stream-tube. Hence:

$$\rho u A = \text{const.} \tag{2.41}$$

Thus the speed of the fluid is inversely proportional to the cross-sectional area of the stream-tube (Example 2.4).

EXAMPLE 2.4 Flow along a stream-tube

An incompressible ideal fluid flows at a speed of $1\ \mathrm{m\ s^{-1}}$ through a pipe of 1 m diameter in which a constriction of 0.1 m diameter has been inserted. What is the speed of the fluid inside the constriction?

Putting $\rho_1 = \rho_2$ and using eqn (2.39), we have $u_1 A_1 = u_2 A_2$, or:

$$u_2 = u_1 \frac{A_1}{A_2} = \left(1\ \mathrm{m\ s^{-1}}\right) \times \left(\frac{1}{0.1}\right)^2 = 100\ \mathrm{m\ s^{-1}}$$

2.10 Energy Conservation in an Ideal Fluid: Bernoulli's Equation

In many practical situations, viscous effects are much smaller than those due to gravity and pressure gradients over large parts of the flow field. We can then ignore viscosity to a good approximation in these regions and derive an equation known as **Bernoulli's equation** (or **Bernoulli's theorem**) for **energy conservation** in a fluid. For steady flow, Bernoulli's equation is of the form:

$$\frac{p}{\rho} + gz + \frac{1}{2}u^2 = \text{const.} \tag{2.42}$$

(For a proof of Bernoulli's equation, see Derivation 2.4)

For a stationary fluid, $u = 0$ everywhere in the fluid, and eqn (2.42) reduces to:

$$\frac{p}{\rho} + gz = \text{const.} \tag{2.43}$$

Eqn (2.43) is the equation for hydrostatic pressure. It shows that the fluid at a given depth z is all at the same pressure p (see Example 2.5).

EXAMPLE 2.5 Hydrostatic pressure

The atmospheric pressure on the surface of a lake is 10^5 N m^{-2}. Assuming the water is stationary, what is the pressure at a depth of 10 m? (Assume $\rho_{water} = 10^3$ kg m^{-3} and $g = 10$ m s^{-2}.)
From eqn (2.43), we have $p_1/\rho + gz_1 = p_2/\rho + gz_2$. Putting $p_1 = 10^5$ N m^{-2} at $z_1 = 0$, and $z_2 = -10$ m, we have $p_2 = p_1 - \rho g(z_2 - z_1) = 10^5 - (10^3)(10)(-10) = 2 \times 10^5$ N m^{-2}.

The significance of Bernoulli's equation is that it shows that the pressure in a moving fluid decreases as the speed increases. The practical importance of this effect is illustrated in Example 2.6 and Exercises 2.32–2.34.

EXAMPLE 2.6 Effect of wind on air pressure

Assuming the pressure of stationary air is 10^5 N m^{-2}, calculate the percentage change in pressure due to a wind of 20 m s^{-1} (assume $\rho_{air} \approx 1.2$ kg m^{-3}).
From eqn (2.42) we have $p_1/\rho + \frac{1}{2}u_1^2 = p_2/\rho + \frac{1}{2}u_2^2$. The change in pressure is given by
$p_2 - p_1 = \frac{1}{2}\rho(u_1^2 - u_2^2) = -\frac{1}{2}(1.2)(20)^2 = -2.4 \times 10^2$ N m^{-2}. Hence, the percentage change in pressure is $-(2.4 \times 10^2)/(10^5) \times 100 \approx -0.24\%$.

Derivation 2.4 Bernoulli's equation for steady flow

Consider the steady flow of an ideal fluid in the control volume shown in Fig. 2.10.

The height, cross-sectional area, speed, and pressure at any point are denoted by z, A, u, and p, respectively. The increase in gravitational potential energy of a mass δm of fluid between z_1 and z_2 is $\delta mg(z_2 - z_1)$. In a small time interval δt the mass of fluid entering the control volume at P_1 is $\delta m = \rho u_1 A_1 \delta t$, and the mass leaving P_2 is $\delta m = \rho u_2 A_2 \delta t$.

In order for the fluid to enter the control volume it has to do work to overcome the pressure p_1 exerted by the fluid. The work done in pushing the elemental mass δm a small distance $\delta s_1 = u_1 \delta t$ at P_1 is $\delta W_1 = p_1 A_1 \delta s_1 = p_1 A_1 u_1 \delta t$. Similarly, the work done in pushing the elemental mass out of the control volume at P_2 is $\delta W_2 = -p_2 A_2 \delta s_2 = -p_2 A_2 u_2 \delta t$ (note change of sign). The net work done is $\delta W_1 + \delta W_2 = p_1 A_1 u_1 \delta t - p_2 A_2 u_2 \delta t$. By energy conservation, this is equal to the increase in potential energy plus the increase in kinetic energy, so that:

$$p_1 A_1 u_1 \delta t - p_2 A_2 u_2 \delta t = \delta mg(z_2 - z_1) + \frac{1}{2}\delta m(u_2^2 - u_1^2)$$

Putting $\delta m = \rho u_1 A_1 \delta t = \rho u_2 A_2 \delta t$ and tidying up, we obtain:

$$\frac{p_1}{\rho} + gz_1 + \frac{1}{2}u_1^2 = \frac{p_2}{\rho} + gz_2 + \frac{1}{2}u_2^2$$

Finally, since points P_1 and P_2 are arbitrary it follows that:

$$\frac{p}{\rho} + gz + \frac{1}{2}u^2 = \text{const.}$$

everywhere along the stream-tube.

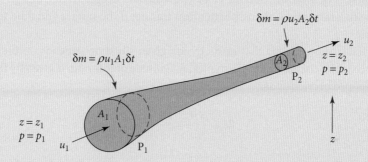

Fig. 2.10 Control volume for Bernoulli's equation.

2.11 **Dynamics of a Viscous Fluid**

In general, the motion of a viscous fluid is more complicated than that of an inviscid fluid. Consider a fluid flowing over a flat surface. Owing to strong forces of attraction between the fluid and the surface, the fluid next to the surface is at rest. The velocity u increases with distance y into the fluid stream in a *thin* boundary layer containing a large velocity gradient, until it attains the velocity of the bulk of the flow. The viscous shear force per unit area in the fluid is proportional to the velocity gradient, i.e.

$$\frac{F}{A} = -\mu\frac{du}{dy} \tag{2.44}$$

where μ is known as the **coefficient of dynamic viscosity**. The drag is given by the shear force at the surface.

A viscous fluid can exhibit two different kinds of flow regime: **laminar flow** and **turbulent flow**. For a boundary layer in laminar flow (Fig. 2.11 (a)), the fluid slides along distinct stream-tubes and is predominantly stable, but in turbulent flow the motion is disorderly and

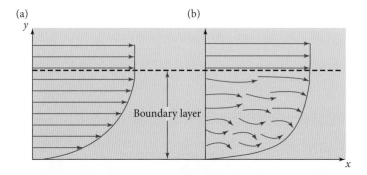

Fig. 2.11 Velocity variation in viscous boundary layer in (a) laminar and (b) turbulent flow.

more unstable (Fig. 2.11 (b)), and the velocity gradient is steeper. While this gives more friction than with laminar flow, there is better heat transfer if the surface and fluid temperatures are different.

The particular flow regime that exists in any given situation depends on the ratio of the inertial force to the viscous force. The typical magnitude of this ratio is given by the **Reynolds number**, defined previously as:

$$Re = \frac{\rho UL}{\mu} = \frac{UL}{v} \tag{2.45}$$

where U, L, and $v = \mu/\rho$ are the characteristic speed, the characteristic length, and the *kinematic viscosity* of the fluid, respectively. For the flow over a surface, L is the thickness of the boundary layer and U is the velocity of the bulk of the fluid. Re is named after Osborne Reynolds, who conducted pioneering experiments on flow in pipes. He discovered that flows at small Re are predominantly laminar, whereas flows at large Re contain regions of turbulence.

For fluid flow in pipes, the flow is mainly laminar when $Re < 2300$ or turbulent for $Re > 4000$, and is transitional for intermediary values. The pressure drop Δp is related to the average velocity of flow by the Darcy-Weisbach empirical equation:

$$\Delta p = f_D \frac{\rho l v^2}{4b} \tag{2.46}$$

where l is the length of the pipe, b its radius, ρ is the density of the fluid, and f_D is the *Darcy friction factor*. The mean velocity of the flow v, is given by the volume rate of flow Q divided by the cross-sectional area of the pipe, and f_D is a function of the Reynolds number and the surface roughness of the pipe (both dimensionless quantities). When the pipe is smooth and the flow is laminar, $f_D = 64/Re$, but for turbulent flow f_D is larger and is evaluated using empirical relations. For laminar flow, eqn (2.46) can be rearranged in terms of Q to give *Poiseuille's equation*

$$Q = \pi \frac{\Delta p b^4}{8 \mu l} \tag{2.47}$$

EXAMPLE 2.7 Reynolds number

Estimate the Reynolds number for:

(a) treacle flowing over a plate ($v = 10^{-1}$ m^2 s^{-1}, $U = 10^{-2}$ m s^{-1}, $L = 10^{-2}$ m), and

(b) air flowing around a jet aircraft ($v = 1.5 \times 10^{-5}$ m^2 s^{-1}, $U = 3 \times 10^2$ m s^{-1}, $L = 10$ m).

The Reynolds numbers in each case are given by:

(a) $Re = \dfrac{UL}{v} \approx \dfrac{10^{-2} \times 10^{-2}}{10^{-1}} \ll 10^3$

(b) $Re = \dfrac{UL}{v} \approx \dfrac{3 \times 10^{-2} \times 10^1}{1.5 \times 10^{-5}} \gg 10^3$

Hence in (a) the flow is predominantly laminar but in (b) the flow contains regions of turbulence.

Another important aspect of the Reynolds number is that two different flows with the same Reynolds number, i.e.:

$$Re = \frac{\rho_1 U_1 L_1}{\mu_1} = \frac{\rho_2 U_2 L_2}{\mu_2}$$

exhibit *geometrically similar* behaviour. This has important engineering applications because it implies that results obtained from tests on a small scale can be applied to a full-scale model with the same Reynolds number (see Exercise 2.40).

We can derive the above algebraic form of Re from the following dimensional consider-ations. Consider a fluid flowing with speed U through a cross-sectional area A of order L^2, where L is some characteristic length (e.g. the diameter in the case of flow around a cylinder). The mass flowing per second is $\sim \rho U L^2$, so the inertial force (i.e. the rate of change of momen-tum) is $\sim \rho U L^2 \times U = \rho U^2 L^2$. Also, from eqn (2.44) the viscous force $\sim \mu A U/L \sim \mu U L$. Hence the ratio of the inertial force to the viscous force is of order:

$$Re = \frac{[\text{inertial force}]}{[\text{viscous force}]} \approx \frac{\rho U^2 L^2}{\mu U L} = \frac{\rho U L}{\mu} = \frac{UL}{v} \tag{2.48}$$

Viscous effects in a real fluid are important in determining the flow around objects. Fig. 2.12 shows the flow around a cylinder for (a) an inviscid fluid and (b) a viscous fluid. For inviscid flow the velocity fields in the upstream and downstream regions are symmetrical. It follows that the corresponding upstream and downstream pressure distributions are also symmetrical, so the net force exerted by the fluid on the cylinder is zero! This startling result is in contradiction with common experience and is an example of **d'Alembert's paradox**.

For a body immersed in a viscous fluid, the component of velocity tangential to the sur-face of the fluid is zero at all points on the surface of the body. At large Reynolds numbers ($Re > 4 \times 10^3$), the viscous force is negligible in the bulk of the fluid but is very significant in a **viscous boundary layer** close to the surface of the body. Rotational components of flow known as **vorticity** are generated within the boundary layer. At a certain point (known as the

(a)   (b)

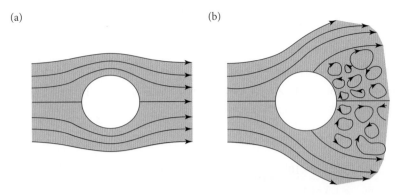

Fig. 2.12 Flow around a cylinder for (a) an inviscid fluid and (b) a viscous fluid.

separation point) the boundary layer becomes detached from the surface and the vorticity is discharged into the body of the fluid. The vorticity is transported downstream of the cylinder in the **wake**. As a result, the pressure distributions on the upstream side and the downstream side of the cylinder are not symmetrical in the case of a viscous fluid, so the cylinder experiences a net force in the direction of motion, known as the **drag force**.

When the flow in the boundary layer is turbulent, rather than laminar, the higher velocity close to the surface means that the separation occurs later and the extent of the turbulent wake is less and so is the drag. (This is why golf balls have dimples in them to ensure that the boundary layer is turbulent.) Owing to the symmetry of the velocity field above and below the cylinder, there is no component of force normal to the direction of the flow past the cylinder. However, for a spinning cylinder, a force (called the **lift force**) arises at right angles to the direction of flow due to the circulating airflow.

2.12 **Lift and Circulation**

It is possible to explain lift using inviscid fluid dynamics by introducing the concept of **circulation**. In order to understand circulation, it is helpful to begin by considering why a spinning ball swerves sideways as it flies through the air, an effect well known to golfers! Spinning creates an imbalance in the pressure on either side of the ball, and generates a net force at right angles to the direction of motion. This is known as the **Magnus effect** and is illustrated in Fig. 2.13.

We consider an inviscid fluid that is both passing over and rotating around a stationary cylinder, rather than flowing over a rotating cylinder, since it illustrates the same effect and is more like the flow over a stationary aerofoil. Fig. 2.13 (a) shows the flow of a uniform stream incident on a cylinder. The velocity profile is symmetrical on the upper and lower surfaces, so the resulting pressure distribution is also symmetrical and there is no net sideways force on the cylinder. Fig. 2.13 (b) shows an inviscid fluid rotating around a cylinder, with a circumferential velocity u_θ that varies inversely with distance r from the centre of the cylinder, i.e.

$$u_\theta = \frac{\Gamma}{2\pi r} \tag{2.49}$$

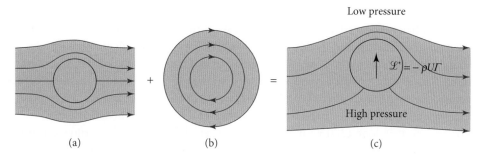

Fig. 2.13 Magnus effect: (a) flow of a uniform stream around a cylinder; (b) rotating flow around a cylinder; (c) superposition of (a) and (b).

where Γ is a constant called the **circulation** (see Fig. 2.13 (b)). Note that the circulation is positive if counter-clockwise. Superposing the velocity profiles shown in Figs. 2.13 (a) and (b) produces a velocity field in which the fluid moves faster on the upper side than on the lower side (Fig. 2.13 (c)). It follows from Bernoulli's equation (eqn (2.42)) that the pressure is smaller on the upper side than on the lower side, so that a net upward force is exerted on the cylinder at right angles to the incident stream. Its magnitude per unit length of cylinder is $\mathcal{L}^* = -\rho U \Gamma$ (see Exercise 2.46*).

2.13 **Flow over an Aerofoil**

While the circulation in the flow shown in Fig. 2.13 arises because the cylinder is spinning, a similar flow pattern occurs over an aerofoil, as shown in Fig. 2.14.

The flow can be described by the flow of a uniform stream over the aerofoil plus a clockwise circulating flow. If the circulation is Γ, then the lift force per unit length in the upward direction is given by the **Kutta–Joukowski lift theorem**:

$$\mathcal{L}^* = -\rho U \Gamma \tag{2.50}$$

This expression is in fair agreement with experimental observation, despite the fact that the inviscid theory allows the fluid to slip over the surface of the body. Eqn (2.50) also describes the lift force acting on a spinning cylinder.

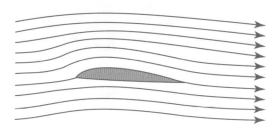

Fig. 2.14 Flow over two-dimensional aerofoil.

The physical justification for the concept of circulation around an aerofoil arises from the way that vorticity is generated when an aerofoil starts to move from rest. In the early stages, vorticity is generated around the leading edge, which is swept towards the trailing edge and then shed downstream in the wake, leaving an equal and opposite rotational flow around the aerofoil. This is why aircraft have to wait on the runway to allow time for the shed vortices generated by the previous aircraft to disperse. The effect of viscosity is therefore to produce circulation, and the lift force on the aerofoil can then be derived from inviscid theory.

For an infinitely long (2D) aerofoil there is no net deflection of the airflow an infinite distance downstream, but for a finite length (3D) aerofoil there is one, as only a finite extent h of the air above and below the aerofoil is involved in the flow. This is the situation for the flow over an aircraft wing, and can be modelled by considering the flow over an infinite stack of uniformly-spaced 2D aerofoils.

Consider the control volume $A_1A_2B_2B_1A_1$ enclosing unit length of a single aerofoil, as shown in Fig. 2.15 The streamlines Ψ_1 and Ψ_2 separate the flow passing over neighbouring aerofoils, a constant vertical distance h apart. The vertical planes A_1A_2 and B_1B_2 are chosen to be in regions of roughly uniform flow. Since there is no mass flow across a streamline, it follows that $u = U$, where U is the horizontal velocity upstream and u the horizontal component downstream. There is also no momentum transfer across the streamlines Ψ_1 and Ψ_2, and, by symmetry, there is no pressure gradient in the vertical direction on the streamlines Ψ_1 and Ψ_2.

There is flow downwards, and the rate of change in momentum of the fluid in this direction (by Newton's third law) will give rise to a reactive force upwards on the aerofoil. We will define f_y as the force acting on unit length of aerofoil, and positive when upward, as shown in Fig. 2.15.

The mass flow per second crossing B_1B_2 is ρUh, so f_y is given by:

$$f_y = \rho Uhv \tag{2.51}$$

The circulation Γ is given by the closed integral $\Gamma = \oint_C u.ds$, where C is taken to be the contour

$C = A_1A_2B_2B_1A_1$. Noting that the contribution along A_1A_2 is zero, and the contributions along the streamlines Ψ_1 and Ψ_2 cancel one another, the only non-zero contribution is along B_2B_1, given by:

$$\Gamma = -hv \tag{2.52}$$

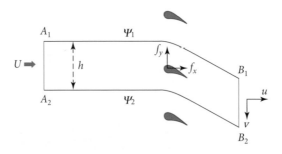

Fig. 2.15 Two-dimensional flow through a cascade of aerofoils.

Hence:

$$f_y = -\rho U \Gamma \qquad (2.53)$$

We see that this lift is the same as that given by the **Kutta–Joukowski lift theorem**, eqn (2.50), so lift can also be described in terms of the rate of change of vertical momentum and Newton's third law, as well as in terms of circulation.

2.14 **Lift and Drag Forces on an Aerofoil**

Since the aerofoil imparts a component of velocity v downward, the kinetic energy of the flow has increased. So power must be supplied, equal to $F_D U$, where F_D is called the induced drag force, and is given by:

$$F_D U = \frac{1}{2} \rho U h s v^2 \qquad (2.54)$$

where $\rho U h s$ is the mass of the air that is swept downwards with a velocity v. The effective area of air affected by the wing is equal to that of circle of diameter s, so $hs = \pi s^2/4$. The total force upward on the aerofoil of length s equals the weight of the aeroplane mg, so $f_y s = mg$. Combining these relations and eqn 2.51 with eqn 2.54 yields:

$$F_D = \frac{2(mg)^2}{\rho \pi s^2 U^2} \qquad (2.55)$$

The total drag force on an aeroplane is the sum of the induced drag and normal drag $\mathcal{D}$ that arises from the effects of viscosity. The normal drag of an object in a fluid will depend on the fluid density, kinematic viscosity, its frontal area A, and on the relative speed of the object through the fluid U. For a blunt object where the impinging air comes essentially to rest then the rate of change of the momentum of the fluid would be $\rho A U^2$ and would be the magnitude of the drag $\mathcal{D}$ on the object. Using dimensional analysis, the drag can be expressed as:

$$\mathcal{D} = \frac{1}{2} C_D \rho A U^2 \qquad (2.56)$$

where the drag coefficient C_D is dimensionless and depends on the Reynolds number Re of the flow, as well as on the shape and surface texture of the object. For a blunt object at large Re, C_D is about unity and has little dependence on the viscosity, and hence on Re, while for a streamlined object viscosity is more significant.

Likewise, dimensional analysis shows that the lift on an aerofoil is of the form

$$\mathcal{L} = \frac{1}{2} C_L \rho A U^2 \qquad (2.57)$$

where the lift coefficient C_L is a dimensionless function of the Re number. From eqn (2.56) the normal drag depends on U^2 while the induced drag depends on U^{-2}, so there is an optimum speed when the total drag is a minimum, and this determines the speed at which passenger aeroplanes fly (see Exercise 2.44).

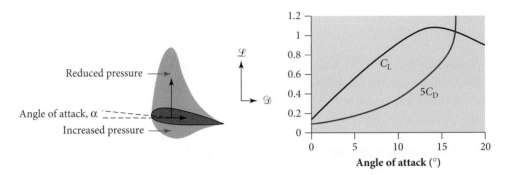

Fig. 2.16 Lift and drag coefficients and pressure distribution for an aerofoil.

According to inviscid flow theory, for flow over an aerofoil with a sharp trailing edge (see, for example, Acheson 1990) and an angle of attack α, there is only one value of the circulation such that the velocity is finite at all points on the surface of the aerofoil, given by:

$$\Gamma = -\pi U c \sin \alpha \tag{2.58}$$

where c is the width (or chord) of the aerofoil. Combining eqns (2.53) and (2.58), we can write the lift force on a single aerofoil of length s in the form

$$\mathcal{L} = f_y s = -\rho U s \Gamma = \pi \rho U^2 s c \sin \alpha \tag{2.59}$$

where we have put $\mathcal{L} = \mathcal{L}^* s$. Eqn (2.59) is of the same algebraic form as eqn (2.57), obtained from simple dimensional analysis. The essential difference is that eqn (2.59) gives an explicit expression for the dimensionless lift coefficient, i.e. $C_L = (2\pi s c / A) \sin \alpha = 2\pi \sin \alpha$. This predicts that C_L equals unity at an angle of attack of 9°, which is close to what is observed (see Fig. 2.16).

Birds are able to control the lift and drag forces by changing the shape of their wings, the ruffle of their feathers, and the angle of attack of their wings relative to the incident flow. Humankind has copied nature in designing the shape of an **aerofoil** for aircraft wings and turbine blades. For small angles of attack (tilt of wing to the horizontal), the pressure distribution on the upper surface of an aerofoil is significantly lower than that on the lower surface, resulting in a vertical lift force on the aerofoil.

Fig. 2.16 also shows the variation of the lift and drag coefficients C_L and C_D with angle of attack for a typical aerofoil, [NB For ease of visualization the drag coefficient has been enlarged by a factor of 5.] The lift and drag coefficients C_L and C_D cannot be determined by dimensional analysis because they depend on non-dimensional parameters such as the Reynolds number Re, the shape of the body, and the surface roughness. In practice, C_L and C_D are obtained from wind tunnel tests on model shapes or from numerical models of the flow.

SUMMARY

- The equations governing heat transfer are described; in particular, the rate at which energy is radiated from a surface of unit area is given by the Stefan–Boltzmann law: $\mathcal{P}_e = \varepsilon \sigma T^4$

- The first and second law of thermodynamics are explained, and the Carnot efficiency $\eta_c = 1 - T_2 / T_1$ is defined.

- The use of entropy, exergy, enthalpy, and the Gibbs energy in understanding processes is discussed.

- The basic physical properties of fluids are described, and Bernoulli's equation:
$$p / \rho + gz + \frac{1}{2}u^2 = \text{const. is derived.}$$

- The lift forces on aerofoils, and the Kutta–Joukowski theorem are explained.

 ## FURTHER READING

Acheson, D.J. (1990). *Elementary fluid dynamics*. Clarendon Press, Oxford. Good mathematical introduction to fluid mechanics.

Blundell, S. and Blundell, K. (2006). *Concepts in thermal physics*. Oxford University Press, Oxford. Very good textbook on thermal physics.

Cengel, Y. and Boles, M. (2014). *Thermodynamics: An engineering approach*, 8th edn. McGraw Hill. Provides a large number of engineering examples.

Douglas, J.F., Gasiorek, J.M., and Swaffield, J.A. (2001). *Fluid mechanics*. Prentice-Hall, Englewood Cliffs, NJ. Textbook on fluid mechanics; good discussion of dimensional analysis and of turbines.

Ingram, W. (2010). *A very simple model for water vapour feedback on climate change*. Q. J. R. Meteorol. Soc. 136: 30–40. Clear explanation for the general size of water vapour feedback.

Jaffe, R. and Taylor, W. *The Physics of Energy*, Cambridge University Press, Cambridge (2018). Detailed and clear discussion of the background physics.

Taylor, F.W. (2005). *Elementary climate physics*. Oxford University Press, Oxford. Good introduction.

http://pubs.acs.org/doi/10.1021/acs.jchemed.5b00333 Good explanation of why combustions are always exothermic, yielding about 418 kJ per mole of O_2.

www.chem1.com/acad/webtext/thermeq/index.html Useful introduction to the thermodynamics of chemical equilibrium.

 ## EXERCISES

2.1 Derive the form of eqn (2.5) using dimensional analysis. Estimate the characteristic timescale for heat to conduct through a heat shield of thickness 1 cm ($\rho = 5 \times 10^3$ kg m^{-3}, $k \approx 10^{-1}$ Wm^{-1} °C^{-1}, $c \approx 10^3$ J kg^{-1} °C^{-1}.)

2.2* Consider a uniform bar $x \geq 0$ which is initially at temperature $T = 0$, where ρ, c and k are the temperature, density, specific heat and thermal conductivity of the material, respectively. Suppose that the temperature at $x = 0$ is raised to $T = T_0$ for times $t \geq 0$. To a good approximation, we can suppose that the temperature is zero for distances

$x \geq s(t)$, and assume a simple quadratic function of the form $T(x,t) = T_0 \left[1 - \dfrac{x}{s(t)} \right]^2$, for the temperature profile between $x = 0$ and $x = s(t)$.

Assuming that $s(t) = \sqrt{12\kappa t}$, where $\kappa = \dfrac{k}{\rho c}$ is the diffusivity, show that (a) the heat flux $-k\dfrac{\partial T}{\partial x}$ is zero at $x = s(t)$, and (b) the heat flux $-k\dfrac{\partial T}{\partial x}$ at $x = 0$ equals the rate of increase of stored heat $\rho c \displaystyle\int_0^{s(t)} \dfrac{\partial T}{\partial t}\, dx$ in the bar between $x = 0$ and $x = s(t)$.

How long does it take for the temperature at some point x to reach $\dfrac{1}{2}T_0$? Compare your answer with that from the exact solution $T(x,t) = \mathrm{erfc}\left(\dfrac{x}{2\sqrt{\kappa t}} \right)$, where erfc is the complementary error function.

2.3* Find out how far a room would need to be below the surface of the Earth for the seasonal variation in temperature to be reduced to 10%. Model the summer–winter variation in temperature by assuming the surface temperature of the Earth varies as $T = A \cos \omega t + T_0$, and try a solution to the heat conduction equation $\dfrac{\partial T}{\partial t} = \kappa \dfrac{\partial^2 T}{\partial x^2}$, where $\kappa = \dfrac{k}{\rho c}$, of the form: $T = A \exp(-\beta x)\cos(\beta x - \omega t) + T_0$

The specific heat of soil $c \approx 1000$ J K^{-1} kg^{-1}, its thermal conductivity $k \approx 1$ W m^{-1} K^{-1}, and its density $r \approx 1600$ kg m^{-3}.

2.4 Re-do the calculation in Example 2.1 for a tube with a diameter of 10 mm, with the rest of the data unchanged.

2.5 Using the empirical correlation for turbulent flow in a pipe, $Nu = \dfrac{0.5 f\, Re\, Pr}{1 + 2 Re^{-1/8}(Pr - 1)}$, where $f \approx 0.08 Re^{-1/4}$, calculate the Reynolds number Re required to give a Nusselt number of $Nu = 100$ for a fluid with a Prandtl number $Pr = 3.5$.

2.6 A counter-flow heat exchange consists of two concentric copper tubes with wall thicknesses $d = 0.5$ mm and lengths 1.5 m, and diameters 15 mm and 21 mm. The hot water flow through the inner tube and the counterflowing cold water flow in the annulus between the inner and outer tubes are both equal to 2 litres per hour, and their input temperatures are 80 °C and 20 °C, respectively. Calculate their output temperatures when thermal equilibrium has been established. Assume the outer cylinder is well insulated. The Nusselt number when there is a constant heat transfer rate and the water flow is laminar is equal to 4.36. The thermal conductivities of water and copper are 0.58 and 400 W m^{-1} K^{-1}, respectively. The specific heat of water is 4180 J kg^{-1} K^{-1}.

2.7 Explain the principles of a water filled copper heat pipe operating between 55 °C and 45 °C.

2.8* A 10 mm diameter heat pipe of length $L = 300$ mm contains water and is lined with a wick of cross-sectional area $A = 50$ mm^2. Darcy's law gives the volume flow rate $\dot{V}$ through the wick when the pressure drop is Δp as $\dot{V} = KA\Delta p/\mu L$, where wick permeability $K = 1.5 \times 10^{-10}$ m^2 and water viscosity $\mu = 55 \times 10^{-4}$ Pa s. The heat and volume flow rate are related by $\dot{Q} = \rho\lambda\dot{V}$, where $\rho = 1000$ kg m^{-3} and latent heat $\lambda = 2380$ kJ kg^{-1}. The maximum capillary pressure $\Delta p = 2\sigma/r$, where water surface tension $\sigma = 69 \times 10^{-3}$ N m^{-1} and wick pore radius $r = 20$ microns. Estimate the maximum heat flow allowed by capillary action, and compare to a solid copper rod with $\Delta T = 10\ °C$, and $k = 393$ $W\ m^{-1}\ K^{-1}$.

2.9 Estimate the power radiated from a black body with a surface area $1\ cm^2$ at a temperature $1000\ °C$.

2.10 In a solar cooker, 400 W of sunlight is focused on the underside of a hotplate of area $0.04\ m^2$. The receiving surface of the hotplate has an emissivity of 0.95 for radiation with a wavelength less than $2\ \mu m$ and 0.05 for wavelengths greater than $2\ \mu m$. When the hotplate has a temperature of $200\ °C$, estimate the net radiative input and compare with that obtained with a surface with a constant emissivity of 0.95.

2.11 Verify the physical dimensions of the Stefan–Boltzmann law for the power per unit area radiated by a body at temperature T, given by

$$\frac{P}{A} = \frac{2\pi^5 k^4 T^4}{15 h^3 c^2}$$

2.12* The power per unit area per unit solid angle per unit wavelength of radiation emitted from a black body is given by Planck's law:

$$u(\lambda, T) = \frac{2hc^2}{\lambda^5} \frac{1}{\exp(hc/\lambda kT) - 1}$$

Show that the wavelength λ_m where the power is maximum satisfies Wien's displacement law: $\lambda_m T = \text{const.}$

2.13 Explain qualitatively what the 'greenhouse effect' is. Why is it enhanced when the concentration of CO_2 in the atmosphere is increased, and what is the role of water vapour? What are the main concerns about the Earth's climate that arise from increased atmospheric GHG concentrations?

2.14* Consider the following simple model to describe solar radiation incident on the Earth, in which the atmosphere is included. A fraction f of the incident solar flux is absorbed by the atmosphere and a fraction A is reflected, the rest being absorbed by the Earth. Assume that the only heat transfer to the atmosphere from the Earth is by radiation and that $f = 0.25$ and $A = 0.3$. Find the temperature T_E of the surface of the Earth for radiative equilibrium.

2.15 Derive equation (2.13) for the radiation outward from the top of the atmosphere and explain its significance.

2.16* Balancing the weight $\rho g \delta z$ of an element of the atmosphere of unit area and thickness δz with the force arising from the pressure difference $-\delta p$ across it, gives the pressure variation in an atmosphere as $dp/dz = -\rho g$. Noting that $d(pV) = Vdp + pdV$, $\rho = M/V$, $C_p = C_v + R$, and $dU = C_v dT$ (for an ideal gas), show that the variation of temperature with height z in an atmosphere where the conduction of heat is negligible (adiabatic) is $dT/dz = -g/c_p$, where c_p is the specific heat at constant pressure ($\sim 1000\ J\ kg^{-1}\ K^{-1}$ for dry air).

2.17* Take the mass of CO_2 above the height z_{ERL} of the emitting radiation level (ERL) in the atmosphere as proportional to $\rho_0 \exp(-z_{ERL}/b)$, where ρ_0 is the concentration of CO_2 at ground level and b is a constant. The intensity of the black body radiation emitted at the height of the ERL for temperatures $T = 200\text{-}300$ K is approximately proportional to T. Show that the shift in the height of the ERL when the CO_2 concentration is increased by a factor of a is proportional to $\ln a$. Assuming a constant lapse rate, deduce that the radiative forcing F is also proportional to $\ln a$.

2.18 Explain qualitatively why the warming of the Earth's surface is proportional to the *cumulative* emissions of CO_2.

2.19 Discuss whether the caloric theory is consistent with (a) Fourier's law of heat conduction and (b) the first law of thermodynamics.

2.20 For an ideal gas, $pV = nRT$, where n is the number of moles of gas. By the first law of thermodynamics (2.17) we have $đQ = dU - đW = nC_v dT + p dV$, where C_v is the heat capacity per mole at constant volume. For the Carnot cycle shown in Fig, 2.6, along the isotherms ab and cd $dT = 0$, and along the adiabatics da and bc $đQ = 0$. Show that $-Q_2/Q_1 = T_2/T_1$ and that the efficiency of a Carnot cycle $\eta_C = 1 - T_2/T_1$.

2.21 Why is it important in a heat engine to have high heat transfer rates across small temperature differences, and how can this be obtained?

2.22* Two systems are in thermal contact. Energy (heat) will flow from one to the other until they are in thermal equilibrium, i.e. at the same temperature. Show that the condition that the entropy S of the combined system is a maximum is given by $\partial S_1/\partial U_1 = \partial S_2/\partial U_2 = 1/T$, where T is the equilibrium temperature and $S = S_1 + S_2$, $U = U_1 + U_2$. Hence show that $\Delta Q = T\Delta S$ and deduce that in a Carnot cycle $Q_1/T_1 = -Q_2/T_2$.

2.23 The number of states available to a molecule in a box at a temperature T is proportional to the volume V of the box. In an isothermal expansion from V_1 to V_2, show using $\Delta S = k \ln(W_2/W_1)$ that $T\Delta S = nkT\ln(V_2/V_1)$, where n is the number of molecules, and that this is equal to the work done by the expansion of the gas.

2.24 Calculate the exergy efficiency of an electrical heater which heats air to a temperature 100 °C above an ambient temperature of 15 °C.

2.25 What is the loss in exergy when a heat transfer of 50 J occurs across a temperature difference of 700 °C to 650 °C when the ambient temperature is 20 °C?

2.26 The temperature in an endothermic reaction is increased. In what direction will this change shift the reaction?

2.27 Estimate the boiling point of water by assuming that the values for $\Delta S° = 0.119$ kJ K^{-1} mol^{-1} and $\Delta H° = 44.1$ kJ mol^{-1} do not change significantly with temperature. Explain qualitatively why vaporization occurs spontaneously above the boiling point.

2.28 The water shift reaction $CO(g) + H_2O(g) \leftrightarrow CO_2(g) + H_2(g)$ is in equilibrium. Will the concentration of H_2 be increased if (a) a catalyst is added (b) the concentration of CO is increased (c) the temperature is increased?

2.29 Estimate the higher heating and lower heating values of ethanol and compare them to the measured values.

2.30 Estimate the heat of combustion of oleic acid whose composition is $C_{18}H_{34}O_2$.

2.31 What happens to the speed of cars when two slow-moving lanes of cars converge into a single lane, assuming the spacing between cars remains constant?

2.32 Why does an open-door swing shut when air blows through the doorway?

2.33 A Pitot tube uses water manometers to measure the pressure in two tubes, one with its open end facing the flow of a fluid (p_s), and the other with its open end normal to the flow (p_0). Show that the speed of the flow U is given by $U = [2(p_s - p_0)/\rho]^{1/2}$ and calculate the speed due to a difference in height of 1 cm between the manometers. The density of the fluid is ρ.

2.34 In a Venturi meter, an ideal fluid flows with a volume flow rate Q and pressure p_1 through a horizontal pipe of cross-sectional area A_1. A constriction of cross-sectional area A_2 is inserted in the pipe and the pressure is p_2 inside the constriction. Verify that the volume flow rate through the meter is given by

$$Q = A_1 u_1 = A \left[\frac{2(p_1 - p_2)}{\rho} \right]^{\frac{1}{2}}, \text{ where } A = A_1 A_2 (A_1^2 - A_2^2)^{-\frac{1}{2}}$$

2.35 Verify that the Reynolds number is a dimensionless parameter.

2.36 Estimate the Reynolds number Re for a body in an air stream, for a characteristic length $L = 10$ mm, $U = 1$ m s^{-1}, $\rho = 1.3$ kg m^{-3}, and $\nu = 10^{-6}$ m^2 s^{-1}.

2.37 Verify that all the terms appearing in Bernoulli's equation (eqn (2.42)) have physical dimensions of the form $L^2 T^{-2}$.

2.38 A fountain shoots vertically upwards with speed u_0. Use dimensional analysis to derive an expression for the maximum height h in terms of u_0 and g.

2.39 A jet of water emerges from an orifice in a dam at a depth h below the water surface. Using Bernoulli's equation, and the fact that the surface of the water in the dam and the jet are both at atmospheric pressure, show that the velocity of the jet on leaving the orifice is given by $u = \sqrt{2gh}$.

2.40 It is desired to examine the flow over a model wind turbine using water instead of air. Assuming the kinematic viscosities ($\nu = \mu/\rho$) of air and water are 1.5×10^{-5} m^{2}s^{-1} and 10^{-6} m^{2}s^{-1}, respectively, and that the model is 100 times smaller than the full size turbine, what is the ratio of the speed in the water to that in air, in order for the Reynolds number to be the same in both cases?

2.41* A viscous fluid flows in the x-direction between parallel plates at $y = 0$ and $y = b$, under the action of a pressure gradient dp/dx = const. Given the momentum equation for the fluid is $\mu \dfrac{d^2 u}{dy^2} = -\dfrac{dp}{dx}$ show that the velocity profile is given by $u(y) = -\dfrac{1}{2\mu} \dfrac{dp}{dx} y(y - b)$.

2.42 Design an experiment to examine the Magnus effect on a rotating cylinder in a flowing stream of water to investigate how the sideways force varies with the angular velocity of the cylinder and the velocity of the stream.

2.43 Estimate the lift on an aircraft with wingspan $s = 10$ m, chord $c = 2$ m, $\alpha = 1°$, flying at 900 m s^{-1}, assume $\rho = 1$ kg m^{-3}.

2.44 By considering the aerodynamic and induced drag on an aeroplane, show that the optimum speed is given by

$$u_0 = \left[\frac{4(mg)^2}{AC_D \pi s^2 \rho^2} \right]^{\frac{1}{4}}$$

where m is the mass, s is the wingspan, C_D is the drag coefficient, A is the frontal cross-sectional area of the aeroplane, ρ is the density of air, and g is the acceleration due to gravity.

Calculate the optimum speed for an airliner of mass 325 tonnes travelling at an altitude of 10 km, where the air density is 0.41 kg m^{-3}. The airliner has a drag coefficient $C_D = 0.06$, a frontal cross-sectional area $A = 110$ m^2, and a wingspan $s = 70$ m.

2.45 Derive an algebraic formula for the drag force, $\mathcal{D}$, on a sphere moving with uniform speed u in a fluid, in terms the radius of the sphere, r, the density of the fluid, ρ, and the coefficient of viscosity, μ.

2.46* The circumferential component of velocity on the surface of the cylinder from the uniform flow with speed U shown in Fig. 2.13 (a) is $u_\theta = -2U \sin \theta$; i.e. it is $2U$ at the top and bottom and zero at the points on axis (NB θ increases counterclockwise). For the circulating flow shown in Fig. 2.13 (b) $u_\theta = \Gamma/2\pi a$, where Γ is the circulation and a is the radius of the cylinder. For the superposition of these flows, Fig. 2.15 (c), $u_\theta = -2U \sin \theta + \Gamma/2\pi a$.

From Bernoulli's theorem, the pressure on the surface is given by $p = k - \frac{1}{2}\rho u_\theta^2$ where k is a constant. Show that the lift force per unit length is given by:

$$\mathcal{L}^* = -\int_0^{2\pi} pa \sin \theta \, d\theta = -\rho U \Gamma.$$

2.47 Write an article of about 100 words for a popular science magazine on the statement that 'without viscosity, birds could not fly and fish could not swim'. (NB Assume the readers have no knowledge of mathematics or fluid mechanics.)

For further information and resources visit the online resources
www.oup.com/he/andrews_jelley4e

3　Energy from Fossil Fuels

→ **Introduction**

The primary fossil fuels—**coal, oil, and natural gas**—are hydrocarbons that originated from the remains of dead plants and creatures over millions of years. They have a high carbon content and are still the dominant source of global primary energy (~86% in 2018).

The global consumption of all fossil fuels—coal, oil and natural gas—has been rising for many decades, from about 40,000 TWh in 1965 to 135,000 TWh in 2018 (see Fig. 3.1). The carbon dioxide released by the combustion of fossil fuels is the main cause of **global warming** and reducing such emissions or taking carbon out of the atmosphere both present an enormous challenge.

The global emission of CO_2 due to the combustion of fossil fuels stalled over the period 2014–2016 at ~35.6 $GtCO_2\ y^{-1}$, raising hopes that the world was starting to get on top of the problem. However, there was a rise of 2.7% by 2018, and although it slowed in 2019, it needs to fall sharply over the coming decades to avoid putting the world at risk of very significant climate change due to further global warming.

After a brief history of fossil fuels (see Box 3.1), we describe the different forms of fossil fuel, their availability, combustion, and the various thermodynamic cycles used in fossil fuel power stations. We also describe the various carbon capture and storage technologies being developed to reduce their emissions.

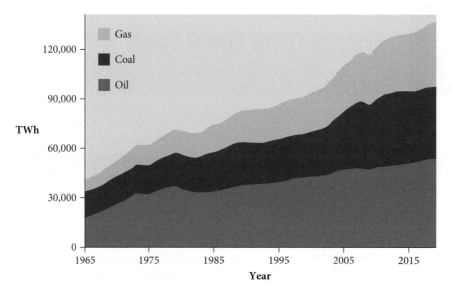

Fig. 3.1 Global consumption of fossil fuels 1965-2019 (OurWorld).

Box 3.1 A brief history of fossil fuels

Fossil fuels have transformed society and provided employment on a vast scale through-out the whole world. For thousands of years, the abundance of forests provided the wood and biomass needed to heat homes and for industry. Some coal, peat, and crude oil were used to provide heat and light, though their contribution was small, and tended to be confined to regions where this fuel was easily obtained. However, in the centuries leading up to the Industrial Revolution in Britain, cities like London were already experiencing a shortage of wood, pollution from coal burning was on the rise, and by the eighteenth century, coal was displacing charcoal in the smelting of iron.

The invention of the Watt steam engine led to a significant increase in the demand for coal, and industrial towns sprang up on both sides of the Atlantic. The high-energy content of coal also made it the fuel of choice for the early steam locomotives of the nine-teenth century, enabling people and goods to move much faster and eventually making horse-drawn transport redundant. It also revolutionized the shipping industry, with coal replacing wind as the means of propulsion—more reliable and requiring less manpower. Coal was also the predominant fuel for the vast number of power stations built in Europe and North America throughout the first half of the 20th century, and in the second half of the century it powered the huge expansion of the Chinese economy. Coal-fired power stations still dominate in the developing countries of SE Asia, where they are seen as national assets in the transformation of their economies.

Although oil had been used for thousands of years, it was not until the middle of the nineteenth century that it began to be exploited on an industrial scale, for making lubri-cation oils, paraffin wax, and kerosene for street lighting. The first oil refineries were built in Scotland, Romania, Canada, and the United States, but the start of the modern oil industry is generally credited to Edwin Drake. He devised a method of protecting the sides of a drill hole in gravel from collapsing, by installing a tube down to the bedrock below. In 1859 in Pennsylvania he struck oil at a depth of 21 m, which yielded 20–40 barrels a day. By 1870, the founding of Standard Oil by John D Rockefeller made him the richest man in the world, and demand grew with the development of internal combustion engine driven cars and machinery in the late nineteenth century.

After the Second World War, the centre of gravity of oil production shifted towards the Middle East, which held the bulk of the world's known oil reserves at the time. In 1960, the oil-rich countries of Iran, Iraq, Kuwait, Saudi Arabia, and Venezuela formed a cartel, the Organisation of the Petroleum Exporting Countries (OPEC), which used its considerable muscle to hike oil prices in the 1970s, creating havoc in western economies. By the 1980s, however, OPEC's grip had begun to weaken, due to political tensions within the OPEC bloc and the overproduction of oil, which destabilized oil prices.

Gas was first produced commercially by the gasification of coal in the early nineteenth century, and initially provided street lighting. The introduction of electric lighting towards the end of the nineteenth century led to coal gas being used for cooking and heating. Natural gas from oil fields was also employed, and by the latter part of the twentieth century it had completely displaced coal gas. Natural gas is increasingly being burned in power plants for the generation of electricity in place of the more polluting coal.

The main reserves of natural gas are in the Middle East, Russia, and the USA. After 2000, Russia emerged as a major exporter of natural gas from Siberia to Europe. In more recent years, the USA has become world's largest producer of oil and gas, made possible by high pressure fracking of its vast shale deposits, causing the balance of oil power to shift away from the Middle East. Canada, China, and Argentina are now looking at how they, too, can exploit their own shale oil and gas reserves. However, overcapacity has created a global glut of oil, causing huge swings in crude oil prices and leading some countries to adopt aggressive measures to protect their own oil interests. Venezuela has been the big loser in this epic struggle of the 'survival of the fittest'.

For the last two hundred years, the energy from fossil fuels has transformed societies across the world. The significant and flexible power (available on demand) that initially coal-fired and later combustion engines could produce has raised the average standard of living very significantly. But the massive increase in their use arising from the huge rise in the world's population and the incessant demand for goods is leading towards dangerous levels of atmospheric carbon dioxide, which has already caused significant global warming and adverse climate change.

Given the pressure to move to renewables in order to avoid CO_2 emissions, the world order established by the fossil fuel industry is looking increasingly shaky. The age of fossil fuels must be replaced by one that is greener and more sustainable. Renewables are now generally cheaper, but it will be a difficult transition as fossil fuels are so enmeshed in global society.

3.1 **Coal**

Coal is a carbon-rich solid that originated hundreds of millions of years ago in vast wetland forests and swamps. The remains of vegetable matter accumulated on the bottom of these swamps and turned into peat, which got covered over and transformed into coal through the effects of pressure and some heat. (N.B. The Earth's geothermal temperature gradient (see Chapter 5 Section 5.7) provides heat that cracks the complex organic molecules and drives off the volatile components.) Coal is graded according to its carbon content: e.g. 60–75% for lignite and 75–90% for bituminous coals. Coal also contains other combustible elements, including hydrogen and sulphur, and non-combustible materials such as water and ash-forming minerals. Coal provided about 27% of primary energy consumption in 2018 and 38.5% of energy for electricity generation (source: IEA).

Before it can be burned in power stations, coal has to be pulverized into a fine powder. For lignite and bituminous coals, the carbon content is often enriched prior to combustion by removing moisture and some pollutants, which adds to the cost of electricity production.

There are major environmental issues associated with coal-fired power stations, notably **global warming**, due to the release of carbon dioxide into the atmosphere (N.B. it has a high carbon intensity of ~1000 kg CO_2eq MWh_e^{-1}), and **acid rain**, from coals with a high sulphur content, which causes damage to buildings. There are also serious **health issues**, including lung cancer, heavy metal poisoning, and radiation exposure from fly ash. The smoke from burning wood, charcoal and coal for heating and cooking is estimated by the World Health Organisation (WHO) to cause around 4 million deaths a year.

Although global coal production has levelled off over the last decade, Fig. 3.2 shows that there was a sharp rise over 2000–2010 in production in the Asia Pacific region, mainly in China.

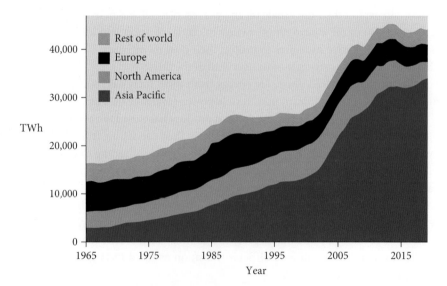

Fig. 3.2 Coal production by region 1965–2019; the sharp rise over 2000–2010 mainly in China and to a lesser extent in India and Australia (OurWorld).

Three countries accounted for 71% of global coal consumption in 2018: China (51%), India (12%) and USA (8%) (see **Case Study 3.1**). Most developed economies are reducing their coal consumption, whereas some emerging economies of SE Asia (e.g. Vietnam and Indonesia) are increasing it. The key question is to what extent China, India and the USA will reduce their coal emissions through the 2020s to help avert excessive global warming.

Case Study 3.1 Coal use in China, India, and the USA

China

Coal is the main source of energy behind China's economic boom in recent decades, and still accounts for about half the world's consumption of coal. Although China has the fourth largest reserves of coal (after USA, Russia and Australia), it increasingly relies on imported coal, which rose by nearly 10% in the first nine months of 2019 (Reuters). China plans to cap coal consumption in 2020, when a staggering 1,100 GW of electric power is expected to be generated from coal. The contribution of coal as a fraction of China's total energy consumption has declined significantly over the past decade, from 72% in 2009 to 59% in 2018. COVID-19 has lowered electricity demand, which may lead to fewer new coal plants but also to fewer renewable projects, due to vested interests in maintaining existing coal-fired generation.

In the draft national 2020 Energy Law, renewables have priority, but the focus is mainly on ensuring uptake of renewable energy through, for instance, electric vehicles and hydrogen production, while only maintaining a steady increase in renewable capacity. Installations of coal-fired power plants continue in the provinces where, even though their capacity factors are low and many lose money, coal is often seen as necessary for reliability, and this is hindering clean growth. China has a stated aim to become a world leader in mitigating climate change, pledging net-zero carbon by 2060, and has invested heavily in renewables and nuclear. By 2020, wind should account for 210 GW and nuclear for 58 GW. In addition, China is now the largest manufacturer of solar power technology, with total installed capacity 175 GW by 2018. (Also, around 500,000 solar panels are installed around the world every day, most being made in China.)

Significant increases in the share of non-fossil energy in China will depend in particular on whether the falling costs for renewables are maintained, on grid capacity, and on adjusting to overseas trade tariffs and the reduction of renewable subsidies in China. Another factor is the possibility of pressure from Western countries for China to cut back on coal-fired power generation, as global temperatures rise and the connection between the burning of fossil fuels and global warming becomes more widely accepted. Even though in September 2020 China announced a net-zero carbon target by 2060, attaining it will be a considerable challenge.

India

Coal has been the primary source of energy behind India's rapid economic growth in the last two decades. India's economy is expected to grow by 8% through the 2020s and, with the second largest population on the planet, there is a pressing need to provide remote rural areas with electricity and to improve the infrastructure of the grid. There is also a need to improve the reliability of electricity supply, especially at peak demand times when voltages tend to fall. Even those consumers that are connected to the grid can experience blackouts for 4–16 hours a day. Although India has the fifth largest coal reserves in the world, most are situated in the north-east of the country and the cost of transportation of coal to where it is needed is significant. Also, the combustion of coal is causing acute air pollution, with 7 of the world's 10 most polluted cities in the world in India. India is also experiencing longer and more intense summer heat waves over the last decade, causing thousands of deaths.

Although coal is expected to remain the dominant fuel for energy consumption for at least the next decade, solar power and wind power are posing an increasing threat to the construction of new coal plants. According to Forbes International, new wind and solar plants were undercutting 65% of existing coal plants in 2018 and it is becoming increasingly difficult to obtain support from banks and insurers for new coal-fired plant. However, handling the variability of renewables is difficult as the poor transmission grid in India makes it hard to distribute electricity from areas where renewable generation is high to those where it is low. This has favoured the continuance of coal-fired power plants, which are dispatchable, particularly those close to mines. The situation is changing with the drop in the cost of batteries starting to make renewable farms with storage able to provide electricity competitively in the evenings when there is a peak in the demand in India, particularly for air-conditioning. However, there has been over-expansion of coal-fired plants in the last few years and considerable pressure for continued government support. It will be a significant challenge for India to eliminate its coal-fired plants, but like China, India could face sanctions on its manufactured goods by western governments, as they take ever more drastic measures to combat global warming.

USA

Coal consumption is a highly contentious issue in the USA, with President Donald Trump promising to revive the coal industry to create jobs in the 'rust belt' and pulling out of the 2015 Paris Climate Change Agreement; a decision that was reversed by President Biden in 2021. Despite this, 24 US states joined the United States Climate Change Alliance to uphold the Paris Agreement. Moreover, the period 2010–2019 saw a 40% drop in coal-fired generation and an increase in the share of natural gas from 23% to 37%, through the exploitation of shale gas by fracking (see Chapter 3 Section 3.3). Wind and solar are also making significant contributions.

Under the Obama administration, the Clean Power Plan represented an attempt to significantly reduce carbon dioxide emissions by 2030. The plan was attacked by the Trump administration for undermining jobs in the fossil fuel economy, but it seems unlikely that this will see a reverse in the long-term decline of coal-fired generation, since market forces tend to dictate in the power sector. And particularly so, following President Biden's commitment in 2021 of reducing the emission of greenhouse gases by at least 50% from 2005 levels by 2030.

3.2 **Crude Oil and Natural Gas**

Crude oil is a hydrocarbon-rich liquid that formed about 65–140 million years ago from the decay of planktonic organisms in ancient seas (e.g. in the Tethys Ocean which became the Gulf of Mexico). These remains settled on the seabed and formed a dense sludge which got buried under layers of sand and mud. Over time, the combined effect of pressure and heat (100–150 °C) at a depth of a few kilometres converted the sludge into crude oil.

There are two basic types of crude oil. **Sweet crude oil** is sulphur-free and is the most economic, whereas **sour crude oil** contains sulphur, which needs to be removed. Oil is refined into petrol (gasoline), jet fuel, diesel oil, fuel oil, asphalt, lubricants, and kerosene, and is used in the manufacture of plastics, fertilizers, cosmetics, foodstuffs, and many other products used in the modern world. While plastics have brought many benefits, their safe disposal is posing a significant challenge.

Only a comparatively small fraction of oil is used for electricity production. Its main advantages as a fuel source are its high-energy density—which means it can also be used for storing energy (e.g. a petrol tank)—ease of transport, and widespread distribution, which have made it difficult for alternative energy to replace it. The combustion of oil derivatives contributes significantly to global warming, providing 34% of primary energy consumption in 2018, more than coal (27%) and natural gas (24%). The largest proven reserves of crude oil are in the Middle East.

Most **natural gas** was formed in organic-rich rock formations containing the fossilized remains of bacterial organic matter, where the temperature was high enough to crack the organic matter to produce natural gas. A small fraction of this gas migrated from source rocks to reservoirs, with a larger part escaping or forming gas hydrate accumulations. The remaining part formed an *unconventional resource* (see Section 3.3). The largest proven reserves of natural gas are in Russia and the Middle East. Natural gas was generally regarded as a waste product in the oilfields of the nineteenth and early twentieth centuries and was burned off ('flaring').

Gas is now regarded as a valuable resource, and is used for electricity generation, especially for meeting peak-load demand, for space heating of buildings, for transportation and in numerous industrial processes. The combustion of natural gas typically emits 50–60% less carbon dioxide than coal. Even so, natural gas contributes significantly to global warming, accounting for 21.3% of primary energy consumption in 2012. Also, fracking sites (and the connecting pipework) leak methane, which is a far more potent greenhouse gas, and undercuts some of the benefit of switching from coal to natural gas.

Combined-cycle gas turbine (CCGT) power stations have a carbon intensity of ~450 kg CO_2eq MWh_e^{-1}, about half that of coal-fired stations. Natural gas is a mixture of methane (CH_4) and smaller proportions of ethane (C_2H_6) and other hydrocarbons, together with water vapour, carbon dioxide, and various impurities. Fig. 3.3 shows the Trainel Kraftwerk Hamm CCGT power plant in Germany whose turbines can have a combined cycle efficiency of ~57.5%.

Fig. 3.3 Trainel Kraftwerk Hamm 800 MW CCGT power plant in Germany.

3.3 **Unconventional Oil and Unconventional Gas**

Conventional resources of oil and gas occur in accumulations trapped below impermeable rocks in porous and permeable reservoirs, from which they can be extracted by pressure depletion. **Unconventional** resources are ones that do not flow naturally to the surface, either because the oil and gas remain in the source rocks (**shale oil** and **shale gas**) or because of bio-degradation to **tar sands**; these resources are widespread but more difficult to extract. Except for periods of high oil prices (e.g. during the oil price shocks of the 1970s), unconventional oils were regarded as uneconomic. But in the last two decades some sources have become economically viable and have made a significant contribution to production.

Shale oil is the oil derived from **oil shale**—a sedimentary rock containing a high proportion of an oily material called **kerogen**—and other **tight rock formations** (tight here means very low permeability to hydrocarbons). In the most common process, the oil shale is brought to the surface, where it is crushed to increase its surface area, and then heated for a long period in an oxygen-free environment to separate the condensable gases, which are condensed to form **shale oil**, and the non-condensable **shale gas**, which is combustible. Shale gas can also be extracted *in situ* by hydraulic fracturing of the oil shale formations (**fracking**), but this is a controversial technology since some experts believe it to be a potential cause of earthquakes. There are also serious environmental issues associated with unconventional oils and gases, notably pollution of groundwater by toxic elements in the waste materials, disposal of spoil, enormous water usage, and larger greenhouse gas emissions than for conventional fossil fuels (see Case Study 3.2: Fracking in the USA).

Another major unconventional source is **tar sands** (also called **oil sands**), which are oil deposits of a mixture of porous sand and clay saturated with bitumen—a very viscous oil. The oil accumulated in reservoirs that were very shallow by seepage to the surface or near surface. There it was degraded by bacteria which preferentially metabolize short chain components of the molecules, leaving the higher molecular weight parts which have a higher density and viscosity. Large deposits are found in Canada and Venezuela, and oil has been extracted from the tar sands in Alberta, Canada; though there are concerns about the impact on the environment. Extraction is energy intensive, since the tar sands have to be heated—about 1 GJ of energy is required to produce a barrel of oil, which can generate about 6 GJ of heat. This can make new oil from tar sands projects uneconomic when the price per barrel drops to around $50 or less. The estimated oil reserves in these two accumulations are about 200 billion barrels each, and together account for approximately 30% of the global reserves of oil.

Other unconventional gas resources include **gas clathrates** (also called **gas hydrates**). These are ice-like substances consisting of light gases contained within a cage of water molecules, which can form at atmospheric pressure, in or below the permafrost layer. Methane-rich clathrates are found on the continental shelves and under permafrost. The amounts are very uncertain but likely to be of order of at least 1000 Gt, equivalent to ~150 times the annual global energy consumption. Attempts at production of gas from hydrates have been made in Alaska, Canada (onshore) and Japan (offshore), but the resource density is low and extraction is difficult—it is necessary to increase the temperature (since the dissociation reaction is endothermic) or drop the pressure in the permafrost or just below the sea bed without producing large quantities of water. Concern has been expressed over whether global warming could cause some clathrates to melt, and thereby release large quantities of methane; however, of the few areas likely to be unstable, few of those are thought likely to release any significant quantities into the atmosphere.

Case Study 3.2 Fracking in the USA

Fracking—the high-pressure extraction of oil and gas from rock formations—is a boom industry in the USA which has transformed the country into a global superpower in terms of oil and gas production. According to the EIA, conventional oil production in the USA declined from a peak of about 9.4 million barrels of crude oil a day (bbd) in the early 1970s to 5bbd by 2008. In the following seven years this trend was reversed with the extraction of oil from shale. In 2018, the USA became the world's largest producer of oil and gas, creating many thousands of new jobs and independence from imported oil from the Middle East. However, fracking is an expensive process, which makes the industry vulnerable to price drops, as in the period 2015–2016. Also, there are huge uncertainties about the size of the reserves of shale oil and gas, and the environmental impact of the fracking process has generated public opposition to the further expansion of the industry. Growth in the US fracking industry has now slowed and raising capital to drill new fracking wells has become more difficult. One factor putting off investors is that the output from a fracking well drops by 70% in the first year of production, compared with a 5% drop for a conventional oil well.

The first commercially successful fracturing procedure was in 1949, when Halliburton Oil Well Cementing Company injected high pressure water into wells in Oklahoma and Texas. The process was improved in 1952 by the Russians, by introducing **proppants** into the fracking fluid—particles which remain behind after the fracturing process and keep the fracture open. The main breakthrough came by combining **horizontal drilling techniques** (which enabled a much larger area of contact to be made with the gas-containing sediments), the use of **slickwater** (chemical additives which alter the properties of the fluid), and **massive hydraulic fracturing** (the employment of much greater volumes of water). Landowners in the USA have the rights to the oil and gas resources below ground, unlike in most other countries. This, together with the necessary technological and geological know-how, and the ability to raise vast sums of capital on US stock markets, gave land owners a huge incentive to exploit their own reserves.

However, there are growing safety and environmental concerns, notably about the vast amount of water used in massive hydraulic fracturing, the risks of contamination of groundwater and the atmosphere by chemicals used in fracking processes, the spillover of slickwater, and the risk of triggering earthquakes. Being a relatively young industry, there is a general lack of reliable data needed to establish a regulatory regime that would be acceptable to all parties. Although the industry is now more closely regulated than it was, there are significant variations from one State to the next. The industry is reluctant to disclose commercially sensitive information which could be used by competitors.

The dramatic rise of oil and gas production in the USA has had a significant impact on global energy markets. In 2014–16 there was a glut of oil which forced the price of oil down and drove some fracking operations out of business. The lower price also had a very negative effect on the economies of those countries which rely on oil and gas production as their main source of income. According to Reuters (2019), lack of investment is forcing shale companies to exist on their operating revenues, resulting in only 7 out of 29 shale producers making a net profit in 2018.

The abundance of cheaper oil and gas in the USA has reduced its dependence on coal, which has led to a large reduction in the emission of carbon dioxide into the atmosphere and has enabled the USA to be much more proactive in negotiating climate change agreements. Also, the wide geographical spread of the shale gas reserves in the US has enabled gas production to be focused at a local level, thereby avoiding the high cost of transporting gas around the country and reduced the price of gas to customers.

From a global warming perspective, although the abundance of shale oil and gas promises short-term economic benefits to the US economy in the next decade, its success poses a risk to the global climate and reduces the incentive to develop low carbon technologies.

Table 3.1 Proven reserves and R/P values for oil, gas, and coal in 2019 (BP2020)

Fuel	Unit	Proven Reserves	Annual Production	R/P (years)
Oil	Gt	244.6	4.90	49.9
Coal	Gt	1069.6	8.10	132
Natural gas	Tm^3	198.8	3.99	49.8

3.4 Fossil Fuel Production and Reserves

The proven reserves of crude oil, natural gas, and coal at the end of 2019 and the reserve to present production per annum ratio R/P are shown in Table 3.1, where the proven reserves are 'generally taken to be those quantities that geological and engineering information indicates with reasonable certainty can be recovered in the future from known reservoirs under existing economic and operating conditions' (BP2020). The R/P ratio gives an estimate of the number of years that the production of the energy source can be maintained at its present rate based on proven reserves. This is therefore an absolute minimum of availability and not a measure of 'how much is left' as significant effort and capital is continually spent in finding new resources and converting resources to reserves (for oil and gas about \$600 billion annually in the mid-2010s). The percentages of the proven conventional reserves of (oil, gas, coal) in 2019 were in North America (14.1, 7.6, 24.1), South and Central America (18.7, 4.0, 1.3), Europe (0.8, 1.7, 12.6), CIS (8.4, 32.3, 17.8), Middle East (48.1, 38.0, 0.1), Africa (7.2, 7.5, 1.4), and Asia Pacific (2.6, 8.9, 42.7). [CIS = Commonwealth of Independent States]

Given the annual production of oil, gas, and coal, we can estimate the global emissions of CO_2.

EXAMPLE 3.1 (a) Estimate the global CO_2 emissions, assuming that ~75% of the oil produced is combusted (~70% in transportation), that the specific carbon dioxide emission of oil, gas, and coal in kg CO_2 per kg fuel are 3.1, 2.7, and 2.1, respectively, and the density of natural gas is 0.8 kg m^{-3}. (b) Estimate the amount of CO_2 released if all the proven fossil fuel reserves were combusted.

(a) Noting that there are 1000 kg per tonne and assuming all the coal and gas are combusted then the total emissions would be $4.90 \times 3.1 \times 0.75 + 3.99 \times 2.7 \times 0.8 + 8.1 \times 2.1 = 37.0$ Gt of CO_2. The largest uncertainty in this estimate comes from the significant variation in the specific carbon emission of coal—the value of 2.1 is an average for coal in the USA. An estimated 36.8 Gt of CO_2 was emitted in 2019, up 4% on 2016 after being approximately constant for 2014, 2015, and 2016.

(b) Substituting the specific carbon emissions into the total proven reserves given in Table 3.1 gives a total of 3244 GtCO_2, which itself is close to the trillion-tonne limit on cumulative carbon emissions (3/11 of CO_2) set to restrict global warming to ~2 ^{0}C. Given that cumulative emissions are already over 2200 GtCO_2, and the large amount of unconventional resources (see below) and undeveloped resources, **fossil fuels are limited by emissions and not resources.**

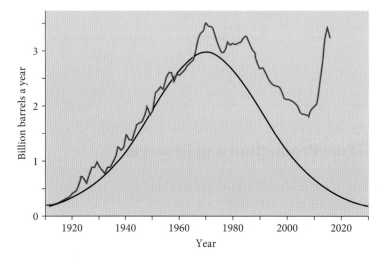

Fig. 3.4 Hubbert's peak prediction (conventional oil) vs. actual oil production in USA, in billion barrels a year (OurWorld).

According to **Hubbert peak theory**, the rate of production of any fossil fuel rises to a maximum (peak), and then falls as new deposits are found and extraction techniques improve, until deposits become scarcer and too expensive to exploit. It was first proposed for oil production in the USA in 1956 by M. King Hubbert, an American geologist, who argued that the amount of oil extracted would follow an approximately sigmoid (logistic) curve and accurately predicted that the US production peak would occur in the early 1970s, as shown in Fig. 3.4.

Hubbert's prediction was based on *conventional* oil accumulations, but the production of shale oil has reversed this decline from the peak in the early 1970s. By 2030 the output from existing oil fields could be down by over 50% and the world will eventually become reliant on unconventional sources of oil, at ever increasing costs of production. The reserve of a resource expands as technology improves, and also depends on its cost and its accessibility (e.g. whether the deposit is under a National Park). The estimated amount of unconventional oil increases the global reserves of oil by approximately a factor of two. From an assessment of 32 countries, the IEA has estimated that shale gas reserves are similar to the proven reserves of conventional sources of natural gas, i.e. ~200 Tm^3.

3.5 **Oil and Fossil Fuel Prices**

The price of fossil fuels is subject not only to the economic laws of supply and demand but also to political pressures. The oil price crisis of 1973 arose after Organisation of Petroleum Exporting Countries (OPEC) declared an oil embargo (October 1973 to March 1974) after the

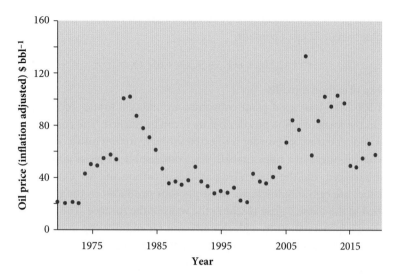

Fig. 3.5 Crude oil price fluctuation during the period 1970–2019 (*Source*: Oil-price).

USA's support of Israel in the Yom Kippur war. The oil price after the embargo was higher than before, as shown in Fig. 3.5. OPEC was set up in 1960 to co-ordinate petroleum policies among its member countries, with the aim of securing a regular supply to consuming countries at a price that gave a fair return on capital investment. However, after the 1973 OPEC oil embargo, the aim has largely been to maximize profits, with periods of very high returns. The member countries are Iran, Iraq, Kuwait, Saudi Arabia, Venezuela, Qatar, Indonesia, Libya, United Arab Emirates, Algeria, Nigeria, Ecuador, Angola, and Gabon.

In 1979-1980, the Iran-Iraq War caused another rise in the price of oil, and during the decade and a half following the 1973 crisis, research and development into alternative energy sources flourished around the world. By 1986, however, oil prices were almost back to their previous levels and funding and interest in alternative energy diminished. More recently, a greater appreciation of the risks from global warming, and the increase in oil prices, has rekindled efforts to develop low-carbon sources of energy. For the other fossil fuels, the cost of gas has tended to follow that of oil while that of coal has historically been more stable.

3.6 **Combustion**

The combustion of methane in natural gas is the exothermic reaction:

$$CH_4 + 2O_2 \rightarrow CO_2 + 2H_2O \qquad (3.1)$$

which produces 55 MJ of heat per kilogramme of methane. The atomic masses of H, C, and O are in the proportion 1:12:16, so the burning of 16 kg of methane releases 44 kg of carbon dioxide.

The combustion of the hydrocarbons in oil follows similar exothermic reactions. For example, the combustion of 72 kg of pentane:

$$C_5H_{12} + 8O_2 \rightarrow 5CO_2 + 6H_2O \tag{3.2}$$

releases 220 kg of carbon dioxide.

A typical 500 MW coal-fired plant consumes around 250 tonnes of coal per hour. Coal needs to be in the form of a fine powder before it can be burned. The coal lumps are ground in large coal mills and the coal powder is injected through nozzles into a combustion chamber (Fig. 3.6), where it burns in a huge fireball. The combustible material undergoes exothermic reactions with oxygen, and the suspended particles of carbon and ash emit and absorb radiation. The flames are optically thick, i.e. most photons produced in the interior of the flame are reabsorbed in the flame before reaching the walls of the furnace. The outer surface of the flame is a close approximation to a black body and radiation is the dominant form of heat transfer to the boiler tubes lining the walls of the combustion chamber. The radiant heat incident on the outer wall of the boiler tubes is conducted through the tube walls and heats the water inside. Over a long period, a solid layer of slag deposit forms on the outer surface of the boiler tubes, which reduces the heat transfer; the slag is removed during plant outages for general maintenance.

The combustion of coal is a complex process, involving:

- the evaporation of water trapped inside the coal (which uses some of the energy content of the fuel);

- the production of combustible gases from the dissociation of coal (notably methane CH_4 and carbon monoxide CO), which react with oxygen and release heat;

- the combustion of solid carbon matter, $C + O_2 \rightarrow CO_2$; thus 12 kg of carbon releases 44 kg of CO_2.

Incomplete combustion of the carbon results in the formation of carbon monoxide:

$$2C + O_2 \rightarrow 2CO \tag{3.3}$$

Since carbon monoxide is poisonous, sufficient air must be injected into the furnace to ensure that the production of carbon monoxide is minimized, which also improves the fuel efficiency. The design of efficient coal-fired furnaces that minimize the production of environmentally harmful gases is an active area of coal technology.

Controlling the emission of sulphur dioxide from fossil fuel power stations is also an important issue since it is a major contributor to **acid rain** (see Section 3.18). There are various ways of tackling the problem, including:

- removing the sulphur prior to combustion by coal scrubbing and oil desulphurization;

- gasification of coal under pressure with air and steam to form gases which can be burned to produce electricity;

- flue gas desulfurization, in which the waste gases are scrubbed with a chemical absorbent (e.g. limestone).

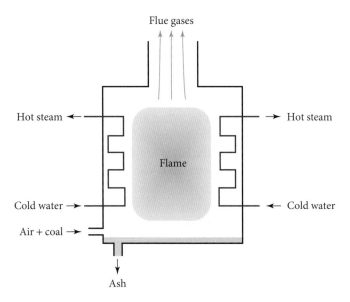

Fig. 3.6 Coal-fired combustion chamber.

Although these processes benefit the environment, they typically add about 10% to the overall cost of electricity, so power companies need to be given incentives to incorporate such measures, or regulations need to be imposed.

3.7 **Carbon Capture and Storage (CCS)**

One proposal to offset the impact of carbon dioxide on global temperature is to develop **carbon capture and storage** (CCS) technologies. Another is **carbon capture and utilization** (CCU), where CO_2 is used in the production of building materials (e.g. through conversion to magnesium carbonate) and fuels, for enhancing plant (e.g. algae) growth, or in the chemical or beverage industry. However, to implement CCS or CCU (collectively called CCUS) on a global scale presents enormous technical, economic, and political challenges.

The main sources of CO_2 are power stations and industries which burn fossil fuels. Since the concentration of CO_2 in the atmosphere decreases rapidly with distance from a point source, the best place to capture the gas is at the source itself. The three main methods of capture under consideration are:

- **post-combustion capture**: scrubbing CO_2 from the flue gases exiting the combustion chamber;

- **pre-combustion capture**: converting the fuel into a mixture of CO_2-rich gas and H_2, prior to combustion;

- **oxyfuel capture**: burning the fuel in oxygen rather than air, producing a stream of CO_2-rich gas.

Technologies exist for the first two processes, but not on a scale necessary to deal with the emissions from a large power station or industrial plant. One process proposed for post-combustion capture is to pass the flue gases through an amine solution in which the CO_2 dissolves, with the nitrogen and other gases passing through. When the amine solution is saturated, the flue gas is switched to another tank of amine while the first tank is heated to release the dissolved CO_2, which is then compressed and liquefied. The liquid CO_2 is pumped to where it will be stored, e.g. in an underground aquifer. A plant at Mongstad in Norway employing this technology opened in 2012 but closed in 2013 following health-related concerns about the use of amines.

A technology for pre-combustion capture is the integrated gasification combined cycle (IGCC) plant. In this process a pulverized coal–water slurry is burned in oxygen from a cryogenic (liquid oxygen) air separator unit to produce synthesis gas or 'syngas', a mixture of carbon monoxide and hydrogen, which is then passed through a shift converter. There, the syngas reacts with superheated steam over a catalyst, to produce hydrogen and carbon dioxide. The principal reactions are:

$$3C + O_2 + H_2O \rightarrow 3CO + H_2 \quad \text{Syngas production}$$
$$CO + H_2O \rightarrow H_2 + CO_2 \quad \text{Shift reaction}$$

In a conventional IGCC plant, the syngas is sent to a combined cycle gas turbine (CCGT; see Section 3.12). Adding a shift converter produces hydrogen and a relatively pure CO_2 stream that can be captured, but makes the plant quite complex and expensive. The efficiency when using hydrogen as a fuel is also reduced, since ~20% of the heat produced is taken away by the steam generated in the combustion process. The final efficiency of IGCC-CCS is about 35%, similar to that of a good post-combustion CCS plant.

The oxyfuel route is still at the research and development stage. The plant is like an air-fired pulverized coal boiler, but with O_2, and involves recycling a portion of the resultant flue gases. The remainder is rich in CO_2, which then has to be captured.

All three processes require significant energy input; this 'energy penalty' adds 15–40% to the cost of electricity generated. After capture, the CO_2 is treated (desulphurization, dehydration and compression) and then sent by pipe or by ship to a storage location.

The long-term storage of CO_2 is known as **carbon sequestration**. The main line of attack is:

Geological storage (geo-sequestration), by direct injection of liquid CO_2 into underground geological formations, including:

- natural gas formations and crude oil fields (see Case Study 3.3);
- heavy oil fields in enhanced oil recovery (EOR), in which CO_2 is pumped into the field to reduce the oil's viscosity so that it flows out;
- un-mineable coal seams, displacing the methane trapped in the coal;
- saline reservoirs.

Suitable geological storage sites have been identified (IPCC) to store all the CO_2 produced in the USA for the next 500 years, but not all countries have suitable sites close to hand.

Direct injection of gaseous CO_2 into the oceans has been proposed, either at depths of more than 3000 m, where it liquefies and forms 'lakes' of liquid CO_2 on the ocean floor, or at depths of between 1000 m and 3000 m where the CO_2 then rises towards the surface and dissolves, forming carbonic acid. However, currently it would be in conflict with the London Protocol and the Convention for the Protection of the Marine Environment of the North East Atlantic (OSPAR Convention). [Air capture of carbon dioxide either using chemical absorbers, biomass, or accelerating the natural reaction of CO_2 with minerals to form bicarbonates, is considered in Chapter 12 Section 12.7.1 Negative emission technologies.]

It is clear from the above that CCS technologies are expensive, in particular for the capture process, and will probably require internationally binding agreements on a carbon price before they are widely employed. Furthermore, there are serious environmental issues associated with certain technologies, e.g. leakage of carbon dioxide from underground storage sites, and acidification of the sea. In 2019, there were 51 large scale CCS projects, 19 in operation, 28 in development and 4 under construction. 24 were in the Americas, 12 in Europe, 12 in Asia Pacific and 3 in the Middle East, and of those operating 2 were power and 17 industrial projects (Global CCS Institute). The amount captured in 2019 was 25 million tonnes out of an estimate total capture capacity of 40 million tonnes per year– about 0.1% of the current global emission. However, many of these projects use captured carbon dioxide to pump oil from oil wells, so their net benefit in terms of global warming is questionable (see Case Study 3.3 below).

Case Study 3.3 Geological storage of CO_2

In August 2019, CO_2 injection started at the Gorgon natural gas processing plant on Barrow Island off the Western Australian coast. When fully operational it will be the world's largest dedicated geological depository, storing 4 million tonnes of CO_2 per annum. The world's first dedicated store was the Sleipner carbon capture and storage project (Norway), which was started in 1996 and sequesters about 1 million tonnes of carbon dioxide a year. Discovered in 1974, the Sleipner gas field is situated in the North Sea about 250 km from the coast of Norway. It produces natural gas and light oil condensates from sandstone structures about 3km below sea level. Nearly 10% of the gas is carbon dioxide, which needs to be separated from the natural gas.

Instead of releasing it to the atmosphere and contributing to global warming, the carbon dioxide is injected in liquid form into a porous and permeable layer of rock about 3000 m below sea level, known as the Utsira Sand. The distribution of liquid carbon dioxide in the Utsira Sand is regularly monitored by seismic reflection analysis, which shows a distinct plume of carbon dioxide liquid trapped between layers of shale. The project has provided powerful support for the technical feasibility of using suitable geological formations for the long-term storage of carbon dioxide. Even if only 1% of the Utsira Sand were used, it could absorb 50 years' worth of carbon dioxide emissions from 20 coal-fired 500 MW power stations.

The Boundary Dam Carbon Capture Project, built in 2014 and situated in Saskatchewan Province, Canada, is a post-combustion process retrofitted to an aging 140 MW coal-fired power station. The removed carbon dioxide is then pumped into nearby oilfields to recover oil. It was assumed that the plant would remove 90% of the carbon dioxide produced by the power station, amounting to around a million tonnes of carbon dioxide a year. In reality, only about 50% of carbon dioxide is removed, and disappointing performances have also been reported at the Petra Nova in Houston, Texas, and other CCS plants.

The World Coal Association promotes the idea of carbon capture for coal-fired power processes as means of transforming the industry to a 'clean coal technology' that could make a significant contribution to reducing global warming. In the light of its poor performance there seems little financial incentive to invest in CCS technology just to keep the coal industry alive, especially given that electricity from renewables is now cheaper than that from coal-fired plants in many countries around the world.

3.8 Costs for Carbon Capture and Storage (CCS)

The major challenge for carbon capture is reducing the costs of capture, especially from the relatively low concentration streams of carbon dioxide from gas-fired power stations and industrial plants. For electricity generation, the cost of CCS has been estimated as $60–$100 per tonne of CO_2, making the cost of carbon capture 50–80% of the total cost of generation! For gas-fired power stations, post-combustion capture is the cheapest method.

CO_2 emissions are ~0.35 tonne MWh^{-1} for a gas combined cycle plant and ~0.85 tonne MWh^{-1} from a coal-fired plant, and the corresponding additional costs are ~$20–$35 per MWh and ~$50–$85 per MWh, respectively. When these are added to the LCOE estimates in Fig. 3.6, we can see that CCS adds a considerable cost penalty for fossil fuel generation.

3.8.1 Outlook for Carbon Capture and Storage

Progress on CCS up to 2015 was much slower than expected due to technical difficulties, lack of suitable sites, a lack of governmental drive to provide economic incentives, and public opposition. However, more recent progress in the USA in particular is starting to indicate that CCS could make some contribution to combating global warming. The IEA has promoted 6 $GtCO_2$ capture and stored per year by 2050, but at the rate of progress so far 2.5 $GtCO_2$ per year might be possible with an effective carbon price set and long-term policy support to enable the required investment. There are also liability and regulatory issues that need to be resolved. Carbon capture and utilization (CCU), an example of which is given in Chapter 11 Section 11.2.3, may offer some effective mitigation and avoids any problems of storage.

Even though carbon capture may not achieve more than ~10% of the reduction in emissions needed by 2050 to reduce global warming to acceptable levels, it could be the most cost-effective way to decarbonize some processes, in particular some in industry (see Chapter 11 Section 11.2.3). Focusing research and development on these may be a good policy to pursue, rather than on electricity generation, for which renewables are increasing competitive. We will take the accessible potential as equivalent to 10,000 TWh of carbon free heat energy (see Exercise 3.3). We will discuss these issues more in Chapter 12 where we will also consider the possibility of air-capture, which is the capture of CO_2 directly from the air, as a mitigation strategy for global warming.

3.9 Thermodynamics of Steam Power Plants

In a conventional thermal power plant, the working fluid is water and, at different stages in a **closed thermodynamic cycle**, the working fluid changes phase from water, to a two-phase mixture of water and steam, to dry steam, and finally back to water. The change of phase gives a high heat transfer rate (see Chapter 2 Section 2.2.2). Some knowledge of thermodynamics and the thermal properties of water and steam is essential to calculate the power output from such a cycle, as we now describe.

The most convenient thermodynamic variables for describing thermal power plants are temperature T and entropy s. The T–s diagram for water and steam is shown in Fig. 3.7.

There are three distinct regions of interest:

  I water;

  II two-phase mixture of water and steam;

  III dry steam.

The bell-shaped curve represents the **phase boundary**. The solid blue lines are **isobars** (constant pressure), and the dashed lines in region II are lines of constant **steam quality** x, i.e. the fraction by mass of steam in the two-phase mixture, defined by eqn (3.5) below.

To illustrate how to interpret the T–s diagram, consider the process of boiling water in a kettle. Since the fluid remains at atmospheric pressure throughout the heating process, we

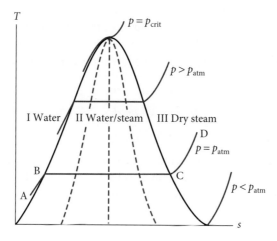

Fig. 3.7 T–s diagram for water and steam (not to scale).

follow the isobar ABCD (at a pressure of 1 bar). Along AB water is heated from cold to the boiling point 100 °C. Water starts to boil at point B, but the temperature remains at 100 °C along BC as the fluid absorbs the latent heat of evaporation. By point C, all the water has been converted into dry steam. (This is an idealization of what happens in practice, since some water droplets may still exist for a while beyond point C.) The line CD represents **superheated fluid** (since the temperature of the steam is above the boiling point), and the temperature of the dry steam rises at constant pressure as more heat is supplied, though the kettle should have switched off before getting this far!

The properties of water (Region I) and dry steam (Region III) can be obtained directly from **steam tables**. However, in order to use the steam tables in the **two-phase region** (Region II), some further explanation is necessary. Consider a mass of water m at B that turns into an equal mass of dry steam at C. At any point along the isobar BC the mixture of water and steam contains a mass m_f of water and m_g of steam (where the subscripts refer to fluid and gas, respectively). The total mass of the mixture is $m = m_f + m_g$. If v_f is the specific volume of liquid water at B, and v_g the specific volume of dry steam at C, then the total volume of the mixture is $V = V_f + V_g = m_f v_f + m_g v_g$. Hence the specific volume v of the water–steam mixture is given by:

$$v = \frac{V}{m} = \frac{m_f v_f + m_g v_g}{m} = \frac{(m - m_g)v_f}{m} + \frac{m_g}{m}v_g = \left(1 - \frac{m_g}{m}\right)v_f + \frac{m_g}{m}v_g \tag{3.4}$$

The ratio:

$$x = m_g/m \tag{3.5}$$

is called the **steam quality** and represents the proportion by mass of steam in the mixture. Likewise, $(1 - x)$ represents the proportion by mass of water in the mixture. Thus, the steam quality at B is $x = 0$ (all water) and at C is $x = 1$ (all steam). Substituting for x from eqn (3.5) in eqn (3.4) gives the specific volume of the mixture as:

$$v = (1 - x)v_f + xv_g \tag{3.6}$$

The numerical values of the coefficients v_f and v_g are given by the steam tables. Eqn (3.6) can then be used to determine the specific volume of the mixture for any particular value of the steam quality x.

The corresponding values of u, h, and s in the mixture are similarly of the form:

$$u = (1 - x)u_f + xu_g \tag{3.7}$$
$$h = (1 - x)h_f + xh_g \tag{3.8}$$
$$s = (1 - x)s_f + xs_g \tag{3.9}$$

and the numerical values of the coefficients u_f, u_g, h_f, h_g, s_f, s_g are also given in the steam tables.

To illustrate how to use the T–s diagram to solve practical problems, we now consider a steam power plant operating in a **Carnot cycle** (Example 3.2).

EXAMPLE 3.2 Steam power plant operating in a Carnot cycle

A steam power plant operates in the Carnot cycle shown in Fig. 3.8. The boiler is at $T_1 = 352$ °C, $p = 170$ bar and the condenser is at $T_2 = 30$ °C, $p = 0.04$ bar. Calculate (a) the efficiency of the cycle, (b) the heat input to the boiler, (c) the work done on the turbine, (d) the heat output in the condenser, and (e) the fraction of heat used in a complete cycle, using the steam table data below.

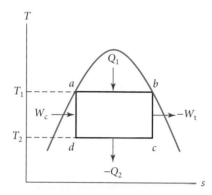

Fig. 3.8 *T–s* diagram for a steam plant operating in a Carnot cycle.

Steam table data

T (°C)	p (bar)	h (kJ kg^{-1})		s (kJ kg^{-1} K^{-1})	
		h_f	h_g	s_f	s_g
30	0.04	126	2556	0.436	8.452
352	170	1690	2548	3.808	5.181

[Note the convention that heat and work are positive when input, and are negative when output; see Chapter 2 Section 2.4.]

(a) From eqn (2.17) the efficiency of the cycle with the temperature expressed in Kelvin is

$$\eta_c = 1 - \frac{T_2}{T_1} = 1 - \left(\frac{273 + 30}{273 + 352} \right) \approx 0.52$$

(b) The boiler operates at constant pressure, so we can use Section 2.5.2 (i) to calculate the heat input Q_1 from the change in enthalpy along *ab*. Thus

$$Q_1 = h_b - h_a = 2548 - 1690 = 858 \text{ kJ kg}^{-1}$$

(c) Since *bc* is an adiabatic in a Carnot cycle, we can use Section 2.5.2 (ii) to calculate the work $-W_t$ done by the turbine (NB Work is negative for output) from the change in enthalpy, i.e.

$$-W_t = h_b - h_c = 2548 - h_c$$

To evaluate h_c we use eqn (3.8), i.e. $h_c = (1 - x_c)h_f + x_c h_g$, where x_c is the steam quality at c. In order to determine x_c we use the fact that the expansion in the turbine is adiabatic. Equating the entropy at b and c gives $s_c = s_b = 5.181$ kJ kg^{-1} K^{-1}. Using eqn (3.9), yields the following equation for x_c:

$$s_c = (1 - x_c)s_f + x_c s_g = 5.181$$

From the steam table, $s_f = 0.436$ kJ kg^{-1} K^{-1} and $s_g = 8.452$ kJ kg^{-1} K^{-1} on the isobar $p = 0.04$ bar. Hence:

$$0.436 (1 - x_c) + 8.452x_c = 5.181, \text{ which gives } x_c \approx 0.59.$$

We can now evaluate the enthalpy at c as:

$$h_c = (1 - x_c)h_f + x_c h_g \approx (1 - 0.59)(126) + (0.59)(2556) \approx 1560 \text{ kJ kg}^{-1}$$

The work done by the turbine is then given by:

$$-W_t = h_b - h_c = 2548 - 1560 = 988 \text{ kJ kg}^{-1}$$

(d) Point a lies on the phase boundary, where the specific entropy is $s_a = 3.808$ kJ kg^{-1} K^{-1}. Since there is no change in entropy in the compressor, the entropy at d and a are identical. Hence $s_d = 3.808$ kJ kg^{-1} K^{-1}. From eqn (3.9), the steam quality at d is then given by:

$$(1 - x_d)s_f + x_d s_g = 3.808$$

Using the steam table, $s_f = 0.436$ kJ kg^{-1}K^{-1} and $s_g = 8.452$ kJ kg^{-1}K^{-1} on the isobar at $p = 0.04$ bar. Hence:

$$0.436(1 - x_d) + 8.452x_d = 3.808, \text{ yielding } x_d \approx 0.42$$

So:

$$h_d = (1 - x_d)h_f + x_d h_g = (1 - 0.42)(126) + (0.42)(2556) = 1147 \text{ kJ kg}^{-1}.$$

Since the condenser operates at constant pressure, the heat lost ($-Q_2$, as heat is negative if output) to the environment is minus the change in enthalpy, so that:

$$-Q_2 = -(h_d - h_c) = h_c - h_d = 1560 - 1147 = 413 \text{ kJ kg}^{-1}$$

(e) The fraction of heat used in the complete cycle is given by:

$$1 - \frac{(-Q_2)}{Q_1} = 1 - \frac{413}{858} \approx 0.52$$

i.e. the Carnot efficiency derived in part (a).

3.10 Disadvantages of a Carnot Cycle for a Steam Power Plant

Despite the fact that a Carnot cycle yields the maximum possible efficiency for a thermal power plant operating in a closed cycle, it has various disadvantages that make it impractical for a real working fluid like water.

First, a Carnot cycle requires the temperature T_1 of the upper reservoir to be constant. However, it can be seen from the T–s diagram for water/steam (Fig. 3.6) that this can only be achieved by operating the boiler along an isobar in the two-phase region II. In practice, it is not possible to operate the boiler in the dry steam region since the temperature rises along any given isobar in Region III. Hence the upper temperature of the cycle, T_1, is limited to the maximum temperature of the two-phase boundary.

Another problem arises in the turbine. In a Carnot cycle the turbine operates with a two-phase mixture of water and steam (Region II in Fig. 3.6). The momentum of fast-moving water droplets in the mixture damages the turbine blades and shortens their life. Similarly, since the compressor is required to compress a mixture of water and steam into high pressure water, water droplets in the mixture damage the blades of the compressor.

Finally, the volume of steam in the mixture is very large, which means that the compressor needs to be very large and therefore expensive. The combined effect of all these factors makes a Carnot cycle impractical for a steam power plant.

3.11 Rankine Cycle for Steam Power Plants

Fortunately, a thermodynamic cycle exists that overcomes the problems of a Carnot cycle. It is called the **Rankine cycle**, in honour of Thomas Rankine, one of the founders of thermodynamics. We begin by considering the simplest case of a Rankine cycle without reheat.

3.11.1 Rankine Cycle without Reheat

The Rankine cycle without reheat is shown in Fig. 3.9. Unlike a Carnot cycle, all the steam in a Rankine cycle is converted into water in the condenser (*de*) before it enters the compressor. The compressor increases the pressure of the water (*ei*) adiabatically before the water enters the boiler. In modern steam power plants, boilers are normally constructed in three separate sections, where each is made with a different grade of steel, using cheaper steels in the lower temperature sections and more expensive steels in the higher temperature sections. In the **economizer** section (*ia*), water is heated at high pressure until it starts to boil. In the **evaporator** section (*ab*), a two-phase mixture of water and steam is heated at constant pressure until all the water has been converted into dry steam. The dry steam is then heated at constant pressure in the **superheater** section of the boiler (*bc*). Dry steam enters the turbine at high pressure and does work on the turbine blades (*cd*). On leaving the turbine, wet steam enters the condenser (at sub-atmospheric pressure), where it condenses on the cold outer surfaces of a large bank of condenser tubes containing cold water from the external environment. This is a

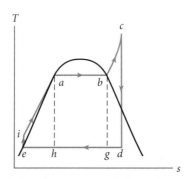

Fig. 3.9 *T–s* diagram for Rankine cycle without reheat (*abcdeia*). *abgha* represents a Carnot cycle with the same lower temperature.

simplification of the situation in a real plant. In practice, there are various complications, e.g. a pressure drop through the boiler (due to frictional losses), the incorporation of a recirculating loop in the economizer section (to take advantage of natural circulation), and possible instabilities in the position of the two-phase boundaries.

To calculate the efficiency of a Rankine cycle without reheat, it is not possible to use the Carnot formula (2.19), since the temperature of the upper reservoir of heat is not constant. The average upper temperature in a Rankine cycle lies somewhere between the lowest temperature in the economizer and the maximum temperature in the superheater. The method of calculating the heat transfer processes in each stage of the cycle and the overall efficiency is shown in Example 3.3.

EXAMPLE 3.3 Thermal power plant operating in a Rankine cycle without reheat

A steam power plant operates in a Rankine cycle without reheat, as shown in Fig. 3.9. The boiler and the condenser are at $p = 170$ bar and $p = 0.04$ bar, respectively. The temperature of the evaporator is 352 °C, the maximum temperature in the superheater is 600 °C, and the temperature in the condenser is 30 °C. Calculate (a) the work done by the compressor, (b) the heat input in the boiler, (c) the work done by the turbine, (d) the heat output in the condenser, (e) the efficiency of the Rankine cycle without reheat, and (f) the efficiency of the Carnot cycle operating between a reservoir at 352 °C and a reservoir at 30 °C.

Steam table data

T (°C)	p (bar)	h (kJ kg^{-1})		s (kJ kg^{-1} K^{-1})	
		h_f	h_g	s_f	s_g
30	0.04	126	2556	0.436	8.452
352	170	1690	2548	3.808	5.181
600	170		3564		6.603

(a) Assuming that water is incompressible, the work done by the compressor per unit mass of water is given by $W_c = v_f (p_i - p_e)$; see Derivation 2.1(b). Inserting $v_f = 10^{-3}$ m^3 kg^{-1} and $p_i - p_e = (170 - 0.04) \times 10^5$ N m^{-2}, we obtain $W_c \approx 17$ kJ kg^{-1}.

(b) We first calculate the enthalpy at the entrance to the boiler (i). Assuming the compressor is adiabatic, the work done is equal to the change of enthalpy, i.e. $W_c = h_i - h_e$, or $h_i = W_c + h_e \approx 17 + 126 = 143$ kJ kg^{-1}.

The boiler operates at constant pressure, so the heat input is given by $Q_1 = h_c - h_i = 3564 - 143 \approx 3421$ kJ kg^{-1}.

(c) The work done by the turbine is $-W_t = h_c - h_d = 3564 - h_d$, where h_d is given by $h_d = h_f(1 - x_d) + h_g x_d = 126(1 - x_d) + 2556 x_d$. To obtain x_d we use the fact that the expansion in the turbine is adiabatic, so that $s_d = s_c = 6.603$.

Hence $0.436 (1 - x_d) + 8.452 x_d = 6.603$, which yields $x_d \approx 0.77$.

Thus $h_d = 126 (1 - 0.77) + 2556 (0.77) \approx 1997$ kJ kg^{-1},

and $-W_t = 3564 - 1997 = 1567$ kJ kg^{-1}.

(d) The heat output in the condenser is $-Q_2 = h_d - h_e = 1997 - 126 \approx 1871$ kJ kg^{-1}.

(e) The efficiency of the cycle is $\eta_R = \dfrac{-W_t - W_c}{Q_{in}} = \dfrac{1567 - 17}{3421} \approx 0.45$.

(f) The efficiency of the Carnot cycle is $\eta_c = 1 - \dfrac{T_2}{T_1} = 1 - \left(\dfrac{273 + 30}{273 + 352} \right) \approx 0.52$.

3.11.2 Rankine Cycle with Reheat

The Rankine cycle *without* reheat does not completely eliminate the production of water droplets with high momentum which damage the turbine blades. To overcome the problem, modern power plants tend to use the **Rankine cycle with reheat**, in which the steam is reheated several times before it leaves the turbine. Fig. 3.10 shows the T–s diagram with three reheat stages. Steam is reheated after leaving the high pressure (HP) turbine before it enters the intermediate pressure (IP) turbine, and is again reheated between the IP turbine and the low pressure (LP) turbine. The effect of reheat increases the overall efficiency of a Rankine cycle and greatly reduces the formation of water droplets.

In order to maximize the efficiency of a steam power plant, it is desirable to operate at as high a temperature as possible in the superheater. However, above about 650 °C, **metal fatigue** becomes significant due to the very high temperatures and pressures that the walls of

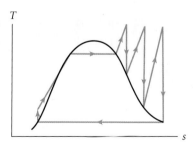

Fig. 3.10 *T–s* diagram for Rankine cycle with reheat.

the boiler tubes have to withstand. **Erosion/corrosion** of the boiler tubes due to the presence of trace chemicals in the water can also be a lifetime-limiting factor. Replacement of boiler tubing in a power plant is a major operation that requires the plant out of service for a considerable period. As a consequence, the maximum operating temperature of a steam power plant is limited to about 650 °C; above this temperature the benefits from improved efficiency are outweighed by the cost of tube replacement and outage costs. Modern plants using Rankine cycles with reheat achieve overall efficiencies of around 40–45%.

3.12 Gas Turbines and the Brayton (or Joule) Cycle

As we observed in Section 3.11.2, the maximum temperature in a steam power station needs to be kept below about 650 °C to avoid excessive metallurgical damage. However, in a **gas turbine**, the gaseous products of combustion are typically at around 1300 °C. The turbine blades are covered by a ceramic coating of low thermal conductivity, which prevents the hot gases from making direct contact with the metal surfaces. Furthermore, the blade assembly is water-cooled so that the temperature of the blades is maintained below the metallurgical limit.

Gas turbines for electricity generation originally evolved from **jet turbine engines**. In such an engine, the thrust arises from the combustion of gaseous fuel and the expansion of the exhaust gases. Since the working fluid does not change phase, a condenser is not involved in the process, which means that the size and cost of a gas turbine plant is less than that of a steam plant with the same power output. Gas turbines operate in a **Brayton (or Joule) cycle**, as shown in Fig. 3.11. It is an **open cycle** but is equivalent to a closed cycle since the atmosphere effectively acts as a heat exchanger which cools the air before it enters the combustion chamber.

Air enters the **compressor** at atmospheric pressure and is compressed to around 10–20 bar (*ab*). It is then mixed with fuel in the **combustion chamber**, (*bc*), where it produces hot combustion gases that do work on the **turbine** (*cd*). The exhaust gases are then vented to atmosphere (*da*). Around the cycle *abcda* the change in internal energy ΔU is zero, so by the first law $\Delta W + \Delta Q = 0$, where $-\Delta W$ is the work done by the system and ΔQ is the heat added to the system.

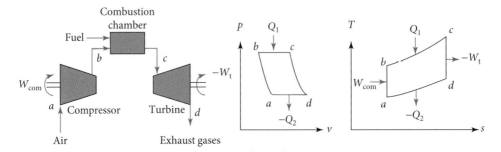

Fig. 3.11 Brayton (or Joule) cycle.

Assuming the compressor and turbine are adiabatic ($\Delta Q = 0$), then the only heat transfers are on bc and da at constant pressure, and $\Delta Q = Q_1 + Q_2 = c_p\,(T_c - T_b) - c_p\,(T_d - T_a)$, where c_p is the specific heat at constant pressure. Hence the efficiency of the cycle is given by:

$$\eta = \frac{-\Delta W}{Q_1} = \frac{\Delta Q}{Q_1} = \frac{(T_c - T_b) - (T_d - T_a)}{T_c - T_b} \tag{3.10}$$

A more useful expression for the efficiency of a gas turbine is given by the formula:

$$\eta = 1 - r^{-\left(\frac{\gamma-1}{\gamma}\right)} \tag{3.11}$$

Where:

$$r = \frac{p_b}{p_a} = \frac{p_c}{p_d} \tag{3.12}$$

is called the **pressure ratio** and $\gamma = c_p/c_v$ is the ratio of the specific heats at constant pressure and at constant volume (see Exercise 3.14).

Gas turbines are relatively low capital cost devices that can be started up quickly and are employed for satisfying sudden surges in electricity demand. Efficiencies of simple gas turbines are up to around 40%.

EXAMPLE 3.4 Gas turbine

Calculate the exhaust temperature, given $T = Ap^{(\gamma-1)/\gamma}$, where A is a constant (see Exercise 3.14), and the efficiency η of an ideal gas turbine operating with a maximum temperature of 1300 K for a pressure ratio $r = 8$. (Assume $\gamma = 1.4$.)

From $T = Ap^{(\gamma-1)/\gamma}$, the exhaust temperature T_d is given by:

$$T_d = T_c r^{-\left(\frac{\gamma-1}{\gamma}\right)} = 1300 \times 8^{-\left(\frac{1.4-1}{1.4}\right)} \approx 718 \text{ K}$$

Equation (3.11) yields the efficiency as $\eta = 1 - r^{-\left(\frac{\gamma-1}{\gamma}\right)} = 1 - 8^{-\left(\frac{1.4-1}{1.4}\right)} \approx 0.45.$

3.13 Combined Cycle Gas Turbine

The overall efficiency of a gas turbine can be increased by feeding the heat of the exhaust gases into a steam power plant. The combination of a Brayton cycle and a Rankine cycle is called a **combined cycle gas turbine** (CCGT), and is shown in Fig. 3.11. The net effect is equivalent to that of a single cycle operating between the upper temperature of a Brayton cycle and the lower temperature of a Rankine cycle. Efficiencies of up to 64% are now possible in CCGT plants (see Fig. 3.12).

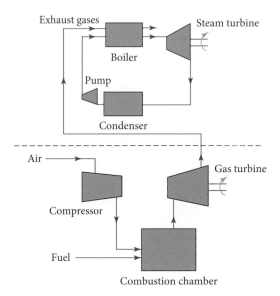

Fig. 3.12 Combined cycle gas turbine (CCGT) generation.

Even greater efficiencies can be achieved in a **combined heat and power** (CHP) cycle. In a CHP plant, the condenser in the steam power cycle is operated at a higher temperature than that in a conventional steam power plant, and the waste heat from the condenser is used to provide **district heating** in the local community. The total efficiency of CHP schemes is typically around 80%. However, the cost involved in installing the pipework and other infra- structure is high, so the application of CHP tends to be limited to industrial complexes or to densely populated urban areas.

3.14 **Efficiency of Power Plant Allowing for Heat Transfer Loss**

By far the most significant thermodynamic loss in systems heated by an external energy source is heat transfer loss, due to the entropy generated in the process of transferring heat from the source to the working fluid, and from the working fluid to the heat sink. Optimization of the total plant cost requires a trade-off between power cycle efficiency and size.

This optimization can be modelled for a power plant operating between T and T_a, where T is the receiver temperature and T_a is the temperature at which heat is rejected, by embedding an ideal engine operating between temperatures lower than T and greater than T_a. The max- imum power output of this 'endoreversible' engine is subject to a trade-off between the heat input $\dot{Q}_{in}$ and the efficiency of the internally reversible cycle, $1 - \dfrac{T_c}{T_h}$, as illustrated in Fig. 3.13.

Assuming Newtonian heat transfer (i.e. linear dependence on temperature differences), the flow rates of heat into the thermodynamic cycle of an endoreversible engine at a mean tem- perature T_h and of heat rejected by the working fluid at a mean temperature T_c are given by:

$$\dot{Q}_{in} = C(T - T_h) \text{ and } \dot{Q}_{out} = C(T_c - T)$$

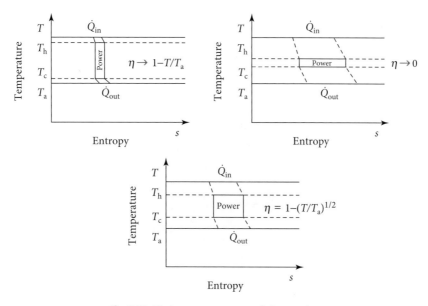

Fig. 3.13 Maximum power output of a heat engine.

where the conductance C is taken to have the same value for the hot as for the cold heat exchanger and is proportional to their size. Assuming that all thermodynamic irreversibilities other than heat exchange processes are negligible (i.e. that the cycle is internally reversible), the thermal efficiency η_{th} is the Carnot efficiency between mean working fluid hot and cold temperatures T_h and T_c; i.e. $\eta_{th} = 1 - \tau$, where $\tau = T_c/T_h$. Differentiating eqn (2.19) wrt time with $Q_{in} = Q_1$ and $Q_{out} = -Q_2$, then $\dot{Q}_{in}/\dot{Q}_{out} = T_h/T_c$ and the power output is given by

$$P = \dot{Q}_{in} - \dot{Q}_{out} = \frac{1}{2}C\left(T - \frac{T_a}{\tau}\right)(1-\tau) \tag{3.13}$$

The value of τ that maximizes P is given by $\tau_{max} = \sqrt{\dfrac{T_a}{T}}$ (see Exercise 3.15*), so the thermal efficiency at maximum power is:

$$\eta_{max} = 1 - \tau_{max} = 1 - \sqrt{\frac{T_a}{T}} \tag{3.14}$$

which is the Chambadal–Novikov–Curzon–Ahlborn value. This value can be a fair approximation for actual power plants (see Exercise 3.16).

3.15 Fluidized Beds

Fluidized beds (Fig. 3.14) provide an alternative means of burning coal, biomass, or municipal waste that reduces the emission of environmentally harmful gases. The appeal of fluidized beds is their ability to cope with a wide range of feedstock and also their simplicity. The

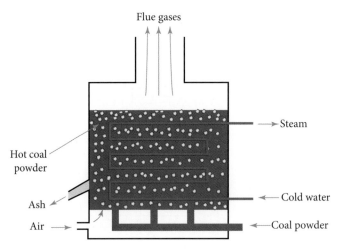

Fig. 3.14 Fluidized bed.

fuel in the bed is in the form of small solid particles, which are suspended in an upward jet of air. Turbulent mixing of the particles with the air results in more complete chemical reactions than in a conventional combustion chamber. Most of the sulphur in coal is removed by using limestone or dolomite to precipitate the sulphur. The generation of NO_x is reduced by operating the bed at a lower temperature than in a conventional combustion chamber. Heat is removed by an array of boiler tubes immersed in the fluidized bed, and the resulting steam is used to drive a steam turbine. Fluidized beds are also used outside combustion in several catalytic processes as a way of mixing reactants with solid catalysts.

The first generation of fluidized beds tended to suffer from erosion/corrosion, due to fine particles and gases damaging the boiler tubes. Later designs have significantly reduced the problem by using a pressurized system to increase the contact between sorbent and flue gas. Furthermore, by the incorporation of a coal gasifier, a fuel gas is produced which provides significant improvements to the efficiency of combined cycle systems.

3.16 **Supercritical and Ultrasupercritical Plants**

Conventional power plants operate at pressures around 170 bar, which is below the critical pressure of 220 bar, above which there is no distinction between liquid water and steam. **Supercritical** and **ultrasupercritical plants** (SCP and USCP) operate above the critical pressure with SCP at ~240 bar and USCP at ~250–300 bar. The higher operating temperature increases the efficiency of the plants: thus a SCP can achieve ~40–42% at ~566 °C and an USCP 43–46% at 600 °C, compared with ~38% at 538 °C for a conventional plant. There is a corresponding reduction in CO_2 emissions per kWh, but CCS is needed for effective reduction. The higher efficiency of SCP and USCP means that the cost penalty of CCS would be less than that with conventional plants without CCS. Materials such as NiCrMoV steels that can withstand very high temperatures are required and there is considerable research and development into new materials, e.g. nickel-base alloys and high boron steels, which can operate at even higher temperatures.

Supercritical carbon dioxide (CO_2) plants are under development in which the working fluid is carbon dioxide at temperatures and pressures above its critical point. These plants operate in a Brayton cycle and, since the density of the CO_2 around the cycle is always high, there is less energy lost in pumping. Moreover, no energy is lost in changing phase, as in a Rankine steam cycle, and high efficiencies are obtained at moderate temperatures where standard materials can be used. Supercritical CO_2 plants would also be very compact and so be cheaper.

3.17 Internal Combustion Engines (ICEs)

About 60% of all the oil produced is used in transportation, nearly all of which is in internal combustion engines (ICEs). In 1876, **Nikolaus Otto**, developed the first practical 4-stroke spark ignition internal combustion piston engine that is now used in petrol (gasoline) cars. The other prevalent ICE is the diesel engine, which is also used for cars but is what usually powers trucks (since it produces a larger torque). In 1897, **Rudolf Diesel** demonstrated the first practical compression ignition internal combustion engine. When shown in Paris in 1900, it ran on peanut oil, but 'diesel' fuel derived from crude oil, was found to be more suitable and readily available; it contained hydrocarbons with typically ~13–16 carbon atoms per molecule, as compared with ~5–12 for petrol.

The diesel engine, like the petrol engine, normally has a four–stroke cycle: intake, compression, expansion, and exhaust. In the case of a diesel engine, only air is taken in and compressed, whereas in a petrol engine it is a fuel–air mixture. This means that the compression ratio of intake to compressed volume, corresponding to the piston being at the bottom and top of the cylinder, respectively, can be much higher in a diesel than in a petrol engine, typically 20:1 compared with 9:1. Compression in a petrol engine has to be limited to avoid auto-ignition, since the temperature of the fuel–air mixture rises as it is compressed. In a diesel engine fuel is injected into the hot (~900 °C) compressed air whereupon it ignites and rapidly expands. (N.B. It is a very fast burn, not an explosion.) As a consequence, the maximum temperature and pressure are higher in a diesel than in a petrol engine, and the efficiency is therefore higher, as would be expected from the maximum theoretical thermodynamic efficiency given by the Carnot formula, $(1 - T_{min}/T_{max})$.

In a modern diesel engine, the air taken into the cylinder is pre-compressed, so that the final maximum pressure is higher. This increase in air (oxygen) allows more fuel to be injected and therefore more power to be obtained. This process, called **turbocharging**, makes the performance of diesel engines higher than that of petrol engines. Their efficiency is ~35%, compared with ~25% for petrol engines, and for large diesel engines their efficiency can be ~45%. However, increasing concerns are being raised about the effects of particulate and NO_x emissions from diesel cars on health and the environment.

3.18 **Environmental Impact of Fossil Fuels**

Besides producing a very significant fraction of GHG emissions (36.8 $GtCO_2$ were emitted in 2019, see Example 3.1), which cause global warming, the combustion of fossil fuels gives rise to pollution through the formation of **acid rain** and the emission of particulates. Acid rain contains high levels of sulphuric acid and nitric acid, caused by sulphur dioxide and nitric oxides, which are emitted in the combustion of fossil fuels, and dissolve in water droplets in clouds. (Volcanic eruptions also release sulphur dioxide). The acid can be incorporated in snow, mists, and dry dusts, which can cause respiratory illnesses in humans and animals. Acid rain can cause serious damage to trees, poison rivers and lakes, and hasten the erosion of building materials These emissions can give rise to serious smog conditions in cities, a situation that has affected China in recent decades due to its greatly increased use of coal (see Fig. 3.15). To reduce the emissions from cars and trucks, **catalytic converters** are used to oxidize carbon monoxide and incompletely combusted hydrocarbons to produce carbon dioxide and water, and to reduce oxides of nitrogen (NOx). The catalyst is generally a mixture of platinum and other precious metals.

Conventional oil is becoming increasingly inaccessible and deep-water drilling has increased the risk of major oil spills. For example, after the Deepwater Horizon platform explosion in the Gulf of Mexico in 2010, oil escaped for three months before the well was secured and ~5 million barrels of oil were released, causing extensive damage to fish, birds, and their habitat. As discussed earlier in this chapter, the extraction of shale gas and shale oil can also have significant environmental effects and, as with all fossil fuels, generate CO_2 on combustion.

Coal mining can also cause serious damage to the environment; in particular strip mining—also called surface, opencast or mountain top removal (MTR). Strip mining involves the removal of layers of earth and rock to expose the seams of coal, sometimes by blasting the tops of mountains apart. It is often used in preference to underground mining, because it requires fewer workers and is more cost effective: it now constitutes ~40% of all coal mines. It can leave landscapes scarred and cause water and soil contamination through the release of

Fig. 3.15 Smog in Jinan, China.

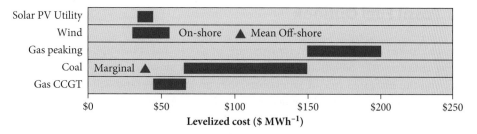

Fig. 3.16 Levelized costs of energy in the USA 2019 (Lazard).

minerals and heavy metals. Most countries require reclamation of sites but often this is slow and ineffective. In the USA in the 70 years prior to 2000, 2.4 million hectares were impacted, much of it forest; and in China, by 2004 coal mining had affected an estimated 3.2 million hectares. While underground mining has less visual impact, it can create large amounts of waste, cause significant subsidence, and affect ground water and streams. Mining also releases methane—a powerful GHG—and globally there are a significant number of smouldering seams of coal that generate polluting smoke. There have also been a large number of deaths from coal mining accidents and related diseases.

3.19 **Economics of Fossil Fuels**

Fig. 3.16 compares the unsubsidized **levelized cost of energy** (LCOE) for fossil fuels with non-fossil fuels in the USA. (Marginal is the cost when plant fully depreciated). Onshore wind and utility scale solar PV are now significantly cheaper than fossil-fuel generation, though this comparison does not take into account any additional cost due to their variability, low power density (MW km^{-2}), or the environmental costs associated with conventional sources. Also note that power provided by gas peaking plants, to meet the peak demand, is considerably more expensive than baseload power. Solar PV and wind power are now close to the marginal cost of coal in some countries.

According to Forbes, 42% of global coal-fired capacity was already unprofitable by 2018. This raises the possibility that a significant chunk of the fossil fuel industry and the fossil fuel reserves, worth between \$1tn and \$4tn, could become 'stranded assets' as the world shifts to a low-carbon economy in the coming decades, a phenomenon known as a **carbon bubble**. One example is the sharp drop in orders for gas turbines in 2017, with GE losing 2/3s of its value in that year. Another is the threat to the controversial Adani coal mine proposed in Queensland, Australia because the perceived markets in SE Asia may no longer exist by the time the mine is ready to export coal. Furthermore, Australia is on target to produce 50% of its electricity from renewables by 2024 and 100% by 2032 despite the Government's declared support for the coal industry.

A consequence of a sudden bursting of the carbon bubble could be a catastrophic global economic downturn, and to avert such a possibility many economists propose the gradual introduction of a carbon pricing policy which takes into account the full environmental cost of fossil fuels.

3.20 **Fossil Fuel Outlook**

China and India currently account for over 50% of global coal consumption. However, with growing concern over pollution and global warming, a slowing expansion in their economies, and the increased role of renewables, they are expected (and will be under increasing pressure) to reduce their dependence on coal over the coming decade. In non-OECD countries the use of renewables needs to increase significantly in order to avoid fossil fuels being used to meet their increasing demand for energy—nuclear power can also help in this respect. Moving from coal-fired to gas-fired power stations, which emit about half the CO_2 per MWh of electricity, will help to reduce emissions, as is already happening in the USA. However, to contain global warming in line with the IPCC recommendations, much tougher curbs on the burning of fossil fuels are needed, helped by ambitious CCS schemes and, in particular, stronger incentives for renewable energy technologies to replace fossil fuel power for electricity generators, transport, and industrial plants.

SUMMARY

- Wind and utility scale solar are now (2020) as cheap or cheaper than fossil fuels for electricity generation, even the marginal cost of coal-fired generation. This lack of competitiveness, together with their environmental costs, are making some fossil fuel reserves increasingly 'stranded assets'.

- The major coal consuming economies will need to make a significant shift to non-fossil energy sources over the coming decade to reduce the impact of global warming.

- The amount of unconventional oil and shale gas approximately doubles the global oil and gas reserves, but the rate at which these reserves can be extracted is limited by their emissions rather than the quantity of resources left in the ground.

- Carbon sequestration is a means of storing carbon dioxide for long periods. Though expensive, it could be cost effective for certain industrial processes, and make a small contribution to limiting global warming.

- The Rankine cycle overcomes the difficulties of a Carnot cycle for steam power plants and yields much higher efficiencies.

- Even higher efficiencies can be obtained in combined cycle gas turbines, which use the waste heat of a Brayton cycle in a Rankine cycle of a steam power plant.

- Fluidized beds can cope with a wide range of feedstock, produce smaller quantities of environmentally harmful gases than conventional combustion chambers, and are now commercially viable.

FURTHER READING

Berkowitz, N. (1997). *Fossil hydrocarbons: chemistry and technology*. Academic Press, San Diego. Comprehensive treatment of fossil hydrocarbons.

Hove, A. (2020). *Current direction for renewable energy in China.* The Oxford Institute for Energy Studies. Interesting analysis of current policy in China.

IEA (2015). *Carbon Capture and Storage: The solution for deep emissions reductions.*

Rackley, S. (2009). *Carbon capture and storage.* Butterworth-Heinemann, Oxford. Comprehensive overview of a wide range of technologies involved in CO_2 capture and sequestration.

Richie, H. and Roser, M. Our World in Data, *ourworldindata.org/energy* (OurWorld).

Rogers, G. and Mayhew, Y. (1992). *Engineering thermodynamics,* 4th edn. Longman, Harlow. Engineering approach to thermodynamics with many practical examples.

Rogers, G. and Mayhew, Y. (1995). *Thermodynamic and transport properties of fluids*, 5th edn. Blackwell, Oxford. Steam table data.

en.wikipedia.org/wiki/Hydraulic_fracturing_in_the_United_States Fracking in the USA

www.bp.com BP Statistical Review of World Energy (2020) (BP2020).

www.carboncommentary.com Chris Goodall newsletter on options to move off fossil fuels.

www.globalccsinstitute.com/resources/global-status-report/ Global CCS Status 2019 (GCC).

www.lazard.com Levelized cost of energy version 13 (Lazard).

www.macrotrends.net/1369/crude-oil-price-history-chart (Oil-price).

There are many websites that quote steam table data for specific thermal conditions.

EXERCISES

3.1 Comment *critically* on the concept of *peak oil*. Do your comments apply equally well to gas and coal?

3.2 Estimates of the reserve to production ratio, R/P, for coal, oil, and gas are 132, 50, and 50 years, respectively. What would the duration of these reserves be if production increased linearly by 15% per decade? Estimate the reserves given in Table 3.1 in EJ and in TWy.

3.3 Justify the values of 3.1, 2.7, and 2.1 kg CO_2 per kg fuel for the specific carbon dioxide emission of oil, gas, and coal, respectively, and that the density of natural gas is 0.8 kg m^{-3}. Is 10,000 TWh y^{-1} of carbon free heat energy a reasonable estimate of the effect of capturing 10% of CO_2 emissions per year?

3.4 Describe the impact on total carbon emissions from the exploitation of shale oil and shale gas.

3.5 Comment on any correlation between investment in renewable technology and the price of oil (see Fig. 3.4).

3.6 Discuss the challenges facing CCS technologies and how much they might allow fossil fuel combustion.

3.7 Investigate what methods are being considered for CCU, and to what extent do they mitigate CO_2 emissions.

3.8 An adiabatic compressor increases the pressure of water from 0.04 bar to 150 bar. Assuming that water is incompressible, calculate the work done per kg of water.

3.9 Using the following steam table data, estimate the work done on an adiabatic turbine by 1 kg of superheated steam entering the turbine at a pressure of 200 bar and a temperature of 600 °C and leaving the turbine at a pressure of 1 bar and a temperature of 100 °C.

Steam table data

T (°C)	p (bar)	h (kJ kg^{-1})
100	1	2676
600	200	3537

3.10 Show that the internal energy u, enthalpy h, and entropy s of a two-phase mixture can be expressed in the form

$$u = (1-x)u_f + xu_g$$
$$h = (1-x)h_f + xh_g$$
$$s = (1-x)s_f + xs_g$$

3.11 Consider a thermal power station operating in a Carnot cycle between an upper reservoir at $T = 400$ °C and $p = 180$ bar, and a lower reservoir at $T = 20$ °C and $p = 0.02$ bar. Calculate (a) the efficiency of the cycle, (b) the heat input in the boiler, (c) the heat output in the condenser, and (d) the work done on the turbine.

Steam table data

T (°C)	p (bar)	h (kJ kg^{-1})		s (kJ kg^{-1} K^{-1})	
		h_f	h_g	s_f	s_g
20	0.02	84	2538	0.296	8.666
400	180	1732	2510	3.872	5.108

3.12 A power station operates in a Rankine cycle without reheat, consisting of (i) an adiabatic compressor, (ii) a three-stage boiler at 200 bar, (iii) an adiabatic turbine, and (iv) a condenser at $p = 0.02$ bar, $T = 20$ °C. The maximum temperature of the boiler is 700 °C. Using the steam table data in the following table, calculate:

i.   the specific work done by the compressor

ii. the heat supplied per unit mass to the boiler

iii. the specific work obtained from the turbine

iv. the efficiency of the cycle

Steam table data

	T (°C)	p (bar)	h (kJ kg^{-1})		s (kJ kg^{-1}K)	
			h_f	h_g	s_f	s_g
Water/steam mixture	20	0.02	84	2454	0.296	8.666
Dry steam	700	200		3806		6.796

3.13 Calculate the exhaust temperature and the efficiency η of an ideal gas turbine operating with a maximum temperature of 1600 K for a pressure ratio $r = 9$. (Assume $\gamma = 1.4$)

3.14* Consider the adiabatic expansion of a perfect gas, then $du + pdv = 0$, or $c_v dT + pdv = 0$. Since $pv = RT$, $pdv + vdp = RdT$. Putting $R = c_p - c_v$ and $v = RT/p$, and eliminating dv, show that $\dfrac{dT}{T} = \left(\dfrac{\gamma-1}{\gamma}\right)\dfrac{dp}{p}$, where $\gamma = c_p/c_v$, and hence that $T = Ap^{(\gamma-1)/\gamma}$, where A is a constant. Apply this result to show that the efficiency η of the Brayton cycle is given by:

$$\eta = 1 - r^{(\gamma-1)/(\gamma)}, \text{ where } r = \frac{p_b}{p_a} = \frac{p_c}{p_d} \text{ is the pressure ratio.}$$

3.15* By maximizing the power P given by eqn. (3.14) with respect to τ, show that the efficiency of an 'endoreversible' heat engine at maximum power is given by eqn (3.15), i.e.

$$\eta_{max} = 1 - \tau_{max} = 1 - \sqrt{\frac{T_a}{T}}$$

How much has the energy available for doing work (the exergy) been reduced by the finite temperature differences in the heat exchangers from the value $Q_1(1 - T_a/T)$ in an ideal heat engine?

Explain why the power is zero when $\tau = T_a/T$ and the efficiency equals that of a Carnot cycle.

By considering eqn (1.1) for the levelized cost of electricity, explain why it is more economic to run at less than maximum power when the cost of fuel is included (ignore O&M costs). Assume the cost of the power plant is proportional to the size of the heat exchanger; i.e. to the conductance C.

3.16 Compare the efficiency of actual power plants, including sub-critical, supercritical, and ultra-supercritical, with the Chambadal–Novikov–Curzon–Ahlborn value (see eqn 3.15) and comment on the agreement.

3.17 What are the advantages of a power plant using super-critical CO_2 as its heat transfer fluid?

3.18 Why are nitrous oxide emissions worse from diesel than from petrol engines?

3.19 Besides their CO_2 emissions, what are the environmental impacts of using fossil fuels?

3.20 What costs need to considered besides the levelized cost of energy when considering two different sources of electricity?

3.21 Why is the amount of fossil fuels determined by their associated emissions rather than by their reserves?

3.22 What are 'stranded assets' and what actions should fossil fuel companies take.

3.23 Discuss the challenges facing China and India in reducing their CO_2 emissions, and to what extent is it in their interests to switch from coal to renewables?

 For further information and resources visit the online resources
www.oup.com/he/andrews_jelley4e

4 Bioenergy

→ Introduction

Bioenergy is a form of energy derived from plant and animal sources. The various materials used to produce bioenergy are collectively known as **biomass**, which includes, for example, wood, sugarcane, corn, palm oil, and forest residues. The burning of biomass or fuels derived from biomass is an important and currently the largest source of renewable energy in the world, providing about 12.4% (46 EJ) of the final energy demand in 2017. Together with hydropower (3.6%), and wind plus solar (2%), the total percentage supplied by renewables was 18%. Traditional biomass (wood, charcoal, dung, and crop residues) contributed about 7.4% or 27.5 EJ and modern biomass provided 5% or 18.4 EJ. The amounts of modern bioenergy, and their percentages of the global demands were: 13.3 EJ (7%) for heating, 3.5 EJ (3%) for transport, and 1.6 EJ (2%) for the world's electricity supply (see Fig. 4.1).

In plant-based biomass, the energy originates from **sunlight**, which is stored as chemical energy in the plant via **photosynthesis**. Traditional biomass has been burnt for cooking and keeping warm for thousands of years and is still widely used in the developing world. However, traditional cook stoves are very inefficient and constitute a serious health hazard.

The next largest consumption of biomass is as forest and crop residues, together with energy crops, as sources for heat and power plants. Energy-rich crops are also grown on a huge scale in some parts of the world, notably in Brazil and USA, to produce liquid and gaseous **biofuels** for transportation. Energy crops require enormous areas of land, and attention needs to be focused on the means of land clearance, planting, fertilizing, harvesting, and the transportation of biomass, to ensure that significant amounts of carbon dioxide are not generated in the process, as replanting the crops only absorbs the CO_2 emitted in their combustion. The biomass can then be a low-carbon source of energy.

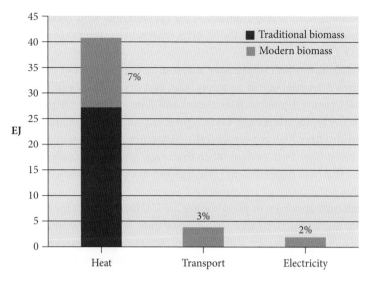

Fig. 4.1 Biomass consumption in 2017 and percentages of global demand supplied by modern bioenergy (REN2019).

We now describe how solar energy is absorbed by plants in photosynthesis and make a rough estimate of typical **crop yields**. We then look at the current and potential contribution of biomass in providing heat and electricity, and in producing liquid biofuels for transportation, as alternatives to the conventional fossil fuels of petrol (gasoline) and diesel.

4.1 **Photosynthesis and Crop Yields**

In **photosynthesis**, plants use the energy of sunlight to convert carbon dioxide and water into oxygen and carbohydrate (sugar):

$$CO_2 + H_2O + h\nu \rightarrow O_2 + [CH_2O]$$

where $h\nu$ represents the energy of the photons in sunlight and $[CH_2O]$ the carbohydrate. The final products of the chemical reaction are about 5 eV per carbon atom higher in energy, which corresponds to around 16 MJ kg^{-1} for pure carbohydrate. The energy per kilogramme depends on the degree of oxidation of the carbon, from zero—when fully oxidized as CO_2—to about 16 MJ kg^{-1} as carbohydrate, with a maximum of 55 MJ kg^{-1} when fully reduced as CH_4.

In plants, photons between the blue and red ends of the visible spectrum ($\lambda \sim 400$–740 nm) are absorbed by the chlorophyll pigment in the leaves of the plant. Green light is not absorbed by these pigments as well as other colours, so it preferentially emerges, after diffuse reflections of white light within a leaf, and is why leaves appear green.

In most plants, the subsequent capture of CO_2 is catalysed by the enzyme RuBisCo and produces intermediary three-carbon molecules (C3), which are converted into carbohydrates—the **Calvin (or C3) cycle**. In some tropical plants, the capture of CO_2 first produces intermediary four-carbon molecules (**C4**) which transport the CO_2 to a separate site deep within a leaf where the RuBisCo catalyzes the production of carbohydrates. An adaption of this C4 mechanism to arid conditions is the **crassulacean acid metabolism (CAM)**. To reduce evaporation from the leaves, the stomata (pores) in the leaves are open only at night to capture the CO_2.

The C4 mechanism is an evolutionary adaption to higher levels of oxygen in the atmosphere. C4 plants are better suited to sunny, hot, and dry regions, and have higher conversion efficiencies than C3 plants under these conditions, whereas C3 plants have a greater efficiency at lower temperatures and lower light levels (temperate regions). **Algae** (which are C3 plants) can also have high conversion efficiencies, producing potentially an order of magnitude more biofuel per hectare than other sources; however, the challenge is to find robust strains of algae that have high efficiency when grown in large quantities, and can be harvested cost effectively.

Even though only about 3% of plant species are C4, they account for around 20% of plant matter, mainly due to the large areas of C4 grasslands. Some of the most productive crops, such as maize and sugarcane, are also C4 plants. For example, **sugarcane**—a C4 plant—can yield around 100 t $ha^{-1} y^{-1}$, though about 70% of the mass is water and the yield of sugar is typically 10 t $ha^{-1} y^{-1}$.

Plants need energy to grow and this is supplied via **cellular respiration**. This process is the opposite reaction to photosynthesis and proceeds by enzyme-catalyzed reactions that convert carbohydrate and oxygen to carbon dioxide and water, which releases about 5 eV per carbon atom. Cellular respiration occurs all the time in humans and is the source of energy for human activity. In plants, respiration occurs mainly at night, but also during the day if water is scarce. Under these conditions, forests can act as a source rather than a sink of carbon dioxide. The percentage ratio of respiratory loss to photosynthetic gain of energy varies considerably, and for major crops is 30-60%.

4.1.1 Efficiency of Photosynthesis

The conversion efficiency of solar energy to biomass energy depends on a number of factors. The most significant is the percentage (49%) of the spectrum of sunlight that falls within the photosynthetically active window (400-740 nm). About 10% of the incident sunlight is reflected off leaves. Of the absorbed light, about 16% is lost as heat from relaxation of higher excited states of chlorophyll. Fig. 4.2 illustrates the factors that contribute to maximizing the **conversion efficiency**.

In C3 plants, a minimum number of eight photons (each with ~1.8 eV, making a total of 14.4 eV) are needed to produce one O_2 molecule and one C atom in a carbohydrate, which stores ~4.8 eV of energy. However, for carbohydrate synthesis in C4 plants, an additional four photons are required, to make the ADP used in the transport of the CO_2.

C3 plants also have a competing process with photosynthesis, in which oxygen, catalyzed by RuBisCo, is used to oxidize carbohydrate with the emission of CO_2. This **photorespiration** increases with temperature and oxygen concentration, which wastes some of the energy generated by photosynthesis. When there is little water, a plant's stomata close in order to reduce evaporation. However, this lowers the CO_2 level while O_2 is still being produced, so photosynthesis decreases and photorespiration increases. In C4 and CAM plants, the RuBisCo is isolated from atmospheric oxygen, so the photorespiration loss is much reduced or stopped.

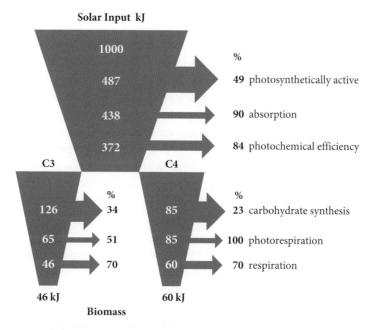

Fig. 4.2 Maximum conversion efficiency in photosynthesis (30°C, 380 ppm CO_2).

This effect, together with the level of CO_2 around the enzyme being around 10 times higher than in C3 plants, means that C4 plants are much better adapted to hot dry climates.

In C3 plants the loss from photorespiration (at 30°C, 380 ppmv CO_2) is 49%, while the minimum loss in C4 plants is zero. Taking the minimum ratio of respiration to photosynthesis as 30%, yields maximum efficiencies of 4.6% for C3 and 6.0% for C4. The rate of photorespiration decreases with temperature and below about 22°C the efficiency of C3 equals that of C4.

The final factors to consider are the light intensity and the concentration of CO_2. Respiration still occurs at night-time when there is no light. In daytime the light level at which photosynthesis balances respiration (including photorespiration) is called the **compensation point**. The net assimilation of energy by a plant equals the gain through photosynthesis minus the loss through respiration. As the light level increases, there comes a point when the net assimilation rate is no longer proportional to the light level but saturates.

Saturation is caused because most chlorophyll molecules gather light and only a small fraction are where the photoreactions occur. At high light levels the energy input to the reaction centres exceeds the energy taken away in the production of carbohydrate, and the excess is lost. The net assimilation rate saturates at a lower level in C3 plants than in C4 plants, due to the photorespiration loss which rises as more O_2 is being produced. Increasing the concentration of CO_2 in the atmosphere will reduce photorespiration in C3 plants, and at high concentrations (at 30°C > ~650ppm) C3 will be more efficient than C4 due to its higher carbohydrate synthesis efficiency.

Further conversion of carbohydrates to oils occurs when certain plants ripen and these oils provide a more compact form of energy storage—typically ~38 MJ kg^{-1} [see Box 4.6 at the end of the chapter].

4.1.2 Crop Yields

Saturation reduces the conversion efficiency of a plant to well below the maximum value. Below about 25% of full intensity, photosynthesis increases linearly with light level, but above 25% it drops off and saturates at around 50%. The highest values for C3 and C4 crops observed for sunlight intercepted by the plants' leaves in a growing season are 2.4% and 3.7%, respectively. Since growing environments are generally not optimal, typical observed conversion efficiencies (ε_c) are closer to 1%, with an average of 0.7% for C3 and 1.2% for C4.

The annual dry weight yield per hectare (10^4 m^2) of a crop Y is proportional to the fraction of sunlight intercepted by the leaves (ε_i), the fraction of solar radiation in the growing season (f), the fraction of biomass harvested (η), the conversion efficiency (ε_c), and the annual amount of solar energy per hectare (S). The **total yield** is then given by the product:

$$Y = S \times f \times \varepsilon_i \times \varepsilon_c \times \eta \tag{4.1}$$

In sunny regions, $S \sim 2000$ kWh m^{-2} or 72 TJ ha^{-1}. The growing season varies significantly, as does the harvested fraction, and we will take 0.5 as a rough estimate for f and η. (NB Root mass is generally small at ~20%). The fraction intercepted by the leaf canopy with good planting is $\varepsilon_i \approx 0.9$. Assuming $\varepsilon_c = 1\%$ then gives $Y \approx 160$ GJ ha^{-1}. So the yield of harvested biomass (e.g. sugar from sugarcane), which requires ~16 GJ t^{-1}, is approximately:

$$\text{Biomass yield} \approx 10 \text{ t ha}^{-1}\text{y}^{-1} \cong 160 \text{ GJ ha}^{-1}\text{y}^{-1} \cong 5 \text{ kW}_{th} \text{ ha}^{-1} \equiv 0.5 \text{ MW}_{th} \text{ km}^{-2} \tag{4.2}$$

Note that a *continuous* thermal power of 0.5 MW$_{th}$ would generate typically 0.2 MW$_e$ of electrical power, since the thermal efficiency of a power station is of order 40%.

In practice, the yield and energy content depend on the type of plant and the growing conditions. The above estimates are only ball-park figures, but they at least provide a rough estimate of the area of land required to produce biomass for a certain electrical power output. The following example illustrates just how large an area of land is needed.

EXAMPLE 4.1

Estimate the area of land required to supply fuel for a 2 GW biomass-fired power station. Take the capacity factor to be 60%. Compare with the area of land needed for solar PV.

Equation (4.2) gives an estimate of 5 kW ha^{-1}, or 0.5 MW per square kilometre. At a capacity factor of 60%, the average power output is 1200 MW of electricity. From a thermal power plant of 40% efficiency, this would require 3000 MW of thermal power, and would need 6000 square kilometres of land; i.e. an area of 77 km $\times$ 77 km, to grow the crops!

This is equivalent to approximately **600 sq km for 1 TWh per year**, compared with about 15 sq km for solar PV (see Table 1.2).

4.2 Biomass Resource

The net primary production (NPP) of terrestrial biomass is about 125 Gt annually, and the total mass of live biomass is around 1000 Gt. As a source of energy, at 18 MJ per kg, some

2300 EJ, or 630,000 TWh, of biomass is produced a year, which corresponds to about six times the world's final energy demand. A similar amount of biomass is produced annually in the oceans as on land, but the amount alive is only about 1-2% of that on land, because its mean life span is much shorter. While fish biomass is used as food, providing about 3% of our total calories, virtually no aquatic biomass is used for energy due to its inaccessibility.

At present some 12% (1.5 Gha) of the world's land surface (13.4 Gha) is cropland, 26% grazing lands, 30% forests, about 2% urban area, and 30% other largely uninhabitable land. Of the biomass used by humans (~230 EJ y^{-1}), some is used as material, some provides energy (46 EJ y^{-1}) either as traditional or modern biomass, some is not harvested, some discarded, and about 130 EJ y^{-1} is processed for food.

4.2.1 Biomass for Food

Of the biomass for food (130 EJ y^{-1}) about 20% ends up as food for consumption. The efficiency of production of vegetables is about 67% while that of meat is only 4%, with beef, pigs and poultry accounting for over 90% of the meat eaten. Meat makes up ~15% of food intake, but about a third of cropland is used for livestock feed production. The amount of feedstock to food eaten for beef is 36% to 1.6%, while for pigs and chicken it is an order of magnitude less. This is because ruminants have a much lower reproduction and relative growth rate to pigs and poultry. Cattle also account for most of the enteric methane given off annually, which amounts to about 5.5% of energy related emissions—in total, livestock account for about 15% (FAO).

Ruminants obtain about half their food from grazing and half from cropland; for instance, of Brazil's deforested land about 70% is used for grazing, with soybean fields grown for feed in much of the rest. A lot of cultivatable land is taken up by feeds for ruminants, for which there will be competition both to grow vegetables, as well as the meat to feed the increasing world population, and also to grow plants to act as carbon sinks or cellulosic crops for bioenergy, or to preserve or restore biodiversity. So, moving to a more vegan than omnivorous diet could save land and biodiversity, as well as GHG emissions.

4.2.2 Traditional Biomass for Cooking and Heating

Some 25 EJ, or 7000 TWh, is supplied each year in the form of heat from traditional biomass. In the developing world, about 2.6 billion people have no access to clean cooking but instead rely on traditional biomass, coal, or kerosene, and these can also provide space heating. The air pollution associated with these primitive forms of cooking and heating is very damaging to health, with around 4 million premature deaths a year—women and children being the most affected. The pollution is caused by the incomplete combustion of the fuel and the generation of microscopic (PM2.5) particles that penetrate far into the lungs.

There have been attempts worldwide to introduce improved cooking stoves (ICS) over the last three decades, with China in the 1980s and early 1990s introducing 130 million; and, in particular since 2010 when the global alliance for clean cook stoves was set up. But while ICS are more efficient than the traditional stoves, and hence save money and time in collecting fuel (often a couple of hours a day), most of them have been shown to only have limited health benefits, since the combustion is still incomplete, and effective ones are expensive (see Box 4.1).

Box 4.1 Improved cook stoves (ICS)

Besides cooking, a traditional three-stone hearth with wood provides light and heat, and it is easy to use and cheap to make. The smoke can be utilized to preserve food; it also deters mosquitoes and flies, and its tar deposits can protect thatched roofs. However, the smoke is very damaging to health, and the incomplete combustion lowers the efficiency to typically 10–15%. Adding a chimney helps, but the problem of community air pollution persists and they are expensive. Traditional charcoal burners use a simple metal box plus grate. The closeness of the cooking pot and higher temperature than with wood gives efficiencies of ~20%, but combustion is incomplete and CO levels can be high.

Three types of ICS that attempt to avoid these problems are shown in Fig 4.3.

For wood burning, the **rocket design** has an insulated short chimney, which improves the draught and hence combustion of the fuel, and heat transfer is improved by directing the hot exhaust gases around the cooking pot, as shown. It significantly reduces fuel use and emissions by about 50%.

For charcoal burning, the **Kenyan Ceramic Jiko** (KCJ) design has proved very popular and is cheap. To improve insulation and optimize the combustion, the stove has an hour-glass shape with a ceramic liner in the top half with perforations to let the ash fall through. Fuel consumption is improved over conventional charcoal stoves by 30-50% and the emissions of particulate matter and toxic gases are reduced. However, the health benefits of the rocket and Kenyan Ceramic Jiko stoves are limited and better biomass combustion is really needed.

An example of a clean biomass stove is the **gasifier**, shown in Fig. 4.3 (c), in which biomass is first converted by heat to a gas in a reduced oxygen environment (pyrolysis), which leaves a carbon-rich residue called **char**. The evolved gas is mixed with a heated secondary supply of air and combusted. The primary supply of air can be controlled to either burn the char or to conserve it. Gasifiers for cooking can be very efficient and clean, particularly with an added fan, but are relatively new and still expensive and their penetration in the market is very small.

Cost is often a barrier to clean cooking, but there can also be a reluctance to change old habits, and in some neighbourhoods, the fear of theft can put off investment in new technology. Still, the introduction of simple credit schemes that use mobile phones is giving access to good stoves to a few on lower incomes.

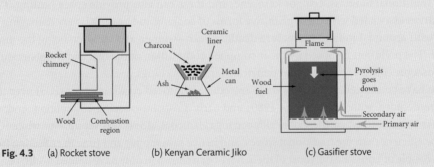

Fig. 4.3   (a) Rocket stove   (b) Kenyan Ceramic Jiko   (c) Gasifier stove

Fig. 4.4 The Haiti Dominican Republic border.

While unregulated harvesting of wood can cause devastating environmental damage, as can be seen in the deforestation in Haiti compared with the preserved forest in the Dominican Republic (Fig. 4.4), the harvesting of wood for cooking fires has been found not to significantly reduce the sustainability of woodland.

This means that ICS have not greatly reduced CO_2 emissions, which has lowered funding and development of ICS through carbon offsetting.

In 2018 the Global Alliance for Clean Cookstoves acknowledged that, though it hopes that the cost of clean biomass stoves can be brought down through further research and mass production, affordable clean cooking can often best be achieved using liquified petroleum gas (LPG) or propane. Stoves burning these gases, which burn cleanly, are now being widely promoted and are used by about 30% in the developing world. While there is some warming from the emitted CO_2, which is partly offset by the reduction in black carbon from wood or charcoal burning, their use saves many lives. However, the introduction of carbon pricing would put up the cost of LPG and jeopardize access to supplies.

Access to clean cooking differs markedly across the developing world, with percentages in 2018 for China 71%, India 49%, and sub-Saharan Africa (SSA) 17% where the vast majority gather biomass for cooking. There has been a gradual decline in the percentage globally (43–39% during 2010-2017) of people without clean cooking facilities, who use mainly wood as their fuel—but the situation in sub-Saharan Africa is slightly worsening. Charcoal is very widely used in SSA, particularly in urban districts. It is preferred to wood due to its compactness, ease of use, higher temperature, and somewhat cleaner burning, though LPG is now being promoted.

Ideally, heating without any pollution or GHG emissions is required. Besides fan assisted gasifier biomass stoves, cooking with biogas produced from the anaerobic digestion of organic matter (see Section 4.3.1) is clean. However, while it has a low LCOE, the digester is expensive and uptake is small with only about 1.5% of the global market, mainly in China with a few in India and Latin America. China is now promoting larger-scale deployment of biogas plants. Biogas is well suited to agricultural regions where feedstock is readily available, but culturally it is not always acceptable. Its take up in sub-Saharan Africa has been poor, where often it is too expensive and labour intensive, as resources are not close by.

Solar energy is also clean but the main disadvantages of using reflectors are the requirement to cook around midday, which is not always convenient, and no ability to store energy simply. As a result, penetration is very low at ~0.1%. Larger scale units can produce steam or hot oil, which can provide some heat storage, and are used for institutional cooking in India, but are too expensive for households. Electric cookers though are an increasingly attractive option and are used in 5% of urban homes in the developing world.

Global electrification reached 89% in 2017, and has been fast expanding into rural regions with both rooftop solar PV and grid extensions. Induction cookers are efficient (~90%) and relatively cheap and solar panels are now increasingly affordable. The main barrier is in the cost of battery storage, but this is falling fast. Electric cooking looks like the most sustainable long-term solution, but with LPG (or biogas where available) cooking as an interim measure.

4.3 Biomass for Heat and Power

Biomass is being increasingly used to provide heat, power and fuel on an industrial scale, and in 2017 modern biomass provided 18.4 EJ. Fig. 4.5 shows the various processes for converting raw biomass material into heat, power, fuel and fertilizer.

The main use of modern biomass is for the production of heat and its most efficient utilization is for **combined heat and power (CHP)**, where efficiencies can be as high as 80%. CHP

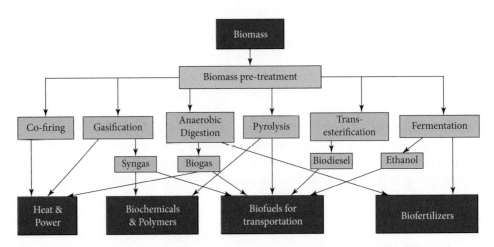

Fig. 4.5 Energy conversion processes (adapted from European Biomass Industry Association: Biomass processing technologies).

plants can be very effective off-grid or where the biomass supply is close by. The development of fluidized-bed combustion plants with higher efficiencies has resulted in the greater use of biomass and waste products in power and heat generation. Using biomass to provide industrial heat is an important way of exploiting this resource and has significant potential. The cost of transporting biomass can be high, because of its low-energy density, so small bio-power plants that can utilize locally produced feedstock are often preferred.

For heat and power there are two main sources of biomass: from agricultural and municipal wastes and from energy crops. These have rather low-energy content per kilogramme compared with fossil fuels and relatively low density, making them bulky and expensive to transport. Economic use for energy production therefore generally requires the biomass source to be readily available, e.g. waste dump or factory residue. For this reason, bioenergy production is currently often combined with crop production or as a useful way of disposing of organic waste, both municipal and agricultural. For a greater supply of biomass for heat and power, large areas dedicated to energy crops such as willow or switch grass would be needed.

The combustion of traditional biomass is the largest source of bioenergy but another widespread source is to take advantage of natural decomposition of organic waste by anaerobic bacteria to produce **biogas**, which is mainly methane. Biogas is used in developing countries for heating and cooking to a small extent and in industrialized countries for small power units.

4.3.1 Biogas from Anaerobic Digestion

Anaerobic digestion is the decomposition of organic matter in the absence of air by bacteria. Bacteria break down the organic matter and produce a gas consisting of methane (~65%) and carbon dioxide (~35%), with traces of other gases. The gas has a calorific value of ~17–25 MJ m^{-3} (STP, standard temperature and pressure, which is 0 °C and 100 kPa $\equiv$ 1 bar) and the conversion efficiency is typically 40–60%. Suitable biomass supply is sewage and industrial sludge, waste water, animal manure, landfill and energy crops. For municipal waste, the organic fraction of the waste needs to be separated out, which can be a costly procedure. The biochemical processes involved are explained in Box 4.2.

Box 4.2 Biochemical processes in anaerobic digestion

Anaerobic digestion consists of several processes in which organic matter is transformed by anaerobic bacteria to a gas consisting mainly of CH_4 and CO_2 and a residue called **sludge** or **biosolids**, which can be used as a soil improver. Anaerobic digestion results in a significant fraction of the original biomass energy being stored in the methane, since the reactions only give out a small amount of heat. These reactions are catalyzed by enzymes (complex proteins) whose action depends on their molecular shape. These enzymes increase the reaction rate by reducing the energy required to start a chemical reaction.

Anaerobic digestion consists of three stages: hydrolysis, acidification, and methane production. The large molecules in the organic matter are initially broken down by hydrolysis, which is a reaction that causes molecules to break apart when water is added, i.e. AB + HOH → A-OH + H-B. The hydrolysis of cellulose and protein produces fatty acids, amino acids, and glucose. These are then converted to organic acids, such as ethanoic (acetic) and butanoic acid, by acidogenic and acetogenic (acid and acetate forming) bacteria, and H_2 and CO_2 are produced, e.g.

$$C_6H_{12}O_6 \text{ (glucose)} + 2H_2O \rightarrow 4H_2 + 2CH_3COOH \text{ (ethanoic acid)} + 2CO_2. \qquad (4.3)$$

In the final stage (methanogenesis), bacteria digest the products of the acidification stage and produce methane via reactions such as:

$$CH_3COOH \rightarrow CH_4 + CO_2; \quad 4H_2 + CO_2 \rightarrow CH_4 + 2H_2O. \qquad (4.4)$$

The acidification and methane production stages are symbiotic because the H_2 and acetate inhibit the bacteria which produce them. Hence, their consumption by the methane-producing bacteria is beneficial. If the production is complete, then the overall process would amount to:

$$C_6H_{12}O_6 \rightarrow 3CH_4 + 3CO_2 \qquad (4.5)$$

The efficiency of the conversion is very high, with around 90% of the stored energy in glucose being stored in the methane produced. The methane can be burnt to provide heat or it can be converted to **syngas** (carbon monoxide plus hydrogen) by reaction with steam: the steam reforming process:

$$CH_4 + H_2O \rightarrow CO + 3H_2 \qquad (4.6)$$

Syngas can be converted into long chain hydrocarbons by the Fischer-Tropsch process (see Section 4.4.2) and these can be used as fuel or the feedstock for chemicals.

EXAMPLE 4.2

Calculate the energy efficiency of conversion of carbohydrate to methane in anaerobic digestion.

The molecular weights of glucose ($C_6H_{12}O_6$) and methane (CH_4) are 180 and 16, respectively, so 180 kg of glucose converts to 48 kg of methane. The heats of combustion of glucose and methane are ~16 MJ kg^{-1} and ~55 MJ kg^{-1}, respectively. Hence, the stored energy in 180 kg of glucose is ~2880 MJ, and in 48 kg of methane it is ~2640 MJ. The conversion efficiency ε_{AD} is then given by:

$$\varepsilon_{AD} \approx 2640/2880 \approx 92\%.$$

Anaerobic digestion occurs naturally, e.g. in compost heaps, and is the source of **marsh gas**. It takes place in landfill sites over a period of years, with typically peak methane production after 10 years, but only a few weeks in purpose-built digesters, where the process occurs at higher temperatures (30–60 °C). In digesters, the residue can be used as a fertilizer. In comparison with *aerobic* digestion, in which organic matter is converted to a residue plus carbon dioxide and water (as in combustion but at much lower temperatures), anaerobic digestion produces considerably less residue (~5–10 times less) as well as a gas that can be used to produce power. Anaerobic digestion also occurs in cows and is a source of a significant amount of methane in the atmosphere; CH_4 production from cows and other ruminants is estimated to account for **about 5.5 % of the global warming** associated with greenhouse gas emissions!

Anaerobic digestion is used in Asian villages, where the biogas is burnt for heating and cooking. About half the use is in China, where an estimated 10% of households (~50 million) have biogas. Elsewhere the technology for larger scale anaerobic digestion is well developed.

An anaerobic digestor consists of a feedstock holder, digestion tank with mixing system, biogas and residue recovery, and where necessary, e.g. northern Europe, fitted with heat exchangers to maintain the optimum temperature for the bacteria to produce the biogas (~30–60 °C). Small digesters are mainly used for heat production while the larger units are used for electricity generation, with power outputs of ~1 MW (see Fig. 4.6).

In the EU, the dumping of biodegradable waste in landfill sites is being reduced. Although the methane can be utilized to provide energy, recovery is variable with typically only 50% being used. Since methane is a very potent greenhouse gas—some 28 times more than carbon dioxide—reducing its emission is extremely important. Moreover, there can be pollution from leachates from biodegradable waste in the landfill sites.

Global biogas and biomethane production provided ~1.5 EJ in 2018, while its estimated sustainable potential is 30-45 EJ y^{-1} (IEA Outlook 2020—World Biogas Association 2019),

Fig. 4.6 1 MW anaerobic digestion plant in Wales, turning food waste into electricity (Photo courtesy of Biogen, UK).

equivalent to about a 10% reduction in global GHG emissions. Biogas could therefore make a significant low carbon contribution and give improved waste management, reduce reliance on traditional biomass for cooking and heating, together with co-benefits of residue as fertilizer and reducing deforestation. There is an even larger potential resource if low grade pastureland is considered, as shown in the case study on using CAM plants.

Case Study 4.1 Potential of CAM crops as a bioenergy resource

CAM plants require typically about 5 times less water per unit of dry biomass produced than C4 plants (and about 10 times less than C3), and can also store substantial amounts of water. This water efficiency enables them to thrive on semi-arid land that is unsuitable for food crops due to insufficient rainfall. It has been estimated that there are globally between 2 billion and 5 billion ha of semi-arid land, distributed widely in both the developing and the developed world. Only a small percentage of this would supply a substantial amount of energy. At 10 t ha^{-1} y^{-1}, and a conversion efficiency of 50%, 10% of this area would produce 16–40 EJ y^{-1}, which compares with a global final energy consumption of ~400 EJ in 2018.

As their lignin content is low, CAM plants can also be used for biogas production, which requires much less sophisticated technology for the conversion process than for bioethanol production. (N.B. Lignin impairs anaerobic digestion.) Two CAM plants with good yields, a high tolerance to drought, and a reasonably high percentage ratio of dry mass to water, are **Opuntia ficus-indica** (see Fig. 4.7) and **Euphorbia tirucalli**. Both can be coppiced for many years and be harvested mechanically. Due to their high water content, the digestate from their anaerobic digestion is estimated to contain around 100 tonnes of nutrient-rich water per hectare, which could be used to enhance the productivity of the land.

Opuntia is grown for forage quite widely and is digested quickly with a typical yield of biomethane of ~325 litre kg^{-1}. The sparser data on Euphoria suggest a yield of around 260 litre kg^{-1}. These numbers translate to anaerobic digestion efficiencies of 64% and 52%, respectively. The energy density of the crops is ~18 MJ kg^{-1}.

Taking the global area of semi-arid land available to be 2.5 billion hectares, the amount of electrical energy that 10% of this area could produce from these crops is close to the 5 PWh produced globally by natural gas each year. Table 4.1 summarizes the energy potential from AD using these two CAM plants.

A significant fraction of the global semi-arid land is unsuitable, due to the terrain, unavailability of electrical power infrastructure, and the need to preserve ecosystems (biodiversity). Much of the most suitable land is probably used already for grazing but a recent study of eight sub-Saharan African countries found that ~10% of the semi-arid and arid land was available for energy crops. As shown in Table 4.1, this land could provide much-needed electricity in this part of the developing world. As biomethane, it could also provide a much-needed source of clean energy (in compressed gas cylinders) for cooking and heating.

Fig. 4.7 Opuntia ficus-indica.

Table 4.1 Global energy potential from AD of CAM crops

Contributions	Opuntia	Euphoria
Dry tonnes ha^{-1} year^{-1}	12	20
Equivalent thermal power (W m^{-2})	0.68	1.14
Gas yield (biomethane litre kg^{-1})	325	260
Energy yield (MJ kg^{-1})	11.59	9.28
Efficiency of AD	64%	52%
Efficiency of biomethane to electricity	41%	41%
Electrical power (W m^{-2})	0.18	0.24
PWh from 10% semi-arid land (2.5 10^8 ha)	3.9	5.2

The cultivation of CAM plants on semi-arid land for biogas production can be mechanized and therefore does not require much labour, and on land that is little used for food. The harvested crops (or biomethane) can also be stored, so the energy is available whenever needed. A mixture of CAM and conventional crops could give a farmer greater resilience against changing market conditions. The economic cost is not yet determined. The distance that the energy would need to be transported, as electricity or as fuel, may be a significant factor. These crops could also be an important source of biomass for the chemical industry that would avoid competition with food production, and the emissions arising from the use of oil.

4.3.2 **Biomass Combustion and Gasification**

As we have seen for traditional biomass, energy can be produced by direct combustion or by combustion after gasification. In order to use biomass on an industrial scale, **pelletization** of the biomass prior to combustion is required before either co-firing with fossil fuels or direct combustion. Co-firing gives a better efficiency than biomass alone plants, and up to 30% biomass can be added to the feed of a pulverized coal-fired power station without significantly reducing its efficiency. In 2018 in the EU close to 200 TWh of electricity was generated by biomass combustion, and globally 581 TWh (2.1 EJ). In the UK, the Drax power station, which supplies 5% of the country's demand, has converted four of its six generating units over the last decade to use pelletized wood instead of coal. This has helped significantly in reducing the carbon intensity of UK electricity (see Exercise 4.9).

Gasification plants are currently not competitive due to their high investment, maintenance, and operating costs; however, gasification has the potential to be cheaper and more efficient. Combining the production of heat and power (CHP) gives the highest efficiency and for power plants with an output of more than 2 MW, steam (Rankine) cycle generators are used. For outputs in the range of 0.2–2 MW, organic Rankine cycle (ORC) plants are preferred, while for small generators less than about 100 kW Stirling engines have the best potential (see Box 4.3 Stirling Engine).

Box 4.3 Stirling engine

The **Stirling engine** was conceived in 1816 by the Revd Dr Robert Stirling. In his engine, a gas is sealed in a cylinder and alternately heated and cooled, which drives a piston that in turn drives a generator. The heat supply is external, and the cold side can be an air-cooled heat exchanger. The internal pressures are lower than those in a steam engine. Unfortunately, it is slow to warm up, is less compact than a steam engine, and requires precise machining. As a result, it was never commercially competitive in his lifetime.

However, it is quiet, reliable, efficient, has a completely external heat supply and has no emissions. As a result, it has already found important applications in submarines and in space. It is also used to provide low-carbon electricity with biomass providing the energy input. Typical generator sizes are 10–25 kW and such systems can provide electricity in arid rural regions far from the grid because **they do not require any water supply**, and can be combined with thermal storage to generate electricity at night.

An ideal Stirling cycle is represented on a T–V diagram in Fig. 4.8. The sealed gas is moved from the heated to the cooled cylinder through a porous matrix of thin tubes or wire gauze, called a **regenerator**, which acts as a temporary heat store or supply (Steps 1 and 3). The basic principle of the cycle is that more work is produced when a gas expands from V_1 to V_2 at T_h than it takes to compress the gas from V_2 to V_1 at T_c, where T_h and T_c are the temperatures of the hot and cold sides of the sealed engine, respectively. Stirling's brilliant innovation was to store and reuse the heat transferred during the constant volume parts of the cycle (Steps 1 and 3).

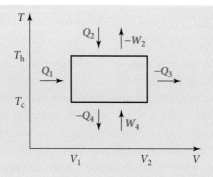

Fig. 4.8 Stirling cycle.

The values for the heat flows Q_1 and Q_2 are given by:

$$Q_1 = \alpha nR(T_h - T_c), \quad Q_2 = -W_2 = nRT_h \ln(V_2/V_1) \tag{4.7}$$

where α ($\alpha = C_v/R$) for a monatomic gas is 1.5 and for a diatomic gas is ~2.5, and n is the number of mole. $\Delta U = Q_2 + W_2 = 0$, as we are assuming an ideal gas, so $U = U(T)$. Likewise

$$-Q_3 = \alpha nR(T_h - T_c), \quad -Q_4 = W_4 = nRT_c \ln(V_2/V_1) \tag{4.8}$$

Assuming the regenerator is 100% efficient (i.e. assuming all of $-Q_3 = Q_1$ is absorbed by the regenerator and then subsequently absorbed by the gas), then the efficiency ε is given by:

$$\varepsilon = (-W_2 - W_4)/Q_2 = (T_h - T_c)/T_h = \varepsilon_C \tag{4.9}$$

where ε_C is the efficiency of a Carnot cycle. Actual Stirling engines are more complicated thermodynamically, but very good thermal efficiencies (~40%) can be achieved, with regenerator efficiencies of more than 95% possible.

Gasification can be brought about by anaerobic digestion as described previously, and also by burning the biomass in a reduced supply of air. One of the earliest ways that biomass was converted to produce a more useful fuel was the burning of wood with a reduced air supply to yield charcoal. Charcoal burns at a much higher temperature than wood and the higher temperatures that could be achieved using charcoal significantly advanced the extraction of metals from ores. Modern methods of biomass conversion by thermochemical processes concentrate on the gaseous and liquid products and in particular on gasification.

The first industrial use of gasification was in the production of **coal gas** or **producer gas** from heating coal in the presence of steam. With the advent of natural gas this production ceased, but a similar process is involved in the gasification of biomass. The different thermochemical processes that occur are described in Box 4.4. The gas that results from burning biomass in a reduced air supply at relatively low temperatures (700–1000°C) consists of CO, H_2, CO_2, CH_4, and nitrogen (from the air), and is called producer gas. At high temperatures (1200–1600°C), (bio) syngas, which is mainly CO and H_2 and less hydrocarbons, is made.

Box 4.4 Thermochemical processes in gasification

Various thermochemical processes occur simultaneously in a gasifier. We will consider wood as an illustration. Drying of the wood occurs at ~150 °C, which drives off the water as steam. The dried wood is decomposed by heat in the absence of air. This occurs between ~150 °C and ~700 °C and produces gases, in particular CO and H_2, liquids (oils), and a solid residue—charcoal or biochar when used as a soil additive and as means of sequestering carbon. These products are then oxidized by reacting with the oxygen in air supplied in just sufficient quantity to produce CO_2 and H_2O and heat by the exothermic reactions (ΔH values are per kmol, where a kmol of carbon, for example, is 12 kg):

$$C + O_2 \leftrightarrow CO_2 \quad \Delta H = -394 \text{ MJ}; \quad H_2 + \frac{1}{2}O_2 \leftrightarrow H_2O \quad \Delta H = -286 \text{ MJ} \qquad (4.10)$$

which occur between 700 °C and 2000 °C. The heat raises the temperature and these gases are reduced by the following reactions:

$$CO_2 + C \leftrightarrow 2CO \quad \Delta H - +173 \text{ MJ}; \quad C + H_2O \leftrightarrow CO + H_2 \quad \Delta H = +131 \text{ MJ} \qquad (4.11)$$

$$C + 2H_2 \leftrightarrow CH_4 \quad \Delta H = -75 \text{ MJ}; \quad CO_2 + H_2 \leftrightarrow CO + H_2O \quad \Delta H = -41 \text{ MJ} \qquad (4.12)$$

The products of the *endothermic* reactions, i.e. those requiring heat (ΔH positive), are favoured at higher temperatures, by Le Chatelier's principle, while the *exothermic* reactions are at lower temperatures. The result is that the production of syngas is favoured over producer gas at the higher temperatures.

It also means that it is important that the temperature should not fall too far in the reduction reactions, otherwise the yield of CO and H_2 is reduced. By altering the conditions, hydrogen production can be enhanced and this can be used in fuel cells to generate electricity (see Chapter 11 Section 11.4.3).

Producer gas has a low calorific value of 1000–1200 kCal m^{-3} at STP, which is equivalent to 4.2–5.0 MJ m^{-3} (STP). However, producer gas burns cleanly with low emissions. The total conversion efficiency can be 60–70%. Each kilogramme of air-dried biomass (10% water content) yields about 2.5 m^3 of producer gas equivalent to ~12 MJ kg^{-1}, consistent with our rough estimate of 16 MJ kg^{-1} of biomass. (Biomass has a higher output if it is low in water content since water requires 2.3 MJ kg^{-1} to evaporate.) Producer gas can be used to provide heat, or in internal combustion engines, and in gas turbines for electricity generation. Syngas has a higher heating value of 13 MJ m^{-3} for equal parts CO and H_2 and can also be used in combustion engines.

Liquid fuels in the form of oils can also be derived from heating biomass in the absence of air, a process called **pyrolysis**, but so far this has been less economic, producing fuels with too low a calorific value to substitute for gasoline or diesel. The producer gas can also be purified to produce syngas, which can be used to synthesize chemicals and fuels such as methanol, biomethane, and biodiesel, using catalysts.

4.3.3 **Municipal Solid Waste**

Industrialized countries produce a large amount of household waste—around 1 tonne per year per household, which globally amounts to ~1.3 billion tonnes or ~1 kg per person per day. The typical energy content of municipal solid waste (MSW) is ~10 MJ per kg, so the global energy resource is ~13 EJ. Only ~11% of this waste is treated in around 2000 waste-to-energy (WTE) plants worldwide, with ~70% going to landfill and 19% recycled. The global production of electricity and heat in 2014 was 1.32 EJ. The combustion of municipal waste in combined heat and power plants (CHP), with the use of the heat for electricity production and the rejected heat for space heating, is a useful way of disposing of this waste, particularly with the decreasing availability of landfill sites.

The waste often needs to be processed before it can be used as fuel. Although the fuel is mainly organic the combustion is not carbon neutral, because some of the material is derived from fossil fuels: typically, 20–40%. This analysis, called a **life-cycle analysis**, calculates the amount of CO_2 (and other gaseous emissions) per kWh of energy produced (see Section 1.5). While the burning of agricultural wastes produces less than 30 g per kWh, municipal solid waste (MSW), also called energy from waste (EfW), produces around 360 g per kWh compared with ~970 g per kWh from coal and ~450 g per kWh from a natural gas CCGT power plant.

Besides providing heat and power, biomass can provide fuels which reduce our dependence on oil-based fuels such as petrol (gasoline) and diesel. We now look at how biomass has been used for transport fuel production and its future potential for such fuels.

4.4 **Liquid Biofuels**

While sustainable electric power can be provided by other renewables, biomass is a direct renewable source of carbon-based fuels and chemicals. Biofuels produced in a country can provide energy security and economic development, thereby reducing petroleum imports and protection against international political disturbances that affect oil supplies. There is also the environmental benefit from reduced CO_2 emissions. These considerations encouraged many countries to grow biofuels. In Europe, biodiesel was encouraged by the European Commission and vegetable oil production rose, particularly in the first decade of this century. In 2018, globally about 3.8 EJ of biofuel was produced, made up of around 112 Gl of bioethanol, 35 Gl of biodiesel, and 6 Gl of renewable diesel—up from 26 Gl of ethanol and 1.5 Gl of biodiesel in 2004 (Gl gigalitres), see Fig 4.9.

However, the growth in biofuel production in recent years has slowed down (see Fig. 4.9), due to a growing concern over its impact on food production and on the environment, and also their cost relative to fossil fuels. While energy and agricultural policies can be coupled with energy crops, aiding rural development as well as energy resources, a study in 2016 found that biofuels were using 2-3% of the world's cropland and water supply. This area could supply food for some 280 million people. Biofuels only provide 3% of the transport demand, so biomass on cropland could not meet the global demand for both food and fuel. It has also been found that the clearance of land for energy crops for biofuels can cause a significant release of CO_2, offsetting their low emissions.

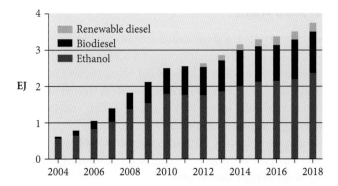

Fig. 4.9 Global biofuel production 2004–2018 (REN2019).

Furthermore, the economic competitiveness of biofuels has fallen in the last few years, due to the large fall in the price of oil (see Fig. 3.5). In the US, the significant increase in the production of shale oil has led to a secure source and reduced dependence on oil imports, and this has adversely affected the US biofuel industry. However, in Brazil, biofuels contribute about 8% to its domestic energy use—mainly in transport. It is important to understand the factors that can make a biofuel competitive, and the limits to its contribution. In 2018, biofuels provided ~3% of the ~110 EJ used in transport worldwide.

A useful measure of the efficacy of biofuels in replacing fossil fuels is given by the **fossil energy replacement ratio** (FER):

$$\text{FER} = \text{Energy supplied to customer/ Fossil energy used} \tag{4.13}$$

where the energy used is that in production. When the FER is close to unity, the biofuel is essentially only displacing fossil fuel use. The FER is also referred to as the energy budget or energy balance.

The two most important liquid biofuels are **bioethanol** and **biodiesel**, the first being derived from sugar-containing plants and the second from oil-containing plants. We look first at how the biological fermentation of sugar and starch containing plants is used to produce bioethanol.

4.4.1 **Bioethanol**

The glucose in sugarcane or sugarbeet, or in starch, which contains α–glucose polymers, can be easily extracted. However, since these feedstocks for bioethanol require good quality soil and plentiful amounts of water, they can be in competition with food crops. They are the only ones currently commercially developed and are called **first-generation biofuels**. The largest producers of bioethanol are the USA (primarily from corn) and Brazil (from sugarcane).

Bioethanol from sugar feedstocks

Sugar from sugar-containing plants can be directly fermented by yeast or bacteria, which reduce the carbohydrate to ethanol and produce CO_2:

$$C_6H_{12}O_6 \rightarrow 2C_2H_5OH + 2\,CO_2 \quad \Delta H = -\,0.4 \text{ MJ kg}^{-1} \tag{4.14}$$

Since the heat released is very small, nearly all the energy stored in the sugar is stored in the ethanol. Ethanol has a much higher heat of combustion 30.5 MJ kg^{-1} than that of glucose (15.6 MJ kg^{-1}). It is sufficiently high that it can be used as a substitute for gasoline, which has a heat of combustion of ~45 MJ kg^{-1}. Almost half the weight of glucose (molecular weight, MW = 180) is converted to carbon dioxide (MW = 44) and the maximum conversion efficiency by weight to ethanol is 51%.

The FER for bioethanol from sugarcane is good with a FER of ~8. Also, the waste product, **bagasse**, can be used to provide heat, which contributes to the good FER.

Brazil is the largest producer of bioethanol from sugarcane, providing a significant source of fuel for cars and reducing its dependence on oil. Bioethanol has been described as a sustainable advanced biofuel since the associated carbon dioxide emissions are relatively low: up to 90% smaller compared to gasoline on direct emissions, but only about 60–70% when change in land use emissions are included.

Case Study 4.2 Bioethanol from Brazilian sugarcane

Brazil originally started producing ethanol from sugarcane in the late 1920s. The programme expanded following the oil price shocks in the 1970s, and by 2019 it was using about 8.5 million ha, 1.0% of the total arable land, and ethanol consumption was 32.8 billion litres. Brazil adopted a policy that all cars must be able to run on a mixture of gasoline and bioethanol, and in 2019, the percentage of ethanol used by cars and light goods vehicles was close to 50%. The price of the bioethanol blended fuels has been competitive with that of gasoline. The majority of the bioethanol is produced in the subtropical south-eastern region of the country, far away from the Amazon rainforest. Roughly half the sugarcane is used to make bioethanol and the rest for sugar; the relative amounts can be adapted to market conditions.

Sugarcane is a perennial grass that produces cane stalks 3–4 m in height and about 5 cm in diameter (see Fig. 4.10). It is a C4 plant, very efficient at photosynthesizing (up to ~2%), and grows in tropical or temperate regions where there is plentiful rainfall (> 60 cm yr^{-1}). Genetic improvements to the sugarcane varieties can make them more resistant to disease and improve productivity. The yield of sugarcane is around 65 t ha^{-1}, and every six years the sugarcane is replanted from cuttings. About 15% of the plant's mass is sugar and 15% is fibre, the remainder being water, producing ~10 tonnes of sugar per hectare. After the cane juice containing the sugar has been extracted, the residual fibre (bagasse), is used as a fuel in the production of the alcohol. The energy ratio, FER, is good at 8–10 with a typical breakdown per tonne of sugarcane being: agricultural phase 190 MJ, industrial processing 40 MJ, ethanol 2000 MJ, and bagasse 300 MJ; giving an FER of 10.

Although most of the growth of sugarcane plantations is on degraded and pasture lands, it is important that this does not indirectly cause deforestation and an associated large release of carbon dioxide, as cattle grazing land is expanded. Another concern is pollution from the burning of the crops prior to harvesting, which is used to

Fig. 4.10 Sugarcane plantation.

kill poisonous animals and remove the sharp leaves, to make it easier to cut the stalks manually. Mechanized harvesting avoids the need for burning, but it has a big impact on employment, since there are about 1 million people employed in the sugarcane business, many of them migrant workers, with about half as sugarcane cutters.

The huge abundance of pastureland and favourable temperature and rainfall conditions, together with the required infrastructure, make Brazil a special case for the production of bioethanol from sugarcane. It is clearly possible for Brazil to provide a significant fraction of its fuel for transport and to do so with relatively low carbon dioxide emissions, and without a damaging impact on food production. However, care is needed to avoid short-term economic hardship due to mechanization, and to avoid causing any significant release of carbon dioxide through indirect changes in land use.

Bioethanol from starch feedstocks

In the production of bioethanol from starch, which is contained in corn, the glucose is in the form of a biopolymer. The glucose molecules are linked together by α–glycosidic bonds (see Box 4.6 at the end of the chapter). These bonds are easily broken apart using human and animal enzymes. The enzymes catalyse the decomposition by hydrolysis of starch to glucose, which can then be fermented to produce bioethanol. A considerable amount of energy, though, is required in the production of the corn.

The corn produced in the US is mainly used to provide animal feed, with some for bioethanol production. Energy is needed in producing the bioethanol and on the farm, for planting, harvesting and for making fertilizers. The resulting FER for corn-ethanol is positive but small, with an FER of ~1.2–1.4; as a result, the corn-ethanol does not reduce CO_2 emissions significantly. However, it does reduce the amount of gasoline and hence oil required, though about 50% more bioethanol is required for the same stored energy.

However, US government subsidies for corn-ethanol have meant that farmers have used more of their corn crops to produce bioethanol, leaving less for animal feed and concern over less corn for food and the effect on its price. So, while corn-based ethanol can displace oil, a better solution would be a glucose-containing feedstock that is both cheap to produce and process, not in competition with food production, and has a good FER.

Moreover, the amount of bioethanol produced in the US in 2018 ($\sim$6 $\times$ 10^{10} litres yr^{-1}) is only equivalent to ~7% by energy (~11% by volume) of the gasoline that was consumed. To produce another 7%, which would be ~1.3 EJ yr^{-1} of bioethanol, would require ~1.8 $\times$ 10^7 ha using corn (at 3400 litre ha^{-1}, see Table 4.2). This area is ~11% of the US cropland, so there would be a conflict between food and fuel. Ideally, we would like a plant that can grow in marginal land that is unsuitable for food crops. Starch-rich plants that will grow on degraded soil, such as the cassava plant and sweet sorghum, are possibilities. The large group of cellulose-rich plants, such as switch grass or miscanthus, could also provide a suitable feedstock. However, extracting the glucose from cellulose is less straightforward than from starch; cellulosic feedstocks are called **second-generation biofuels**.

Bioethanol from cellulosic feedstocks

Cellulosic feedstocks, such as wood and grasses, contain mainly cellulose, hemicellulose, and lignin, and are also called lignocellulosic-based feedstocks. Lignin is a biopolymer rich in phenolic components that confer stiffness and make up about a tenth to a quarter of the biomass. It is the part of a plant that fossilizes and eventually becomes coal. Switch grass, for example, is a deep-rooted perennial that prevents soil erosion and can restore degraded land. Switch needs only a small amount of fertilizer or pesticide and uses water efficiently. As a result, the costs of production can be low.

Cellulose, which is the largest component of the biomass (40–60%), is a biopolymer (polysaccharide) of glucose, as is starch, and consists of bundles of long chains of glucose molecules bonded together by β-glycosidic linkages (see Box 4.6 at the end of the chapter). The fibre bundles are strong due to a high level of hydrogen bonding between the glucose chains and are resistant to cleavage. The other component of the biomass, hemicellulose, (20–40%), interlinks the cellulose and is mainly made up of the 5-sugar xylose. The hemicellulose and lignin enclose the cellulose bundles and protect them from microbial attack.

The hydrolysis of cellulose to glucose is less straightforward than in starch containing plants, due to the hemicellulose and lignin which encase the cellulose. Pre-treatment with dilute acid, combined with heat and pressure, is used to separate the hemicellulose and lignin and expose the cellulose for hydrolysis. The hydrolysis can be acid-catalyzed but this process is expensive since it may require pressure vessels and significant amounts of energy.

Enzyme hydrolysis is being pursued as a potentially cheaper method. This process was first noticed in World War II when a fungus was found that attacked cotton clothes and tents. The fungus was providing cellulose enzymes. The process requires low temperatures and can be carried out at atmospheric pressure.

However, making enzyme hydrolysis cost effective has proved to be much more difficult than anticipated. In 2007 a target of 16 billion gallons of cellulosic bioethanol by 2022 was thought possible in the US and set as part of the Energy Independence and Security Act. However, production in 2017 was a mere 10 million gallons and corn grain was by far the dominant source of bioethanol. It appears that much research is needed before cellulosic ethanol is a cost-effective resource, and cellulosic feedstock may best be used to make materials currently produced from petroleum.

4.4.2 Biodiesel from Plant Oils

Biodiesel can be made from plant oils, and its production grew rapidly in the first decade of this century. This rise was driven by the same concerns over emissions and availability of petrol and diesel that led to the growth in bioethanol production. The global output in litres of biodiesel grew from 1.5 billion in 2003 to 19 billion in 2010. However, growth has slowed since, with 30 billion in 2014 and 41 billion in 2018, caused mainly by concern over their impact on the environment, and also on food production when food crops such as palm trees are used. China has encouraged the planting of jatropha as it grows on marginal land. Use of 100% biodiesel occurs in some European countries (such as Germany), and also as a blend (5–25%) throughout North America and Europe. Biodiesel is made by the chemical transesterification of vegetable oils from oilseed crops, such as rapeseed and sunflower, or from other sources such as waste cooking oil.

Rudolf Diesel demonstrated his engine in Paris in 1900 using peanut oil, but the availability and cheapness of diesel fuel derived from fossil fuel oil meant that vegetable oils were not used. While diesel engines can run on pure vegetable oils, their viscosity is rather high and transesterification produces a lower viscosity fuel that starts more easily. Fatty acid methyl esters (FAME) are the product and biodiesel is also called FAME. The process involves a relatively simple reaction of the oil with either methanol or ethanol using sodium or potassium hydroxide as a catalyst; the chemical processes involved are described in Box 4.5. The efficiency of the process is high (>97%) and requires ~10% by weight of alcohol; and the resultant FER is quite good at about 3.2.

Renewable diesel from biomass

Diesel engines may require modification to run on FAME, and alternative processes, which produce hydrocarbons that can replace the fossil fuel, are gaining ground, and in 2018 produced 6 Gl out of the 41 Gl of biodiesel. These are called drop-in fuels. One such process is the treatment (hydroprocessing) of vegetable oils that produces hydrocarbons more similar to those found in diesel derived from fossil fuels. The triglyceride molecules (see Box 4.5) in the vegetable oils are converted to hydrocarbons by reacting the oil with hydrogen at a raised temperature and pressure in the presence of a catalyst. The product is called renewable diesel.

A summary of the main processes involved in producing liquid biofuels from biomass is given in Fig. 4.11. Of particular importance is the yield of biofuel per hectare and we will see that this limits how much can be produced easily without impacting on normal agricultural use of the land.

Box 4.5 Transesterification of plant oils

Triglycerides can be used neat in diesel engines but better starting is obtained in cold weather by lowering the viscosity. This can be done by mixing the triglyceride with a solution of methanol and sodium hydroxide, the sodium hydroxide acting as a catalyst:

$$
\begin{array}{cccc}
CH_2OOR_1 & & \text{catalyst} & CH_2OH \\
| & & \downarrow & | \\
CHOOR_2 & +\ 3CH_3OH & \Leftrightarrow\quad 3CH_3OOR_x\ + & CHOH \\
| & & & | \\
CH_2OOR_3 & & & CH_2OH \\
\textit{Vegetable oil} & \textit{Methanol} & \textit{Biodiesel} & \textit{Glycerine}
\end{array}
\tag{4.15}
$$

The methanol and sodium hydroxide form sodium methoxide plus water and the sodium methoxide then successively converts the triglyceride to methyl esters plus glycerine. The first step can be represented by:

$$NaOH + CH_3OH \rightarrow NaOCH_3 + HOH \tag{4.16}$$

$$HOH + NaOCH_3 + (-CH_2OOR_1) \rightarrow (-CH_2OH) + CH_3OOR_1 + NaOH. \tag{4.17}$$

This process continues until all three methyl esters have been formed. The catalyst, sodium hydroxide, is not used up in the reaction, and the triglycerine molecules (glycerine esters) have been converted to methyl esters, hence the description of the process as a *transesterification*.

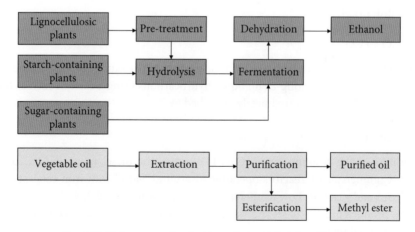

Fig. 4.11 Main processes involved in producing biofuels from biomass.

Biodiesel from microalgae

Microalgae are unicellular organisms, unlike macroalgae or seaweeds that are multicellular, typically containing 20-50% of their mass as oil in the form of triglycerols (see Box 4.6 at the end of the chapter). They can be grown in salty or waste water on arid land where they would not be in competition with food crops. They also can grow much faster than land crops and can produce up to 20× more oil per hectare, with estimated yields of 5000–15,000 litres per ha per annum. However, generally high oil yield is at the expense of growth rate.

Their production can be either on open ponds or in closed (transparent) bioreactors. Open ponds are susceptible to contamination, and the microalgae are more expensive to harvest. Closed systems are easier to control but are more expensive. Genetically modified strains, which may be more robust, are being considered, but as microalgae play a very important role in regulating marine food chains, the concern is that they would upset the delicate balance of species in the environment if they were released. (Algae are also a primary source of omega-3 fats, and all fish get their omega-3 content from eating algae.) Pyrolysis of the microalgae, rather than extraction of their oil, has been suggested as a more cost-effective way of producing bio-oil, but after many years of research, producing oil from them is still not commercially competitive, particularly when the cost of oil is low.

Renewable fuels from (bio)gas-to-liquid (Fischer-Tropsch (FT) process)

The gaseous fuels from biomass are either biogas from anaerobic digestion, or producer gas or syngas from gasification. After cleaning to remove impurities, the upgraded products are CH_4, CO and H_2 and these can be converted to liquid fuels or chemicals by the Fischer-Tropsch (FT) process. The feedstock for the FT process needs to be syngas with a ratio of H_2 to CO of about 2. Syngas can be made from methane by the steam reforming reaction:

$$CH_4 + H_2O \leftrightarrow CO + 3H_2 \tag{4.18}$$

The water shift reaction:

$$CO + H_2O \leftrightarrow CO_2 + H_2 \tag{4.19}$$

can be used to alter the ratio of H_2 to CO in the gas input. The output of the FT process is the polymerization of CH_2 into long-chain liquid hydrocarbons using a catalyst such as the transition metal iron. It can be approximately described by:

$$nCO + (2n+1)H_2 \rightarrow C_nH_{(2n+2)} + nH_2O \tag{4.20}$$

Hydrocarbons in the gasoline range (C5-C11) and in the diesel range (C12-C20) can be produced. Commercial FT reactors operating at around 340°C with iron catalysts produce mainly olefins and gasoline. Low temperature FT reactors at around 230°C use iron or cobalt catalysts to produce mainly diesel and waxes. Refining of the products is generally needed to obtain the required fuel.

The selectivity of the process can be improved by using bi-functional catalysts; these have two sites that catalyze two different reactions. In the FT process a metallic site will catalyze

Table 4.2 Typical values of FER, biofuel yield, suitability for cultivation on degraded land, water requirement, and time to replace the CO_2 associated with converting land to biofuel production, for a number of bioethanol and biodiesel feedstocks

Feedstock	FER (location)	Biofuel yield (litre ha^{-1} y^{-1})	Cultivation on degraded land?	Water requirement	Replacement of CO_2 (y)
Corn	1.34 (USA)	3400	No	high	~50
Sugarcane	8 (Brazil)	6000	No	high	~20
Rapeseed	2.3 (EU)	1000	No	high	~50
Cassava	9 (Thailand)	~3000	Yes	low	~0
Jatropha	6 (Thailand)	530	Yes	low	~0
Palm oil	9 (Malaysia)	3750	No	high	~100
Switchgrass	5 (USA)	2800	Yes	low	~0

Source: WorldBank2009

syngas to methanol and an acidic site methanol to hydrocarbons. An example is Cr_2O_3-ZnO for the metallic and ZSM-5 zeolite for the acidic component of the catalyst. Together they produce liquid hydrocarbons in the gasoline range. The reactions are:

$$nCO + 2nH_2 \leftrightarrow nCH_3OH$$
$$nCH_3OH \rightarrow (CH_2)_n + nH_2O \quad \text{Overall } 2nCO + nH_2 \rightarrow (CH_2)_n + nCO_2 \qquad (4.21)$$
$$nCO + nH_2O \leftrightarrow nCO_2 + nH_2$$

As well as fuels, chemicals can be made using the FT process, and with biomass as the feedstock these are low carbon products.

4.4.3 Liquid Biofuel Yields and Energy Budgets (FERs)

We expect roughly 10 t ha^{-1} y^{-1} of dry carbohydrate and, as the maximum conversion efficiency by mass to alcohol is ~50%, then assuming ~40% in practice gives an annual yield for bioethanol of about ~4 t ha^{-1} or 5000 litres ha^{-1}. Typical values for the FER and biofuel yields from corn, sugarcane, rapeseed, cassava, jatropha, palm oil, and switchgrass are shown in Table 4.2. For switchgrass, the lignin is a by-product that cannot be fermented, but (like bagasse from sugarcane) it can be used as a fuel, and this helps to improve the FER values.

EXAMPLE 4.3

A palm tree plantation produces 3 tonnes of oil per hectare per annum. (a) What area of plantation would be required to displace 5×10^{11} litres of gasoline per annum (i.e. the US annual consumption) with biodiesel? (b) What would be the amount of carbon emissions displaced? (Density of gasoline is 0.73 kg per litre.)

(a) Gasoline has a LHV of ~43.5 MJ kg^{-1}, so there are ~32 MJ per litre. The energy content of 5×10^{11} litres of gasoline is therefore 1.6×10^{13} MJ. An area A ha of palm trees would produce $3000A$ kg of oil per year. The conversion efficiency to biodiesel is close to 100% and biodiesel has a LHV of ~38 MJ kg^{-1}. The efficiency of diesel cars is ~ 4/3 times that of gasoline cars. So the area of plantation required is given by:

$$3000 \times 38 \times A = 3/4 \times 1.6 \times 10^{13} \text{ so } A = 105 \times 10^6 \text{ ha} = 1.05 \times 10^6 \text{ square kilometres}$$

(c.f. land area of USA ~ 9.6×10^6 square kilometres)

(b) Assume that gasoline combusts like octane. Example (2.3) shows that 114 kg of octane produces 352 kg of CO_2. This is equivalent to $(12/44)352 = 96$ kg of C. So 1 kg of octane produces 0.84 kg of carbon on combustion. Petrol has an FER of 0.83, and so, assuming the fossil fuel energy used in its production is from gasoline, 1 kg of petrol produces roughly 1 kg of carbon. A volume of 5×10^{11} litres of gasoline has a mass of 3.7×10^{11} kg. This amount would produce ~ 0.37 Gt of carbon.

Taking the FER of biodiesel as 3.2, the fossil fuel energy used in its production is $3/4 \times 1.6 \times 10^{13}/3.2 \approx 3.8 \times 10^{12}$ MJ, which would be provided by $3.8 \times 10^{12}/43.5$ kg = 0.09 Gt of gasoline, equivalent to ~0.09 Gt of carbon. So:

$$C_{\text{displaced}} \approx 0.37 - 0.09 \approx 0.28 \text{ Gt of carbon per year.}$$

4.5 Environmental Impact of Biomass

Whether biomass is really a renewable and carbon-neutral source of energy has become the subject of debate in recent years. While it can be regarded as a renewable source of energy provided the carbon dioxide released into the atmosphere due to burning is reabsorbed by the planting of new vegetation, the associated emissions, particularly with land clearances as the areas required are very large, can be considerable. Moreover, the timescale for equilibrium to be established is typically several decades, and it is important to consider the complete cycle for any bioenergy process to establish whether it is really carbon neutral. Also, the combustion of biomass produces NO_x and particulate pollution.

Given the need to make drastic reductions in carbon dioxide emissions through the 2020s in order to limit global warming to safe levels in later decades, the continued burning of biomass, unless carefully resourced and replenished, could make a dire situation even worse. And deforestation is already a major concern, with soy cultivation a major factor in the Amazon basin. The climate change committee in the UK concluded in 2018 that as sustainable biomass supplies are limited, new large-scale biomass power plants should be equipped with carbon capture and storage.

Plants and soils are huge stores of carbon, and hold about 2.7 times the amount of carbon in the atmosphere. The demand for palm oil for biodiesel led to increased deforestation in Indonesia and Malaysia, and the draining of large areas of peatlands. This caused decomposition of the peat and led to fires that released huge amount of CO_2. Furthermore, the clearance of vegetation before planting new crops can result in a significant release of carbon bound up in the soil—the increased aeration of the soil speeds up the decomposition of organic matter with the emission of CO_2. The biomass planted on the cleared land can eventually replace this lost carbon if the fossil-fuel replacement ratio is greater than 1, but the replacement time can be many decades. Table 4.2 shows that the estimated replacement times for the conversion of grassland, savannah, tropical forest, and peatland, range from about 20 to 400 years. In contrast, the conversion of marginal cropland results in no significant release, which is a strong argument that only these areas should be developed for bioenergy.

The use of residues and wastes can also have little environmental risk and there is considerable potential from forestry and timber waste, such as black liquor, which is a by-product of the wood-pulping process. CHP is a particularly effective way of using biomass in the Nordic countries. The combustion of wood also gives much less SO_2 than coal and hence less acid rain. However, some crop residues are used as a soil improver, and large areas of energy crops may reduce biodiversity, though forestry energy crops can have a greater variety of wildlife and flora than arable or pastureland. It should be noted that biocrops, like all crops, can be vulnerable in bad weather, so keeping reserves and a range of biomass supplies are prudent.

An important contribution could come from genetic engineering (GE), i.e. modifying the DNA of a plant or animal to change the yield, resistance to disease, or tolerance to drought. For example, a strain of rape seed has been produced that is herbicide resistant, allowing the farmer to use broad-spectrum herbicides to clear a rape field of weeds. There is concern, though, that the development of herbicide resistant weeds may be promoted by the use of such GE modified crops.

The production of biomass could be limited by the availability of water. Currently about a quarter of the world's population is living in countries, including India, Pakistan, and Botswana, that are close to running out of water. In these countries water withdrawals are over 80% of the average water availability, which leaves them vulnerable to fluctuations caused by droughts and high demand. Extreme water stress can be experienced in many countries as worldwide water availability is very uneven. Agriculture is responsible for some 70% of the demand and water efficiency needs to be improved to cope with the increasing demand from rising populations, living standards, and urbanization. Also, climate change is making precipitation more variable and droughts more common. The MENA countries of North Africa and the Middle East, which are hot and dry, are at risk, as is India where their groundwater resources, which are mainly used for irrigation, are severely stretched. Widespread investment in grey (treatment and reuse) and green (wetlands and watersheds) infrastructure is needed.

4.6 Global Potential and Economics of Biomass

At present, the main feedstocks for biomass heat and power generation come from forestry, agricultural and municipal residues and wastes. Large amounts of these resources remain untapped, and many countries hope that biomass will make a significant contribution in reducing their GHG emissions. Typically, to be competitive with fossil fuels, the supply needs to be readily available, as in Sweden where biomass and wastes provided in 2013 almost a quarter of their energy supply.

Bioenergy can generate baseload or dispatchable electricity cost-competitively. Over the last decade, the LCOE has been about US$0.04 kWh^{-1} in India, US$0.05 kWh^{-1} in China, and \$0.085 kWh^{-1} in Europe and North America. However, the sharp fall in the price of solar PV has meant that it is more economic when producing both heat and power, as in CHP. Its dispatchability can also help its competitiveness. For biofuels, the feedstock price is the main cost, but the by-products, which can be used for animal feed, or, in the case of bagasse, for fuel provide some offset. In the US and Brazil, which account for 84% of global ethanol production and 26% of biodiesel production in 2017, the production costs were competitive with gasoline and diesel production costs when oil prices were around US\$55 and US\$100 per barrel, respectively. Taxes and credits affect the prices at the pump, and, in both countries, there is support for first generation biofuels. But in Europe there has been a shift away from food-based to second generation or advanced biofuels.

In a recent critical assessment by Searle and Malins (2015) of the global biomass potential, in which food production and natural forests are protected, they calculated that the area of land available for energy crops in 2050 that is not forest, wetlands, tundra, desert, cropland, or pastureland is 0.93 Gha, and that using a much higher area would be unsustainable. A typical yield on this land is 7.5 t ha^{-1} y^{-1} with an energy per kg of 19 MJ. Of this area almost 80% would be cost-effective at a carbon price of \$30 but governance issues around environmental protection in some countries would reduce the output by about 25%. The average energy conversion efficiency for using the biomass for CHP and for biofuels is estimated as 75%.

The resulting estimates of the maximum potential in 2050 were 10–20 EJ yr^{-1} for biofuel, 20–40 EJ yr^{-1} for electricity, and 10-30 EJ yr^{-1} for heating, approximately equivalent to 10,000–25,000 TWh. What was not considered were land use emissions from the planting of the energy crops, water availability, or biodiversity loss, and the need to keep land for vegetation to act as a carbon sink to reduce atmospheric CO_2 levels.

4.7 Biomass Outlook

Making the right choice of biomass and its location are vital if biomass is to be an effective low carbon energy source that does not adversely affect food supply, biodiversity, and the world's forests. Regulation and certification can also help, but it is essential that complete LCAs are made and required for all biomass projects. The use of biomass for food is already causing a significant perturbation of the natural carbon cycle and of ecosystems. Improving the efficiency of agricultural land in developing countries would help accommodate the demands for global biomass for energy and for food production. Vertical farming is also being considered as a means of providing food security in an increasingly urbanized world; however, the associated energy demand is high.

As it can be stored and available on demand, bioenergy is also a useful complement to the variable sources of wind and solar. However, large areas of suitable and accessible land required are not widely to hand. Where biofuels can be produced economically and sustainably, they might make a significant contribution in aviation, as a substitute for jet fuel, and in shipping as a replacement for oil. Biomass could also be a sustainable source of carbon for the chemical industry, and can bring employment opportunities. Overall a realistic contribution is likely to be around 15,000 TWh by 2050, but policy support will need to be strong for this to actually happen sustainably.

SUMMARY

- Traditional biomass provides about 7.5% of global final energy demand, mainly for cooking and heating in the developing countries. However, the associated air pollution is very damaging to health and alternatives are being promoted.

- Modern biomass is a significant source of energy that could realistically supply 10–20 EJ yr^{-1} for biofuel, 20–40 EJ yr^{-1} for electricity, and 10–30 EJ yr^{-1} for heating by 2050 with strong policy support—a total of around 15,000 TWh per year.

- A large area of land is required to produce a significant quantity of biomass for power or biofuel (~4.5 GWh y^{-1} km^{-2} or ~0.5 MW$_{th}$ km^{-2}) and care is needed to avoid adverse impacts on the environment, food production, land use, and climate change.

- Biofuels may best be reserved for aviation, shipping, and trucks.

- The availability of water will become an increasing concern as demand is already close to the sustainable supply in many areas.

FURTHER READING

Cushion, E., Whiteman, A., and Dieterle, G. (2009). *Bioenergy development*. The World Bank. siteresources.worldbank.org/INTARD/Resources/Bioenergy.pdf. Very good review of the environmental issues surrounding the use of biomass (WorldBank2009).

Mason, M. et al. *The potential of CAM crops as a globally significant bioenergy resource: moving from 'fuel or food' to 'fuel and more food'*. Energy Environ. Sci. 8 (2015) 2320.

Peake, S. (ed.) (2018). *Renewable energy*, 4th edn. Oxford University Press, Oxford. Useful overview of bioenergy.

Renewables 2019 Global Status Report; *www.ren21.net/gsr-2019/* (REN2019).

Searle, S. and Malins, C., *A Reassessment of global bioenergy potential in 2050*. GCB Bioenergy (2015) 7, 328–336, doi: 10.1111/gcbb.12141.

Wirsenius, S., The Biomass Metabolism of the Food System (2003); *onlinelibrary.wiley.com/doi/10.1162/108819803766729195*.

World Bank, Landscape report (2014). *Clean and improved cooking in Sub-Saharan Africa*.

Zhu, X et al. *What is the maximum efficiency with which photosynthesis can convert solar energy into biomass?* Current Opinion in Biotechnology 19 (2008) 153–159.

www.fao.org/home/en Food and Agriculture Organization of the UN (FAO).

www.iea.org/fuels-and-technologies/bioenergy Bioenergy.

www.unwater.org/publications/world-water-development-report-2019

EXERCISES

4.1 Estimate the land area required to grow willow that would provide 1 GW of power in a region where the annual solar radiation is 1500 kWh m^{-2}. Assume an efficiency of 0.5%.

4.2 Describe how energy is stored in plants.

4.3 Quantify how much agricultural land would be released if the world consumption of meat was halved and discuss the implications.

4.4 Investigate vertical farming as a way of increasing food production.

4.5 What are the challenges of using biomass for cooking in the developing world.

4.6 Investigate the cost of cooking using electricity from a solar PV panel plus a battery compared with that using propane.

4.7 A household disposes of 300 kg of domestic waste per year, 90% of which is carbohydrates and 10% of which is inert, in an anaerobic digester. The family uses the methane for cooking on an open fire. Estimate the number of litres of water that could be boiled annually. Assume: 16 MJ per kg energy content of carbohydrates, 92% efficiency of conversion to methane, and 5% efficiency of heating on an open fire. (Specific heat of water is 4.2 kJ kg^{-1} °C^{-1}.)

4.8 Discuss why some hydrocarbon fuels are being promoted for cooking in the developing world.

4.9 Estimate the reduction in CO_2 emissions from the full conversion of the Drax power station in the UK to combusting pelletized wood.

4.10 Compare C3, C4, and CAM plants and discuss what types of plant are most suitable for biofuel production.

4.11 India has a land area of 2.97×10^6 square kilometres, 57% of which is cropland. If 5% of India's cropland were dedicated to producing jatropha plants, estimate the annual production of biodiesel. Compare your estimate with the global use of oil for transport and comment.

4.12 Is it realistic to replace the 300 million tonnes of oil used by international shipping each year with biodiesel?

4.13 What are the difficulties with producing bioethanol by enzymatic hydrolysis of cellulosic feedstock?

4.14 If all cars in the US were to run with gasoline blended with 20% bioethanol produced by enzymatic hydrolysis of cellulosic feedstock, estimate the annual reduction in CO_2 production in tonnes. Assume: FER $= 4$, energy content ethanol 21 MJ litre^{-1}, gasoline 32 MJ litre^{-1}.

4.15 Estimate what size of forest would need to be planted to absorb the carbon dioxide produced by a 3 GW$_e$ coal-fired power station. Is this a practical way to combat greenhouse gas emissions?

4.16 In a Stirling engine the regenerator has an efficiency of ε_R, i.e. the heat input Q_1 required) equals $(1 - \varepsilon_R)|Q_3|$. Show that the efficiency ε_S of the Stirling engine operating between temperatures T_h and T_c is given by

$$\varepsilon_S = \varepsilon_C / [1 + (1 - \varepsilon_R)\alpha\varepsilon_C / \ln(V_2/V_1)],$$

where ε_C is the Carnot efficiency $(T_h - T_c)/T_h$ and $\alpha = 1.5$.

Calculate the efficiency for $T_h = 325$ °C, $T_c = 75$ °C, $\varepsilon_R = 0.5$, and $V_2/V_1 = 5$.

4.17 Describe how anaerobic digestion of a CAM crop coupled with the Fischer-Tropsch process could be used to make aviation fuel, and discuss whether this would be economically viable.

4.18 Discuss the extent to which energy derived from the following sources is carbon neutral: (a) short rotation coppice willow; (b) jatropha; (c) corn; (d) sugarcane; (e) cassava; (f) palm oil.

4.19 (a) Estimate the annual reduction in CO_2 emissions if a 1 GW_e coal-fired power station were replaced by a MSW plant. (b) How much waste would be required per year by the MSW plant? Assume a 30% efficiency for a municipal solid waste plant.

4.20 A 1 GWe coal-fired power station is converted to use biomass as fuel. Estimate the area of land required for energy crops. (a) What is the reduction in CO_2 and in C emissions per year? (b) What area would be required for an annual reduction of 1 Gt of carbon?

4.21 Calculate the amount of carbon emitted when 1 litre of biodiesel is burnt. Take the composition of biodiesel to be 100% cetane (hexadecane) $C_{16}H_{34}$. Assume a density of 0.88 kg per litre.

4.22 Estimate the savings in CO_2 emissions per litre when biodiesel is used instead of diesel, if the source of biodiesel is from jatropha?

4.23 Does the production of nitrous oxides and of particulates limit the use of biofuels in urban areas?

4.24 In parts of South America the yield of sugar from sugarcane is 1600 tonnes per square kilometre per year. The sugar is fermented into ethanol via the reaction

$$C_6H_{12}O_6 \rightarrow 2C_2H_5OH + 2CO_2.$$

Calculate the area of sugarcane required to produce sufficient ethanol to displace 2×10^{10} litres of petrol per year. What would be the resulting reduction in carbon emissions, measured in tonnes of carbon dioxide per year? (Assume petrol is pure octane.)

4.25 Should the Brazilian bioethanol from sugarcane program be adopted globally?

4.26 A strain of algae is developed that can produce 900 GJ $ha^{-1} y^{-1}$. What area is required to displace 400 million tonnes of conventional diesel?

4.27 What is the best way to increase the contribution of biomass to global energy production?

4.28 Discuss whether the CO_2 associated with converting land to biomass production, and the resulting loss of land and water for food production, make biomass an unsustainable energy resource.

Box 4.6 Energy storage in plants

Plants synthesize carbohydrates from CO_2 and H_2O by using the energy from sunlight (photosynthesis, see Section 4.1). The carbohydrates are part of a plant's structure and provide a store of energy. The simplest carbohydrates are sugars or monosaccharides, which have the composition $(CH_2O)_n$. Glucose, $C_6H_{12}O_6$, is the commonest plant sugar and is called a 6-sugar since it contains six carbon atoms. The glucose molecule can exist in several forms, in which the atoms have different bonding and orientations, called 'structural isomers'; in particular, as α-glucose and β-glucose. These two hexagonal ring

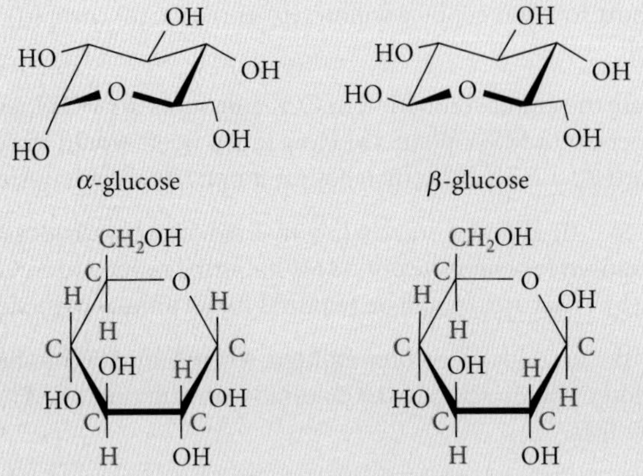

Fig. 4.12 α- and β-glucose.

forms are illustrated in Fig. 4.12. The upper illustrations indicate that the hexagonal rings are not actually flat in nature. Also, the carbon and hydrogen symbols are omitted from the upper diagrams for clarity.

Glucose forms a disaccharide by a condensation reaction:

$$C_6H_{12}O_6 + C_6H_{12}O_6 \rightarrow C_{12}H_{22}O_{11} + H_2O, \tag{4.22}$$

that is the reverse of a hydrolysis reaction. For example, maltose is formed by linking one α-glucose and one β-glucose and cellubiose by linking two β-glucose molecules. Further condensation reactions convert glucose to polysaccharides, in particular to starch and cellulose. These biopolymers store energy and provide bulk and structure in a plant.

The differences in the bonding in starch and cellulose significantly affect their structure, with starch much more amorphous than cellulose which forms fibrous bundles. Amylose, which is a component of starch, is a polymer of α-glucose molecules linked by α-**glycosidic bonds**, while cellulose is a polymer of β-glucose molecules linked by β-**glycosidic bonds**. The linkages in amylase and cellulose are illustrated in Fig. 4.13.

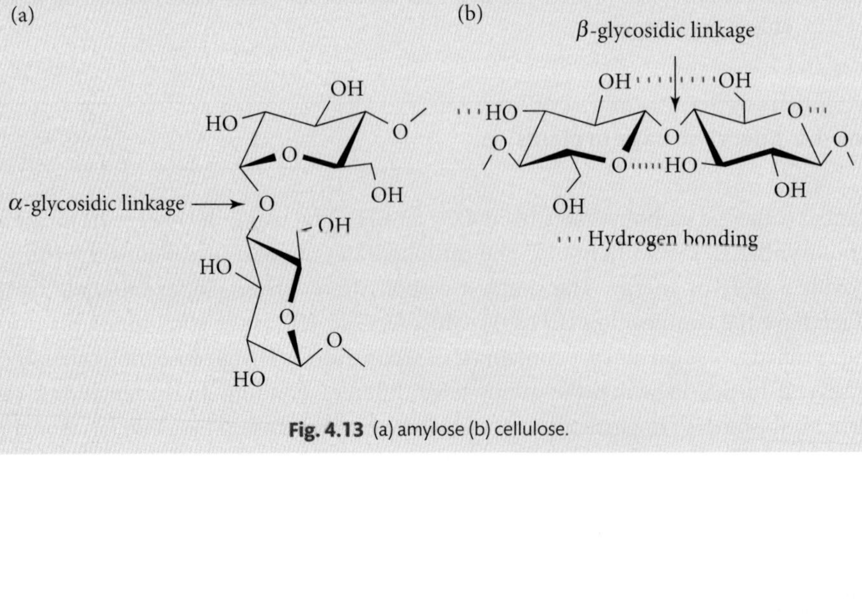

Fig. 4.13 (a) amylose (b) cellulose.

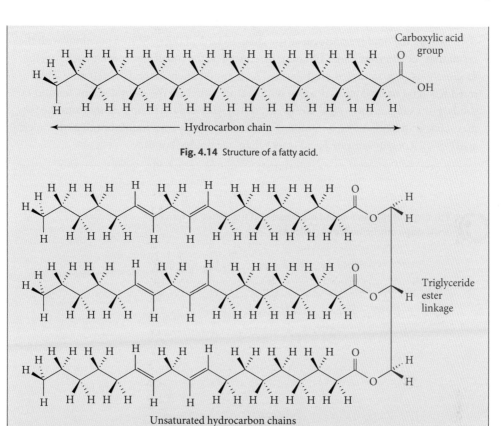

Fig. 4.14 Structure of a fatty acid.

Fig. 4.15 An unsaturated triglyceride: trilinolein.

The hydrogen bonding gives increased stability and leads to long straight chains. These chains can hydrogen bond with each other giving rise to strong micofibrils.

The sugars are a store of energy with the carbon in a state of partial reduction; on combustion ~16 MJ kg^{-1} is released. A more compact form of storage is afforded by further reducing the sugars to form fatty acids whose structure has the form shown in Fig. 4.14. (The H of the carboxylic group is easily ionized, which makes the molecule an acid.)

The commonest length of the hydrocarbon chain lies within 12 and 24 carbon atoms. These molecules have a very low solubility in water, and they and their derivatives are called **lipids**. The carbon is almost fully reduced and has the highest ratio of H:C when there are no double bonds between the carbon atoms—such fatty acids are called saturated. The heat of combustion is therefore much higher than for carbohydrates with typically 38 MJ kg^{-1} released.

A common storage molecule is a triglyceride, which is a fatty acid ester: three fatty acids joined to a glycerol molecule with the removal of three water molecules. This transformation of carbohydrate to triglyceride occurs when certain plants ripen (e.g. olives). The reverse process occurs when a seed starts to grow (germination) with the hydrolysis (uptake of water) of 1 gramme of oil producing ~2.7 grammes of carbohydrate. Shown in Fig. 4.15 is an unsaturated triglyceride: trilinolein.

In this fat, the double bonds cause the chain to kink at the indicated positions with the result that the intermolecular bonding is reduced as the molecules cannot pack together so closely as in a saturated triglyceride. As a result, such naturally occurring unsaturated fats tend to be liquids at room temperature, whereas the saturated fats tend to be solids. For example, olive oil is composed of triglycerides made up mainly from the isomers oleic acid (55%–85%), a monounsaturated acid (only one double bond) and linoleic acid (~9%), a polyunsaturated acid (more than one double bond).

For further information and resources visit the online resources
www.oup.com/he/andrews_jelley4e

5   Solar Thermal and Geothermal Energy

→ **Introduction**

As we saw in Chapter 1 Section 1.2.2, the world uses a significant amount of its energy in the form of heat, particularly in thermal power plants, industrial processes, internal combustion engines, and in heating homes. Much of this energy is presently supplied by fossil fuels, and in this chapter we describe two renewable low-carbon sources of heat that could help in displacing their use: solar thermal and geothermal.

Solar energy is by far the largest source of renewable energy; in about one hour the Earth receives the same amount of energy from the Sun that is consumed globally in one year; or more precisely, only ~0.015% of the Sun's energy reaching the Earth's surface would have provided the global average primary energy consumption of 576 EJ in 2016. Sunlight can be converted to electricity using photovoltaic (PV) cells, and this increasingly important source of electricity is discussed separately in Chapter 8. Solar energy can also be used as a direct source of heat (solar thermal), or when concentrated in a thermal power plant to generate electricity (concentrated solar thermal power-CSP), as well as providing heat for industrial processes or for producing chemicals.

Although a smaller resource than solar energy, the thermal energy that potentially can be extracted from the Earth per year is also very significant at around 10 TWy. This geothermal energy, though, has to be extracted from hot rocks often deep within the Earth, hence making this process economic is very challenging. Easier and cheaper to extract is the heat in the ground near the Earth's surface, using heat pumps. This source is often also referred to as geothermal, although nearly all of the thermal energy extracted from near the surface is replaced by solar energy.

5.1 **Solar Thermal Energy**

Solar thermal energy is the direct conversion of electromagnetic radiation from the Sun into heat, via a **solar thermal collector**. It is used especially in colder climates to provide supplementary space heating and for water heating for domestic and commercial buildings, by mounting solar panels on Sun-facing surfaces.

The two most common designs of solar panels are the flat plate collector and the evacuated tube collector. In the **flat plate collector** (Fig. 5.1 (a)), the incident solar radiation is absorbed by a flat plate with a black coating. Heat is conducted along the plate to pipes attached to the plate, and is transferred to a fluid (e.g. water or antifreeze) inside the pipes. Flat plate collectors lose a considerable amount of heat to the environment in cold and windy conditions, and the efficiency is further reduced by condensation.

The **evacuated tube collector** (Fig. 5.1 (b)) consists of an array of parallel tubes, with a reflecting surface on the rear of the panel. The incident solar radiation penetrates an outer glass tube and is absorbed on the blackened outer surface of a metal **heat pipe** (see Chapter 2 Section 2.2.2), which is surrounded by a vacuum barrier to reduce heat loss to the environment. The heat pipe contains a fluid which is heated and evaporates from contact with the hot inner surface of the tube; the vapour rises due to buoyancy forces, and the heat is transferred to a heat exchanger. The cooled vapour/liquid from the heat exchanger is then returned to the lower end of the heat pipe, to complete the cycle.

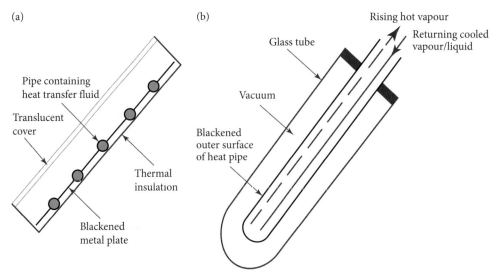

Fig. 5.1 (a) Flat plate collector (b) Evacuated tube collector.

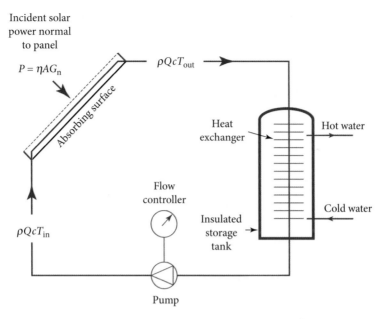

Fig. 5.2 Solar thermal energy system for domestic hot water.

The schematic design of a typical domestic solar thermal energy installation is shown in Fig. 5.2. The working fluid is contained in a closed circuit. Circulation can be just by natural convection without a pump: a **thermosyphon** system. Heat from the solar panel is stored in a thermally insulated tank, which provides hot water for radiators and hot water taps. It should be noted that a back-up heating system is usually required for cloudy days and periods of higher demand. However, the capital cost of a domestic solar thermal installation is offset by a significant reduction in water heating bills, typically around 50–80%.

The temperature of the fluid leaving the solar panel is determined by energy conservation. Equating the power absorbed by the collector to the rate at which heat is absorbed by the fluid, we have:

$$P_{\text{solar}} = \eta A G_n = \rho Q c T_{\text{out}} - \rho Q c T_{\text{in}} \tag{5.1}$$

where η is the efficiency of the collector, $A (\text{m}^2)$ is the area of the collector, $G_n (\text{W m}^{-2})$ is the solar energy per unit area per second normal to the surface of the collector, $\rho (\text{kg m}^{-3})$ and $c (\text{J kg}^{-1} \text{K}^{-1})$ are the density and specific heat of the fluid, and $Q (\text{m}^3 \text{s}^{-1})$ is the volume flow rate.

Rearranging eqn (5.1), we obtain the output temperature as:

$$T_{\text{out}} = T_{\text{in}} + (\eta A G_n)/\rho Q c \tag{5.2}$$

The thermal performance of the heat transfer fluid can be improved by around 20% by adding nanoparticles; e.g. carbon-based, CuO, or Al_2O_3. The optical absorption can also be increased so that no solid surface absorber is needed. However, in both the direct absorption and the conventional solar thermal collectors, the long-term stability of the nanofluids is a concern and further research is needed.

EXAMPLE 5.1

Calculate the time taken to heat 1000 litres of water to $T_{out} = 333$ K, given that $T_{in} = 293$ K, $\eta = 0.6$, $A = 20$ m², $G_n = 500$ W m⁻², $\rho = 10^3$ kg m⁻³, $c = 4 \times 10^3$ J kg⁻¹ K⁻¹.

From eqn (5.2), we have:

$$Q = \frac{\eta A G_n}{\rho c (T_{out} - T_{in})} = \frac{(0.6)(20\,\text{m}^2)(5\times10^2\,\text{Wm}^{-2})}{(10^3\,\text{kg m}^{-3})(4\times10^3\,\text{J kg}^{-1}\text{K}^{-1})(40\text{K})} \approx 3.75 \times 10^{-5}\,\text{m}^3\text{s}^{-1} \tag{5.3}$$

Hence, the time taken to produce 1000 litres of hot water is given by:

$$t \approx \frac{1000 \times 10^{-3}\,\text{m}^3}{(3.75 \times 10^{-5}\,\text{m}^3\text{s}^{-1})} = 26670\,\text{s} = 7\,\text{hours}\,25\,\text{minutes}$$

5.1.1 Global Solar Thermal Heating Capacity

By the end of 2019 the total global capacity of solar thermal water heaters, nearly all glazed, was estimated to be 479 GWth, providing about 389 TWh of heat annually, corresponding to savings of 0.135 GtCO₂. About 70% were evacuated tube collectors with flat plate 23%. The market for small scale heating systems especially in China and Europe is under pressure from heat pumps. But nonetheless, around 60% of global installations are small scale thermosyphon systems, and the number of megawatt-scale systems for district heating and industrial applications is growing. The share of the market at the end of 2018 was China 70% (see Fig. 5.3), Europe 12%, and the Americas 7%, accounting for 89% of the installed capacity.

Solar thermal systems are used in the food, beverage, and mining industry, and in recent years to supply heat to greenhouses. Photovoltaic thermal systems combine electricity and heat generation and grew by 9% in 2019 with an installed capacity of 606 MWth and 208 MWpeak, with 58% in Europe. Solar thermal energy can also be used for cooling by using an air-source heat pump (ASHP) (see Section 5.2) that uses heat rather than electricity to drive it. In a water and lithium bromide solar cooler, for example, water enters an evaporator, which is at low pressure, and boils at ~5°C. The change in phase extracts heat through a heat exchanger from the building, and the water vapour is absorbed in a lithium bromide solution. The diluted lithium bromide is pumped to a vessel where it is heated by solar radiation

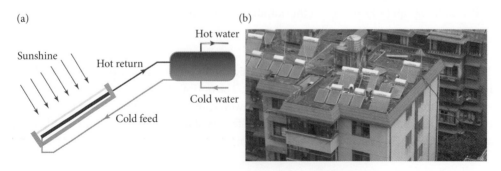

Fig. 5.3 (a) Thermosyphon system (b) Solar water heaters mounted on rooftops in China.

and the water evaporated off. The water vapour is condensed, transferring its heat to the surrounding air, and both the water and concentrated lithium bromide solution are returned to the evaporator. An absorption cooler's coefficient of performance is about four times less than that of a vapour compression system (i.e. ~0.8), but its energy input is heat, which can be freely available, as with solar energy, or using waste heat.

Globally about 20% of final energy is for heating, cooling, and water heating in buildings, approximately 20,000 TWh, so solar thermal is providing about 2%. Solar thermal technologies can provide industrial heating (process heat) and cooling, and the market is expanding. IRENA estimates that its contribution by 2050 could be around 3000 TWh y^{-1} (~10 EJ y^{-1}), about equally split between industrial applications and homes, saving about 0.8 $GtCO_2$ a year by 2050.

5.2 Ground- and Air-source Heat Pumps

Decarbonizing the large heat demand in buildings will be challenging. While it is estimated that solar thermal, modern biomass, and geothermal could supply by 2050 about 3000, 2000, 1000 TWh y^{-1}, respectively, this is only roughly 30% of the expected demand (~20,000 TWh y^{-1}). The heat required is at moderate temperatures, and while **radiant heating** using clean electricity can be a cost-effective way to provide low-carbon localized heating, a very efficient low-carbon technology to provide space heating that fits in well with renewable electricity is a heat pump.

5.2.1 Principle of a Heat Pump

A **heat pump** is a heat engine that uses work to pump heat either into or out of a building. When used to pump heat to or from the ground or air, the process takes advantage of the fact that the ground, and to a lesser extent the air, is maintained at a relatively constant temperature by heat from the Sun (there is also a very small contribution from geothermal energy when using the ground as a source). The temperature of the ground is reasonably constant below a depth of ~10 m (~10°C in the UK). The amount of work required to pump a quantity of heat across temperature differences of 20–40°C between the inside of the house and the ground or air is typically a factor of around three times less than the heat transferred. Ground-source pumps draw heat from the ground during the winter, so it can be important to pump heat back into the ground during the summer, which occurs when a heat pump is used to both air-condition and heat a building, in order to avoid the ground freezing. The operation of a heat pump is based on the same principle as a refrigerator.

In an air-source heat pump, the following processes take place when heating a room at temperature T_2 with heat extracted from the outside air at a temperature T_1 (see Fig. 5.4). In this case the evaporator is outside and the condenser is inside, and $T_2 > T_1$.

(a) **Along 3-4:** A fluid at high pressure, initially near the temperature T_2, passes through a narrow tube which throttles the flow (the expansion valve) and causes a large drop in pressure. The boiling point of the fluid is lowered and some of the fluid evaporates, which cools the fluid (as the latent heat of evaporation comes from the fluid itself).

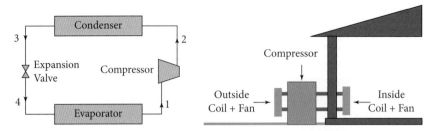

Fig. 5.4 Components and layout of an air source heat pump.

(b) Along 4-1: The cold fluid is piped through a heat exchanger coil (called the evaporator) where it warms up, as heat flows from the surrounding air at temperature T_1 to the cold fluid. This causes the cold fluid to evaporate more and to end up as a cold vapour.

(c) Along 1-2: The cold vapour is then compressed to high pressure, during which process it heats up.

(d) Along 2-3: The hot vapour under high pressure is then passed through a heat exchanger (called the condenser) during which it condenses back to a fluid, releasing heat to the surrounding air and cooling to near the temperature T_2, and the cycle is then repeated.

We can very roughly approximate the process by a Carnot cycle where during (b) heat $|Q_1|$ is extracted from the outside air at a temperature T_1, during (d) heat $|Q_2|$ is expelled to the room air at a temperature T_2, and during (c) and (a) work W is done by the compressor. From the first law of thermodynamics, eqn (2.10), $W = |Q_2| - |Q_1|$, and for a Carnot cycle $|Q_2|/T_2 = |Q_1|/T_1$ (eqn (2.19)), so:

$$|Q_2| = T_2 W / |T_2 - T_1| \tag{5.4}$$

Thus, a factor $T_2/|T_2 - T_1|$ more heat would be transferred to the room than the work done by the compressor.

To provide cooling for a building, the flow of the fluid is reversed and the heat exchanger inside the room is now the evaporator, and the one outside the house is the condenser. Heat is extracted from the room air at temperature T_2 when cold fluid is pumped through the evaporator, step (b), and is transferred to the outside air at temperature T_1 when the hot vapour is passed through the condenser, step (d); in this case $T_1 > T_2$.

In the process, heat $|Q_2|$ is either extracted or transferred to the building, the ratio $|Q_2|/W$ for a heat pump is called the **coefficient of performance** (COP), and, for an ideal heat pump the COP $= T_2/|T_2 - T_1|$; e.g. for heating a building with $T_2 = 37°C = 310$ K and a ground temperature of $T_1 = 6°C = 279$ K, $\Delta T \equiv |T_2 - T_1| = 31$ K and we have COP $= 10$. The actual COPs for heating or cooling units for buildings typically lie between 2.5 and 4 (HP2018), as shown in Derivation 5.1 at the end of the chapter. Even allowing for the inefficiency of the heat pump, much less energy is required to heat buildings compared with using direct heating units like electric resistance heaters or gas fires.

Ground-source heat pumps tend to be more efficient than air-source, because the ground is at a more stable and suitable temperature, being generally warmer in the winter and colder in the summer than the surface air temperature, but their capital cost is generally higher. For such pumps, water or a water–antifreeze mixture is circulated through pipes that are buried in the ground at a depth of typically 30–120 m. There they extract or transfer heat to the ground, depending on whether the pipe temperature is lower or higher than that of the ground. Inside the buildings, ducts are used to transport the hot or cool air throughout the building. The heat pump can also be used to provide domestic hot water.

5.2.2 Outlook for Heat Pumps

Currently, heat pumps supply about 300 TWh of heat a year but this could increase to 4000 TWh a year by 2050. This would, as we have seen, only take about a third of that amount of electricity. There has been a shift to air-source heat pumps (ASHPs) as their efficiency has improved and their cost decreased, and they are increasingly being used for new domestic buildings, though ground-source heat pumps (GSHPs) are well suited for large buildings, though their cost is generally higher, as deep holes have to be drilled.

Rivers running underground below cities, such as the Tyburn and Fleet under London, could provide good sources for heat pumps in large city buildings. Coils could be placed in these subterranean rivers, which would avoid the cost of drilling boreholes. The technology is already employed in Stuttgart to heat the Baden-Württemberg state ministry using the underground river Nesenbach. While seasonal performance factor (SPF is the average COP over a season) values of three or more, and good reliability under a wide range of conditions, are expected, there have sometimes been installations with a poor SPF of around 2 when retrofitting homes (see Section 11.1.7).

Most of the sources and sinks of heat for heat pumps are sustainable, so if non-fossil fuel electricity is used to run the pumps, the process is renewable. Many, but not all, countries treat heat pumps as a renewable source, and heat pump installations are eligible for renewable support. Presently, approximate one-third of electricity is non-fossil, and the global carbon intensity ($475 \text{ gCO}_2 \text{ kWh}^{-1}$ in 2018) is predicted to fall as more low-carbon generation comes on line.

Currently, high temperature heat pumps can heat air up to 120°C and produce steam at 165°C. They can replace boilers in several industrial processes, and R&D is expected to extend their operating range, and their efficiency by up to 50%. Costs are expected to fall as a result, which will also be helped by the learning effect through greater market penetration. Combining heat pumps with thermal storage can reduce the peak demand for energy.

The global demand for space heating or cooling accounts for approximately half of the world's energy consumption in buildings, with a fast-increasing demand in developing countries, and most of this is provided now by fossil fuel combustion. In their transition scenario IRENA estimate some 14 EJ y^{-1} from heat pumps by 2050, and we will take 4000 TWh y^{-1} as the accessible potential.

5.3 Concentrated Solar Thermal Power (CSP) Plants

The first commercially operated concentrated solar thermal power (CSP) plant was a 75 kW (100 hp) pumping engine powered by five parabolic troughs in Al Meadi, Egypt, in

1913, which was designed by an American, Frank Schuman. The plant had a total aperture area of around 1200 m^2 that generated saturated steam at up to 250°C. The British and German governments planned similar units to provide energy in their colonies. However, the need for energy in a readily transportable form during the First World War drove a rapid expansion in oil exploration, and CSP almost disappeared until the first oil crisis in the early 1970s.

The oil crisis revived interest in CSP, and the first commercial plants began operation in California in the 1980s. An early example was at Barstow in California, where a large array of mirrors was used to direct the Sun's rays onto a tank on top of a central tower. Oil and then molten salt (at over 500°C) were used to transfer heat to a boiler for a conventional thermal power plant that produced 10 MW. The molten salt was also used as a store of heat. Another large plant in the Mojave Desert in California used parabolic trough collectors to provide heat for an 80 MW plant from ~500,000 m^2 of collector area. A tank containing oil was heated to ~390°C, and an average conversion efficiency of 18% was achieved.

However, the decrease in the cost of fossil fuels in the 1990s and the withdrawal of incentives stifled any significant growth until 2006 when government initiatives in Spain and in the USA revitalized the market. By 2010 there was globally 1 GW of capacity with a further 15 GW being planned, with parabolic trough systems accounting for the greatest share of the CSP market. However, since then growth has been affected by the sharp fall in PV costs and only recently has investment revived; its flexibility when combined with thermal energy storage (TES) is a significant advantage.

CSP requires direct sunlight to be concentrated by mirrors or lenses onto small areas (see Fig. 5.5), and on clear days, which are often found in hot, semi-arid regions typically lying between latitudes 15° and 40° N and S of the equator, about 80–90% of the solar radiation reaches the Earth directly, without significant scattering. The intensity is also higher at high altitudes where the amount of absorption and scattering of sunlight by aerosols in the atmosphere is reduced.

The reason for concentrating sunlight is to raise the input temperature of a thermal power plant high enough that electricity can be generated economically. Parabolic trough

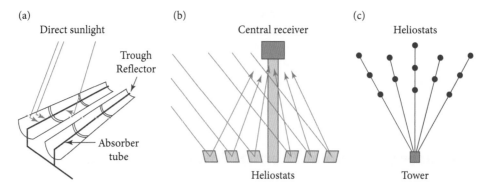

Fig. 5.5 Parabolic trough and solar tower/central receiver plants and heliostat layout.

reflectors typically concentrate light 30-80 times, while solar tower plants 200–1000 times and so have higher thermal efficiencies that can offset their high capital costs. The relation between concentration and input temperature is explained in Box 5.1 CSP thermodynamic fundamentals.

Box 5.1 CSP thermodynamic fundamentals

The maximum possible concentration is $(1/\sin\alpha)^2 \approx 46,000$, where $\alpha = 4.65$ mr is the half-angle subtended by the Sun on Earth, while for optical elements that only concentrate in one plane, like a parabolic trough reflector, the maximum concentration is given by $1/\sin\alpha \approx 220$. However, in practice the concentration is generally restricted by material properties to less than a thousand.

The radiative power absorbed by a receiver is that concentrated on to it less the amount radiated away (given by the Stefan-Boltzmann law, eqn (2.8)), so it can be written in the form:

$$\dot{Q}_r = \alpha GA_c - A_e \varepsilon \sigma T^4 = GA_c \eta_r \tag{5.5}$$

where G is the unconcentrated irradiation incident on the optical concentrators, A_c is the input aperture area of the concentrators, A_e is the area of the receiver, α is the absorptivity for solar radiation, ε is the infrared emissivity of the absorber, σ is the Stefan–Boltzmann constant, and η_r is the receiver efficiency given from eqn (5.5) by

$$\eta_r = \alpha - \beta T^4 \tag{5.6}$$

The radiative loss coefficient β depends on the concentration ratio $C = A_c/A_e$ of the concentrator, according to the formula:

$$\beta = \varepsilon\sigma/GC \tag{5.7}$$

The input power can therefore be increased by decreasing the radiative loss by reducing the infrared emissivity by using a selective emitter, by increasing the absorptivity or concentration, or by any combination of these.

We now determine the concentration for an optimal absorber temperature by considering the thermal exergy. The **thermal exergy** (see Chapter 2 Section 2.6) or available energy flow out of the receiver is given by:

$$\dot{X} = GA_c \eta_r \eta_a \tag{5.8}$$

where $\eta_a = 1 - T_a/T$ is the Carnot efficiency between the absorption temperature T and the temperature of the ambient environment, T_a. The concentration required to make the absorption temperature optimal is obtained by maximizing eqn (5.8) with respect to T for given G, A_c, and C. Solving for C then yields:

$$C = \varepsilon \sigma T^4 (4T - 3T_a) / \alpha G T_a \tag{5.9}$$

For a tower plant α and ε might both equal about 0.9, and $G = 900$ W m^{-2}. For receivers made from nickel alloys the maximum input temperature is around 900 K, while for cobalt alloys 1100 K. Taking $T_a = 300$ K, C_{max}(nickel) = 372 and C_{max}(cobalt) = 1076.

For a parabolic receiver plant with $\alpha = \varepsilon$ and with an input temperature of 400°C, $C_{max} = 77$, though this could be lowered by reducing the emissivity of the absorber tubes.

5.3.1 Solar Tower (Central Receiver) Plants

CSP central receiver (solar tower) plants use large arrays of flat or nearly flat mirrors (heliostats) to track the Sun and reflect the solar radiation onto a central receiver (Fig. 5.7 (b)). The heat absorbed by the receiver is used to generate power either directly or indirectly using a thermal store. The receivers in solar tower plants can be heated to higher temperatures than those in parabolic trough plants, which gives higher efficiencies and a lower cost for thermal energy storage, as there is a greater operating temperature range.

Either steam or molten salt is used as the heat transfer fluid in the receiver. Steam uses conventional technology and, when superheated under high pressure, the steam can be at ~500°C; it can also be hybridized with a natural gas fired boiler. When molten salt is used, it is pumped from one tank to another via the receiver, where it is heated to ~565°C. When steam is required, the salt is pumped from the hot tank through a heat exchanger, to the warm tank (kept at 285°C, which avoids the salt solidifying), decoupling the electricity generation from solar energy collection (see Fig. 5.6). A typical average annual efficiency for a solar tower plant for conversion of solar to electrical energy is about 18%.

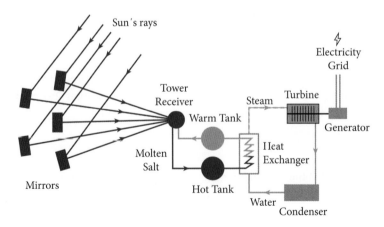

Fig. 5.6 Schematic of a concentrated solar power plant.

A commonly used array is the radial stagger layout illustrated in Fig. 5.5 (c), developed by Lips and Vant-Hull to minimize the blocking of reflected light and shading. The heliostats closest to the tower have the least shading, those due north (south) in the northern (southern) hemisphere have the least inclination (cosine) loss, whilst those farthest from the tower require the greatest tracking accuracy: for heliostats 300 m away an error of $\pm 0.1°$ corresponds to a movement of the Sun's image of ± 0.5 m. For a large plant there can be tens of thousands of heliostats (see Fig. 5.7).

Recent solar-only Brayton cycle projects have investigated using supercritical CO_2 as the working fluid. This has the advantage that compression work requirements are very low, minimizing compressor costs and parasitic losses, and the efficiency for the same temperature is higher than with steam (see Chapter 3 Section 3.15). Shouhang and EDF announced in 2019 that they are to test a supercritical CO_2 cycle in a small 10 MW CSP with storage plant at Dunhuang in China.

Fig. 5.7 shows part of the 150 MW Noor III solar tower plant and the 200 MW Noor II parabolic trough plant in Morocco, both with **seven hours** of storage.

5.3.2 **Parabolic Trough Plants**

The solar field contains rows of collectors (Fig. 5.5 (a)) aligned on a north-south axis that track the Sun from east to west during the day. The heat transfer fluid, usually a synthetic oil, is heated to a temperature of ~390°C and is pumped through a heat exchanger to either generate high-pressure superheated steam (~100 bar 370°C) or to heat molten salt that is pumped from one tank to another. The steam is used to drive a turbine generator to produce electricity.

Fig. 5.7 Noor II parabolic trough and Noor III tower plants (Xinhua/Alamy Live News).

Wet, dry, or hybrid cooling towers condense the steam, the choice affecting water usage and cost. When the Sun is not shining, the molten salt can be pumped through the heat exchanger to heat the oil, which is then used to generate steam. A typical average annual efficiency for a parabolic trough power plant for conversion of solar to electrical energy is about 15%.

In CSP parabolic trough plants, many use slumped glass mirrors. Alternatives are to employ coated anodized aluminium sheet, or polymer film mirrors, which are lighter but are generally less resistant to abrasion. Self-cleaning mirrors are also being investigated as a way of reducing costs. The operating temperatures of parabolic trough plants have historically been limited to around 400°C by the thermal degradation temperature of the oils used as heat transfer fluids (HTFs) between the solar field and the steam generator. Using molten salt HTFs would raise the operating temperature and hence the efficiency. However, they introduce new challenges associated with pumping and maintaining salts above their melting temperatures at all times.

5.3.3 **Solar Receivers**

The main receiver types are illustrated in Fig. 5.8. In CSP trough plants, such as SEGS I to IX, developed in the Mojave Desert between 1984 and 1991, the linear absorbers usually comprise a metallic tube with a selective emissivity coating, which is for solar wavelength absorptance enhancement as well as longer wavelength emittance reduction. This is contained within an evacuated glass Dewar, to which anti-reflection coatings may be applied.

In central receiver steam plants, a cavity receiver is usually employed. The cavity receiver concept arises from the trade-off between maximizing heat exchanger surface area (and hence thermal conductance) on the fluid side whilst minimizing thermal losses due to thermal emission from the absorber surface. Concentrated radiation passes through an aperture into a cavity containing the absorbing elements. Radiation usually enters from below to minimize convective losses. The effective area over which radiative heat loss can occur is limited to the area of the aperture, as the absorbing elements reabsorb most of the thermally-emitted photons.

A volumetric receiver has been developed to optimize the transfer of concentrated solar radiation to gases to enable higher temperatures to be reached. Radiation is absorbed onto a metallic or ceramic porous structure over which a gaseous heat transfer fluid passes and is

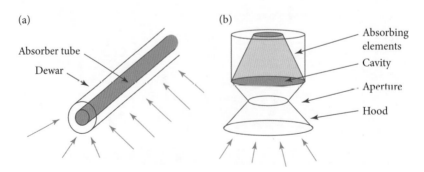

Fig. 5.8 (a) Trough absorber (b) Cavity receiver (CSP2015).

simultaneously heated. Particle receivers are also being considered in which falling particles are heated to over 720°C to both improve thermal efficiency and to provide high-temperature storage.

5.3.4 CSP Thermal Energy Storage (TES)

Thermal energy may be stored at the high temperatures required to achieve useful 'round-trip' conversion efficiencies in a number of ways. All existing CSP plants that incorporate TES utilize the stored heat to improve plant capacity factors on the same site that it is collected. Therefore, energy density is only of indirect importance through economic cost. The size of the solar field is larger than that required to generate full capacity, i.e. a solar multiple (SM) greater than 1. The excess size both allows for maximum generation to be maintained when the solar intensity is lower by a certain amount, and for excess solar heat to be stored to provide for generation when the sun is down.

The great majority of operational high-temperature thermal storage systems employ sensible heat storage. Most of these use molten salts, the most common being a non-eutectic salt mixture of 60 wt% sodium nitrate and 40 wt% potassium nitrate that has become known as 'solar salt'. Typical energy densities (sensible heat per kg) of molten salts are up to 200 kJ kg^{-1}.

5.3.5 Economics of CSP

Before 2015, adding storage to a CSP plant was often uneconomic as penetration of variable renewable generation was relatively small. But since then hardly any project is without storage, which raises the capacity factor (CF), gives more flexibility, and lowers the levelized cost of energy (LCOE). In the period 2010–2019 the cost per kW of output fell from around $9000 to $5800 and CF increased from about 30% to 45%. These improvements are reflected in the LCOE falling from an average 35 to 18 cents per kWh in 2019; in regions of high direct normal irradiance (DNI) prices in 2019 could be around 14 cents per kWh. The LCOE are similar for parabolic trough as for solar tower plants. The operations and maintenance (M&O) costs are about 0.02 to 0.04 US$ kWh^{-1}.

The success of the recent CSP plus storage NOOR III project in Morocco has increased confidence in this technology, and interest rates on capital for new projects have decreased and the lifetimes of power purchase agreements (PPAs) increased. For a new plant in Dubai, due online in 2021, interest rates have fallen from around 7.5% to 4% and the lifetime lengthened from 25 y to 35 y, which improves the capital recovery factor (CRF) from 0.09 to 0.054. This improvement is a large part of why the PPA price is 7.4 cents kWh^{-1}, while the LCOE of a comparable plant in 2019 was 14 cents kWh^{-1}. The LCOE is a proxy for a PPA, which is an agreed price that a utility will pay for generated electricity.

Another project in Morocco, Noor Midelt I, which was announced in 2019, has 800 MW capacity in a location with a DNI of 2360 kWh per square-metre per year. It is a hybrid solar farm, with trough CSP as well as PV capacity and 5 hours of storage. Operation is expected in 2022 and the PPA is for 25 y at 7.1 cents kWh^{-1}. The project has reduced costs by economies of scale, improvements in installation, design efficiencies, and sharing of infrastructure.

The plant will have a flexible output and its storage can smooth out any variability in sunshine. The cost of storage is around €25 per kWh (thermal) ~ $100 per kWh (electricity) and is cheaper than that of Li-ion batteries, currently (2020) around $150 per kWh, and it is hoped that the cost of thermal storage might be reduced to $15 per kWh (thermal).

5.3.6 Environmental Impact of CSP

CSP is a renewable energy generator with low associated carbon emissions, and therefore helps to combat global warming. Condensing systems can use water which in hot, dry regions can be scarce, but dry cooling systems can be deployed, though these may raise costs. There have been concerns over bird deaths and glare from the intense solar beams that tower plants utilize. High solar flux in the region near receivers corresponding to heliostat stand-by positions has been identified as a cause, and limiting regions of high flux can reduce this risk. As with wind turbines, there is a slight risk to birds from CSP, but the wider environmental danger from global warming needs to be kept in mind.

5.3.7 Outlook for CSP

In the 2000s it was thought that concentrated solar power plants would be able to generate electricity more cheaply than photovoltaic systems. The average cost of electricity from CSP plants in 2010 was around US$0.35 kWh^{-1}, similar to that from PV, but the very sharp fall in PV costs since then has meant that CSP plants lost out and their growth has slowed. Global capacity was ~0.5 GW in 2008, ~4.4 GW in 2014, and 6.6 GW in 2019.

The recent sharp fall in PPA prices to around 7c kWh^{-1} for projects starting in early 2020s could well lead to more investment in CSP plus thermal storage and help, particularly in countries with good DNI, with the integration of PV. New projects are planned for Dubai, China, Chile, and South Africa. The reduction in costs of batteries, though, for large-scale storage, either of Li-ion or of flow batteries, may well affect its competitiveness.

In 2018 and 2019 around 0.66 GW capacity globally was added each year, and IRENA in their global energy transformation report (2018) estimated that global CSP installed capacity could be ~600 GW by 2050 generating about 2000 TWh per year. This would require the average installed annual capacity to increase about thirtyfold. It would be a small but useful contribution to renewable energy generation.

5.4 Thermoelectric Generation (TEG)

An interesting area of research is into **solar thermoelectric generation** (STEG), where the drive is to improve thermoelectric generation efficiency. Electricity generation relies upon the **Seebeck effect**, and the efficiency depends on the temperature difference across the material through the Carnot efficiency and through a figure of merit ZT of the material. T is the absolute temperature and Z is defined as $\alpha^2 \sigma / k$, α is the Seebeck coefficient, σ is the electrical conductivity, and k is the thermal conductivity, all of which are temperature-dependent. The aim is to increase $\alpha^2\sigma$, called the **power factor**, and decrease k. Recent improvements in k have come about through using nanostructures.

The conversion efficiency η is given by:

$$\eta = \eta_{\text{Carnot}}(F-1)/(F+T_h/T_c) \tag{5.10}$$

Where $F = \sqrt{(1+ZT_{\text{ave}})}$, T_h, T_c, and T_{ave} are the hot and cold surface temperatures, and average temperature, respectively, and $\eta_{\text{Carnot}} = (T_h - T_c)/T_h$ is the Carnot efficiency. Typically, efficiencies of 20% of η_{Carnot} can be obtained. A good material for temperature differences of around 100°C is bismuth telluride, which has a ZT of ~1–1.5, though MgAgSb-based materials (skutterudites) have been found to give a comparable performance. Enhancement of the power factor of materials may be possible through control of their fabrication process. TEGs can be used to generate electricity from waste heat, and in space applications.

To obtain greater efficiency, optical concentration can be used to increase the temperature difference. Good heat exchange, both at the hot and cold face, are important for maximizing this difference. Also, a thermoelectric material that has good ZT at high temperature can be combined with one that has good ZT at low temperature. The materials may either be segmented or cascaded, and they must be compatible. The use of selective emitters is important to reduce radiation losses. Using segmented skutterudite and bismuth telluride up to 12% efficiency has been achieved across a temperature difference of 541°C. The high temperatures would enable TEG to be a topping cycle with an existing CSP technology, and it may prove to be economically viable as a stand-alone technology.

5.5 Solar Chimneys

Another method proposed of utilizing solar energy in areas of strong solar radiation, and where land is readily available, is to construct a large **solar chimney.** Warm air is produced under a large area of glass and is drawn up the high chimney. A prototype was built in Spain in the 1980s and produced 50 kW with a collector 240 m in diameter. The chimney was 195 m tall and 10 m in diameter.

A large, ambitious project that was planned to be built in Australia was a giant solar tower some 1000 m tall with a diameter of 150 m; the collector would have been about 5000 m in diameter. Black plastic pipes full of water placed on the ground under the glass would heat up during the day and give out their heat to the air during the night, providing output throughout 24 hours. The output power would have been 200 MW, sufficient power for ~200,000 homes. The conversion efficiency is ~2%, but, even where land is cheap and there is a lot of sunshine, the cost per kWh was estimated at ~$0.15 per kWh and has not proved competitive, particularly with the low cost of PV.

The principle of the solar chimney is to utilize the pressure difference arising at the top and bottom of a chimney when the air in the chimney is hotter (T_i) than the outside air (T_o). Then the pressure drop within the chimney is smaller than outside, since the hot air is less dense. These differences in pressure drive air in at the bottom and out at the top, and this process, called the **stack** or **chimney** effect, is used to provide natural ventilation in buildings. The overall pressure difference Δp driving the air flow is given by:

$$\Delta p_o = (\rho_o - \rho_i)gh = \rho_i gh(T_i - T_o)/T_o \tag{5.11}$$

We can estimate the flow of gas up the chimney using Bernoulli's principle (eqn (2.42)).

Consider the air outside as stationary, then the speed within the chimney u will be given by $\Delta p/\rho = \frac{1}{2}u^2$. Energy will be lost because of turbulence as the air enters the chimney, and empirically the velocity u will be given by:

$$u = C_T (2\Delta p/\rho)^{\frac{1}{2}} = C_T \left[2(\Delta T/T_o)gh \right]^{\frac{1}{2}} \tag{5.12}$$

where C_T is a constant <1.

We can obtain an approximate estimate of the power P generated by a solar chimney using this value for u as the wind speed incident on a wind turbine. Its power P is given by:

$$P = \frac{1}{2}\varepsilon A\rho u^3 = 0.1D^2 \left[2(\Delta T/T_o)gh \right]^{\frac{3}{2}} \tag{5.13}$$

where $\rho = 1.25$ kg m^{-3}, the efficiency $\varepsilon \sim 0.4$ for a wind turbine (see Chapter 7 Section 7.7), $\pi D^2/4$ is taken as the cross-sectional area of the chimney, of diameter D, and we have taken $C_T \sim 0.8$. For the extremely high solar chimney that was proposed in Australia, $h \sim 1000$ m and $D \sim 150$ m, and if we take $\Delta T = 30$ K and $T_o = 300$ K, then $P \sim 200$ MW.

This estimate is rather rough, as the model that predicts the speed of air flow up the chimney is very approximate. However, it does show how a considerable amount of power could be generated by a very tall solar chimney, but the cost so far has proved prohibitive.

5.6 Ocean Thermal Energy Conversion (OTEC)

The oceans collect a huge amount of solar radiation. As a result, a vast amount of heat energy is stored within the top 100 m of the oceans, where the temperature is ~20–25°C higher than deep below the surface in the Tropics. Ideas on the exploitation of this resource, called **ocean thermal energy conversion** (OTEC), have been developed (see Box 5.2). The small

---

Box 5.2 Principles of ocean thermal energy conversion (OTEC)

The oceans cover a huge area of the Earth and absorb an enormous amount of energy per day. Approximately 1000 MW falls on a square kilometre of sea. In tropical waters, the result is a small temperature difference of ~20–25°C between the surface and water below ~1000 m. Using this temperature difference to drive a heat engine is the basis of OTEC. The temperature of the top ~100 m of the sea is roughly constant. It then

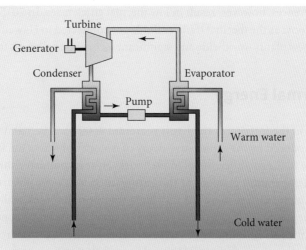

Fig. 5.9 Schematic of a possible OTEC system.

decreases until it is ~5°C at ~1000 m, and below that depth it remains approximately the same.

A schematic of a possible OTEC system is shown in Fig. 5.9. From the second law of thermodynamics there is a limit on the output power P_{out} given by:

$$P_{out} < \eta_{Carnot} P_{flow} \tag{5.14}$$

where η_{Carnot} is the Carnot efficiency,

$$\eta_{Carnot} = (T_h - T_c)/T_h = \Delta T/T_h \tag{5.15}$$

and P_{flow} is the heat flow. If the pump draws a volume Q of seawater per second, and the specific heat capacity and density of the seawater are c and ρ, respectively, then

$$P_{flow} = \rho c Q \Delta T \tag{5.16}$$

So, the output power is limited to:

$$P_{out} < (\rho c Q/T_h)(\Delta T)^2 \tag{5.17}$$

Note the quadratic dependence on ΔT, which favours regions with large ΔT. Seawater has $\rho \approx 1000$ kg m^{-3}, $c \approx 4200$ J kg^{-1} °C^{-1}, so $\Delta T = 20$ °C and $P_{out} = 1$ MW requires $Q >$ 0.18 m^3 s^{-1} (180 litres per second).

Such a system would need large and expensive equipment: big pumps and very efficient heat exchangers. Encrustation by marine organisms (bio-fouling) is a serious problem. The manufacture of long, large-diameter cold water pipes is also difficult. Dredging up the cold water from the deep will release some CO_2 dissolved in the seawater. But the amount is less than 1% of that generated by burning coal to produce the same amount of energy. An OTEC system, however, may affect the climate and marine life of the region. OTEC is starting to become economically competitive in some tropical regions.

temperature differences, however, result in low thermal efficiencies. In 2015 a 100 kW OTEC plant, using ammonia as the thermal transfer fluid, was connected to Hawaii's electricity grid, though there are few plants worldwide and significant technical challenges to be met.

5.7 **Geothermal Energy**

The temperature in the interior of the Earth is around 4000°C. It is maintained by the generation of heat produced by the radioactive decay of isotopes of heavy nuclei and the heat of crystallization due to the solidification of molten rock. Heat is conducted through the mantle, and the average temperature gradient at the Earth's surface is typically around 30°C km^{-1}. So, within 10 km of the surface there is a very large volume of hot rock with temperatures up to 300°C, which represents a vast store of thermal energy. The average continental heat flux is about 65 kW km^{-2}, so an estimate of the sustainable geothermal resource is ~10 TW$_{th}$ or 350 EJ y^{-1} from the flux through the world's land area of ~150 million km^{2}.

Traditionally, geothermal plants have been located near naturally occurring sources, and, as it is difficult to transport heat long distances, the use of geothermal heat is usually restricted to where the source is close to the demand. Geologically active parts of the world such as Iceland, California, Italy, and New Zealand are close to the interfaces between tectonic plates. Here the magma is closer to the surface and the crust can be fractured, allowing cold water to seep through to hot rock and escape as high pressure steam or a mixture of steam and hot water, or just hot springs. These naturally occurring steam jets (**geysers**) and hot springs (up to 350°C) provide a ready source of thermal energy; in Iceland, 9 out of 10 houses rely on it. Up to about 180°C, geothermal energy is primarily used for district heating, industry, and agricultural purposes, but above ~100°C it can be used to generate electricity. Fig. 5.10

Fig. 5.10 The Krafla Geothermal Plant in Iceland (*Source*: Ásgeir Eggertsson—CC BY-SA 3.0, https://commons.wikimedia.org/w/index.php?curid=5244162).

shows the Krafla geothermal plant in Iceland, which has a ~300 MW generating capacity and a ~130 MW capacity for hot water.

Geothermal energy that is used to provide heat is called direct use; in 2019, 95 TWh was used for heating swimming pools and baths, district heating, crop and food drying, and other applications. The use of heat pumps is sometimes classified under the direct use of geothermal, even though nearly all of the extracted energy is replaced by solar. We will now look at the different types of geothermal power plants.

5.7.1 Geothermal Power Plants

Power plants that just use steam to drive a turbine generator are called dry plants, and a notable early example is the Geyser power plant in California. Most (~2/3) of the power-producing hydrothermal sources, though, contain a mixture of water and steam at high pressure and with temperatures >180°C. When the resource is tapped and the pressure is lowered, the water boils producing steam (flashes). The steam is separated and used to run a turbine generator, and the remaining hot water can be flashed again in what are called flash steam plants. Finally, there are plants with binary cycles that can use sources in the range 100–180°C. In these, a low-boiling-point fluid (e.g. pentane or butane in an organic Rankine cycle plant) is heated via a heat exchanger, and the vapour produced is used to drive a turbine to generate electricity.

These traditional geothermal plants provide a small amount of energy, ~0.5% of the global electricity supply. However, much more thermal energy can be 'mined' from below the surface. There are two basic types of rock formation that are suitable for mining geothermal heat: **aquifers** and **hot dry rocks**. An **aquifer** is a layer of porous rock trapped between layers of impermeable rock, e.g. a layer of sandstone tens of metres in thickness. (Aquifers close to the surface provide vast reservoirs of rainwater, which are extracted by water authorities.) Aquifers at depths of 2–3 km are typically at 60–90°C. Cold water is injected at some point in the aquifer through a borehole. The water flows through the aquifer and absorbs heat from the porous rock (see Section 5.7.2), and the hot water is removed through a second borehole and can be used for heating.

In **hot dry rock** heat extraction, which is an example of an **enhanced geothermal system** (EGS), water is pumped at high pressure through narrow gaps in hot rock formations (see Section 5.7.3). Hot regions of crust close to the surface (1–5 km) are generated primarily through the movement of land (plate tectonics) and associated uplift, rather than through radioactivity and thermal conduction. This advection of heat can bring material with temperatures of hundreds of degrees within 10 km of the surface and give rise to higher than average temperature gradients of around 50°C per km. The rock is generally in extension as a result. In these regions, rock at 250°C can be found at ~5 km depth, which is accessible by drilling. The vast energy resource in hot dry rock has resulted in considerable research in enhanced geothermal systems that could extract this energy.

We first look at the extraction of heat from a hot aquifer before describing the status of EGS systems applied to hot dry rock.

5.7.2 Heat Extraction from an Aquifer

In aquifer extraction, heat is removed from porous rock situated between layers of impermeable rock. Heat conduction from the impermeable rock above and below the aquifer is usually negligible over the timescale for heat extraction from an aquifer. For simplicity, we consider a simple one-dimensional fluid flow model for heat removal from an aquifer (see Fig. 5.11). The

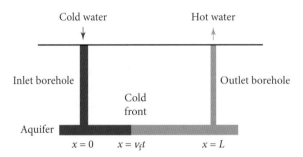

Fig. 5.11 Twin borehole system for heat extraction from aquifer.

actual flow field in the plane of the aquifer is two-dimensional (see Exercise 5.19), but a one-dimensional model gives a useful first approximation and illustrates the main physical features.

Cold water is injected at the inlet borehole $x = 0$, and hot water is extracted at the outlet borehole $x = L$. Suppose that the aquifer is initially at temperature T_1 and the cold water at the inlet is at temperature T_0. Hence the heat available per unit volume from the rock is $\rho_r c_r (T_1 - T_0)$. The power output of the system is the product of the heat per unit volume gained by the water, $\rho_w c_w (T_1 - T_0)$, and the volume flow rate Q, i.e.

$$P = \rho_w c_w (T_1 - T_0) Q \tag{5.18}$$

As cold water flows through the aquifer it absorbs heat from the hot porous rock. A 'cold front' moves from the inlet borehole to the outlet borehole over the lifetime of the system. The speed of the cold front v_f is given by:

$$v_f = \lambda v_w \tag{5.19}$$

where $\lambda = \rho_w c_w / [(1 - \phi) \rho_r c_r]$ is a non-dimensional parameter (see Derivation 5.2) and $v_w = Q/A$ is the bulk velocity of the water in the aquifer, where A is its cross-sectional area. ϕ is the fraction by volume occupied by the pores, known as the **porosity**. The **lifetime** of the system is the time taken for the cold front to reach the outlet borehole, and is given by

$$t_{life} = \frac{L}{v_f} = \frac{(1 - \phi) \rho_r c_r A L}{\rho_w c_w Q} \tag{5.20}$$

Hence, for a long lifetime it is desirable to have low porosity ϕ, large heat capacity per unit volume $\rho_r c_r$, large cross-sectional area of aquifer A, low volume flow rate Q, and large spacing L between the two boreholes. Since the total amount of thermal energy available from the system, $E = \rho_r c_r (T_1 - T_0) A L$, is fixed, the choice of Q and the lifetime of the system are determined by the economics of the system.

Finally, in order to obtain a given volume flow rate Q it is necessary to apply a pressure drop Δp between the boreholes. By **Darcy's law**, the volume flow rate Q through a slab of porous rock of cross-sectional area A and thickness L is given by:

$$Q = KA \frac{\Delta p}{L} \tag{5.21}$$

where the pressure drop $\Delta p = \rho g \Delta h$ is given in terms of a difference in height, the hydraulic head Δh, and K is a constant known as the **hydraulic conductivity**. Poiseuille's law for laminar flow through a pipe of diameter d and area $A = \pi d^2/4$:

$$Q = A \frac{\Delta p}{L} \frac{d^2}{32\mu} = A \frac{\Delta h}{L} \frac{d^2}{32} \frac{g\rho}{\mu}$$

where μ is the viscosity of the fluid, relates K to the **permeability** of the rock $d^2/32$ (i.e. the gaps in the rock) and the properties of the fluid $g\rho/\mu$. Rearranging eqn (5.21), the pressure drop required for a given volume flow rate Q is given by:

$$\Delta p = \frac{QL}{KA} g\rho \tag{5.22}$$

Derivation 5.2 Velocity of the cold front

Suppose the cold front (see Fig. 5.11) moves a distance δx in a time interval δt. The volume of rock exposed is $\delta V_r = (1 - \phi)A\delta x$, where A is the cross-sectional area of the cold front. The amount of heat removed from the element is given by $\delta h_r = \rho_r c_r (T_1 - T_0)(1 - \phi)A\delta x$. The volume of water passing through the element in a small time interval δt is $\delta V_w = Q\delta t$, and the heat gained per unit volume by the water is $\delta h_w = \rho_w c_w (T_1 - T_0)Q\delta t$. By energy conservation, the heat lost by the rock is equal to the heat gained by the water, so that:

$$\rho_r c_r (T_1 - T_0)(1-\phi)A\delta x = \rho_w c_w (T_1 - T_0)Q\delta t$$

Rearranging, we obtain the velocity of the cold front as:

$$v_f = \frac{dx}{dt} = \frac{\rho_w c_w}{(1-\phi)\rho_r c_r} \frac{Q}{A} = \lambda v_w$$

EXAMPLE 5.2 Heat extraction from an aquifer

A sandstone aquifer at 70°C is 20 m thick and 100 m wide. The density, specific heat, porosity, and hydraulic conductivity are 2.3×10^3 kg m^{-3}, 1000 J kg^{-1}°C^{-1}, 0.02, and 2×10^{-5} m s^{-1}, respectively. Estimate the volume flow rate needed to generate a power output of 1 MW, the lifetime of the system, and the pressure drop required for a borehole separation of 1 km. (Assume the water at inlet is at 10°C, $\rho_w = 10^3$ kg m^{-3}, $c_w = 4000$ J kg^{-1}°C^{-1}.)
From eqns (5.18), (5.20), and (5.22) we have:

$$Q = \frac{P}{\rho_w c_w (T_r - T_w)} = \frac{10^6}{(10^3)(4\times10^3)(70-10)} \approx 4\times10^{-3}\,\text{m}^3\text{s}^{-1}$$

$$t_{\text{life}} = \frac{(1-\phi)\rho_r c_r A}{\rho_w c_w Q} L \approx \frac{(1-0.02)(2.3\times10^3)(10^3)(20\times100)}{(10^3)(4\times10^3)(4\times10^{-3})}(10^3) \approx 2.8\times10^8\,\text{s} \approx 9\ \text{years}$$

$$\Delta p = \frac{QLg\rho}{KA} = \frac{(4\times10^{-3})(10^3)(9.81\times10^3)}{(2\times10^{-5})(2\times10^3)} \approx 10^6\,\text{Nm}^{-2} = 10\ \text{bar}$$

5.7.3 Enhanced Geothermal Systems (EGS)

EGS refers to a variety of ways of enhancing the extraction of geothermal energy, from increasing the yield of conventional hydrothermal sources to extracting heat from hot dry rock at around 5 km depth, where there are very large amounts of thermal energy stored under all continents of the world. At these depths the rock temperature can be sufficiently high to make extracting thermal energy economic, and good regions are where temperatures of ~250°C occur at depths of ~5 km. For hot rocks the depths must be less than ~10 km, otherwise stopping holes or cracks sealing under the intense pressures and high temperatures is very difficult. The resource potential in the US has been estimated at over 300 GWe, about a third of the total current capacity. We will concentrate on EGS in hot, dry rocks. A schematic of a hot dry rock power plant is shown in Fig. 5.12.

The method of extracting the thermal energy is to fracture the subsurface rock to make it permeable and to extract the energy in a similar way to other geothermal sources by flowing water through the hot rock. Typically, there will be an injection well and an extraction well separated horizontally with the region between fractured. The main difficulty is that the rock has a low thermal conductivity and a correspondingly low thermal diffusivity of $\kappa_r \sim 10^{-6}$ m^2 s^{-1}. The characteristic distance that heat will diffuse in a time t is $(\kappa_r t)^{1/2}$ (see eqn 2.5 and Exercise 5.23), which for t equal to 10 years is only 18 m; the heat flow is therefore very small.

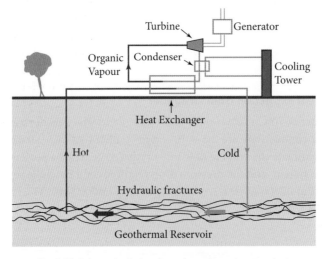

Fig. 5.12 Schematic of a hot dry rock geothermal power plant.

The challenge is in making a network of fractures that has sufficiently large surface area that the temperature of the water rises significantly to give good thermal efficiency. The fractures must also have narrow gaps so that the water flow rate per channel is restricted, as it is the ratio of heating rate to water flow rate that determines the temperature rise. The thermal power extracted is proportional to the flow rate and the temperature rise ΔT, and the maximum efficiency of conversion to electrical power is given by the Carnot efficiency $\Delta T/T_{hot}$; i.e. the electrical power is proportional to $(\Delta T)^2$, so it is very important that the flow rate is correct.

The first plant was in Fenton in the US, at a depth of 3 km and a temperature of 195°C, and operated during the 1970s–1990s. In that time, it demonstrated the concept with a power output of 5 MWth, and research on EGS expanded, with pilot plants in Europe at Soulz-sous-Forêts in France, where there is a 5 MWe plant, in Australia, and elsewhere. After 40 years, though, EGS is still under development and costs are currently uncompetitive.

The main expense is in drilling, typically through hard rock, to the depths required, which can be over 60% of the cost of a project. The drilling costs increase almost exponentially with depth. A breakthrough in drilling technology (from exponential $ per m to linear $ per m) would make EGS more cost-competitive—using lasers to soften the rock prior to drilling may be effective. It has also proved difficult to fracture the rock reliably, with interference sometimes occurring from natural fissures, which tend to be vertical when the rock is in extension, and which can short-circuit the induced fractures and cause loss of injection fluid. However, the development of techniques to extract oil and gas from shale deposits (fracking) has led to considerable improvements in subsurface evaluation, and therefore in reducing the uncertainty as to the nature of the rock formation and rock temperature.

Fracturing the rock can be done by hydraulic fracking, in which a mixture of chemicals and sand is added to water (as done for shale oil and gas extraction), which is pumped down the injection well at high pressure. The high pressure stimulates fissures to form. It is important to check for any seismic activity and so reduce the risk of any serious seismic event, such as the one in Basel that resulted in an EGS project being cancelled in 2009.

5.7.4 Geothermal Energy Production and Economics

In recent years direct use has increased by nearly 8% a year to 117 TWh, with China, Turkey, Iceland, and Japan accounting for 75% of direct use. For electricity generation, growth has been at around 5% per year to 95 TWh with the leading countries by installed capacity: the US, Indonesia, the Philippines, Turkey, New Zealand, and Mexico. The total installed capacity at the end of 2019 was 13.9 GW, giving an average capacity factor of 78%.

In places where there are high-temperature resources available, the economics for electricity generation with modern plants can be competitive with fossil-fuel generation. This applies to binary plants in several cases, but costs are dependent on location, the temperature of the resource, and the size of plant. In 2019, the global levelized cost of energy from geothermal power plants was $73 MWh^{-1}. Exploratory drilling is expensive and the financial risks can be high, as the outcome is often unknown. Drilling and plant construction costs can make investment high, with typically costs for binary plants higher than for steam and flash plants, but operational costs are quite low.

5.7.5 **Environmental Impact of Geothermal Power**

Geothermal power is a relatively harmless technology with little environmental impact. There is generally little carbon dioxide emission (except during the drilling process), although there is sometimes a release of H_2S gas, and the water already trapped in aquifers can contain toxic heavy metals. As a consequence, it is usual to employ heat exchangers to keep the extracted water separate from that used for district heating or electricity generation. Life-cycle assessments (LCA) for flash plants are < 50 g CO_2eq kWh^{-1} and for planned EGS plants < 50 g CO_2eq kWh^{-1}. In EGS, although the possibility of contaminating drinking water supplies is small, it is important to monitor fluid pathways. There has also been concern over induced seismic activity.

5.7.6 **Potential and Outlook for Geothermal Power**

Geothermal energy is a large, mostly untapped source of energy, with only a small fraction of the estimated available resource currently being exploited. The technical potential for electricity production from depths of less than 3 km is ~30,000 TWh y^{-1}, and from less than 10 km some 10 times more, while the potential for heat from resources with T< 130°C is estimated as up to 300 EJ y^{-1}. The heat extracted, both for direct use or for power, can be fully or partly replenished over a long time by the average continental heat flux of 65 kW km^{-2} that amounts to 350 EJ y^{-1} (10 TW$_{th}$). These numbers can be compared with the total global final energy demand of ~400 EJ in 2019.

However, the estimated economic geothermal generating capacity by 2050 is only a small fraction of its technical potential. IRENA project 162 GW generating 1100 TWh y^{-1} at 0.78 capacity factor, and for heat 2.5 EJ y^{-1} (700 TWh y^{-1}) split approximately equally between industry and buildings. We will take the contribution from geothermal power (electricity and heat) by 2050 as ~1800 TWh y^{-1}.

Its low environmental impact and base-load capability (being always available) make it an attractive low-carbon source of heat and electricity, that also provides energy security. However, currently only readily accessible geothermal resources are economic. To make enhanced geothermal technology more competitive with wind and solar power, research is now focusing on extracting the thermal energy in super-hot (greater than 400°C) rock at depths of greater than 5 km. In 2018, AltaRock signed a partnership with China Coal to evaluate enhanced geothermal's potential to replace coal-fired generation in China. The challenges are significant, but the vast geothermal resource, which is available at all times, makes it important that research and development continues.

    SUMMARY

- Solar thermal hot water heating could save ~0.8 GtCO_2 per year by 2050, providing approximately 3000 TWh y^{-1} (~10 EJ y^{-1}) of low-carbon heat. Solar coolers could also make savings.

- Concentrated solar thermal power (CSP) with storage has considerable potential and its LCOE is becoming more competitive. Its accessible potential for electricity generation is estimated as ~2000 TWh y^{-1} by 2050.

- Heat pumps are expected to be deployed more and more, and could supply ~14 EJ y^{-1} (4000 TWh y^{-1}) by 2050.

- Geothermal energy could supply about 1800 TWh y^{-1} (electricity and heat) by 2050.

FURTHER READING

EHPA (2019). Whitepaper: Heat pumps—Integrating Technologies to Decarbonise Heating and Cooling

IRENA (2019). Innovation landscape brief: Renewable power-heat.

IRENA (2019). Global energy transformation: A roadmap to 2050.

IRENA (2019). Renewable power generation costs.

JASON (2013). *Enhanced geothermal systems*, JSR-13-320, DOE.

Jelley, N. and Smith, T. (2015). *CSP: a review*. IMech, Part A. (CSP2015).

REN21 (2020). Renewables Global status report. Geothermal energy production.

energy.mit.edu/publication/future-geothermal-energy/ MIT report on the future of geothermal energy (MIT2006).

www.cibse.org/networks/regions/north-east/north-east-past-presentations Underwood, C. (2018). Heat pumps and their role in the decarbonizing of heat (HP2018).

www.iea-shc.org/solar-heat-worldwide-2020 Global solar heating and cooling statistics.

archive.ipcc.ch/pdf/special-reports/srren/SRREN_FD_SPM_final.pdf Geothermal energy.

EXERCISES

5.1 A solar concentrator produces a heat flux of 2500 W m^{-2} on the outside of a tube of diameter 50 mm. Water flows through the tube at a rate of 0.015 kg s^{-1}. If the water temperature at the inlet is 15°C, what length of pipe is required to produce water at a temperature of 85°C? (Water at 50°C has $\rho = 990$ kg m^{-3}, $k = 0.64$ W m^{-1} K^{-1}, $c = 4180$ J kg^{-1} K^{-1}.)

5.2 A solar receiver only accepts light falling within 30° of normal incidence. What is the maximum concentration of sunlight that could be achieved?

5.3 Calculate the maximum temperature of an absorber in an evacuated tube solar thermal collector, when the sunlight has an intensity of 800 W m^{-2} and the absorber has an emissivity for thermal radiation of 0.1 and an absorptivity for solar radiation of 0.9.

5.4 Compare solar thermal water heating and heating water by heat pumps. For what tasks are they best suited?

5.5 Explain how air- and ground-source heat pumps work. What are their respective advantages? What factors are important to ensure good performance?

5.6 In an air source heat pump the enthalpy change when the cold fluid is compressed is 50 kJ kg^{-1}, and 200 kJ kg^{-1} when the hot compressed fluid is cooled on passing through the condenser. If the isentropic efficiency of the pump is 0.75, what is the COP of the heat pump?

5.7 Calculate the exergy efficiency of a CSP plant, which has a receiver at 900 K with an absorptivity for solar radiation of 0.9, and a concentration factor of 100, when the emissivity for the thermally emitted radiation is (a) 0.9, (b) 0.1. Take $G = 900$ W m^{-2}, $T_a = 300$ K.

5.8* Show that the concentration required to make the absorption temperature optimal is given by eqn (5.9). Estimate what concentration is needed for an alloy with a maximum input temperature of 1200 K and what is the exergy efficiency? Take a and ε equal to 0.9, $G = 900$ W m^{-2}, and $T_a = 300$ K.

5.9 Discuss the relative merits of aligning a parabolic trough collector east–west compared to north–south.

5.10 Calculate in a solar tower CSP what pointing accuracy is required for heliostats 400 m away from the tower to strike a receiver of size 20 m × 20 m. If there are 80,000 heliostats each of area 2.5 m^2, and reflectivity 0.9, what is the concentration and thermal input power to the power plant for a solar DNI of 900 W m^{-2} when the receiver has an absorptivity of 0.9 for solar radiation and emissivity for thermal radiation of 0.3, and is at a temperature of 550°C?

5.11 Assuming the power plant in a CSP plant is operating at maximum power, calculate the improvement in thermal efficiency expected from raising the temperature of the receiver from 550°C to 720°C. Assume an ambient temperature of 25°C.

5.12 Estimate how many tonnes of solar salt would be required to provide 10 hours of storage for a 100 MWe CSP tower plant operating between 550–300°C and ~25°C. What is the average thermal efficiency? The heat capacity of the salt is 1.5 kJ kg^{-1} K^{-1}.

5.13 Discuss the advantages that storage has for CSP.

5.14 Discuss the potential of CSP to reduce global CO_2 emissions.

5.15 What value of ZT_{ave} would a material need to have for a thermoelectric conversion efficiency of 10% for a temperature range of 250–700°C?

5.16 An ocean thermal energy conversion (OTEC) system is proposed for a region where the temperature difference between the surface and the deep water is 25°C. The pumps have a capacity of 100 litres per second. Estimate the power output if the overall efficiency is 50% of the theoretical maximum.

5.17 A solar chimney is proposed to provide 100 MW of power. The temperature difference of the air inside and outside the chimney is predicted to be 35°C. Estimate the height of the solar chimney required if the diameter of the chimney is 100 m.

5.18 Consider a proposal to extract heat directly from the Earth's core by drilling a cylindrical shaft of radius 1 m and depth 100 km through the Earth's mantle and filling it with copper. Estimate the power output assuming the temperature difference is 1000°C and the thermal conductivity is 3.5×10^2 W m^{-1} K^{-1}.

5.19 Consider an aquifer at an initial temperature of 100°C. The aquifer data are thickness 50 m, width 100 m, density 3×10^3 kg m^{-3}, specific heat 1500 J kg^{-1}°C^{-1}, porosity 0.01, permeability 5×10^{-9} m^3 kg^{-1}. Estimate the required separation of the boreholes and the pressure drop in order to produce an output of 5 MW of heat for 10 years. Assume the water at inlet is at 5°C ($\rho_w = 10^3$ kg m^{-3}, $c_w = 4000$ J kg^{-1}°C^{-1}).

5.20 Derive an expression for the cost of drilling a borehole as a function of depth. Assume that the cost of drilling per metre is independent of depth but the cost of lifting the rock material from the borehole increases linearly with depth.

5.21* Consider a two-dimensional aquifer of thickness d, with an inlet borehole at A ($x = -a$, $y = 0$) and outlet borehole at B ($x = +a$, $y = 0$). The velocity field of the water in the aquifer is given by

$$u(x, y) = \frac{Q}{2\pi d}\left(-\frac{1}{r_1} + \frac{1}{r_2}\right)$$

where

$$r_1 = \sqrt{(x-a)^2 + y^2} \quad r_2 = \sqrt{(x+a^2) + y^2}$$

Derive an expression for the time taken for cold water from the inlet borehole to reach the outlet borehole.

5.22 An EGS 1 GWe power station extracts heat from a volume of 100 km^3 of granite that initially is at a temperature of 200°C. Estimate how long the power station could operate before the rock temperature has fallen to 175°C. (Density of granite = 2750 kg m^{-3}; specific heat of granite = 0.79 kJ kg^{-1}°C^{-1}.)

5.23 Assume the amount of energy that can be extracted from a single channel (crack) over a period of time t is the thermal energy $\pi\kappa_r t L C_r \Delta T$ in the cylindrical volume of rock radius $(\kappa_r t)^{1/2}$ and length L corresponding to a temperature drop ΔT, where C_r is the volumetric specific heat of the rock. For an EGS plant extracting heat from one cubic kilometre of rock at 250 °C for 10 years, estimate the minimum separation of cracks and the corresponding total electrical output, assuming a plant thermal efficiency of 15%. Take $C_r = 2.5 \times 10^6$ J m^{-3} K^{-1}, $\Delta T = 100$°C, and the thermal diffusivity of the rock $\kappa_r = 10^{-6}$ m^2 s^{-1}.

5.24 The cost C of drilling to a depth d is given by $C = A\exp(d/b)$. The temperature of the rock T increases linearly with depth: $T = T_0 + Bd$. If all other costs are independent of depth, show that the minimum cost per unit of thermal energy extracted is when the geothermal reservoir is at a depth of b. Discuss the implications of this result.

5.25 Calculate the optimum depth to minimize the cost per unit of electricity generated, given the cost of drilling and the variation of temperature with depth, and other costs, are as given in Exercise 5.24. Approximate the efficiency of conversion of thermal to electrical energy by $\Delta T/2T_0$, where $\Delta T = T - T_0$.

5.26 Discuss the potential of EGS to provide low-carbon electricity and heat.

Derivation 5.1 Practical performance of heat pumps

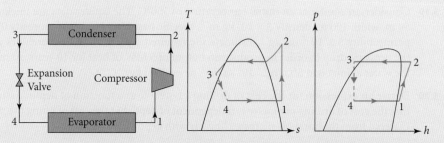

Fig. 5.13 The components and the *T-s* and *p-h* diagrams of a heat pump.

The various parts of a heat pump cycle shown in Fig. 5.13 are as follows. Along 3-4, the working fluid in the heat pump expands. As noted in Chapter 2 Section 2.5.2, when a thermally well-insulated fluid passes through a restriction into a region of lower pressure, without doing any work, its enthalpy ($h = u + pv$) remains constant. As the refrigerant liquid expands and vaporizes, passing through the expansion valve (3 → 4), its molecules lose kinetic energy overcoming the strong intermolecular forces within the liquid, and the temperature of the fluid drops: the product pv increases at the expense of the internal energy u. [It is impractical to expand a liquid through a turbine as its performance is affected by a mixture of liquid and vapour; also, little work would be generated (lines of constant s almost vertical), so a throttle is used; in addition, a turbine would be more expensive.]

The region just before 3 is where the fluid is subcooled to ensure that the fluid is entirely liquid, which is necessary for efficient working of the expansion valve. The liquid-gas mixture at 4 then passes through the evaporator taking heat from its surroundings. Just before 1 is where the gas is superheated to ensure that the fluid is entirely in its gaseous form to avoid damaging the compressor, as liquid is incompressible. After compression, the hot gas passes through the condenser (2 → 3) where heat is given out to its surroundings.

Heat from the subcooling can be used in a separate heat exchanger to superheat the gas before it enters the compressor to reduce energy loss. Typical values for the temperatures, pressures and specific enthalpies of the refrigerant R410A are given in the table below.

		$T\,°C$	p Bar	h kJ/kg		$T\,°C$	p Bar	h kJ/kg
Air source								
	(3)	45	28	270	(2)	75	28	468
	(4)	−12	5.8	270	(1)	−7	5.3	420
Ground source								
	(3)	45	28	270	(2)	65	28	454
	(4)	3	8.5	270	(1)	7	8.5	425

The evaporator temperature is lower for an air-source heat pump because the air temperature is more variable and can be below 0 C. The heat flow is determined by the temperature difference, thermal conductivity, and flow rates.

Neglecting the small pressures drops that occur in practice across the evaporator and condenser, the heat absorbed and rejected by the working fluid takes place at constant pressure. As shown in Chapter 2 Section 2.5.2, the heat transfer is then given by the change in enthalpy. The work done in an adiabatic compression is also given by the enthalpy change. But actual compressors have significant losses that are given by their isentropic efficiency η_C, which is the ratio of the actual work to the work under isentropic conditions. Typically for heat pumps $\eta_C \approx 0.65$.

The coefficient of performance of the pump is given by the ratio of the heat rejected by the condenser along 2-3, $(h_2 - h_3)$, divided by the work done on compression, $(h_2 - h_1)/\eta_C$ i.e.

$$\text{COP}\left(\text{heating}\right) = \eta_C (h_2 - h_3)/(h_2 - h_1)$$

Substituting values gives COP (heating) = 2.7 for the air-source heat pump, and COP (heating) = 4.1 for the ground-source pump.

To obtain good performance requires the correct size of the heat pump—it should not be for the worse case, i.e. not risk averse, as part loading reduces its COP. The sizing of the ground source array is also important. The deployment of underfloor heating is better than using normal radiators. The performance is also increased by good insulation of the house, and by setting thermostatic controls correctly.

In the UK, when the air temperature is below 7 °C the refrigerant is below 0 °C in the evaporator, which causes water vapour in the air to freeze. There is then the need to run the heat pump in reverse to defrost the evaporator, which reduces the COP by a factor of about 0.9 over a year. The seasonal performance factor (SPF) of a pump is the average of the COP over a year.

High temperature (supply temperature > 60 °C) heat pumps can use conventional radiators. When using CO_2 as the refrigerant in a transcritical cycle, where the high pressure (p_2) is above and the low pressure (p_1) below the critical point, they can have high efficiencies; for example, a Mitsubishi Electric CO_2 ASHP has a SPF of 3 when producing hot water at 65 °C. While both the refrigerants CO_2 and R410A do not damage the ozone layer, CO_2 has a much lower global warming potential of one compared to roughly 2000 for R410A. Hybrid systems, which combine a heat pump with another form of heating, can also provide high temperatures.

 For further information and resources visit the online resources
www.oup.com/he/andrews_jelley4e

6 Hydropower, Tidal Power, and Wave Power

List of Topics

- ☐ Power from dams
- ☐ Weirs
- ☐ Water turbines
- ☐ Micro-hydro
- ☐ Pumped storage
- ☐ Economics of hydropower
- ☐ Hydropower potential
- ☐ Outlook for hydropower

- ☐ Tidal power
- ☐ Tidal resonance
- ☐ Tidal barrage and lagoon
- ☐ Tidal stream energy
- ☐ Outlook for tidal power
- ☐ Wave energy
- ☐ Wave power devices
- ☐ Outlook for wave power

Introduction

In this chapter we investigate three different forms of power generation that exploit the abundance of water on Earth: hydropower, tidal power, and wave power. **Hydropower** taps into the natural cycle of:

solar heat → seawater evaporation → rainfall → rivers → sea

It is an established technology which accounted for 16.2% of the global electricity production of 26 730 TWh in 2018, making it by far the largest source of renewable energy. The energy of the water is either in the form of potential energy (reservoirs) or kinetic energy (e.g. rivers). In both cases, electricity is generated by passing the water through large water turbines.

Tidal power is a special form of hydropower that exploits the bulk motion of the tides. Tidal barrage systems trap seawater in a large basin, and the flow of water out (and in) through low-head water turbines generates electricity. Marine currents arise from the tides and also from global oceanic circulation, such as found in the Gulf Stream. In many ocean areas the currents are too slow to exploit. But higher current speeds are found in straits between islands and the mainland, and around headlands, and in recent years, rotors have been developed that can extract the kinetic energy of these underwater tidal currents.

Wave power is a huge resource that is largely untapped. The need for wave-power devices to be able to withstand violent sea conditions has been a major problem in the development of wave-power technology. The energy in a surface wave is proportional to the square of the amplitude, and typical ocean waves transport about 30–70 kW of power per metre width of wave-front. Large-amplitude waves generated by tropical storms can travel vast distances across oceans with little attenuation before reaching distant coastlines. Most of the best sites are on the western coastlines of continents between the 40° and 60° latitudes, above and below the Equator.

6.1 **Hydropower**

The power of water was exploited in the ancient world for irrigation, grinding corn, metal forging, and mining. **Waterwheels** were common in Western Europe by the end of the first millennium; over 5000 waterwheels were recorded in the Domesday Book of 1086, shortly after the Norman conquest of England. The early waterwheels were of the undershot design (Fig. 6.1 (a)) and very inefficient. The development of overshot waterwheels (Fig. 6.1 (b)) and improvements in the shape of the blades to capture more of the incident kinetic energy of the stream led to higher efficiencies (~66%).

A breakthrough occurred in 1832 with the invention of the **Fourneyron turbine**, a fully submerged vertical-axis device that achieved efficiencies of over 80%. Such turbines were used in 1895 to generate electricity on the US side of the Niagara Falls. Fourneyron's novel idea was to employ **fixed guide vanes,** which directed water radially outwards into the gaps between **moving runner blades**, as shown in Fig. 6.2. Moreover, the head was not limited to the diameter of the waterwheel (as in overshot wheels), since the water was contained in a pipe. Many

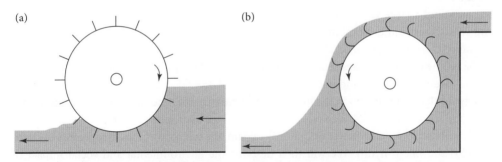

Fig. 6.1 (a) Undershot and (b) overshot waterwheels.

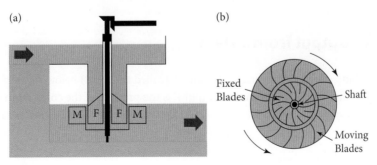

Fig. 6.2 Fourneyron water turbine (a) vertical section, where M and F denote the Moving and Fixed Blades (b) horizontal section through blades.

designs of water turbines incorporating fixed guide vanes and runners have been developed since. Modern water turbines are typically over 90% efficient. Many large installations use a type of turbine designed by James Francis in 1848. In this turbine, water is directed inwards (rather than outwards as in the Fourneyron turbine shown in Fig. 6.2) onto blades attached to the rim of a shaft, which turns an electrical generator. The power is determined by the height of the fall and the volume of water flowing.

Hydropower is now the largest renewable source of power, producing 4306 TWh of electricity in 2019, and delivering ~62% of renewable generation in 2018 (see Fig. 1.8). Concern over climate change due to CO_2 emissions from fossil fuel combustion is one of the drivers for developing hydropower in recent years. Besides being a renewable low-carbon source of energy with a typical carbon footprint of ~55 $gCO_2eq\ kWh^{-1}$ (mainly related to decomposition of submerged vegetation and to concrete and steel used in construction), hydropower is a relatively cheap source of power, particularly when the dam is also built for flood control or for water storage. It is also a well-established and adaptable source of power: it can turn on quickly, and can therefore meet fluctuations in demand as well as providing base-load. Many plants will operate for more than 50 years. Some hydropower stations are dedicated to supplying the power for very energy-intensive operations such as the production of aluminium, and hydropower is also used in the production of silicon for solar panels.

There are three main types of hydropower plant systems: **dams or reservoir plants** (HPP), **run-of-river** (ROR) schemes, and **pumped storage plants** (PSP). RORs use a very small reservoir (pondage) or none at all, while dams have a large reservoir, and PSP have both a reservoir and a lower water supply, typically a river. ROR plants without pondage are very sensitive to water flow variations, while those with pondage are less so. As ROR systems do not impound a large volume of water, their environmental impact is generally small. HPP and ROR plants can be combined in cascaded systems. For example, on the River Durance in France there are 5 cascaded HPPs which produce 7 TWh annually and can provide 2 GW to the grid within 10 minutes. Pumped storage plants (PSP) provided globally in 2019 about 94% of electricity storage and had a total capacity of 158 GW. They can be used to provide energy storage that can help in the integration of the variable renewable power supplies from wind and solar farms. PSP is also called pumped storage hydropower (PSH).

We will first look at the power output from a dam, the type of turbines used, before describing PSP and micro-hydro. We will then describe the environmental and social impact of hydropower and its economics, before discussing the outlook for hydropower.

6.2 Power Output from a Dam

Consider a turbine situated at a vertical distance h (called the **head**) below the surface of the water in a reservoir (Fig. 6.3). The power output P is the product of the conversion efficiency η, the potential energy per unit volume $\rho g h$, and the volume of water flowing per second Q, and is given by:

$$P = \eta \rho g h Q \tag{6.1}$$

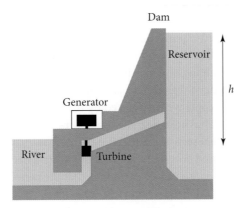

Fig. 6.3 Hydroelectric plant.

Alternatively: the head of water gives rise to a pressure ρgh at the turbine entrance of area A, and so exerts a force $F = \rho ghA$ on the water. If the water is flowing at a speed v then its power is $Fv = \rho ghAv = \rho ghQ$, since $Q = Av$. Eqn (6.1) then follows after allowing for the conversion efficiency η.

Note that the power output depends on the product hQ. Hence, a high dam with a large h and a small Q can have the same power output as a ROR installation with a small h and large Q. The choice of water turbine for a particular location depends on the relative magnitude of h and Q (see Section 6.4.1).

EXAMPLE 6.1

Estimate the power output of a dam with a head of 50 m and volume flow rate of 20 m^3 s^{-1}. (Assume $\eta = 1$, $\rho = 10^3$ kg m^{-3}, g $= 10$ m s^{-2}.)

From eqn (6.1) we have $P = \eta \rho ghQ \approx 1 \times 10^3 \times 10 \times 50 \times 20 \approx 10$ MW.

Case Study 6.1 Aswan High Dam (Egypt)

Completed in 1970, the Aswan High Dam (*Saad el Aali* in Arabic) lies just north of the border between Egypt and Sudan, and is fed by water from the upper reaches of the River Nile. It contains a vast reservoir known as Lake Nasser, with a surface area 5250 km^2. The Aswan Low Dam, 7 km downriver, came close to being overtopped in 1946, emphasizing the need for urgent action. Prior to the construction of the two dams, the Nile flooded in the rainy seasons, providing fresh water for irrigation and valuable nutrients which sustained the population for many thousands of years—about 95% of the population of Egypt lives within 20 km of the Nile. However, in some years no flooding occurred, leading to drought and famine, whereas in other years there was excess flooding and major crop damage. The vast storage capacity of the reservoir (around 132 km^3), together with the ability to control the flow of water through the dam, means that Egypt is now safe from droughts and flooding. In addition, the Aswan High Dam provides about 2 GWe of hydroelectric power, about 3.5% of the national generating capacity, from 12 Francis-type turbines (see Section 6.4). Also, the Nile is now navigable through-out the year, boosting shipping and tourism, and the area of land for growing crops along the Nile and in the delta region has been increased significantly. On the negative side, more than 100,000 people were displaced by the construction of the Aswan High Dam, the quality of the soil has been degraded so that Egyptian farmers now need to use artificial fertilizers, and the reservoir is a breeding ground for parasites. Sediment reten-tion has also caused increased erosion in the Nile delta and a decline in the local fishing industry. Nonetheless, the consensus of opinion is that the benefits which have accrued from the building of the Aswan High Dam have outweighed the disadvantages.

6.3 Measurement of Volume Flow Rate using a Weir

For power extraction from a stream it is important to be able to measure the volume flow rate of water. One particular method diverts the stream through a straight-sided channel contain-ing an artificial barrier called a **weir** (Fig. 6.4). The presence of the weir forces the level of the fluid upstream of the weir to rise. The volume flow rate per unit width is related to the height of the undisturbed level of water y_{min} above the top of the weir by the formula (see Derivation 6.1):

$$Q = g^{\frac{1}{2}} \left(2 y_{min}/3\right)^{\frac{3}{2}} \tag{6.2}$$

(NB Throughout Chapter 6 the y-axis denotes the vertical direction.)

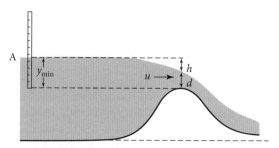

Fig. 6.4 Flow over a broad-crested weir.

Derivation 6.1 Flow over a broad-crested weir

Consider a point A on the surface of the water upstream of the weir where the level is roughly horizontal (i.e. $h = 0$ in Fig. 6.4) and the velocity u_A. Towards the weir, the level drops and the speed increases. For a broad-crested weir, we can ignore the vertical component of velocity and express the volume rate of water per unit width in the vicinity of the crest in the form:

$$Q \approx ud \tag{6.3}$$

where d is the depth of the water near the crest. Using Bernoulli's equation (eqn (2.42)), noting that the pressure on the surface is constant (atmospheric pressure), we have $\frac{1}{2}u^2 - gh \approx \frac{1}{2}u_A^2$. Hence, if the depth of the water upstream of the weir is much greater than the minimum depth over the crest of the weir, then $u_A^2 \ll u^2$ and $u \approx (2gh)^{\frac{1}{2}}$. Substituting for u in eqn (6.3) we obtain $d \approx Q/(2gh)^{\frac{1}{2}}$.

The vertical distance from the undisturbed level to the top of the weir is $y = d + h$. Substituting for d we have:

$$y = Q/(2gh)^{\frac{1}{2}} + h \tag{6.4}$$

The first term on the right-hand side of eqn (6.4) decreases with h, but the second term increases with h. y is a minimum when $dy/dh = 0$, i.e. $-Q/(8gh^3)^{\frac{1}{2}} + 1 = 0$, or:

$$h = \left(Q^2/8g\right)^{1/3} \tag{6.5}$$

Finally, putting h from eqn (6.5) in eqn (6.4), yields $y_{min} = (3/2)\left(Q^2/g\right)^{1/3}$, so that:

$$Q = g^{\frac{1}{2}}\left(2y_{min}/3\right)^{3/2}$$

which is known as the **Francis formula**.

6.4 **Water Turbines**

When water flows through a waterwheel the water between the blades is almost stationary. Hence the force exerted on a blade is essentially due to the difference in pressure across the blade. In a **water turbine**, however, the water is fast-moving and the turbine extracts kinetic energy from the water. There are two basic designs of water turbines: impulse turbines and reaction turbines. In an **impulse turbine**, the blades are fixed to a rotating wheel and each blade rotates in air, apart from when the blade is in line with a high-speed jet of water. In a **reaction turbine**, however, the blades are fully immersed in water and the thrust on the moving blades is due to a combination of reaction and impulse forces.

An impulse turbine called a **Pelton wheel** is shown in Fig. 6.5. In this example there are two symmetrical jets, and each jet imparts an impulse to the blade equal to the rate of change of momentum of the jet. The speed of the jet is controlled by varying the area of the nozzle using a **spear valve**. Thomas Pelton went to seek his fortune in the Californian Gold Rush during

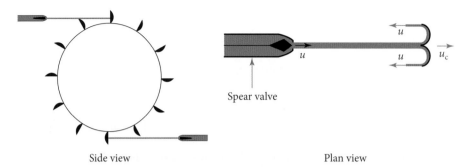

Spear valve

Side view   Plan view

Fig. 6.5 Impulse turbine (Pelton wheel).

the nineteenth century. By the time he arrived on the scene the easy pickings had already been taken, and the remaining gold had to be extracted from rocks that needed to be crushed. Impulse turbines were being used to drive the mills to grind the rocks into small lumps. Pelton observed the water striking the turbine blades and realized that not all the momentum of the jets was being utilized. He deduced that some momentum was being lost because the water splashed in all directions on striking the blades. He redesigned the cups so that the direction of the splash was opposite to that of the incident jet. This produced a marked improvement in efficiency, and Pelton thereby made his fortune.

To calculate the maximum power output from a Pelton wheel, we consider a jet moving with velocity u and the cup moving with velocity u_c. Relative to the cup, the velocity of the incident jet is $(u - u_c)$ and the velocity of the reflected jet is $-(u - u_c)$. Hence the total change in the velocity of the jet relative to the cup is $-2(u - u_c)$. The mass of water striking the cup per second is ρQ, so the force on the cup (the rate of change of momentum) is given by:

$$F = 2\rho Q(u - u_c) \tag{6.6}$$

The power output P of the turbine is the rate at which the force F does work on the cup in the direction of motion of the cup, i.e.

$$P = Fu_c = 2\rho Q\left(u - u_c\right)u_c \tag{6.7}$$

To derive the maximum power output, we put $dP/(du_c) = 0$, which gives $u_c = \frac{1}{2}u$. Substituting in eqn (6.7) then yields the maximum power as:

$$P_{max} = \frac{1}{2}\rho Q u^2 \tag{6.8}$$

Thus the maximum power output is equal to the kinetic energy incident per second.

As in the Fourneyron turbine (see Section 6.1), modern **reaction turbines** use **fixed guide vanes** to direct water into the channels between the blades of a **runner** mounted on a rotating wheel (see Fig. 6.6). However, the direction of radial flow is inward. (In the Fourneyron turbine, the outward flow caused problems when the flow rate was either increased or decreased.)

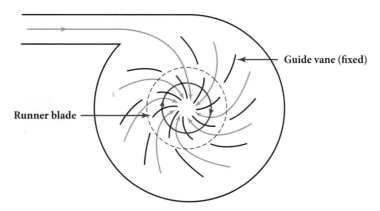

Fig. 6.6 Schematic plan view of a reaction turbine; the runner blades have an aerofoil shape.

The most common designs of reaction turbines are the **Francis turbine** and the **Kaplan turbine**. In a Francis turbine the runner is a spiral annulus whereas in the Kaplan turbine it is propeller-shaped. In both designs the kinetic energy of the water leaving the runner is small compared with the incident kinetic energy.

The term 'reaction turbine' is somewhat misleading in that it does not completely describe the nature of the thrust on the runner. The magnitude of the reaction can be quantified by applying Bernoulli's equation (eqn (2.42)) to the water entering (subscript 1) and leaving (subscript 2) the runner, i.e.

$$p_1/\rho + \frac{1}{2}q_1^2 = p_2/\rho + \frac{1}{2}q_2^2 + E \tag{6.9}$$

where E is the energy per unit mass of water transferred to the runner. Consider two cases: (a) $q_1 = q_2$, and (b) $p_1 = p_2$. In case (a), eqn (6.9) reduces to

$$E = (p_1 - p_2)/\rho \tag{6.10}$$

i.e. the energy transferred arises from the difference in pressure between inlet and outlet. In case (b), E is given by:

$$E = \frac{1}{2}(q_1^2 - q_2^2) \tag{6.11}$$

and the energy transferred is equal to the difference in the kinetic energy between inlet and outlet. In general, we define the **degree of reaction** R as:

$$R = \frac{p_1 - p_2}{\rho E} = 1 - \frac{\left(q_1^2 - q_2^2\right)}{2E} \tag{6.12}$$

(see Example 6.2).

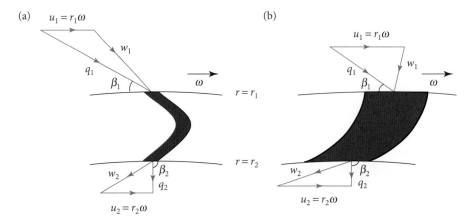

Fig. 6.7 Velocity diagrams for (a) an impulse turbine; and (b) a reaction turbine.

The velocity diagrams in the laboratory frame of reference for an impulse turbine and a reaction turbine are shown in Figs. 6.7 (a) and (b), respectively. The symbols u, q, and w denote the velocity of the runner blade, the absolute velocity of the fluid, and the velocity of the fluid relative to the blade. Fig. 6.7 shows the velocity triangles on the outer radius of the runner, $r = r_1$, and the inner radius, $r = r_2$. The runner rotates with angular velocity ω, so the velocity of the blade is $u_1 = r_1\omega$ on the outer radius and $u_2 = r_2\omega$ on the inner radius.

The torque on the blade is $T = \rho Q(r_1 q_1 \cos \beta_1 - r_2 u_2 \cos \beta_2)$.

Putting $r_1 = u_1/\omega$ and $r_2 = u_2/\omega$, the work done per second is given by:

$$P = T\omega = \rho Q(u_1 q_1 \cos \beta_1 - u_2 q_2 \cos \beta_2) \tag{6.13}$$

which is **Euler's equation**, and relates the power output from a turbine to the rate of change of angular momentum of the fluid between the inlet and the outlet of the turbine.

Equating the incident power due to the head of water h from eqn (6.1) to the power output of the turbine, we have:

$$\eta \rho g h Q = \rho Q(u_1 q_1 \cos \beta_1 - u_2 q_2 \cos \beta_2) \tag{6.14}$$

We define the hydraulic efficiency as

$$\eta = (u_1 q_1 \cos \beta_1 - u_2 q_2 \cos \beta_2)/gh \tag{6.15}$$

The maximum efficiency is achieved when the fluid leaves the runner at right angles to the direction of motion of the blades, i.e. when $\beta_2 = \frac{1}{2}\pi$ so that $\cos \beta_2 = 0$. Eqn (6.15) then reduces to:

$$\eta_{\text{max}} = u_1 q_1 \cos \beta_1/gh \tag{6.16}$$

EXAMPLE 6.2

Consider a particular reaction turbine in which the areas of the entrance to the stator (the stationary part of the turbine), the entrance to the runner, and the exit to the runner are all equal. Water enters the stator radially with velocity $q_0 = 2$ m s^{-1}. The stator curves and narrows so that the water leaves the stationary vanes of the stator at an angle $\beta_1 = 10°$ with an absolute velocity $q_1 = 10$ m s^{-1}. The velocity of the runner at the entry radius $r = r_1$ is u_1 in the tangential direction, and is such that the velocity of the water w_1 relative to the runner is in the radial direction. On leaving the runner, the total velocity is q_2 in the radial direction. Given that the head is $h = 11$ m, calculate the degree of reaction and the hydraulic efficiency.

The volume flow rate is $q_0 A_0$ into the stator, $w_1 A_1$ into the runner, and $q_2 A_2$ out of the runner. Since $A_0 = A_1 = A_2$ it follows by mass conservation that $q_0 = w_1 = q_2$. The energy transfer per unit mass E is given by eqn (6.13) divided by ρQ. Since the total velocity q_2 leaving the runner is in the radial direction, we have $\beta_2 = \frac{1}{2}\pi$. Putting $q_1 \cos \beta_1 = u_1$, then $E = u_1^2$. Also, the square of the total velocity is $q_1^2 = u_1^2 + w_1^2 = u_1^2 + q_2^2$, since w_1 and q_2 are equal and radial. Hence the degree of reaction is $R = 1 - (q_1^2 - q_2^2)/2E = 1 - u_1^2/(2u_1^2) = 1/2$.

The hydraulic efficiency is $\eta_{max} = u_1 q_1 \cos\beta_1 / gh \approx 0.90$.

6.4.1 Choice of Water Turbine

The choice of water turbine depends on the site conditions, notably the head of water h and the volume flow rate Q. Fig. 6.8 indicates the type of turbine which is most suitable for any particular combination of head and volume flow rate. The different designs of turbine are characterized by having different values for the ratio r/R, where πr^2 is the inlet area and R the radius of the turbine blade tip, and for the ratio v_b/v_w, where v_b is the speed of the blade tip and v_w is the speed of the water at the input to the turbine.

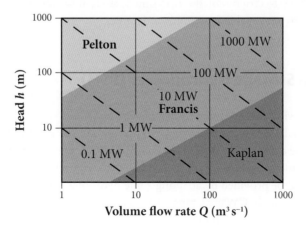

Fig. 6.8 Choice of water turbine in terms of head h and volume flow Q (*Source*: Boyle 2004).

Impulse turbines are suited for large h and low Q, e.g. fast-moving mountain streams. Kaplan turbines for low h and large Q (e.g. ROR sites), and Francis turbines for large Q and large h (e.g. dams).

6.5 **Micro-hydro**

For small or isolated communities with a local supply of flowing water, **micro-hydropower installations** (typically 5–100 kW) can provide an economical source of electricity. The choice of turbine depends on the volume flow rate and head of water (see Section 6.4), and also on the environmental impact, cost, and reliability. In a typical micro-hydropower installation in a mountainous area, water from a high-level stream could be diverted through a pipe (the penstock) to a Pelton turbine, with a valve to control the rate of flow of water through the turbine. Alternatively, in the case of water flowing over a weir or a waterfall, a simple turbine could be used to convert the kinetic energy into electricity. The electricity can be stored in batteries as a reserve for periods of increased demand. Micro-hydropower schemes avoid the need for dams or connection to transmission networks, have virtually zero carbon footprint, and can enable small rural communities to be self-sustaining. One such example is Dyffryn Crawnon, in the Brecon Beacons National Park in Wales, where 15 kW is generated, using a turgo wheel, from a flow of 17 litre s^{-1} and a head of 130 m, enough to sustain a community of about 25 homes.

There are numerous micro-hydro installations in the developing world, e.g. in remote communities in the Andes and Himalayas, and in hilly parts of the Philippines, Sri Lanka, and China. Micro-hydro schemes such as the **Garman water turbine** can be locally manufactured, and can also complement solar power, since river flows are typically highest in the winter when solar insolation is lowest. When not directly near a community, a local grid can be used to transmit the power, with the voltage being stepped up to reduce losses—an example is in Peru where a 10 kV grid is used. In response to fluctuations in demand, the supply can be adjusted by using a load controller or by varying the water supply. Such systems can enable communities to have electricity for the first time. Electricity enables lighting, communications, and the web to be available for schools and homes.

6.6 **Pumped Storage Hydropower (PSH)**

Dams can also provide energy storage, enabling the better use of more variable sources of energy, and pumped storage hydropower (PSH) plants are the most cost-effective form of large energy storage available. In 2019 an estimated 158 GW of generating capacity was provided by pumped storage, providing about 94% of global electricity storage, and considerably more capacity could be provided by adapting some hydropower plants. The main countries with large PSH capacities in GW are China 30.3, Japan 27.6, USA 22.9, Italy 7.7, Germany 6.4, Spain 6.1, France 5.8, Austria 5.6, India 4.8, and South Korea 4.7.

A seawater pumped storage plant near the north-western tip of the Spencer Gulf in South Australia is being planned, as fresh water resources are scarce. It would have sufficient storage to generate 225 MW for up to 8 hours, mitigating pressures on peak demand when temperatures soar. Seawater PSH has great potential due to the large number of sites. Another possibility for PSH is to use underground reservoirs, e.g. old mines, or undersea systems.

PSHs were generally used to pump water up from the lower reservoir to the upper one at night when electricity prices were low, and generate during the day when prices were higher. But increasingly as the penetration of renewables grows, PSH is being used to smooth their variations in output. Variable speed turbines, pioneered in Japan, can provide frequency control and grid stability and a number of schemes are under development in Europe. The volumes and differential heights of the PSH reservoirs determine their storage capability, and that is an important parameter. In Spain, PSHs can buffer several hours of low generation from renewables, while in the UK it is for much less.

The importance of PSH plants in enabling greater penetration of renewables is currently undervalued, and the economics of PSH are being affected by the availability of cheap renewable electricity during the day, which is reducing the price differential between day and night. However, at high penetrations, storage will be very helpful in stabilizing the operation of a grid. Further details of pumped storage are given in Chapter 10 Section 10.17.

6.7 Environmental and Social Impact of Hydropower

The main economic and climatic advantages of hydropower are low operating costs, minimal impact on the atmosphere, quick response to sudden changes in electricity demand, and a long plant life—typically 40 years or more before major refurbishment. However, the capital cost of construction of dams is high, and the payback period is very long. There are also serious social and environmental issues to be considered when deciding about a new hydroelectric scheme, including the displacement of population, sedimentation, changes in water quality, impact on fish, and flooding. A notable example is the Three Gorges Dam on the Yangtze River in China, originally envisaged in 1919, with the first plans being drawn up in the 1930s (see Fig. 6.9). These plans were revived in the 1980s as a way of providing electric power as well as reducing the risk of flooding: 18 million people were displaced and 33,000 killed following a flood in 1954. However, while the dam has a maximum output of 22.5 GWe, 1.3 million people were displaced by the project, and there are concerns over the increase in landslides around the dam.

It has been estimated that 30–60 million people worldwide have had to be relocated because of hydropower. Resettlement is a sensitive issue that needs careful consideration and engagement with the people affected. Projects can also affect indigenous peoples, and these projects must ensure that they are consulted, their traditions respected, and that their society benefits. As a result of these concerns, proposed hydropower plants often provoke controversy, and in some countries public opposition to hydropower has stopped all construction except on small-scale projects; this opposition was noticeable in the last few decades of the last century. Also, dams sometimes collapse for various reasons, e.g. overspilling of water, inadequate

Fig. 6.9 Three Gorges Dam in China (*Source*: www.stema-sustems.nl).

spillways, foundation defects, settlement, slope instability, cracks, erosion, or freak waves from landslides in steep-sided valleys around the reservoir. As with nuclear plants, the risk of major accidents is small but the consequences can be catastrophic. Dams have burst, causing considerable loss of life; notably the failure in 1975, following extreme rainfall, of the Banqiao Dam in China that caused ~26,000 fatalities from the flooding alone, and ~145,000 subsequently from famine and epidemics.

Other problems have beset plans to harness one of the largest single hydropower resources in the world; it lies in sub-Saharan Africa, where an estimated 600 million people are without electricity. Near the mouth of the Congo River are the Inga falls, a series of rapids during which the river drops a great height—about 100 metres, which is the height of a 30-storey building. The river has the second largest flow of water in the world, after the Amazon, and the total potential output is almost twice that of the Three Gorges dam. This is enough to power enormous industrial growth, and to lift many millions of people out of poverty. But realizing this potential has been severely hampered by wars, corruption, decades of social and political unrest, and massive cost overruns.

In order to reduce the environmental impact and the consequences of dam failure, the question arises as to whether it is better to build a small number of large reservoirs or a large number of small ones. Although small reservoirs tend to be more acceptable to the public than large ones, they tend to need a much larger total reservoir area than a single large reservoir providing the same volume of stored water.

A strong argument in favour of hydropower is that its production of greenhouse gases or acid rain gases is small, except during construction. However, while this is generally correct, with median emissions of ~55 $kgCO_2$ per MWh, which are about a tenth of that from fossil fuel generation, there is considerable variation dependent on the area and temperature of the water reservoir. Typically, those where the ratio of the area of water in the reservoir to the annual electrical output (km^2 per GWh) is relatively low, have the lower CO_2 emissions. These emissions arise from the aerobic decomposition of the submerged vegetation. Decomposition can be anaerobic, particularly if the reservoir is hot and thermally stratified, and then methane emissions can be significant, though their climatic effect is relatively short term. The difference in reservoir areas explains, for instance, why the largest number of dams on the Amazon with high carbon emissions occurs in Brazil, which is predominantly low-lying, while those with low carbon emissions are mainly in the mountains in Bolivia, Ecuador, and Peru.

It is important to balance the generation of energy, the social and environmental impacts, and the various demands for water in planning any new hydropower development to ensure a sustainable scheme.

6.7.1 Rainfall Requirement for Large Hydropower Plants

The amount of rainfall needed to maintain the level in a dam's reservoir can be estimated from the size of the catchment area, the head of the dam, and the power output. For the Three Gorges dam, the catchment area $A \sim 10^6$ km^2, the head $h \sim 100$ m, and the continuous output power $P \sim 10$ GWe (which is 22.5 GWe at $\sim 40\%$ capacity factor). The power P is given by eqn (6.1):

$$P = \eta \rho g h Q$$

Taking the efficiency $\eta = 90\%$, then $Q = 10^{10}/(0.9 \times 1000 \times 9.81 \times 100) \sim 10^4$ m^3 s^{-1}. Since $Q = A \mathrm{d}z/\mathrm{d}t$, where $\mathrm{d}z/\mathrm{d}t$ is due to rainfall, then $\mathrm{d}z/\mathrm{d}t \sim 10^{-8}$ m s^{-1} or ~ 0.3 metres per year, which is roughly the amount of rainfall required in a region for a large hydropower plant.

Although climate change could affect water resources, modelling suggests that there would be little overall effect on the global resource, but local impacts could be quite significant and would need to be evaluated. Climate change is already causing larger variations in precipitation and more run-off from glaciers, which are causing modifications and changes in the operation of plants.

6.8 Economics of Hydropower

The economic case for any hydropower scheme depends critically on how future costs are discounted (see Chapter 1 Section 1.5.7). Discounting reduces the benefit of long-term income, disadvantaging hydropower compared with quick payback schemes such as CCGT generation (see Chapter 3 Section 3.13). Hydropower schemes therefore tend to be funded by governmental bodies seeking to improve the long-term economic infrastructure of a region rather than by private capital, where the competitive increasingly deregulated power market has tilted the balance away from capital-intensive projects towards plants with rapid payback of capital. However, the introduction of guarantees by the World Bank has enabled private-sector finance for hydropower schemes to be forthcoming.

When the cost of a new project is large, refurbishing existing plants may be a more cost-effective way to increase capacity. Another cost arises at the end of the effective life of a dam, when it needs to be decommissioned. The issue of who should pay for the cost involved in decommissioning is similar to that for nuclear plants: should it be the plant owners, the electricity consumers, or the general public? On the positive side, production costs for hydropower are low because their lifetimes are long: typically, 40–80 years, the resource (rainfall) is free, and operation and maintenance costs are minimal. However, the efficiency of a hydroelectric plant tends to decrease with age due to the build-up of sedimentation trapped in the reservoir. This can be a life-limiting factor because the cost of flushing and dredging may be prohibitive.

There have also been many dams that have been more expensive than originally budgeted for. Smaller projects, such as run-of-river schemes, where advantage is taken of a natural fall in water level to avoid a large dam and reservoir, are cheaper. The lack of a large reservoir means that their output is more susceptible to changes in rainfall, but their impact tends to be less disruptive for the environment. Climate change is already causing droughts and larger variations in rainfall in some regions, and raising concern in countries which are reliant on hydropower.

The capital costs of construction vary considerably and have risen about 36% since 2010, arising from more expensive development conditions, notably in Asia. In 2019, costs for newly commissioned plants were around $1000 per kW to $4000 per kW, with the average $1700 per kW, and operations and maintenance costs (O&M) around 2% of the investment cost; their average capacity factor (CF) was 48%. When the maximum output of a plant can be considerably higher than its average, so as to meet a surge in demand, its CF tends to be lower. Also, the availability of hydropower is affected by rainfall and so will have a seasonal dependence. Almost 90% of new plants in 2019 generated electricity at prices less than from new fossil fuel plants, and the average price was US$0.047 kWh^{-1}.

This value is for plants with capacities greater than 10 MW and corresponds to an interest rate of 9% on an investment of $1700 per kW, a CF of 48%, an O&M of 2.3% and a lifetime of 40 years (see Chapter 1 Section 1.5.7). About 45% of the capital cost is on civil works and 33% on the procurement cost of equipment for the powerhouse. For smaller projects (<10 MW), the average investment cost is about $2100 per kW. The typically low LCOE means that hydropower is often one of the cheapest forms of power and contributes significantly in a number of countries around the world, including for instance Norway, Canada, New Zealand, China, Paraguay, Brazil, and Angola.

6.9 Global Hydropower Potential

The global technical potential of hydropower is estimated to be about 15,000 TWh y^{-1}, and at a 45% capacity factor (4000 full load hours), would correspond to 3750 GW of installed capacity compared with 1308 GW installed in 2019.

The theoretical potential of hydropower is determined by the annual runoff of precipitation and its fall in metres (see Exercise 6.16). The estimate of technical potential is about a

Table 6.1 Regional hydropower technical potential and percentage untapped

	North America	Latin America	Europe	Africa	Asia	Australasia Oceania
Technical potential ($TWh\,y^{-1}$)	1659	2856	1021	1174	7681	185
Percentage untapped	61	74	47	92	80	80

Source: (IEA2012).

third of the theoretical, and as Table 6.1 shows, there are considerable untapped resources in Asia, ~6000 TWh, Latin America, ~2000 TWh, North America, ~1000 TWh, and Africa, ~1000 TWh per year.

6.10 Current Global Hydropower Capacity

The 10 countries with the largest installed hydroelectric capacity (in GW) in 2019 were: China 356, Brazil 109, USA 103, Canada 81, India 50, Japan 50, Russia 50, Norway 33, Turkey 29, and France 26, and accounted for two thirds of the global capacity. The eight largest sites are shown in Table 6.2 along with their capacity factors (CFs). The significant difference in CFs between the Three Gorges Dam and the Itaipu hydroelectric plants reflects typically low rainfall for several months in the Three Gorges area, while the variation in flow of the Paraná river, which lies on the border of Brazil and Paraguay and feeds the Itaipu dam, is relatively small.

Mountainous countries like Norway and Iceland are virtually self-sufficient in hydropower, but in countries where the resource is less abundant, hydropower is mainly used to satisfy peak-load demand, though pumped storage is becoming increasingly important. It can also act as a backup supply, help to smooth fluctuations in supply, and maintain the output frequency.

In 2019, the global total capacity reached 1308 GW, and the average annual growth seen in the period 2015–2019 was 2.1%. The global hydropower capacity provided ~4306 TWh,

Table 6.2 The largest sites for hydropower

Country	Site	Capacity (GW)	CF(%)
China	Three Gorges	22.5	50
Brazil/Paraguay	Itaipu	14	84
China	Xiluodu	13.9	45
Brazil	Belo Monte	11.2	40
Venezuela	Guri Dam	10.2	60
Brazil	Tucurui	8.4	57
USA	Grand Coulee Dam	6.8	34
China	Xiangjiaba	6.4	54

which corresponds to an average capacity factor of 37.5%. The largest additions in GW were 4.9 in Brazil, 4.2 in China, and 1.9 in Laos, accounting for 70% of new capacity. By the end of 2020, the global hydropower capacity was 1332 GW.

6.11 Outlook for Hydropower

The resource in the EU is quite well exploited but the potential for growth elsewhere is considerable, particularly in Asia, Latin America, and Africa, and will be helped in some of these regions by low labour costs. The IEA anticipate generation in 2050 in TWh in Africa reaching ~350, Central and South America ~1500, with greater than half in Brazil, Asia ~ 3000, with about half in China and a quarter in India, North America ~850, Europe ~900, and Russia + Eurasia ~500, of which about three-quarters in Russia. A total generation of around 7000 TWh per year. Refurbishment of plants can be a cost-effective way to increase capacity. The output of plants with reservoirs is flexible, which is increasingly important in accommodating the variability of wind and solar farms.

Pumped storage hydropower (PSH) is also important for integrating renewables, and there is an increasing emphasis on PSH worldwide to meet the growing penetration of solar and wind. For example, in Chile, financing is being arranged for a 300 MW seawater PSH to complement a 561 MW solar farm in the Atacama Desert, and in Australia a 250 MW 1500 MWh PSH project is being planned to provide flexibility for solar PV farms. The IEA predict that PSP capacity, which was 158 GW in 2019, will increase 3–5-fold to 400–700 GW by 2050. Interconnectors to hydropower plants can also help balance supply and demand. An example is the 240 km power line connection between Denmark and Norway, that has a capacity of 1 GW and allows electricity to be exchanged between Danish wind farms and Norwegian hydropower. PSH could also help with seasonal variations in supply.

The capacity of many systems could be raised by up to 20%, which may be more cost-effective and socially acceptable than large new projects. Research on minimizing the environmental impact of hydropower is particularly important. Whole life-cycle analyses on the impact of projects are necessary to ensure, for example, that water availability for agriculture is not significantly affected. But these concerns must be balanced against the importance of providing energy in raising the standard of living.

About 30% of the global potential has been developed, and the IEA estimates that global capacity will grow to almost 2000 GW by 2050, mainly in developing countries, and provide just over 7000 TWh per year. This would make a very important contribution to the supply of low-carbon energy, and would also help in integrating renewables into grid supplies.

6.12 Tidal Power

Tidal power exploits the bulk motion of the tides. Tidal power has the advantage over other forms of alternative energy of being predictable. For conventional tidal power generation, it is necessary to construct huge tidal basins in order to generate useful amounts of electricity.

The most substantial tidal power plants are at La Rance (France 240 MW) and Sihwa Lake (South Korea 254 MW), commissioned in 1966 and 2011, respectively. Also, in recent years an alternative technology for exploiting strong tidal currents has been under development using underwater rotors.

Tides occur because the gravitational pull of the Moon on the oceans on the near side of the Earth is slightly larger, and on the far side is slightly smaller, than that necessary to keep the Earth and Moon orbiting each other. As a result, the ocean level is slightly higher on either side (and a bit lower in between), and the weight of raised water compensates for the difference in pull. As the Earth rotates, these bulges cause two tides a day, with the variation in sea level—called **the tidal range**—about 0.5 to 1.0 m in mid-ocean (see Derivation 6.3 at end of the chapter). The tidal bulges do not keep pace with the rotation of the Earth (see Exercise 6.30), so the tides lag behind the position of the Moon, with the amount of lag depending on the latitude. The interval between high tides is approximately 12 hours 25 minutes (see Exercise 6.20). The presence of continents and bays significantly disturbs the tides and can enhance their range.

The pull of the Sun increases or decreases the range by about 20%, depending on its relative position, to give spring and neap tides. However, where a natural oscillation of sea water has a period matching that of the rotation of the Earth, the range can be much larger. This happens for the tides in the Atlantic basin, and also in some inlets and estuaries. If a bay has a funnel shape, then it further enhances the range of the tide, since the decreasing width causes an incoming tide to rise up and an outgoing one to drop down. This **tidal resonance** and enhancement is important, since it makes those locations with large tidal ranges ideally suited for tidal power plants.

6.13 **Tidal Resonance**

The tidal range varies in different oceans of the world, due in part because of tidal resonance. Fig. 6.10 shows water of depth h oscillating in a shallow basin of length L. The surface of the water is horizontal at both ends, since there is no flow there, and the displacement in the middle is zero (node). The water goes up and down at the ends with a frequency f_b given by $f_b = c/\lambda$, where c is the speed of shallow water waves and $\lambda = 2L$ is the wavelength of the oscillation.

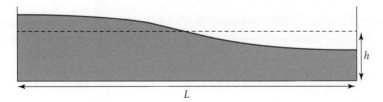

Fig. 6.10 Water of depth h oscillating in a shallow basin of length L.

From dimensional analysis (see exercise 6.31), the speed c of a water wave of wavelength λ in a sea of uniform depth h_0 is given by $c = f(\lambda/h_0)\sqrt{gh_0}$. When $\lambda \gg h_0$, called a

shallow water wave, we expect the speed to be independent of λ, so $f(\lambda/h_0)$ is a constant, and is equal to one (see exercise 6.22); i.e.

$$c = \sqrt{gh_0} \tag{6.17}$$

When the tidal frequency is close to f_b the natural oscillation in the basin builds up and the range (i.e. the difference between the maximum and minimum displacements at the ends of the basin) increases. For example, the Atlantic Ocean has a width of around 4000 km and an average depth of about 4000 m, so the speed of a shallow water wave, eqn (6.17), is about $c = \sqrt{gh_0} = \sqrt{10 \times 4000} = 200$ m s^{-1} and $f_b \sim 2.5 \times 10^{-5}$. The tidal frequency is about 2.2×10^{-5} s^{-1}, which is close to f_b, so the time taken for the tidal bulge to make the round trip, reflecting off both shores, is about the same as the tidal period, so the amplitude builds up; along the Atlantic shore the tidal range is amplified from ~0.5 m to ~3 m.

River estuaries can also exhibit large tidal resonance if the length and depth of the estuary are favourable. An oscillation in an estuary is represented by the oscillation from the middle to one end of the basin, shown in Fig. 6.10: there is no flow at the head, but water flows in and out of the mouth of the estuary. The length of estuary for tidal resonance to occur is therefore a quarter of the wavelength of shallow water waves in the depth of the estuary.

EXAMPLE 6.3

The Bristol Channel (UK) has an average depth of ~80 m and the length of the continental shelf in the channel is ~300 km. Show that a tidal resonance enhancement of the tides at the head of the channel is expected.

The frequency of the natural oscillations f_b in the channel is $c/4L$. The speed of shallow waves in the channel is $c = \sqrt{gh_0}$, so $c = 28$ m s^{-1} and $f_b = 28/(4 \times 3 \times 10^5) = 2.3 \times 10^{-5}$ s^{-1}. Since this is close to the tidal frequency of 2.2×10^{-5} s^{-1}, tidal resonance is expected. The actual situation is more complicated, since the variable width and depth of the channel also have an effect on the tides.

6.14 Tidal Barrages

The earliest exploitation of tidal power was the tidal mill, created by building a barrage across the mouth of a river estuary. Seawater was trapped in a tidal basin on the rising tide and released at low tide through a waterwheel, providing power to turn a stone mill to grind corn. **Tidal barrages** for electricity generation use large low-head turbines and can operate for a greater fraction of the day.

The first large-scale tidal power plant in the world was the tidal barrage built in 1966 at La Rance in France. It generates 240 MW using 24 low-head Kaplan turbines with a capacity factor of 28%. A more recent barrage is the 254 MW Sihwa power plant in South Korea, completed in 2011 (see Fig. 6.11) with a capacity factor of 25%. Various proposals during the last

Fig. 6.11 Sihwa tidal barrage power plant in South Korea. Credit: Topic Images Inc./Getty.

century to build a large-scale tidal barrage scheme for the River Severn in the UK were turned down due to the large cost of construction, public opposition, and the availability of cheaper alternatives—the last time being in 2010 when the estimated cost of £30 billion was deemed too expensive to justify. For these reasons, the focus has switched to smaller projects involving tidal lagoons and tidal streams, though there are plans for some tidal barrage plants.

6.15 Tidal Lagoons

An alternative to a barrage is to create a closed basin in the estuary known as a **tidal lagoon**. The wall of a tidal lagoon does not extend across the whole channel, so the environmental effects are lessened and the impact on fish and navigation is reduced. Also, by restricting the tidal lagoons to shallow water, the retaining wall can be low, and cheaper to build.

Case Study 6.2 Swansea Tidal Lagoon project

The Severn Estuary has a tidal range of 10.5 m, making it the second highest in the world. The old idea of a barrage across the entire Severn Estuary has always suffered from the enormous capital cost involved, but smaller and much cheaper projects are now being proposed, involving self-contained tidal lagoons. The first such project is a 320 MW tidal lagoon at Swansea Bay, with a 9.5 km seawall projecting up to 4 km out to sea. The generators would comprise 16 low-head turbines, which would produce enough electricity for about 150,000 homes, most of which would be close to the site and thus minimize transmission losses. Construction would take four years, and, once built, the lagoon would have a low carbon footprint. It would also stimulate the local economy and encourage tourism. The visual impact would be similar to that of a breakwater wall, and the impact on fish and marine mammals is likely to be minimal, since the turbines are large enough to enable them to pass through unscathed.

The key issue is the high capital cost at around £1.3 billion. The developers argue that this should be seen as a pilot project, with larger lagoons to follow along the Severn Estuary, generating up to 8% of the UK's electricity, with a life-cycle of 120 years. The initial price of electricity, close to £100 per MWh, would be expected to fall for later projects due to the learning effect. An independent government-commissioned review in 2017 backed the scheme, but in 2018 the government rejected the proposal on the grounds that it would be too expensive compared to offshore wind and nuclear power. However, with the withdrawal of support for building two major UK nuclear plants in early 2019, interest in the Swansea Bay lagoon project has revived. The firm behind the original proposal believes that the project could be built without government support, and that prospects are helped by the plan to add floating solar panels to the lagoon, thereby boosting the output by more than a third.

6.16 **Power from a Tidal Barrage or Lagoon**

A rough estimate of the average power output from a tidal barrage or lagoon can be obtained from a simple energy balance model by considering the average change of potential energy during the draining process (ebb tide). One mode of operation is for the basin to fill during the last ~3 hours of the flood tide (from mid to high tide), then to raise the height in the basin by pumping for the next ~3 hours (see Exercise 6.25). The situation is shown in Fig. 6.12. The basin is drained through the turbines for the next ~6 hours: during the first ~3 hours the difference in level between the water in the basin and outside the barrage is somewhat larger than $h/2$ and then it decreases to zero over the next ~3 hours. The result is that a total mass of

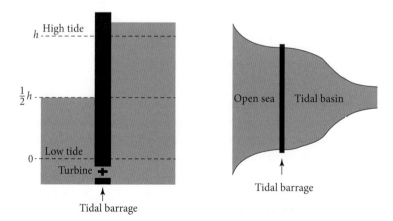

Fig. 6.12 Tidal barrage, with the height in the tidal basin raised by pumping.

water in the tidal basin of area A, $m \approx \frac{1}{2}\rho Ah$, where h is the tidal range, falls a distance $\approx h/2$.

The loss in potential energy is $\sim \frac{1}{4}\rho gAh^2$. Hence the average power output is:

$$P_{\text{ave}} = (\rho gAh^2)/4T \tag{6.18}$$

where T is the interval between successive tides, i.e. the tidal period. In practice, the power varies with time according to the difference in water levels across the barrage and the volume of water allowed to flow through the turbines. Also, the operating company would seek to optimize revenue by generating electricity during periods of peak-load demand when electricity prices are highest.

An important issue is whether it is better to use conventional turbines that are efficient but operate only when the water is flowing in one particular direction, or less efficient turbines that can operate in both directions (i.e. for the incoming tide and the outgoing tide).

EXAMPLE 6.4

Estimate the average power output of a tidal basin with a tidal range of 12 m and a tidal basin area of 520 km^2 (i.e. the Severn Barrage) when operating during the ebb tide.

Substituting in eqn (6.18), noting that the tidal period is $T \approx 4.5 \times 10^4$ s ($T \approx 12.5$ h), the average power output is:

$$P_{\text{ave}} = \frac{\rho gAh^2}{4T} \approx \frac{10^3 \times 10 \times 520 \times 10^6 \times 12^2}{4 \times 4.5 \times 10^4} \approx 4.2 \text{ GW}$$

6.17 Tidal Current (Stream) Plants

In particular locations (e.g. between islands) there may be strong tidal currents that transport large amounts of kinetic energy. In recent years, various devices for extracting this energy have been proposed. These devices are essentially underwater versions of wind turbines. The first large-scale commercial tidal stream generator, **SeaGen**, was installed in Strangford Lough in Northern Ireland in 2008. It has a generating capacity of 1.2 MWe with a capacity factor of 75–80%, operating on both the flood and ebb tide.

For isolated underwater turbines far below the surface, the power generated can be estimated as for a wind turbine (see Chapter 7 Section 7.5), with the maximum fraction of the kinetic energy in the flow extracted being given by the **Betz limit** of 16/27, ~59%, of the amount of kinetic energy flowing through the cross-sectional area of the turbines. However, for turbines in a channel, where the water flow is driven by a tidal oscillation out of phase at either end, generating a head of water, the reactive force of the turbines will cause the flow through the channel to decrease. On equating the forces acting on the mass of water in the channel to its acceleration (see Derivation 6.2), the maximum average power P_{max} that can be extracted is given by:

$$P_{max} = \gamma \rho g a Q_{max} \tag{6.19}$$

where $\gamma = 0.22 \pm 0.02$, a is the amplitude of the sinusoidal variation with angular frequency ω in the head driving the flow, and Q_{max} is the maximum flow rate through the channel when no turbines are deployed. Note that this expression is algebraically similar to that for the power output from a dam, $P = \eta \rho g h Q$ (eqn (6.1)), where the head h is fixed and the efficiency η is typically ~0.9 (see Example 6.2).

Derivation 6.2 Tidal current potential of channels

For a channel that is short compared with the wavelength of the tide (typically hundreds of kilometres) connecting two basins with different elevations, the equation of motion for the mass of water in a channel of cross-sectional area A and length L is:

$$\rho A L \frac{\partial u}{\partial t} = \rho A g a \cos \omega t - F_{turb} - k^* u^2 \tag{6.20}$$

where a is the amplitude of the sinusoidal head driving the flow with angular frequency ω, u is the speed of the water flow, $k^* u^2$ is the natural friction in the channel, and F_{turb} is the additional frictional force caused by the presence of turbines. When the natural friction parameter k^* is large, the acceleration term in the above equation is small compared with the head and can be neglected to a first approximation. Also, since the volume of water flow per second is $Q = uA$, the above equation (6.20) can then be written in the form

$$\rho A g a \cos \omega t \cong F_{turb} + kQ^2$$

where $k = k^*/A^2$. When no turbines are present, the maximum flow Q_{max} is given by $Q_{max}^2 = \rho A g a / k$. The power extracted by the turbines $P = F_{turb} u = F_{turb} Q/A$, so:

$$P = (\rho A g a \cos \omega t - kQ^2)Q/A$$

This has a maximum, P_{max}, when $Q^2 = \frac{1}{3}(\rho A g a/k)\cos \omega t = \frac{1}{3}Q_{max}^2 \cos \omega t$, so the maximum average power P_{max} is given by:

$$P_{max} = \frac{2}{3^{3/2}} \rho g a Q_{max} \left\langle \left(\cos \omega t \right)^{3/2} \right\rangle = \gamma \rho g a Q_{max}$$

which is eqn (6.19). The parameter $\gamma = 0.22$, since $\left\langle \left(\cos \omega t \right)^{3/2} \right\rangle = 0.56$ over half a tidal period (where $\langle X \rangle$ is the average value of X). Consideration of the magnitude and power law dependence of the frictional drag on the flow rate Q gives a 10% uncertainty in γ. Also, the increased flow resistance caused by the turbines could increase the head slightly and thus the power extracted (Garrett and Cummins 2005).

In the majority of designs the axis of rotation of the turbine is horizontal, and the device is mounted on the seabed, or is suspended from a floating platform or is on one submerged. Before installation, the tidal currents for any particular location need to be measured to depths of 20 m or more in order to determine the suitability of the site. The first generation of prototype kinetic energy absorbers have been operated in shallow water (i.e. 20–30 m) using conventional engineering components. Later generations are likely to be larger and more efficient, and use specially designed low-speed electrical generators and hydraulic transmission systems. An array of tidal current turbines under construction is the MeyGen project (see Case Study 6.3).

Case Study 6.3 Pentland Firth MeyGen project

In the UK, work is underway on the MeyGen project, by a consortium led by Atlantis, to install turbines in the Inner Sound in the Pentland Firth. Through this 3 km wide channel, between the island of Stroma and the northern coast of Scotland, there are currents of up to ~5 m s^{-1} at peak spring tide. Each turbine has three blades that sweep out an 18 m diameter area, and there is pitch and yaw control, so individual blades can be turned to face the flow at the right angle. One is a SIMEC Atlantis 1.5 MW turbine, the other three are Andritz Hydro 1.5 MW turbines.

The turbines look like small horizontal wind turbines, HAWTs (see Fig. 6.13 and Chapter 7 Fig. 7.3), and are mounted on a ballasted tripod that is held down by its own weight on the seabed, in water whose depth is ~35 m. They reach their rated capacity at a flow speed of 3 m s^{-1}, at which speed their blades are rotating at ~10 times per minute. Their rated power output P_R is given by:

$$P_R = \frac{1}{2} C_P \rho A u_R^3 \tag{6.21}$$

where C_P is the power factor, ρ the density of water, A the area swept out by the turbine blades $\pi D^2/4$, D is the diameter of the rotor, and u_R is the flow speed at the rated capacity of the turbine. Substituting values for $D = 18$ m, $\rho = 1030$ kg m^{-3} (seawater), and $P_R = 1.5 \times 10^6$ W, makes $C_P = 0.42$. This value for the power factor is similar to those obtained for wind turbines. (NB The Betz limit of 16/27 for wind turbines does not apply to water turbines when the flow is confined by the channel and neighbouring turbines.) It is planned to use anti-fouling paint to reduce fouling by barnacles and weeds.

Each nacelle of the AR1500 turbines contains a permanent magnet generator (see Chapter 10 Box 10.1) and a gearbox, and the output varies in voltage and frequency as the tidal current changes. The power conditioning equipment that produces a fixed frequency and voltage output is located onshore, to reduce the complexity of the underwater generator.

The first phase of the project with four 1.5 MW turbines began operation in 2018 and in 2019 the turbines generated 13.8 GWh, operating with a capacity factor of 26%. In 2019 it was announced that an additional 80 MW of capacity would be added in the next phase of the project. When finally completed there will be an array of 269 turbines mounted on the seabed, with an output of ~400 MW. The predictability of the tides will simplify integration of the generated power into the grid.

Fig. 6.13 MeyGen AR1500 18 m diameter tidal stream turbine.

6.18 **Ecological and Environmental Impact of Tidal Power**

Tidal stream devices (such as SeaGen in Strangford Lough in Northern Ireland and Meygen in Scotland) can be deployed with minimal environmental and ecological effect. The danger to fish is minimal because the blades rotate fairly slowly (typically about 10 revolutions per minute). However, the installation of a tidal barrage can have a major impact on both the environment and ecology of the estuary and the surrounding area. The installation of locks is required to allow vessels to pass through, and the water flow through turbines must be carefully designed to enable the safe passage of fish.

The tidal regime may be affected downstream of a tidal barrage and the water quality in the basin altered, since the natural flushing of silt and pollution is impeded, affecting fish and bird life. In the case of the River Severn in the UK, a barrage would cause the river flow to decrease, which would alter the unique habitat and tend to *increase* biodiversity.

On the positive side, there are the benefits arising from improved flood protection, new road crossings, marinas, and tourism. Tidal lagoons (see Section 6.14.2 and Case Study 6.2) that only partially block an estuary reduce the environmental impact significantly.

6.19 Tidal Resource

The global resource from tidal energy can be estimated from the power dissipated by tides in the Earth's shelf seas of ~2.5 TW out of a total dissipation of ~3.5 TW. The energy is carried by tidal waves in the oceans into coastal waters where tides and tidal currents, which are potential and kinetic energy stores of tidal energy, respectively, are generated.

However, only a small percentage is in regions with sufficient tidal range (> ~5 m) or tidal current speed (> ~2 m s^{-1}) to be economically exploitable by tidal barrages or by tidal streams. For tidal streams, the UK has a particularly high fraction of the global resource. A UK Government report in 2011 (based on the method described in Derivation 6.2) estimated that the UK tidal stream technical resource was ~25 TWh y^{-1}, with the Pentland Firth, a channel connecting the Atlantic to the North Sea with currents of ~3 m s^{-1}, accounting for ~35% of the UK resource. The amount of tidal energy dissipated on the European continental shelf is ~200 GW, ~10% of the global dissipation, with about half the European tidal stream resource estimated to be around the UK. Hence, scaling these estimates would give a global tidal stream resource of around 500 TWh y^{-1}, similar to that from tidal barrages. The global tidal energy technical potential is estimated at about 1200 TWh per year.

6.20 Economics of Tidal Power

Large tidal barrages have the economic disadvantages of large capital cost and long construction times. On the other hand, they have long plant lives (over 100 years for the barrage structure and 40 years for the equipment) and low operating costs. So far, the La Rance plant in France and the Sihwa plant in South Korea are the principal large barrage schemes in operation, which can produce electricity very economically at €40–€120 per MWh and €20 per MWh, respectively.

However, the high upfront costs and environmental impacts of barrages have been a major block to expansion, making it difficult to attract the necessary investment to lower the LCOE through the learning effect. There have been plans for other barrages around the world but their large cost in comparison with wind and solar power is making them increasingly uncompetitive. Plans for tidal lagoons, such as the Swansea Bay project, that combine power generation with other benefits, may have more success.

The economics of small tidal current devices (kinetic energy absorbers) has the attraction that they can be installed on a piecemeal basis, thereby reducing the initial capital outlay. They also have a predictable output, and there is no visual impact. The long-term economic potential and environmental impact of such devices will become clearer after trials on various designs, notably in Canada, China, France, Ireland, Japan, South Korea, Spain, the UK, and the USA. In 2018, a UK report concluded that tidal stream has the potential to reach £150 per MWh by 100 MW installed, reducing to £80 per MWh by 2 GW.

6.21 Outlook for Tidal Power

There is a larger global economic potential for tidal stream plants than for tidal range plants, and most development is on tidal stream devices, several being around Europe, North America, China, and Japan, where in late 2020 SIMEC Atlantis installed an AR500 turbine off Naru Island. These currently (2020) have only a few MW capacity and it is uncertain how much tidal power will contribute by 2050. However, with strong support it is possible that almost half of the technical potential at 500 TWh per year might be generated, though the increasingly low cost of solar and wind power will make this a challenge.

6.22 Wave Energy

The waves on the surface of the sea are caused mainly by the effects of wind. The streamlines of air are closer together over a crest of a wave, so the air moves faster. It then follows from Bernoulli's theorem (see Section 2.10) that the air pressure is reduced, causing the amplitude to increase and waves are generated. As a wave crest collapses the neighbouring elements of fluid are displaced and forced to rise above the equilibrium level (Fig. 6.14).

The motion of the fluid beneath the surface decays exponentially with depth. About 80% of the energy in a surface wave is contained within an eighth of a wavelength of the surface. Thus, for a typical ocean wavelength of 100 m, this layer is about 13 m deep. We now derive an expression for the speed of a surface wave using intuitive physical reasoning. When the

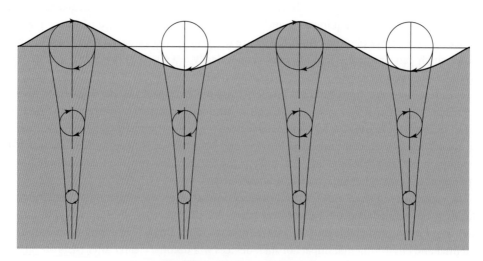

Fig. 6.14 Surface wave on deep water.

depth of water is much greater than the wavelength of the wave, the water particles follow closely circular trajectories, as shown in Fig. 6.14.

Consider a surface wave on deep water and choose a frame of reference that moves at the wave velocity c (to the right in Fig. 6.14), so that the wave profile remains *unchanged with time*. Since the pressure on the free surface is constant (i.e. atmospheric pressure) and the flow is stationary in the moving frame of reference, we can use the *steady* form of Bernoulli's equation (eqn (2.42)):

$$u_t^2 = u_c^2 + 2gh \tag{6.22}$$

where u_c is the velocity of a particle at a wave crest, u_t is the velocity of a particle at a wave trough, and h is the difference in height between a crest and a trough. If r is the radius of a circular orbit and τ is the wave period, then

$$u_c = (2\pi r/\tau) - c, \quad u_t = -(2\pi r/\tau) - c, \quad h = 2r \tag{6.23}$$

Substituting for u_c, u_t, and h from eqn (6.23) in eqn (6.22), and putting $\lambda = c\tau$, we obtain the wave speed as:

$$c = \sqrt{(g\lambda/2\pi)} \tag{6.24}$$

It follows from eqn (6.24) that the wave speed increases with wavelength, so surface waves are **dispersive**. In practice, the wave profile on the surface of the sea is a superposition of waves of various amplitudes, speeds, and wavelengths moving in different directions. The *net* displacement of the surface is therefore more irregular than that of a simple sine wave. As a result, for a wave power device to be an efficient absorber of wave energy in real sea conditions, it needs to be able to respond to random fluctuations in the wave profile.

The total energy E of a surface wave per unit width of wave-front per unit length in the direction of motion is part potential and part kinetic energy. The potential energy is proportional to the height that the mass of water m is raised from a trough to a peak of the wave, and so is proportional to mga, where a is the amplitude of the wave. Since the mass m is proportional to the density ρ and to a, the potential energy is $k\rho ga^2$, where k is a constant of proportionality. The contribution of the kinetic energy equals that of the potential energy (by equipartition of energy), and the constant k is a quarter (see Exercise 6.34), so the total energy E is given by:

$$E = \frac{1}{2}\rho ga^2 \tag{6.25}$$

The dependence of wave energy on the *square* of the amplitude a has mixed benefits. Doubling the wave amplitude produces a fourfold increase in wave energy. However, too much wave energy poses a threat to wave power devices and measures need to be taken to ensure they are protected in severe sea conditions.

The power P per unit width in a surface wave is the product of E and the group velocity c_g, given by:

$$c_g = \frac{1}{2}\sqrt{(g\lambda/2\pi)} \tag{6.26}$$

(see Exercise 6.33). Hence the incident power per unit width of wave-front is:

$$P = \frac{1}{4}\rho g a^2 \sqrt{(g\lambda/2\pi)} \tag{6.27}$$

In mid-ocean conditions the typical power per metre width of wave-front is 30–70 kW m^{-1}.

EXAMPLE 6.5

Estimate the power per unit width of wave-front for a wave amplitude $a = 1$ m and wavelength of 100 m.

From eqn (6.27), the power per unit width of wave-front is:

$$P = \frac{1}{4}\rho g a^2 \sqrt{\frac{g\lambda}{2\pi}} \approx \frac{1}{4} \times 10^3 \times 10 \times 1^2 \times \sqrt{\frac{10 \times 10^2}{2 \times 3.14}} \approx 32 \text{ kW m}^{-1}$$

6.23 Wave Power Devices

Although the first patent for a wave power device was filed as early as 1799, wave power was effectively a dormant technology until the early 1970s, when the global economy was hit by a series of large jumps in oil prices (see Chapter 1 Section 1.1). Wave power was identified as one of a number of sources of alternative energy that could potentially reduce dependence on oil. It received financial support to assess its technical potential and commercial feasibility, resulting in hundreds of inventions for wave power devices, most of which were dismissed as either impractical or uneconomic. The main concerns were whether wave power devices could survive storms, and their capital cost. Through the 1980s, publicly funded research for wave power virtually disappeared as global energy markets became more competitive. However, in the late 1990s interest in wave power technology was revived as a result of increasing evidence of global climate change and the volatility of oil and gas prices. A second generation of wave power devices has since emerged which are better designed and have greater commercial potential.

In general, the key issues affecting wave power devices are:

- survivability in violent storms;
- vulnerability of moving parts to seawater;
- capital cost of construction;
- operational costs of maintenance and repair;
- cost of connection to the electricity grid.

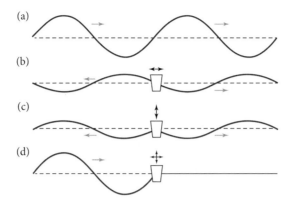

Fig. 6.15 Waves incident and generated by a wave power device orientated perpendicular to the direction of the wave.

We now describe different types of wave power device and examine how they operate and how they address the above challenges:

Linear absorber: A wave incident on a wave power device causes it to oscillate and extract energy and also to generate waves. Figure 6.15 illustrates schematically a wave power device oriented perpendicular to the direction of the wave, i.e. parallel to the wave crests; such a device is called a **terminator**. Figure 6.15 (a) shows a plane wave travelling in the positive x-direction without the device; Figures 6.15 (b) and (c) show the waves generated when the device oscillates forward and backward (pitch) and up and down (surge), respectively; Figure 6.15(d) shows the result when an incident plane wave causes the device to oscillate both ways, each generating waves with *half* the amplitude of the incident wave. The resulting superposition cancels the ongoing part of the incident wave and allows all the incident energy to be extracted. This example illustrates the principle that a good absorber is also a good generator of waves.

Point absorber: For an absorber which is small compared with the wavelength of the wave, the absorber will generate a circular outgoing wave—with the same wavelength and frequency as that of the incident wave—as it oscillates up and down. Depending on their relative phase β, part of the circular wave travelling in close to the same direction as the incident wave can interfere and reduce the amplitude and hence the energy of the incident wave after the absorber. The reduction with distance downstream from the absorber is compensated for by an increase in its width, such that a constant amount of power is extracted from the incident wave (see Exercise 6.36). Some of this power is used to generate the circular wave, and the net power transferred to the absorber divided by the incident power per unit width defines the capture width of the absorber. Its maximum value for an absorber that generates an isotropic outgoing wave is shown in Exercise 6.36 to be $\lambda/2\pi$. This shows that the maximum apparent size of an object interacting with a wave is given by the wavelength rather than the physical extent of the object.

6.23.1 **Floating Devices**

In the early 1970s, public interest in wave power was aroused by a novel device known as the **Salter duck** (Fig. 6.16). The device floated on water and rocked back and forth with the

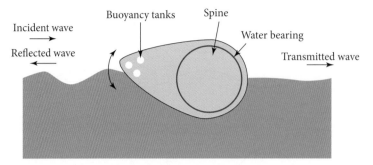

Fig. 6.16 The Salter Duck, a terminator device.

incident waves and had a cam-shaped cylindrical cross section. Its rear surface was circular so that the waves generated behind it by its rocking motion would be small. The front surface profile followed the circular trajectories of water particles, so that most of the incident wave energy was absorbed with only minimal reflection and transmission. Efficiencies in excess of 80% were achieved. The complete system envisaged a string of Salter ducks several kilometres in total length, aligned parallel to an incident wave-front. A spinal column, of 14 m diameter, would have used the relative motion between each duck and the spine to provide the motive force to generate power. The device was designed to be used in the Atlantic Ocean for wavelengths of order 100 m, but never got beyond small-scale trials owing to lack of funding in the 1980s when governmental support for wave power in the UK was dropped in favour of wind and nuclear power. Nonetheless, the Salter duck provides a useful benchmark for comparing the efficiencies of all other wave power devices.

Terminator devices need to be very strong to withstand the full force of large waves, and these forces can be reduced by using converters that are aligned with the wave direction (attenuators).

The **Pelamis** (Fig. 6.17) was an offshore wave power device designed to operate in depths of 50 m or more. As waves moved along the device, the 35 m long, 4 m diameter segments rocked back and forth, and their relative motion activated hydraulic rams which pumped high-pressure oil that powered electrical generators. Although Pelamis had successfully demonstrated that it could work, the project folded in 2014 after failing to raise further capital.

Fig. 6.17 The Pelamis, an attenuator device.

6.23.2 **Spill-over Devices**

TAPCHAN (TAPered CHANnel) is a Norwegian system in which sea waves are focused in a tapered channel on the shoreline. Tapering increases the amplitude of the waves as they propagate through the channel. The water is forced to rise up a ramp and spill over a wall into a reservoir about 3–5 m above sea level (Fig. 6.18 (a)). The potential energy of the water trapped

(a) (b)

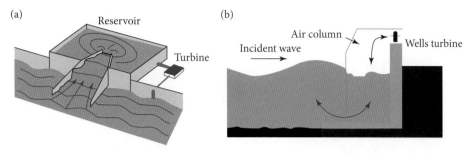

Fig. 6.18 (a) Tapered channel (TAPCHAN) (b) Oscillating water column (OWC).

in the reservoir is then extracted by draining the water back to sea through a low-head Kaplan turbine. Apart from the turbine, there are no moving parts and there is easy access for repairs and connections to the electricity grid. Unfortunately, shore-based TAPCHAN schemes have a relatively low power output and are only suitable for coastal sites where there is a deep-water shoreline and a low tidal range of less than about a metre. To overcome these limitations, a floating offshore version of TAPCHAN called **Wave Dragon** is under development at the Milla Fjord site near Milford Haven in Wales. It consists of two large wave reflectors which focus sea waves towards a ramp; water overtopping the ramp is collected in a large reservoir and discharged through low-head turbines. A combined offshore wave-wind park using the Wave Dragon as a stable floating platform may be a route to delivering cost competitive LCOE.

6.23.3 Oscillating Water Columns

The oscillating water column (OWC) uses an air turbine housed in a duct well above the water surface (Fig. 6.18 (b)). The base of the device is open to the sea, so that incident waves force the water inside the column to oscillate in the vertical direction. As a result, the air above the surface of the water in the column moves in phase with the free surface of the water inside the column and drives the air turbine housed in a duct. The speed of air in the duct is enhanced by making the cross-sectional area of the duct much less than that of the column.

A key feature of the OWC is the design of the air turbine, known as the **Wells turbine**. It has the remarkable property of spinning in the same direction irrespective of the direction of air flow in the column! Unlike conventional turbine blades, the blades in a Wells turbine are symmetrical about the direction of their motion (Fig. 6.19). Relative to the blade, the direction of air flow is at a non-zero angle of attack α. The net force acting on the blade in the direction of motion is then given by:

$$F = \mathcal{L}\sin\alpha - \mathcal{D}\cos\alpha \tag{6.28}$$

where $\mathcal{L}$ and $\mathcal{D}$ are the lift and drag forces acting on the blade. It is clear from the force diagram in Fig. 6.24 (b) (not to scale: $\mathcal{D} \ll \mathcal{L}$) that the direction of the net force is the same, irrespective of whether the air is flowing upwards or downwards inside the air column.

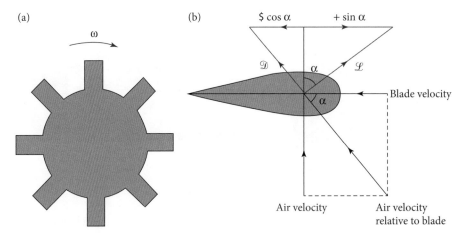

Fig. 6.19 Wells turbine: (a) plan view of blades (b) velocity and force triangles in frame of reference of a blade.

The shape of the blade is designed to maximize the net force on the blade, and the operational efficiency of a Wells turbine is around 80%. At low air velocities the turbine absorbs power from the grid in order to maintain a steady speed of rotation, while for large air velocities the air flow around the blades is so turbulent that the net force in the direction of motion of the blade becomes erratic and the efficiency is reduced.

OWCs can be integrated into buoys, and both small and large floating OWCs have been studied, though near and on the shoreline OWCs are the most developed. A notable example is the 16 Wells turbine OWC plant incorporated into a breakwater at the Mukriku harbour in northern Spain. Each turbines generates 18.5 kW, and it has been operating successfully since 2011. A 500 kW bottom-standing nearshore OWC was successfully installed in 2015 off Jeju Island, South Korea. The OWC's simplicity continues to attract considerable interest as a wave energy converter.

6.23.4 Submerged Devices

Submerged devices have the advantage of being able to survive despite rough sea conditions on the surface. One device whose performance is encouraging is the AW-Energy of Finland **WaveRoller** design shown in Fig. 6.20 (a), which is a submerged oscillating wave surge converter that operates at depths of 8–20 m and ~0.3–2 km from the shore. A single panel unit has a capacity of 0.5–1 MW and has a capacity factor of 25–50% dependent on wave conditions. The panel remains fully submerged during normal operation, but its ballast tanks can be emptied so that the panel floats to the surface for maintenance. A hydraulic system that is hermetically sealed transfers the back and forth kinetic energy of the panel to an electrical generator. There are plans (2020) to deploy units off Portugal, Mexico, and in Southeast Asia.

Another fully submerged converter, shown in Fig. 6.20 (b), is the Australian company Carnegie's **CETO 6**, which is designed to operate 2–3 m underneath the surface. A flattened sphere some 20 m in diameter and 5 m thick contains the generator, tethered to the ocean

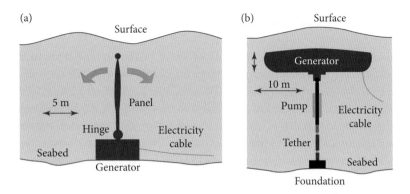

Fig. 6.20 (a) WaveRoller oscillating wave surge converter (b) CETO 6 submerged 'point' absorber.

floor. As a wave passes over, the device moves in a roughly elliptical path, and causes a piston to move up and down and pump fluid that is used to power the electrical generator. Being a 'point' absorber, it absorbs energy from whichever direction the wave is travelling. As of 2020, the company, after recapitalization and restructuring, is progressing its development of this technology.

A device with a similar operating principle to CETO 6 is the **Seabased** wave energy converter. A series of surface buoys are attached to linear generators deployed on the seabed, where they are anchored using a heavy plate that avoids the need for work on the seabed. The electric current varies both in amplitude and frequency, and a special unit inverts and transforms the current to 50/60 Hz AC. A 1 MW array at Sotenäs, off Sweden, started operation in 2016. Another submerged device under development in the UK is the **Wavesub**.

Other types of submerged devices exploit the change in pressure below the surface when waves pass overhead: the pressure is increased for a wave crest but decreases for a wave trough. An example of this type of device is the **Archimedes wave swing**. The AWS is a submerged air-filled chamber (the 'the floater effect'), 9.5 m in diameter and 33 m in length, which oscillates in the vertical direction in response to the action of the waves. The motion of the floater energizes a linear generator tethered to the seabed. In 2020, Malin Renewables in Scotland secured a contract to build a half-scale device.

6.24 Environmental Impact of Wave Power

As with most forms of alternative energy, wave power does not generate harmful greenhouse gases. Opposition to shore-based sites could be an issue in areas of scenic beauty, on account of the visual impact (including the connections to the electricity transmission grid), and the noise generated by air turbines in the case of OWCs. The visual impact is much less significant for offshore devices, but all devices can have an impact on the marine ecology. The noise and vibration during operation can affect fish, and there can be an alteration in sea currents and wave regimes that can affect sedimentation and other users. However, wave power can provide low-carbon electricity and contribute to local development opportunities.

6.25 **Potential Resource of Wave Power**

Some 60% of the global population live within 60 km of a coastline, and it has been estimated that an average 2 TW of wave power is incident on coastlines, with 270 GW off Europe, 40 GW off the UK, 280 GW off Australia, 190 GW off Chile, 220 GW off USA, 320 GW off Asia, and 320 GW off Africa. The technical potential is some 4000 TWh y^{-1}. In deep oceans, waves can travel thousands of kilometres with little loss of energy until they reach depths of around 200 m, when the interaction with the seabed reduces the power in the waves. Nearshore, in depths of ~20 m, maximum wave heights are ~15 m and wave conditions are more stable, c.f. >30 m far-offshore, making nearshore deployment safe from extreme waves and also reducing the cost of transmission to shore. Winds are predominantly from the west between 30° and 60° latitude (see Section 7.2), and result in the highest wave powers off the western coasts of continents in temperate latitudes. The power in the waves can be forecast more than 48 hours in advance, making grid integration easier.

6.26 **Economics of Wave Power**

The main challenges for wave power are to reduce the capital costs of construction, but to be still able to withstand extreme conditions at sea, and to generate electricity at competitive prices. Wave power is generally regarded as a high-risk technology. Even the largest floating devices are vulnerable in freak storms: every 50 years in the Atlantic Ocean there is a wave with an amplitude about 10 times the height of the average wave, so any far-offshore floating device must be able to withstand a factor of 100 times the average wave energy.

Submerging the devices can provide an effective means of defence, and this, together with OWCs where appropriate, appears to be emerging as the preferred technology, though the industry has yet to converge on a limited set of wave energy converters. Another factor to consider is that the frequency of incident sea waves is only about 0.2 Hz, much lower than the frequency of 50–60 Hz for electricity transmission. Although this not a difficult electrical engineering problem, the challenge is to find cost-effective solutions. The high-risk and high-upfront capital costs for developing new devices has led to the establishment of wave test centres in Scotland, Japan, China, the USA, and elsewhere, where prototypes can be tested and developed before large sums need be committed. Once an approach is established, then costs can fall through economies of scale and 'the learning effect'. An estimate by the Strategic Initiative for Ocean Energy in 2013 is for ~8 US cent kWh^{-1} when global capacity is 50 GW, but currently devices are still under development.

6.27 **Outlook for Wave Power**

It is likely to take one or two more decades to gather sufficient operational experience for wave power to compete with other alternative energy technologies, particularly given the low cost of solar PV and wind. But in several parts of the world, e.g. Australia, it is an excellent resource and one that could make a significant contribution provided there is sufficient investment for the price to fall through learning. The potential contribution from wave power by 2050 is estimated at 150 GW installed capacity, which at 40% capacity factor would deliver around 500 TWh per year. In the longer term, an estimate of an eventual global potential for wave power to provide a few per cent of total electricity production does not seem unreasonable, as part of a diverse mix of alternative energy sources.

 SUMMARY

- Hydroelectric installations have a high capital cost but low operational costs. Large dams can provide relief from flooding, but their environmental impact can be a concern, and there can be significant social, safety, and economic issues.

- Hydropower is a large resource of low-carbon energy. The global technical potential from hydro is estimated to be 15,000 TWh, and the global installed capacity by 2050 ~2000 GW, equivalent to about 7000 TWh per year.

- Tidal power is an underdeveloped technology, mainly due to its high capital cost and environmental impact, but tidal stream arrays are looking attractive in areas of good resource. The global technical potential of tidal power is around 1200 TWh y^{-1} with the potential contribution by 2050 estimated to be 500 TWh.

- Wave power is a large natural resource with tens of kilowatts per metre width of wavefront. The global resource is about 17,500 TWh per year, with a technical potential of some 4000 TWh y^{-1}. Oscillating water columns and submerged devices, which can be deployed much more widely, appear to be emerging as the preferred technologies. The potential contribution from wave power by 2050 is estimated to be 500 TWh y^{-1}.

 FURTHER READING

Farley, F.J.M. (2012). Far-field theory of wave power capture by oscillating systems. *Phil. Trans. R. Soc. A* 370: 278–287. Clear derivation of capture width.

Garrett, C. and Cummins, P. (2005). The power potential of tidal currents in channels. *Proc. R. Soc. A* 461: 2563–72. Excellent analysis of tidal stream energy extraction.

Gunn, K. and Stock-Williams, C. (2012). Quantifying the global wave power resource. *Renewable Energy* 44: 296–304.

IEA (2012). Hydropower Technology Roadmap (IEA2012).

IHA (2020). Hydropower Status Report.

IRENA (2019). Renewable Power Generation Costs.

Peake, S (ed.) (2018). *Renewable energy*, 4th edn. Oxford University Press, Oxford. Good qualitative description and case studies.

REN21 Renewables (2019). Global Status Report (GSR2019).

www.ashden.org Micro-hydro projects in developing countries.

theswitch.co.uk/energy/producers/swansea-tidal-power-lagoon Swansea Bay tidal lagoon proposal.

?   EXERCISES

6.1 Check the units to verify the expression $P = \eta\rho ghQ$ for the power output from a dam.

6.2 Estimate the power output of a dam with a head $h = 100$ m and volume flow rate $Q = 10$ m^3 s^{-1}. (Assume efficiency is unity, $\rho = 10^3$ kg m^{-3}, $g = 9.81$ m s^{-2}.)

6.3 Use Euler's turbine equation to estimate the maximum power output of a water turbine operating at 50 Hz such that

$$\rho = 10^3 \text{ kg m}^{-3}, Q = 10 \text{ m}^3 \text{ s}^{-1}, r_1 = 10 \text{ m}, q_1 = 5 \text{ m s}^{-1}, \cos\beta_1 = 0.4$$

6.4 Repeat Example 6.2 with $\beta_1 = 20°$, $q_0 = 6$ m s^{-1}, $q_1 = 32$ m s^{-1}, and $h = 100$ m.

6.5 Assuming the volume flow rate per unit width over a weir is of the form $Q = g^a y_{min}^b$, use dimensional analysis to determine the numerical values of a and b.

6.6* An abrupt change from h_1, u_1 to h_2, u_2 in the depth and speed of a rapidly flowing stream of water in a horizontal channel (a hydraulic jump) can occur when the initial flow speed u_1 is greater than the speed $(gh_1)^{1/2}$ of shallow water waves. Equate the retarding force arising from the mean pressure difference across the jump to the rate of change of momentum, use the conservation of mass flow, and deduce that

$$h_1 u_1^2 - (h_1^2 u_1^2 / h_2) = \frac{1}{2}g(h_2^2 - h_1^2).$$

Show that $h_2 / h_1 = \frac{1}{2}\left[\sqrt{1 + 8Fr^2} - 1\right]$ where the Froude number $Fr = u_1 / \sqrt{gh_1}$.

6.7* The energy flow in a stream of water of depth h and speed u is given by

$$E = \int_0^h \left(p + \frac{1}{2}\rho u^2 + \rho gy\right) u \, dy \text{ where } p = \rho g(h - y).$$ Show that the energy flow in a region

where the depth and speed are h_1 and u_1 is given by $\rho g u_1 h_1^2 + \frac{1}{2}\rho u_1^3 h_1$. Deduce that the energy lost across a hydraulic jump (Exercise 6.6) is given by $\rho g u_1 (h_2 - h_1)^3 / 4h_2$. Calculate the fractional energy loss when the Froude number for the initial (upstream) flow is 20.

6.8 Show that the shape of a waterfall is given by $y = \dfrac{gx^2}{2u^2}$, where u is the horizontal component of velocity and g is the acceleration due to gravity.

6.9 Draw a sketch of an impulse turbine consisting of four jets.

6.10 Verify that the power output of an impulse turbine is a maximum when $u_c = \frac{1}{2}u$, and that the maximum power delivered to the cup is given by $P_{max} = \frac{1}{2}\rho Q u^2$.

6.11 For a head of water h and a cross-sectional area A of the water jet, show that the maximum power P_m from a Pelton turbine can be written as $P_m \propto Ah^{3/2}$. Find the value of the constant of proportionality when P_m is in kW, A is in m^2, and h is in m.

6.12 Explain how a rotary lawn sprinkler works.

6.13 Discuss whether it is better to build a large number of small dams or one large dam.

6.14 Discuss the benefits of micro-hydro systems.

6.15 Discuss what factors should be considered when planning a new hydroelectric plant.

6.16 Discuss who should pay for the cost involved in decommissioning dams when they reach the end of their life.

6.17 Why are pumped-storage systems so important in mitigating carbon emissions?

6.18 A turbine is required to rotate at 6 rpm with a volume flow rate of $5\ m^3\ s^{-1}$ and a head of 30 m. What type of turbine would you choose?

6.19 The average annual rainfall over Asia is ~14 100 km^3 and its average altitude is 950 m. Estimate the theoretical hydropower potential.

6.20 Given that there are two high tides around the Earth at any instant, explain why the interval between successive high tides is 12 hours 25 minutes rather than 12 hours?

6.21 Compare the magnitude of the effect of the Sun on the tides (a) when the Sun and Moon are both on the same side of the Earth, and (b) when the Sun and the Moon are on opposite sides of the Earth ($m_{Sun} = 2 \times 10^{30}$ kg, $m_{Moon} = 7.4 \times 10^{22}$ kg, $d_{Sun} = 1.5 \times 10^{11}$ m, $d_{Moon} = 3.8 \times 10^{8}$ m).

6.22* For the wave shown in Fig. 6.21, the pressure below the surface at a point y above the bottom is approximately given by $p = p_0 + \rho g[h(x,t) - y]$

By considering the force acting per unit area on a slab of water between the planes x and $x + \delta x$, show that $\rho \dfrac{\partial u}{\partial t} = -\dfrac{\partial p}{\partial x} = -\rho g \dfrac{\partial h}{\partial x}$. By equating the difference in the volume per second flowing from x to $x + \delta x$ to the volume displaced per second in the vertical direction, show that $h_0 \delta u = -v \delta x$, and therefore that $h_0 \dfrac{\partial u}{\partial x} = -\dfrac{\partial h}{\partial t}$. Verify the wave equation $\dfrac{\partial^2 h}{\partial x^2} = \dfrac{1}{c^2} \dfrac{\partial^2 h}{\partial t^2}$, where $c = \sqrt{gh_0}$, is the wave speed or phase velocity.

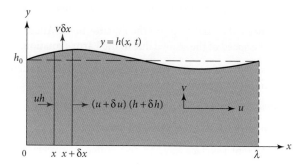

Fig. 6.21 Shallow water wave ($\lambda \gg h_0$).

6.23 Explain the phenomenon of tidal resonance and why it can cause a significant increase in tidal range.

6.24 The travelling wave solution to $\dfrac{\partial^2 h}{\partial x^2} = \dfrac{1}{c^2}\dfrac{\partial^2 h}{\partial t^2}$ is $h = h_0 + a\cos(x \pm ct)$, where a is the amplitude of the wave, with $\cos(x - ct)$ representing a wave travelling to the right and $\cos(x + ct)$ a wave to the left. Show that the amplitude of the velocity flow u_0 and of the travelling wave are related by $h_0 u_0 = ac$, and that the power P per unit width transmitted by a tidal wave, equal to the product of the energy E per unit width per unit length and the group velocity of the wave, is given by $P = \frac{1}{2}\rho g^{3/2}h_0^{1/2}a^2$.

6.25 A tidal barrage is filled during flood tide to a height h above low tide. The level is then raised by pumping to a height $(h + y)$ and held until near low tide, when the water within the barrage is discharged quickly through the turbines. Show that the average net power output is given by $P_{ave} = \rho g A[(h + y)^2 - y^2]/(2T)$. Evaluate the fractional increase in output power using a pump that raises the level from $h = 10$ m to 11 m compared with no pumping. Find the optimum value for y when the efficiency for pumping is half that for generating.

6.26* In a tidal basin power plant, seawater is trapped in a basin of area A at high tide and allowed to run out through a turbine at low tide. The difference in sea level between high and low tides, the tidal range R, varies sinusoidally throughout a month, from a maximum R_S for spring tides to a minimum R_N for neap tides. Show that the maximum mean power P_m produced over the month is $P_m = (\rho A g/4\tau)R_S^2[(3 + 2\alpha + 3\alpha^2)/8]$, where $R_N = \alpha R_S$, τ is the tidal period, g is the acceleration due to gravity, and ρ is the density of seawater. Calculate the mean power for $A = 12$ km^2, $R_S = 6$ m, $\alpha = 0.6$, $\rho = 10^3$ kg m^{-3}.

6.27 What are the advantages and disadvantages of tidal lagoon and tidal stream installations over tidal barrage plants?

6.28 The Pentland Firth is a channel 10.2 km wide with an average depth of 58 m. The maximum flow speed is 3.4 m s^{-1} and the amplitude of the tidal head driving this flow is 1.2 m. Calculate the maximum average power that could be generated by placing turbines in the channel.

6.29* (a) Show by substitution that the profile $h = h_0 + a\cos(kx - \omega t) + b\cos(kx + \omega t)$ satisfies the tidal wave equation $\dfrac{\partial^2 h}{\partial x^2} = \dfrac{1}{c^2}\dfrac{\partial^2 h}{\partial t^2}$.

(b) A uniform channel of length L is bounded at both ends by a vertical wall. Derive the height and velocity profiles of shallow water waves in the channel.

6.30 Show that the speed of a tidal bulge on the Equator in the Atlantic Ocean (depth ~4000 m) is less than the speed of the seabed due to the Earth's rotation.

6.31 Assuming the speed c of surface waves on water of uniform depth depends only on the acceleration due to gravity g, the depth h_0, and the wavelength λ, use dimensional analysis to show that $c = f(\lambda/h_0)\sqrt{gh_0}$. Find the form of $f(\lambda/h_0)$ when $\lambda \gg h_0$ (**shallow water wave**), and when $\lambda \ll h_0$ (**deep water wave**).

6.32 Calculate the speed of a surface wave on deep water of wavelength $\lambda = 100$ m.

6.33 Given that the phase velocity and group velocity of a surface wave are $c = \sqrt{\dfrac{g\lambda}{2\pi}}$ and $c_g = d\omega/dk$, respectively, where ω is the angular velocity and $k = 2\pi/\lambda$, prove that the group velocity is given by $c_g = \dfrac{1}{2}\sqrt{\dfrac{g\lambda}{2\pi}}$.

6.34 Consider a wave with a surface profile of the form $y = a \sin(2\pi x/\lambda)$, as shown in Fig. 6.22. By considering the work done in raising an element of water from $-y$ to y, show that the total potential energy $V = \dfrac{1}{4}\rho g a^2 \lambda$ of the elevated section of fluid.

6.35 Explain the advantages and disadvantages of the Salter Duck wave power device.

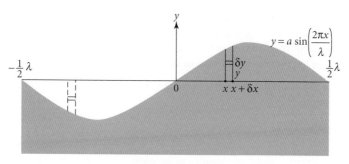

Fig. 6.22 Wave profile.

6.36* Fig. 6.23 shows the interference pattern between the ongoing incident and circular waves at a distance x downstream from a point absorber, when the absorption is greatest. The units are such that the amplitude and power of the incident wave per unit width are unity, and the power per unit width of the circular wave generated by the absorber at a distance x is $a^2\lambda/x$. Away from the incident direction, the oscillations in power per unit width largely cancel out, and it is as if the power per unit width was

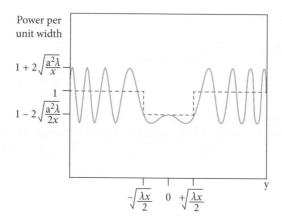

Fig. 6.23 The interference pattern between the ongoing incident and circular waves at a distance x downstream from the absorber, when the absorption is greatest.

given by the dashed line in Fig. 6.23. This has the value of $1 - 2(a^2\lambda/2x)^{1/2}$ over a width equal to $2(\lambda x/2)^{1/2}$.

Show that the total power absorbed is $(2a\lambda - 2\pi a^2\lambda)$, and that this has a maximum value, equal to the maximum capture width as the incident power per unit width is one, of $\lambda/2\pi$.

6.37 An OWC has an air duct of cross-sectional area 1 m^2 and a water duct of cross-sectional area 10 m^2. If the average vertical speed of the water surface is 1 m s^{-1}, calculate the average speed of the air.

6.38 Explain the principles of submerged wave energy converters.

6.39 The maximum width of a wave-front that could be absorbed by a Pelamis type wave energy generator is about $3/k$, where $k = 2\pi/\lambda$ and λ is the wavelength. Estimate the average power output of a wave farm containing 25 such devices moored in an area where the average wave amplitude is 1.0 m and the wave period is 8 s. Take the conversion efficiency of each Pelamis type device to be 70%. The density of water is 1000 kg m^{-3}.

6.40 Discuss the potential for wave power to provide significant amounts of low-carbon electricity.

Derivation 6.3 Equilibrium tidal theory

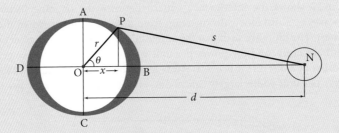

Fig. 6.24 Tidal effects due to the Moon (not to scale).

The force $F = Gm/d^2$, where m is the mass of the Moon, on a unit mass of water at the point P in Fig. 6.24, arising from the centripetal acceleration from the Earth–Moon's rotation, corresponds to a potential $\phi_{rot} = Gmx/d^2$, since the force $F = -d\phi/dx$ acts in the negative x direction. The potential arising from the gravitational attraction of the Moon is given by $\phi_{Moon} = -Gm/s$, where s is the distance from P to point N, the centre of the Moon. For $d \gg r$ we can expand $1/s$ as follows:

$$\frac{1}{s} = \frac{1}{\left[d^2 + r^2 - 2rd\cos\theta\right]^{\frac{1}{2}}} = \frac{1}{d}\left[1 + \left(\frac{r^2}{d^2} - \frac{2r}{d}\cos\theta\right)\right]^{-\frac{1}{2}}$$

$$= \frac{1}{d}[1 + \frac{r}{d}\cos\theta + \frac{r^2}{d^2}(\frac{3}{2}\cos^2\theta - \frac{1}{2}) + \ldots]$$

The potential of unit mass of water due to the Earth's gravitation is $\phi_{Earth} = gh$, where h is the height of the water above its equilibrium level and g is the acceleration due to gravity at the Earth's surface. The total potential of the unit mass of water is given by $\phi = \phi_{rot} + \phi_{Moon} + \phi_{Earth}$. Since $x = r\cos\theta$ and $g = GM/r^2$, where M is the mass of the Earth, and $\phi_{Earth} = GMh/r^2$, ϕ reduces to:

$$\phi = -\frac{Gm}{d}\left[1 + \frac{r^2}{d^2}\left(\frac{3}{2}\cos^2\theta - \frac{1}{2}\right) - \frac{h(\theta)Md}{mr^2}\right]$$

When in equilibrium the surface profile of the water is an equipotential surface, so h must be a function of θ such that:

$$h(\theta) = \frac{mr^4}{Md^3}\left(\frac{3}{2}\cos^2\theta - \frac{1}{2}\right) = h_{max}\left(\frac{3}{2}\cos^2\theta - \frac{1}{2}\right)$$

where $h_{max} = mr^4/Md^3$ is the maximum height of the tide, which occurs at points B and D ($\theta = 0$ and $\theta = \pi$). The tidal range is $3h_{max}/2$. Putting $\frac{m}{M} = 0.0123$, $d = 384\,400$ km, and $r = 6378$ km, we obtain $h_{max} \approx 0.36$ m, and a range of ~0.5 m, which is roughly in line with the observed mean mid-ocean tidal range.

For further information and resources visit the online resources
www.oup.com/he/andrews_jelley4e

7 Wind Power

✔ List of Topics

- ☐ Source of wind energy
- ☐ Global wind patterns
- ☐ History of wind power
- ☐ Modern wind turbines
- ☐ Energy in the wind
- ☐ Maximum efficiency
- ☐ Turbine blade design
- ☐ Optimum tip-speed
- ☐ Power coefficient C_p
- ☐ Design of modern HAWTs
- ☐ Fatigue in wind turbines
- ☐ Control and operation

- ☐ Turbine generators
- ☐ Wind characteristics
- ☐ Power output of turbines
- ☐ Onshore wind farms
- ☐ Offshore wind farms
- ☐ Hybrid wind farms
- ☐ Wind technology developments
- ☐ Environmental impact
- ☐ Public acceptance
- ☐ Economics
- ☐ Wind variability
- ☐ Wind potential and outlook

→ Introduction

The international oil crises of the 1970s and, more recently, concern over global warming have renewed interest in wind power. The global installed capacity has grown at an average annual rate of 15% since 2010 to 651 GW in 2019. In 2019 the wind capacity of 651 GW generated 5.3% of the global electricity production of 26,771 TWh.

The wind is a source of carbon-free and pollution-free energy, and wind power could produce ~35% of the electrical power used globally by 2050 (IRENA2019). Wind power would therefore avoid a considerable amount of fossil fuel use. The modern wind turbine is some 100 times more powerful than the traditional windmills of the seventeenth and eighteenth centuries, and wind farms already generate significant amounts of electricity in several countries, e.g. 47% of Denmark's and 21% of the UK's electricity demand.

We first describe the global wind resources, the energy available in the wind, and how this energy is extracted by a wind turbine. We then consider the issues associated with the siting of turbines, generally as wind farms, that are important for both their output and their environmental impact. We conclude the chapter with a discussion of wind variability and of the economics and potential of wind power.

7.1 **Source of Wind Energy**

The original source of wind energy is radiation from the Sun, which is primarily absorbed by the land and the sea, which in turn heat the surrounding air. Materials absorb radiation differently, so temperature gradients arise causing convection and pressure changes, which result in winds. A simple example is the offshore night-time wind often found on coasts, caused by the sea retaining the heat from the Sun better than the land. On a global scale, the higher intensity of solar radiation at the Equator than elsewhere causes warm air to rise up and cooler air to flow in from the north and south. The direction of a wind is traditionally described in terms of where it comes from, so in the northern hemisphere the warm air rising up from the Equator would give rise to a northerly wind at ground level.

An enormous amount of power resides in the winds, as about 0.5% of the incident solar power of 1.37 kW m^{-2} is converted into wind. The radius of the Earth is approximately 6400 km, so the cross-sectional area receiving solar radiation is about 1.3×10^{14} m^2 and the power in the winds is $\sim 10^6$ GW. This is some 50 times the total global power usage. However, the wind is a diffuse source and it is only practical to harness a very small fraction of this amount (see Section 7.18).

Winds are variable both in time and in location, with some parts of the world exposed to frequent high winds and some to almost no wind. Places where high and low winds occur are, in particular, determined by the effect of the rotation of the Earth. Over distances of tens of kilometres, the Earth's rotation has no significant effect on the direction of a wind; however, over hundreds of kilometres the effect is very noticeable. We will now explain how the Earth's rotation affects the global winds.

7.2 **Global Wind Patterns**

In a simple model, the higher intensity of solar radiation at the Equator would set up a north–south convective flow of air if the Earth were not rotating. However, the Earth's rotation causes a point on the Equator to have a velocity towards the east that is highest at the Equator, decreasing towards the poles. Therefore, a wind moving north or south as seen by an observer on the Equator will initially have a component of velocity towards the east to an observer in space. As the wind moves away from the Equator its distance to the Earth's axis decreases, so its component of velocity towards the east increases. (This is a similar effect to ice skaters spinning faster when their arms are pulled in.) Air initially moving north will therefore reach a northern latitude at a point which is east of its origin. For the observer on Earth the wind appears to be accelerating towards the east, and the apparent force is called the **Coriolis force**.

The wind speed would in principle reach large values by high latitudes, but by latitude 30° the flow becomes unstable. As a result, the north or south motion of the wind is dissipated, and such winds are thereby restricted to within the 30° latitudes. In the northern hemisphere the sinking air near 30° latitude gives rise to the north-east trade winds and the westerly wind belt, which is the prevailing wind over Europe. A map of the resulting global wind patterns expected in this simple model, which ignores the effects of the underlying configuration of

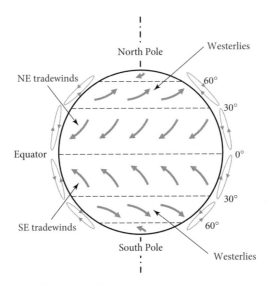

Fig. 7.1 Simplified map of global wind patterns.

oceans and continents, is shown in Fig. 7.1. Notice that there are three regions, called **cells**, in each hemisphere.

In practice, the effects of surface friction and large-scale eddy motions have a big influence, as do seasonal variations, and only the cell nearest the Equator, the **Hadley cell**, is clearly seen. The mid-latitude **Ferrel cell** is quite weak, and the **'polar' cell** is hardly observed. There are many areas where the winds are strong and reliable, and it is in these locations that the energy in the wind can be best exploited, and has been in some places for many centuries (see Box 7.1).

Box 7.1 History of wind power

The first recorded use of wind power was in the tenth century in the Sistan region of Persia, an area where winds can reach speeds of ~45 m s^{-1}, though windmills had probably already been in use in the region for several centuries. The windmills had a vertical axis (see Fig. 7.2 (a)), and were used to pump water and grind grain. Similar windmills were used in China and may have been first developed there. In these vertical-axis machines the wind pushes the sails around with a force, called a **drag** force, dependent on the relative speed of the wind and the sail.

Vertical-axis windmills using a drag force are inherently less efficient than horizontal-axis windmills, which are driven by a **lift** force (see Section 7.6). Horizontal-axis windmills first appeared in Europe in England, France, and Holland around the twelfth century. Their origin is unknown (though it is possible they evolved from the horizontal-axis waterwheel), but such windmills spread rapidly eastward in Europe in the thirteenth century. They were used for grinding corn, pumping water, and sawing wood. The first mills were **post-mills**, where the whole mill swivelled on a post so that it could be manually turned to face the oncoming wind. In later (and larger) mills only the

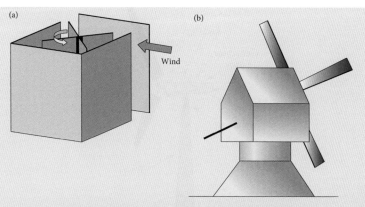

Fig. 7.2 (a) Persian vertical-axis windmill (b) Tower-mill.

top with the sails and windshaft swivelled; these so-called **tower-mills** (Fig. 7.2 (b)) were introduced around the fourteenth century. From experience it was found that more power could be obtained by *twisting* the sails from the root to the tip of the sail.

The use of windmills peaked around the eighteenth century, after which they were displaced by coal-powered steam engines, which were more compact and adaptable and were continuously available. However, windmills continued in regions where the land was sparsely populated, e.g. in the USA, USSR, Australia, and Argentina. In the nineteenth century, small multi-vane windmills were developed in the USA for pumping water, where they became very common. They were eventually displaced with the development of a national electricity grid in the 1930s.

From the late nineteenth century up to the 1960s a number of wind machines were developed for generating electricity and were called **wind turbines** (to distinguish them from windmills). In the early twentieth century, Poul La Cour built turbines using a four-bladed windmill design that produced about 25 kW. The electricity was used to produce hydrogen, which was then used for lighting. These were subsequently displaced by the introduction of diesel engines, but the production of hydrogen as a fuel is now being seriously considered, because hydrogen produces no CO_2 when in use (see Section 11.4.3 on fuel cells). In the late 1930s a massive 1.25 MW two-bladed wind turbine was proposed by Palmer Putman and built in Vermont, USA, by the Morgan Smith Company. Although it ran successfully for a short while, a blade failure in 1945 caused the project to be terminated.

During the Second World War, Denmark used wind energy when oil was not available, though this was only a temporary measure. However, during the international oil crisis in 1973 there was renewed interest in wind power as many countries began looking at sources of alternative energy. In California, concern over high fossil fuel costs led to large-scale investment in wind farms, which was aided by state and federal tax incentives. The technology, though, was not then well developed, and several wind farms were unsuccessful. With the cessation of tax incentives and the fall in oil prices in the mid-1980s, investment in wind power in the USA declined. However, in Europe, particularly in Denmark, support was maintained, and the more recent alarm about global warming has stimulated considerably more interest and investment in wind turbines worldwide.

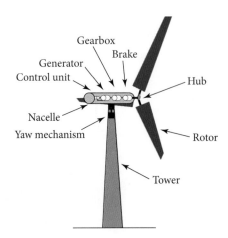

Fig. 7.3 Modern 5 MW horizontal axis wind turbine.

7.3 **Modern Wind Turbines**

7.3.1 **Horizontal-Axis Wind Turbines (HAWTs)**

The vast majority of current designs are HAWTs. The turbine blades are aerofoils, which provide lift forces that drive the turbine. A modern HAWT is illustrated in Fig. 7.3. It consists of a tower on top of which is mounted an enclosure called the **nacelle**, which contains the bearings for the turbine shaft, the gearbox (if used), and the generator. The wind turbine blades (generally three) are mounted on the shaft, and the nacelle is oriented by a drive mechanism—the yaw control—into the wind. The rotor is typically upwind of the tower, in order to prevent the tower from shielding the blades from the wind. Having three blades on a turbine gives a balanced rotation and good performance—having only one blade would be unbalanced, while two give a disconcerting flickering effect as the blades pass the tower, and the marginally improved output with four or more blades is not worth the extra cost.

7.3.2 **Vertical-Axis Wind Turbines (VAWTs)**

There has also been some research and development of VAWTs, in particular the **Darrieus** design shown in Fig. 7.4. In this design the turbine is driven by lift forces generated by the wind flow over the aerofoil-shaped blades. The torque is a maximum when the blades are moving across the direction of wind flow, and zero when the blades are moving parallel to the direction of wind flow (see Exercise 7.10). Unlike a HAWT, the blades are not correctly orientated all the time, which reduces the effective lift-to-drag ratio and the optimal rotational speed (see Section 7.7), which makes for a higher torque for the same power output; as a result, the generator has to be larger and hence more expensive.

On the other hand, VAWTs do not require any yaw mechanism and can be easier to maintain, since the gearbox and generator can be situated at ground level. However, for large machines they have proven to be less cost-effective than HAWTs. Furthermore, in the

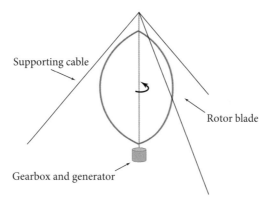

Fig. 7.4 Darrieus vertical-axis wind turbine.

Darrieus design (Fig. 7.4) for a large turbine, cables are required to anchor the top of the rotor, and this limits the mean height of the rotors and loses the advantage of the stronger winds that occur at greater heights. For smaller Darrieus type VAWTs, they are usually taller and narrower than a HAWT of the same output, so there are larger moments on the gearbox and more material is required, and as a result can be more expensive. Small VAWTs, however, can be effective in urban locations where they work well in updrafts and in all wind directions.

We will concentrate on HAWTs, since they are widely used for power generation by electric utilities. We first consider how much energy there is in the wind, and then how HAWTs extract the energy of the wind.

7.4 Kinetic Energy of Wind

The energy of wind is in the form of **kinetic energy**. For a wind speed u and air density ρ, the energy density $\mathcal{E}$ (i.e. energy per unit volume) is given by:

$$\mathcal{E} = \frac{1}{2}\rho u^2 \tag{7.1}$$

The volume of air flowing per second through a cross-sectional area A normal to the direction of the wind is uA. Hence the kinetic energy per second of the volume of air flowing through this area is given by $P = \mathcal{E}uA$, or:

$$P = \frac{1}{2}A\rho u^3 \tag{7.2}$$

Thus, the **power of wind P varies as the cube of the wind speed u**. Hence much more power is available at higher speeds. However, as we explain later, fluctuations in wind speed can cause the output of a wind turbine to vary significantly. A typical wind speed variation is described by the Rayleigh distribution, which is discussed in Sections 7.10 and 7.11.

EXAMPLE 7.1

Calculate the power of the wind moving with speed $u = 5$ m s^{-1} incident on a wind turbine with blades of 100 m diameter. How does the power change if the wind speed increases to $u = 10$ m s^{-1}? (Assume the density of air is 1.2 kg m^{-3}.)
Substituting in eqn (7.2) we have:

$$P = \frac{1}{2} A \rho u^3 = \frac{1}{2} \times (\pi \times 50^2) \times 1.2 \times 5^3 \approx 0.6 \text{ MW}$$

A power of 0.6 MW is sufficient to meet the average electricity usage of about 1000 European households. Doubling the wind speed increases the power by a factor of $2^3 = 8$, so the power would increase to ~8 × 0.6 = 4.8 MW.

7.5 Principles of a Horizontal-Axis Wind Turbine

A wind passing through a turbine loses speed as it pushes the blades round, as power is extracted, which means the wind is slower downwind. The blades are sufficiently wide to extract a good fraction of the wind's power, but not so large that they act as a significant obstruction, as most of the wind would then be diverted around the turbine by the pressure of the air in front of the blades and not push the blades round. The pressure drop Δp across the turbine provides the driving force, and the power P extracted is given by:

$$P = \Delta p A_1 u_1 \tag{7.3}$$

where A_1 is the cross-sectional area of the turbine and u_1 is the speed of the wind at the plane of the turbine. Maximizing the product of pressure drop and wind speed imposes a theoretical maximum efficiency of 16/27 (~59%) for extracting power from the wind, known as the **Betz limit**, which is described in detail in Derivation 7.1.

As the wind flows through a turbine it slows down as some of its energy is transferred to the turbine. The air flow looks like that shown in Fig. 7.5. Upstream, the speed of the wind is u_0 and it passes through an area A_0. By the time the wind reaches the turbine it slows to u_1 and the area of the stream-tube increases to A_1, i.e. the area swept out by the blades of the turbine. Downstream of the turbine the wind's cross-sectional area is A_2 and its speed is u_2. The drop in speed of the wind passing through the turbine, from $u_0 \rightarrow u_1$ and $u_1 \rightarrow u_2$, causes a pressure drop Δp across the turbine (see Bernoulli's principle: eqn (2.42) and Exercise 7.3), which produces a thrust on the turbine blades. The value of Δp is:

$$\Delta p = \frac{1}{2} \rho (u_0^2 - u_2^2) \tag{7.4}$$

The maximum power is generated when downstream of the turbine the wind speed is one-third of the upstream speed u_0 and at the turbine the wind speed is two-thirds of u_0; i.e.

$u_2 = u_0/3$ and $u_1 = 2u_0/3$ (see Derivation 7.1). Under these conditions the power P extracted, using eqns (7.3) and (7.4), is given by:

$$P = \frac{1}{2}\rho A_1 (16/27) u_0^3 \tag{7.5}$$

The power P_w in the wind passing through an area A_1 with a speed u_0 is given by eqn (7.2) as:

$$P_w = \frac{1}{2}\rho A_1 u_0^3$$

Hence, the fraction of the power extracted by the turbine, called the **power coefficient**, C_P, is given by:

$$C_p = P/(\tfrac{1}{2}\rho u_0^3 A_1) \text{ or } P = \tfrac{1}{2}C_p \rho u_0^3 A_1 \tag{7.6}$$

and is $16/27 \approx 59\%$ of the power in the incident wind passing freely through an area equal to that of the turbine, A_1. This limit for the power coefficient C_P of $16/27$ of the incident wind power is usually called the **Betz limit**, after the German physicist who derived it in 1920.

Derivation 7.1 Maximum extraction efficiency

We can obtain an estimate of the maximum efficiency by modelling the turbine as a thin disc (called an **actuator** disc) that extracts energy. Consider the stream-tube of air (see Chapter 2 Section 2.9) shown in Fig. 7.5, that moves with speed u_1 through a wind turbine of cross-sectional area A_1. Upstream of the turbine the cross-sectional area of the stream-tube is A_0 and the air speed is u_0. Downstream of the turbine the cross-sectional area of the stream-tube is A_2 and the air speed is u_2.

Since the turbine extracts energy from the wind, the air speed decreases as it passes through the turbine and the cross-sectional area of the stream-tube increases, as shown in Fig. 7.5. The thrust T exerted on the turbine by the wind is equal to the rate of change of momentum, given by:

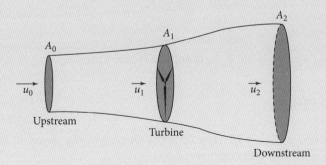

Fig 7.5 Air flow passing through a wind turbine.

$$T = \frac{dm}{dt}(u_0 - u_2) \tag{7.7}$$

where dm/dt is the mass of wind flowing through the stream-tube per second.

The power P extracted is given by the product of the thrust and the air speed at the turbine, u_1, so that:

$$P = Tu_1 = \frac{dm}{dt}(u_0 - u_2)u_1 \tag{7.8}$$

We can also express the power extracted as the rate of loss of kinetic energy of the wind, i.e.

$$P = \frac{1}{2}\frac{dm}{dt}(u_0^2 - u_2^2) \tag{7.9}$$

Comparing eqns (7.8) and (7.9), we require:

$$(u_0 - u_2)u_1 = \frac{1}{2}(u_0^2 - u_2^2) = \frac{1}{2}(u_0 - u_2)(u_0 + u_2)$$

Hence:

$$u_1 = \frac{1}{2}(u_0 + u_2), \text{ or } u_2 = 2u_1 - u_0 \tag{7.10}$$

Also, by mass continuity (see Section 2.9.1), the mass flow per second, dm/dt, passing through the turbine is constant, and given by:

$$\frac{dm}{dt} = \rho uA = \rho u_1 A_1 \tag{7.11}$$

(Note that the changes in pressure are sufficiently small that the density of air ρ is essentially constant; see Example 2.6.)

Substituting for u_2 from eqn (7.10) and for dm/dt from eqn (7.11) in eqn (7.8) yields:

$$P = 2\rho u_1^2 A_1 (u_0 - u_1) \tag{7.12}$$

Let $u_1 = (1-a)u_0$, where a is called the **induction factor**. Then:

$$P = \frac{1}{2}\rho u_0^3 A_1 [4a(1-a)^2] \tag{7.13}$$

The power coefficient C_P, which represents the fraction of the power in the wind that is extracted by the turbine, is given by:

$$C_P = P/(\frac{1}{2}\rho u_0^3 A_1) = 4a(1-a)^2$$

Maximizing P by setting $\mathrm{d}C_P/\mathrm{d}a$ to zero gives the maximum power extracted P_{max} when $a = \frac{1}{3}$; that is,

$$P_{max} = \frac{1}{2}\rho u_0^3 A_1 [16/27]$$

which is ~59% of the power in the incident wind passing freely through an area equal to that of the turbine, A_1.

7.6 Wind Turbine Blade Design

The thrust on the turbine is converted into rotational energy by shaping the turbine blades as aerofoils. A wind turbine is shown in Fig. 7.6 (a), and a section of a blade is shown in Fig. 7.6 (b). The air speed at the turbine is u_1 and the rotational speed of the blade is v (i.e. perpendicular to the direction of air flow). The resultant velocity u_α of the air *relative to the blade* makes an angle ϕ to the direction of the blade, given by:

$$\tan\phi = u_1/v \tag{7.14}$$

The angle of attack of the wind on the blade is α_0. The motion of the wind over the aerofoil section gives rise to a lift force $\mathcal{L}$. The lift $\mathcal{L}$ of an aerofoil is perpendicular to the direction u_α of the air flow, so the thrust T on the aerofoil (neglecting drag $\mathcal{D}$) is given by $\mathcal{L}\cos\phi$ and

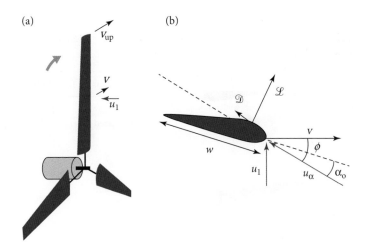

Fig. 7.6 (a) Wind incident on rotating turbine (b) Section of blade at distance r from hub.

the power P developed by $\mathcal{L}(\sin\phi)v$ (force multiplied by velocity). Using eqn (7.14) to substitute for v gives:

$$P = \mathcal{L}(\sin\phi)u_1 \cot\phi = \mathcal{L}(\cos\phi)u_1 = Tu_1 \qquad (7.15)$$

showing that the power developed equals that delivered by the thrust of the wind, when drag is neglected.

The speed v of the blade at a radius r is given by:

$$v = rv_{tip}/R \qquad (7.16)$$

where v_{tip} is the speed of the blade tip and R is the maximum radius of the turbine blade. An important parameter is the **tip-speed ratio** λ, defined as the ratio of the speed v_{tip} of the blade at the tip to the speed u_0 of the incident wind, i.e.

$$\lambda = v_{tip}/u_0 \qquad (7.17)$$

When the Betz condition ($u_1 = 2u_0/3$) is satisfied, then using eqns (7.16) and (7.17) we can express the angle ϕ in terms of r, R, and λ through:

$$\tan\phi = u_1/v = 2R/(3r\lambda) \qquad (7.18)$$

From eqn (7.18) we observe that, for a given radius r, the angle ϕ is only a function of the tip-speed ratio λ, which is why λ is so significant in the design of a wind turbine. The apparent angle of the wind ϕ increases with decreasing radius r (since $\tan\phi \propto 1/r$). Turbine blades are therefore designed with a twist that increases with decreasing r in order for the angle of attack α_0 to remain optimum, and the blade width, w, also increases with decreasing r so that the component of lift generates the thrust required to maintain the Betz condition (see Fig. 7.7).

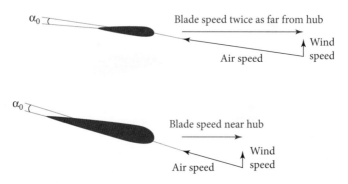

Fig. 7.7 Blade width and twist near hub and twice as far from hub.

Fig. 7.8 Transporting an 88.4 m blade for an 8 MW offshore turbine (*Source*: Courtesy of LM Wind Power).

The calculation of the width of a turbine blade is given in Derivation 7.2. The results for the width and angle of twist of a blade on a turbine with n blades are:

$$\text{Angle of twist} = \phi - \alpha_0 = \tan^{-1}[2R/(3r\lambda)] - \alpha_0 \tag{7.19}$$

$$\text{Width} = w \approx 16\pi R^2/9nr\lambda^2 \tag{7.20}$$

Blades for offshore wind turbines are now enormous—Fig 7.8 shows an 88.4 m long blade being transported by road.

Making the blades flexible is under investigation. This can reduce their weight and cost, and can also reduce the loading on the turbine by reducing the swept area at high wind speeds. Variable length blades are also being considered.

The percentage of the area swept out by the blades that is obstructed by their width—known as the **solidity** of the blades—is ~5%, which makes blades a remarkably cost-effective means of extracting the energy in the wind.

EXAMPLE 7.2

A wind turbine with three blades is operating in a mean wind speed of 11 m s^{-1}. The turbine rotates at 12 rpm. Each blade is 80 m long. Estimate the width at the midpoint and tip of each blade.

The time τ for one revolution of the tip of a blade of length R is

$$\tau = 2\pi R / v_{\text{tip}}$$

so the number n_{rpm} of revolutions per minute (rpm) is

$$n_{rpm} = 60/\tau$$

Therefore the tip speed v_{tip} and tip-speed ratio $\lambda = v_{tip}/u_0$ are

$$v_{tip} = 2\pi R/\tau = 2\pi R n_{rpm}/60 = 2\pi(80)(12)/60 = 100.5 \text{ m s}^{-1}$$

$$\lambda = 100.5/11 = 9.1$$

Using eqn (7.20) gives an estimate of the width of the blade tip:

$$w_{tip} = 16\pi R/(9\lambda^2 n) = 16\pi 80/[9(9.1)^2(3)] = 1.80 \text{ m}$$

At the midpoint, eqn (7.20) shows that the width is twice that at the tip, so

$$w_{mid} = 3.6 \text{ m}$$

Derivation 7.2 Blade design

The effect of a turbine's blade on the wind flow extends over a sufficient distance in the direction of the lift that the blade's reaction on the wind occurs over a significant part of the whole annular area, $dA = 2\pi r dr$ (see Exercise 7.5). The result is that the variation in wind speed across the area swept by the blades of the turbine is relatively small, and u_1 is taken to be the average wind speed at the turbine.

The total thrust dT_n on the annular area dA from n blades equals the rate of change of momentum of the wind. From eqns (7.7) and (7.11) we have:

$$dT_n = (\rho dA u_1)(u_0 - u_2) = \rho u_1^2 2\pi r dr \qquad (7.21)$$

when the Betz condition ($u_2 = u_1/2 = u_0/3$) is satisfied.

This thrust is equal to the sum of the components of the lift $\mathcal{L}$ from each section of blade between r and $r + dr$. (We neglect the effect of drag in this discussion.) The lift $d\mathcal{L}$ is given by (see eqn (2.58))

$$d\mathcal{L} = \frac{1}{2}C_L \rho u_\alpha^2 w dr \qquad (7.22)$$

where α is the angle of attack and w is the width (or chord) of the blade at r. From Fig. 7.6 it can be seen that the speed $u_\alpha = u_1/\sin \phi$. Equating the thrust to the sum of the components of lift and substituting for u_α gives

$$\rho u_1^2 2\pi r dr = nd\mathcal{L}\cos\phi = \frac{1}{2}nC_L\rho u_1^2 w dr \cos\phi/\sin^2\phi \qquad (7.23)$$

In order to satisfy the Betz condition (which maximizes the efficiency of the turbine) the width w of the blade at a distance r from the axis must therefore be given by:

$$w = 4\pi r \tan\phi \sin\phi/(nC_L) \qquad (7.24)$$

The tip-speed ratio $\lambda = v_{tip}/u_0$, so $\tan\phi = u_1/v = 2R/(3r\lambda)$. Substituting for $\tan\phi$ yields:

$$w = 8\pi R \sin\phi/(3\lambda nC_L) \approx 16\pi R^2/9nr\lambda^2 \qquad (7.25)$$

The approximation assumes that $C_L \approx 1$ when the blades are optimal, and that $\sin\phi \approx \tan\phi$, which for $\lambda \sim 10$ and $n = 3$ is accurate to 5% for $r/R > 0.21$.

From eqn (7.20) the width of the blade is approximately proportional to $1/r$, so increases from the tip to the root of the blade (see Figs. 7.6 and 7.7).

The angle of attack α is chosen to give the highest lift-to-drag ratio C_L/C_D and is typically a few degrees. For a modern three-blade wind turbine, λ is often chosen to be about 9. An aerofoil generally has C_L close to 1 at the angle of attack α_0 with the largest C_L/C_D ratio. For $\alpha_0 = 6°$ and $C_L = 1$, the width w of the blade and the shape and angle of twist of the optimum blade of length $R = 60$ m are given in Table 7.1 for a tip-speed ratio $\lambda = 9$. These are calculated using eqns (7.20) and (7.19). For the section of blade near the shaft where $r/R < 0.15$, the area swept out by the blades is less than 2.5% of the total area, and the blades are shaped to give strength and a smooth transition into the hub shaft.

In practice, blade design takes account of drag as well as lift, and the effects of all the sections are calculated using computer programs (blade element theory).

Table 7.1 Width and twist for an optimum turbine blade of length 60 m with $\lambda = 9$, $\alpha_0 = 6^0$

Radius (m)	Width w (m)	Angle (°) Twist $\phi - \alpha_0$	Angle (°) Wind ϕ
15	5.52	10.5	16.5
30	2.76	2.4	8.4
45	1.84	−0.4	5.6
60	1.38	−1.8	4.2

7.7 Dependence of the Tip-speed Ratio for Maximum Power Extraction on the Lift-to-drag Ratio

In the discussion in Section 7.6 of the power extracted P by a wind turbine, we neglected the effect of drag $\mathcal{D}$. This force acts in the direction of the wind at right angles to the lift $\mathcal{L}$

as shown in Fig. 7.6. Although the drag is small compared to the lift, its direction is nearly opposite to that of the blade's motion. Consequently, the effect of the drag is enhanced by a factor of $\sim\lambda$, i.e. the tip-speed ratio, as we will see, and P decreases with increasing λ. For small λ, the energy in the swirling motion of the wake is appreciable and P is less. So, there is an optimum value of λ.

As a result of the drag, the rotational force is reduced and becomes:

$$\mathcal{L}\sin\phi - \mathcal{D}\cos\phi = \mathcal{L}\sin\phi(1 - \frac{1}{k}\cot\phi) \tag{7.26}$$

where $k \equiv C_L/C_D$ is the lift-to-drag ratio of the blade's aerofoil shape. From eqn (7.18)

$$\cot\phi = 3r\lambda/(2R)$$

so the reduction varies with radius r. We can estimate the effective reduction as $(1 - \lambda/k)$ by taking a typical radius r as $2R/3$ so $\cot\phi \sim \lambda$. There is a further reduction for small lambda caused by the swirling motion of the wake (see Derivation 7.3 at the end of the chapter). The overall effect on the power coefficient C_p is given roughly by:

$$C_p \approx [(1-\lambda/k)/(1+5/\lambda^2)]C_{\text{PBetz}} \tag{7.27}$$

and the optimum value of λ is approximately given by:

$$\lambda_{\text{opt}} \approx L - 5/L \text{ where } L = (10k)^{\frac{1}{3}} \tag{7.28}$$

For a modern wind turbine $k \sim 100$, $\lambda_{\text{opt}} \sim 9.5$, and putting $C_{\text{PBetz}} = 16/27$ yields $C_{\text{Pmax}} \approx 0.51$.

EXAMPLE 7.3

A wind turbine is operating in a mean wind speed of 11 m s^{-1}. It rotates at a speed of 15 rpm. Each blade is 65 m long and has an aerofoil section with a lift-to-drag ratio of 100. The blades are optimized for the mean wind speed. Estimate the maximum power coefficient and the power output at the mean wind speed. (Assume the density of air is 1.2 kg m^{-3}.)
From Example 7.2:

$$\lambda = \frac{2\pi R n_{\text{rpm}}}{60u_0} = 2\pi \times 65 \times \frac{15}{60 \times 11} = 9.3$$

substituting in eqn (7.27) with $k = 100$ gives:

$$C_{\text{Pmax}} = 0.51$$

substituting in eqn (7.5):

$$P = \frac{1}{2}C_P\rho u_0^3 A_{\text{turbine}} = \frac{1}{2}(0.51)(1.2)(11^3)\pi(65)^2 \approx 5.41\,\text{MW}$$

7.8 Dependence of the Power Coefficient C_P on the Tip-speed Ratio λ

Fig. 7.9 shows a plot of the variation of C_P with λ for a turbine designed to have maximum efficiency at $\lambda_{\text{opt}} \sim 10$. The width w and twist of the turbine's blades are designed for this tip-speed ratio, so if the wind speed or the rotational angular velocity of the turbine alters, then ϕ and therefore the lift $\mathcal{L}$ and thrust change from their optimal values, and C_P decreases. So, we expect C_P to fall away on either side of its maximum value, as in Fig. 7.9. At low λ the blade stalls (see Section 7.9), and at $\lambda = 0$ no power is extracted. Above $\lambda \approx 13$ corresponds to the induction factor (see Derivation 7.1) $a > \frac{1}{2}$ and is where using the momentum change to derive the thrust on the turbine (eqn (7.7)) is invalid, as it corresponds to a negative final velocity. In this region the thrust rises rather than falls, and corresponds to a turbulent wake behind the turbine blades. Although the thrust T continues to rise in this region, while the speed of the wind u_1 at the turbine decreases, the power imparted to the turbine blades Tu_1 continues to fall. At very high rotational speeds, i.e. very large λ, the turbine blades act like a solid disk deflecting the airflow around the turbine and no power is extracted.

Also shown in Fig. 7.9 is the theoretical maximum C_P curve, which for $\lambda \geq 4$ is close to the Betz limit. At lower λ the curve dips down, due the effect of the swirling motion of the wake (see Derivation 7.3).

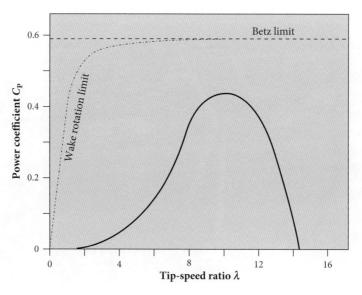

Fig 7.9 C_P–λ curve for high tip-speed rotor.

7.9 **Design of a Modern Horizontal-Axis Wind Turbine**

A modern HAWT is illustrated in Fig. 7.3. Turbines are designed with a large tip-speed ratio λ of ~ 9 to give a higher power coefficient C_P. A larger λ also means a higher shaft speed and hence a lower torque. Lowering the torque allows the use of a smaller gearbox (if used) or smaller generator. Turbines with a large λ also have smaller width blades (see eqn (7.20)), so less blade material is required, which cuts costs. Ensuring that the blades are each wide enough to have sufficient strength is another reason why there are not more than three blades on a modern large turbine. Even then, starting may be difficult, but this problem can be overcome by using a starting motor.

Blades were originally made from wood, and aluminium and steel were employed in later designs. Nowadays though, fibreglass and other composite materials are used due to their high strength and stiffness coupled with low density. For example, for a modern large offshore turbine, Siemens has produced 75-metre-long blades from balsa wood and fibreglass reinforced with epoxy resin. Wood is a natural composite in which cellulose fibres are embedded in lignin. The glass fibres in the fibreglass give it high strength because they are thin enough to be without the flaws that normally weaken glass in bulk. The polymer that encloses them both transfers the loads to the fibres, and also protects them.

As blades are the highest cost component of a wind turbine, much research and development are directed on better composites; for example, combining glass and carbon fibres in a composite blade can give sufficiently improved performance to more than offset the increase in the cost of the blade (see Fig. 7.10).The fatigue properties of the materials used are very important due to the very large number of revolutions that a turbine makes over a 30-year design lifetime—typically of order 10^8.

Fig. 7.10 An 8 MW wind turbine with 88-metre-long carbon and glass fibre composite blades, in Bremerhaven, Germany, where it can provide power for about 10,000 homes (*Source*: Courtesy of LM Wind Power).

When a wire is bent repeatedly it tends to weaken and break. The strength of the wire decreases with the number of times it is bent—an effect known as **metal fatigue**. Likewise, the rotation of turbine blades causes the load experienced by each blade to fluctuate with time. This weakens the blade material after many cycles and has to be allowed for in the design. The fatigue of materials is discussed further in Box 7.2.

The nacelle containing the generator and control mechanisms is mounted on a tower. Most towers have a strong and, for economy, lightweight structure. As a result, their natural frequency of vibration lies below the rotational frequency of the blades. When starting or stopping, the shaft speed is therefore changed quickly to avoid shaking the tower. The size of wind turbines has increased a lot over the last 35 years and Table 7.2 gives typical specifications—the **rated power** is the maximum continuous output power. Fig. 7.11 shows the relative size of a large onshore and offshore turbine.

Table 7.2 Typical wind turbine sizes by date

Year	Rotor diameter (m)	Hub height (m)	Rated power (MW)
1985	20	30	0.08
1995	50	50	0.75
2005	75	80	1.5
2015	125	120	3.0
2020	165	140	9.0*
2025	220	150	13.0*

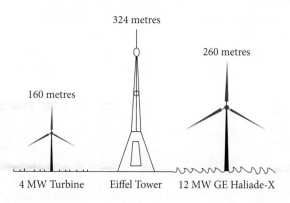

Fig. 7.11 Relative size of large onshore and offshore wind turbines.

Box 7.2 Fatigue in wind turbines

Wind turbine fatigue requirements are particularly severe because the number of load cycles is so high. For a 30-year lifetime, an 80 m diameter turbine operating for ~80% of the time at a $\lambda = 8$ in a wind speed of 10 m s^{-1} will make some 2.4×10^8 rotations. This means that the maximum stresses (force per unit area) must be lower than in other structures to avoid failure through fatigue; see Fig. 7.12.

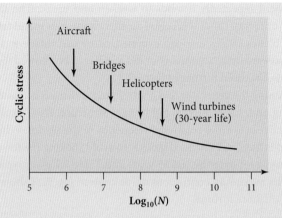

Fig 7.12 Cyclic stress versus $\log_{10}$(cycles to failure) (*Source*: Sand99).

Fatigue failure is the fracture of material after it has been subjected to repeated cycles of stress changes at levels considerably below its initial static strength. The number of cycles to failure decreases as the alternating stress level increases. (The stress level can be characterized by the mean stress or its range, which is equal to $\sigma_i^{max} - \sigma_i^{min}$). Fatigue involves the initiation and growth of cracks in a material under the repeated stress cycles. Discontinuities such as a sharp corner or flaws in the material are prime sites. Wind turbines installed in California in the 1980s suffered blade failures as a result of fatigue not being fully understood.

Fatigue can be quantified by using the Palmgren–Miner linear damage rule (often called **Miner's rule**). This method breaks down the cyclic stresses that a structure undergoes into the number of cycles n_i that occur at each stress level σ_i. The total damage D_M sustained by a structure is then obtained by summing the damage at each stress level, i.e.

$$D_M = \sum_{i=1}^{s}(n_i/N_i) = n_1/N_1 + n_2/N_2 + \cdots + n_s/N_s$$

where N_i is the number of cycles to failure at the stress level σ_i. Miner's rule states that failure will occur when $D_M = 1$, though factors of 2 are often found between predicted and measured lifetimes.

The fatigue strength of a material is the value of the stress level σ_{max} required to cause failure after a specified number of cycles N. The results can be expressed as an S versus N plot (S–N plot), where S is the ratio σ_{max}/σ_0. The stress σ_0 is the static strength of the material. The data can be represented by the equation:

$$\sigma_{max}/\sigma_0 = 1 - b\log_{10}(N) \tag{7.29}$$

where b is a positive constant.

Eqn (7.29) predicts how the stress that can be tolerated decreases with the number of cycles. The results for some fibreglass composite materials are shown in Fig. 7.13. (The data are for tensile stresses with a ratio $R = \sigma_i^{min} / \sigma_i^{max} = 0.1$; data for compressive stresses would also need to be considered.) The good-quality material has a value of $b = 0.10$, i.e. the fatigue strength decreases by 10% for each decade increase in the number of cycles. The material with $b = 0.14$ is poor-quality, as can be seen at $S = 0.2$, where the good quality material has over two orders of magnitude longer lifetime. Fig. 7.13 illustrates the importance of the fatigue performance of the materials used in a wind turbine.

A material with the lowest b coefficient is not necessarily the best, since the static strength is also important. Increased strength could be obtained by having more fibres but with a slight increase in the value of b. Whether this change would give a better performance depends on the fatigue strength required.

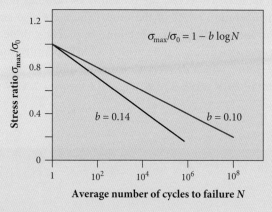

Fig 7.13 S-N curves for two fibreglass composites (*Source*: Sand99).

7.9.1 Turbine Generator Size and Capacity Factor

The ratio of the annual energy yield to that which would be produced at the rated power is called the **capacity factor**. The average capacity factor in 2018 was ~ 0.34 for new onshore and ~ 0.43 for offshore wind turbines (source: IRENA), with some recent (2019) values for offshore turbines as high as 0.55, reflecting the generally better winds offshore. These values correspond to a rated wind speed (corresponding to the rated generator power), about 1.5 times as high as the mean wind speed. There is a balance between the maximum energy output and the capital cost of generators. To take full advantage of periods of high wind speed would require large but expensive generators, which would have a larger output and a smaller capacity factor than a smaller generator. The choice is determined by what will give the lowest generation cost of electricity and is dependent on a number of factors. For example, the cost of offshore foundations has tended to favour larger turbines, but their larger weight pushes up costs and it is not obvious where the economic optimum lies.

Recent innovations have also allowed significantly larger turbine rotor diameters to be installed on the same turbine at low wind-speed sites, which increases the capacity factor and reduces the cost of electricity. For estimation purposes a capacity factor of 0.35 can be taken, but more accurate values can be obtained if details of the wind speed distribution and of the turbines are available (e.g. see Case Study 7.1).

The land area needed for a given output is largely independent of the size of turbine, as larger turbines need to be more spaced out (see Section 7.13), but larger turbines tend to operate in a greater mean wind speed, because their hubs are higher. Fewer turbines are required for the same output, so operation and maintenance costs are reduced; the cost per kWh therefore decreases. The price of infrastructure, such as connecting to the grid, is also reduced. Furthermore, some ridges only allow a single line of turbines, so more power is produced if they are larger. But for a given amount of capital, more smaller units reduce the risk from failures. However, the reliability of turbines is now very good, with turbines available for operation ~97% of the time. (There are also small (~3%) electrical losses between the turbine and the grid connection.) In practice, these considerations have tended to favour larger sizes where possible, with 8-12 MW planned for the early 2020s off-shore and 3-5 MW onshore, where delivery to site often limits the size. The maximum size of an offshore turbine in 2019 was 9.5 MW. Fig. 7.14 shows a 2 MW wind turbine under construction.

Fig 7.14 2 MW wind turbine under construction.

7.10 **Turbine Control and Operation**

In the generation of electrical energy from the wind, a wind turbine needs to be controlled to optimize its output. First the turbine needs to be oriented into the wind, which is achieved by the **yaw control** mechanism. The wind provides the driving torque for the electrical generator and the current flowing in the generator produces an opposing torque, called the **generator torque**. Ignoring friction, the wind torque equals the generator torque in steady operation.

In order to optimize the aerodynamic efficiency, most modern large wind turbines tend to be variable speed variable pitch machines and generate electricity using AC–DC–AC converters (see Box 10.1).

The output power curve for a typical variable speed variable pitch turbine is shown in Fig. 7.15. Also shown is the frequency distribution of a wind following a **Rayleigh distribution** (see Section 7.11) with a mean wind speed of 8 m s^{-1}. Both the speed of rotation and the pitch of the blades can be altered. This facility allows C_P (and hence the output) to be optimized when the wind is above the minimum required to operate the turbine ($u_{\text{cut-in}}$). The **rated wind speed** (u_{rated}) is such that the wind is strong enough to produce the maximum output power of the turbine generator. Generally, only the turbine speed is altered, by changing the generator torque, in the region between the cut-in and rated wind speed to maintain the optimum tip-speed ratio λ. The generator torque can be varied by changing the electrical load. Above u_{rated}, the speed of the turbine is usually maintained at a constant value and the pitch of the blade is adjusted (**feathering**, which reduces the angle of attack and hence the lift), to reduce the wind torque and keep loads on the turbine within safe limits. The turbine can be stopped by application of the shaft brake.

For a turbine running at a constant rotational speed and where the blades have a fixed twist, the angle of attack α increases with the wind speed, because the ratio u_1/v gets larger (see Fig. 7.6). If α becomes too large then the lift drops sharply as the aerofoil stalls. This

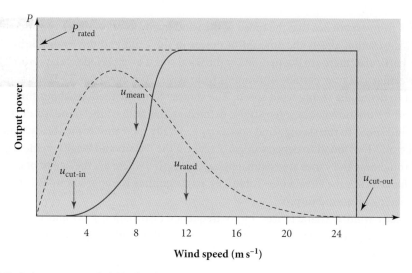

Fig 7.15 Typical output power (solid line) and the Rayleigh wind speed distribution (dot-dash line) for a mean wind speed of 8 m s^{-1}.

occurs at around $\alpha \sim 10°$ for the aerofoil profile shown in Fig. 2.14. This effect, called 'stall' control, provides a means of limiting the torque in high winds.

Turbines are operated typically for 65–80% of the time, depending on demand and on when the wind speed is between $u_{cut\text{-}in}$ and $u_{cut\text{-}out}$. The power density of a wind turbine is defined as the rated output power per unit of swept area and tends to be lower on low wind sites where longer blades are employed giving higher capacity factors, and often on taller towers since the load increase is less in weaker winds, and conversely on high wind sites.

EXAMPLE 7.4

Calculate the average power output of a wind turbine with blades of 85 m diameter operating in wind with a mean speed of 7 m s^{-1}. At this wind speed the power coefficient is 0.45. The rated output power is 2.5 MW when the wind speed is greater than 13 m s^{-1}. What is the power coefficient at a wind speed of 13 m s^{-1}? (Assume the density of air is 1.2 kg m^{-3}). Substituting in eqn (7.6), we have:

$$0.45 = P / \left[\frac{1}{2}(1.2)\left(\frac{1}{4}\pi \times 85^2 \right)7^3 \right], \text{ so } P = 526 \text{ kW}$$

$$C_p = 2.5 \times 10^6 / \left[\frac{1}{2}(1.2)(\pi \times 85^2/4)13^3 \right], \text{ so } C_p(13 \text{ m s}^{-1}) = 0.33$$

In this example the power coefficient at the rated wind speed is about three-quarters of its value at the mean wind speed.

7.11 Wind Characteristics

We know from experience that the speed of the wind in any location varies considerably with time. This variation affects the amount of power in the wind and the loads felt by the turbine. In particular, there are fluctuations in the wind speed over periods of days from changes in the weather and over periods of minutes from gusts. Averages over a ~10-minute period are used to define the steady wind speed. The shorter-term fluctuations about this value are quantified by the turbulence intensity I_T, defined as the ratio of the standard deviation σ_T of the wind speed to the steady wind speed. σ_T generally increases as the steady wind speed increases, and I_T is found to depend in particular on the terrain and height. I_T increases with surface roughness and varies approximately as $[\ln(z/z_0)]^{-1}$, where z is the height of the turbine and z_0 characterizes the terrain (see Table 7.3). Its magnitude is important in determining the fatigue loading on a wind turbine.

The steady wind is characterized by its frequency distribution $f(u)$ and its persistence. Persistence data give, for example, the number of times the wind is expected to blow for more than an hour with a speed greater than u, while $f(u)\Delta u$ gives the fraction of time that the wind speed is expected to be between u and $u + \Delta u$. The persistence of the wind is important in estimating the dependability of the generated wind power. Long periods of low wind affect the

Table 7.3 Surface roughness (z_0) values

Terrain	z_0 (m)
Urban areas	3–0.4
Farmland	0.3–0.002
Open sea	0.001–0.0001

Source: Spera, D. (ed.) *Wind turbine technology*. ASME Press, New York

amount of storage or back-up required on a grid, while long periods of high wind can lead to low electricity prices, transmission congestion or curtailment, and determine the potential for power-to-gas projects.

For sites that have an annual mean wind speed greater than 4.5 m s^{-1} the frequency distribution is often well described by the **Rayleigh distribution**,

$$f(u) = (2u/c^2)\exp[-(u/c)^2] \tag{7.30}$$

where $c = 2\langle u\rangle/\pi^{\frac{1}{2}}$ and $\langle u\rangle$ is the average wind speed. For a Rayleigh distribution the average power in the wind is given by

$$P = \frac{1}{2}\rho A\langle u^3\rangle = 0.955\rho A\langle u\rangle^3 \approx \rho A\langle u\rangle^3 \tag{7.31}$$

The Rayleigh frequency distribution for a mean wind speed of 8 m s^{-1} is shown in Fig. 7.15.

The wind speed u increases significantly with the height above the ground, with the speed $u = 0$ at ground level. Its rate of change decreases with height as the frictional forces decrease, as shown in Fig. 7.16. A commonly used form to describe the dependence of u on height z is:

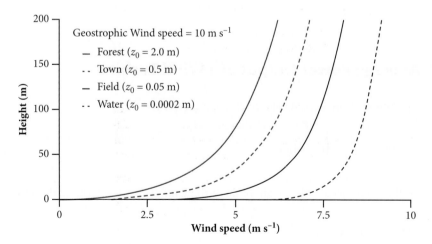

Fig. 7.16 The variation of wind speed with height in various types of location; the geostrophic wind speed is the speed above the boundary layer (DTU).

$$u(z) = u_s (z/z_s)^{\alpha_s} \qquad (7.32)$$

where z_s is the height at which u is measured to be u_s, ideally as close to hub height as possible, and α_s is the wind shear coefficient, which is strongly dependent on the terrain. α_s also generally shows a large variation over a 24-hour period and can change from less than 0.15 during the day to greater than 0.5 at night. This diurnal variation arises because at night the surface temperature drops as the ground loses heat by radiation, giving a stable atmosphere with the lower part cooler than the upper part of the atmosphere. The air is not mixed, and wind shear can be high. After sunrise the ground is heated by the Sun and warms the air in contact, which then rises, causing mixing and reduced wind shear.

The roughness of the terrain is characterized by a surface roughness parameter z_0, which is approximately one tenth of the size of protrusions, e.g. buildings, trees, grass, and some typical values are given in Table 7.3. An approximate parametrization for the dependence of the wind shear coefficient α_s on z_0 for steady wind speeds lying between 6 and 10 m s^{-1} at a height of 10 m is given by the formula:

$$\alpha_s = \frac{1}{2}(z_0/10)^{0.2} \qquad (7.33)$$

For $u_s = 8$ m s^{-1} at a height of 10 m, this relation gives $\alpha_s = 0.32, 0.13$, and 0.05 for $z_0 = 1, 0.01$, and 0.0001 m, respectively. Since a typical hub height for a 2 MW turbine is 80 m, these different wind shears translate into mean hub wind speeds of 15.6, 10.5, and 8.9 m s^{-1}. We can see that, for the same wind speed at 10 m, more power is produced if a turbine is mounted higher, particularly when the wind shear is large. However, large wind shears are often associated with high turbulence and hence more fatigue.

A high mounting position is particularly important for small wind turbines because, to be economic, the average wind speed needs to be at least 5 m s^{-1} and the turbulence to be small for the aerofoils to be effective. To achieve this, heights of 30 m or more are best. Rooftop-mounted wind turbines generally perform poorly due to low wind speeds and large turbulence (see Exercise 7.15).

7.12 Average Power Output of a Wind Turbine

The average power P_w in the wind at a given site is given by:

$$P_w = \frac{1}{2}\rho A \langle u \rangle^3 = \frac{1}{2}\rho A \int_0^\infty u^3 f(u)\, du \qquad (7.34)$$

where u is the wind speed at the height of the turbine hub (we are neglecting the variation in wind speed over the turbine blades) and $f(u)$ is the wind-speed distribution.

The average output power P_o is given by:

$$P_o = \frac{1}{2}\rho A \int_0^\infty C_P(u)u^3 f(u)du \tag{7.35}$$

For a variable pitch variable speed turbine, C_P can be kept quite close to its maximum value over a range of wind speeds about the mean, and then reduced to keep the output at its rated value for high wind speeds. The effect of C_P varying for typical values of the mean and rated wind speeds is to make the integral roughly equal to $0.35\langle u \rangle^3$, where $\langle u \rangle$ is the mean wind speed at the height of the turbine hub. When turbines are in an array there is an array loss factor (~0.9), so the average output of a turbine in a wind farm becomes:

$$P_o \approx 0.15 D^2 \langle u \rangle^3 \tag{7.36}$$

where D is the diameter of the turbine. A typical turbine spacing in a wind farm (see Section 7.13) when space is not at a premium is $8D$ (downwind) by $5D$ (crosswind), so the power density, which is the average output power per unit land or sea area P_{area} of a wind farm, is:

$$P_{\text{area}} \approx 3.75 \times 10^{-3} \langle u \rangle^3 \text{ MW km}^{-2} \tag{7.37}$$

For $\langle u \rangle = 8 \text{ m s}^{-1}$, $P_{\text{area}}(\text{wind}) \approx 1.9 \text{ MW km}^{-2}$.

The typical installed capacity (or rated power) is about 5 MW km^{-2} onshore or offshore, though it varies considerably dependent on conditions. For onshore wind farms, the area required for 1 TWh per year is ~50–90 km^2 while for offshore wind farms it is ~30–60 km^2.

EXAMPLE 7.5

Estimate the power output and capacity factor of a wind farm consisting of 50 4 MW turbines. The turbines' hub height is 100 m and their diameter is 120 m. The wind has an average speed of $u = 7$ m s^{-1} and 8.1 m s^{-1} at a height of 10 m and 60 m, respectively. Deduce the surface roughness parameter z_0.

Substituting in eqn (7.32) we have:

$$u_{60} = u_{10}(60/10)^{\alpha_s}, \text{ so } \alpha_s = \log(8.1/7)/\log 6 = 0.081$$

$$u_{\text{hub}} = u_{60}(100/60)^{\alpha_s}, \text{ so } \alpha_{\text{hub}} = 8.44 \text{ ms}^{-1}$$

From eqn (7.33), $z_0 = 10(2\alpha_s)^5 = 0.0011$

Putting u_{hub} in eqn (7.36) gives an estimate for the power output P_o for each turbine in the array, assuming an array loss of 0.9:

$$P_o = 0.15(120)^2(8.44)^3 = 1.30 \text{ MW}, \quad \text{capacity factor} = 0.325$$

So the wind farm output is estimated to be $50 \times 1.30 \text{ MW} = 65 \text{ MW}$.

We now consider the locations which favour the siting of wind turbines.

7.13 Wind Farms

Good data over a long period of time are required to determine where to site a turbine or a number of turbines, which is then called a **wind farm**. Experience has shown that onshore sites require an average wind speed greater than 6.5 m s^{-1}. Offshore wind speeds are generally higher, and good sites have average wind speeds of ~9 m s^{-1} or more. Wind speed is very important; e.g. the power available in a 12 m s^{-1} wind is twice that in a 9.5 m s^{-1} wind (due to the cubic power law, eqn. (7.2)). Suitable locations include mountain gaps and passes, high-altitude plains, exposed ridges, and coastal areas. For example, where there is restriction in the flow of wind across a plain caused by a hill, the wind speed for a turbine mounted on the top of the hill will be higher than for one mounted upwind on the plain.

A crucial consideration in the case of wind farms is the **turbine spacing**. A spacing of between three and ten rotor diameters, dependent on conditions, is required for the reduced wind speed downwind from a turbine to have been brought up to its original speed by gaining energy from the surrounding wind that is not slowed down. For a farm with a spacing of 7–8D (downwind) by 4–5D (crosswind), an array loss (the amount by which the output of the farm falls below the output of the turbines sited separately) of ~5–10% might be expected. The loss is sensitive to the layout and spacing of the turbines, the size of the array, and the distribution, in both direction and speed, of the wind.

7.13.1 Onshore Wind Farms

When land is at a premium a tighter spacing will generally be used. Some land is needed for the tower and service roads, but the land between the turbines is available for other use, e.g. for grazing animals or growing crops. To avoid **noise** being a problem in the case of onshore turbines, the tip-speed has to be limited to about 80 m s^{-1}, and their siting must not be too close to housing. The **visual impact** of wind turbines can also lead to difficulties in obtaining planning permission, though this can be mitigated if the local community is involved in the planning and receives benefits. A low wind shear reduces the differential loads on turbine blades and hence the fatigue damage, which favours flat sites, though some ridge sites have low wind shear.

The availability of suitable areas away from populations varies between countries and the economics of these sites is affected by their remoteness, since transmission lines are expensive. Turbine installation costs are cheaper on land than at sea, but their size tends to be limited through transport by road. In 2018, the global average rating of onshore turbines was 2.6 MW with rotor diameters of 110 m, and the maximum size built in that year 4.3 MW. The rating of new turbines is expected to be around 5–6 MW by 2025. The average capacity factors increased from 27% to 34% in the period 2010-2018, helped by longer blades and higher hub heights, with around 45% possible by 2050. Considerable land areas are globally available for the expansion of onshore wind farms, but offshore wind farms are looking increasingly attractive in some regions.

7.13.2 Offshore Wind Farms

Offshore sea areas can provide excellent wind conditions: wind speeds are generally higher and more consistent than on land with the result that capacity factors are higher, averaging 43% in 2018, compared with 34% onshore in 2018; and ~60% is expected by 2050. There is also less turbulence and, as a consequence, less fatigue. These areas can be far from the coastline and have little or no visual impact from land, but still quite close to areas of demand as many cities in the world are near the sea. In Europe new offshore turbines averaged 7.2 MW in 2019—with 12 MW turbines under construction—much larger than typical new onshore wind turbines of about 3.1 MW. They can also operate at higher rotational speeds, since noise is less of a concern out at sea. An example of an offshore wind farm is the **Horns Rev 2** offshore wind farm, commissioned in 2011, described in Case Study 7.1.

Case Study 7.1 The Horns Rev 2 offshore wind farm

The Horns Rev 2 wind farm is located in the North Sea, 30 km from the shore at Blåvandshuk in Denmark, where the depth of water is only 9–17 m. There are 91 Siemens SWT-2.3-93 turbines in the farm, which covers an area of 35 km^2 and has a total capacity of 209.3 MW. The cut-in, rated (for 2.3 MW output), and cut-out wind speeds are 4, 13.5, and 25 m s^{-1}, respectively, and the rotor diameter is 93 m.

The output power and power coefficient C_p of the turbine at various wind speeds are shown in Table 7.4.

The variable speed and pitch control of the turbine allows C_p to be close to its maximum value for the most probable wind speeds. The mean wind speed $\langle u \rangle$ at the hub height of 68 m is equal to ~9.7 m s^{-1}, and the annual production in 2010 was 855.5 GWh, equivalent to 1.074 MW of continuous output. This output corresponds to a capacity factor of 46.7%, illustrating the higher capacity factors found offshore than onshore. The average capacity factor (up to December 2018) over 9.3 years was 48%, corresponding to a generation 0.88 TWh per year or an area requirement of 39.8 sq km per TWh, corresponding to a power density of 2.87 MeV km^{-2}.

The average power output is given by:

$$P_o = \int_{u_{cut-in}}^{u_{rated}} P(u)f(u)\mathrm{d}u + P(u_{rated}) \int_{u_{rated}}^{u_{cut-out}} f(u)\mathrm{d}u \qquad (7.38)$$

Table 7.4 Output power of SWT-2.3-93 wind turbine versus wind speed

Speed (m s^{-1})	4	6	8	10	12	14	16	18
Power (kW)	98	376	914	1784	2284	2300	2300	2300
C_p (% Betz limit)	62	71	72	72	54	34	23	16

Source: Siemens data sheet.

where $P(u)$ is the output at a wind speed of u and $f(u)$ is the wind-speed distribution at the turbine hub height. The capacity factor CF of the turbine in a wind farm is given by $\eta_{farm}P_o/P(u_{rated})$, where η_{farm} is the array loss factor, typically 0.9. Approximating $P(u)$ in the range $u_{cut-in} \leq u \leq u_{rated}$ by $P(u) \propto (u - u_{cut-in})$, and $f(u)$ by a Rayleigh wind-speed distribution (eqn (7.30)), the capacity factor of the Horns Rev 2 wind farm is estimated to be 48.2% (Exercise 7.17*). An analytic expression for the capacity factor, calculated (see Exercise 7.17*) by assuming $P(u) \propto u^2$ in the range $0 \leq u \leq u_{rated}$, is

$$CF \approx \frac{\eta_{farm}}{x_r^2}\left[1-\exp(-x_r^2)\right] \tag{7.39}$$

where $x_r = u_{rated}/c$ and $c = 2\langle u\rangle/\pi^{1/2}$ is the parameter in the Rayleigh distribution formula eqn (7.30). Though only approximate, eqn (7.39) illustrates the dependence of CF on both the rated and mean wind speeds, and is a more reliable estimator for the output of a wind farm than just assuming an average capacity factor of 35%. This value for CF can be combined with the area of the farm A_{farm}, the number of turbines $N_{turbine}$, and their rated power P_{rated} to give an estimate of the power per unit area P_{area} of the wind farm:

$$P_{area} \approx \frac{N_{turbine}\eta_{farm}P_{rated}}{A_{farm}x_r^2}\left[1-\exp(-x_r^2)\right] \tag{7.40}$$

For the Horns Rev 2 wind farm, eqn (7.39) gives $CF = 46.2\%$ and eqn (7.40) gives $P_{area} = 2.93$ MW km^{-2} c.f. 2.87 MW km^{-2} observed, while eqn (7.37) gives $P_{area} = 3.42$ MW km^{-2}. (For comparisons with other wind farms, see Exercise 7.18.)

Planning permission is generally much easier to obtain for offshore sites than for onshore ones, particularly in countries with high population densities. Where there are restrictions on land-based expansion, investment and research is now aimed at expansion offshore. Turbines can cause electromagnetic interference, and planning permission has been refused for some sites where a wind farm would affect radar coverage. Access to the electrical grid is important, and some of the windiest sites are also some of the most remote.

Costs for offshore farms have fallen sharply since 2014 and some of those planned for the early 2020s will produce electricity at grid parity, excluding transmission costs (see Section 7.16). The larger size of turbines (>10 MW) has helped to offset the higher tower construction costs; the marine environment can be harsh, so structures need to be strong and corrosion resistant. Towers have foundations in deep water and are more expensive than on land, which has favoured shallower near-shore sites. For less than 10 m depth, a large heavy base (gravity base) can be used; up to ~40 m a monopile is suitable, and for 20–50 m a tripod/jacket can be deployed. Most offshore farms off Europe use monopile foundations; these are steel pipes (which by 2018 were some 10 m in diameter), driven about 20 m or more into the seabed. There are plans for 12 m diameter piles supporting 15 MW turbines in up to 50 m depth at heights above sea level of around 100 m. These wind farms require undersea cables to land, which are costly. High voltage direct current (HVDC) is favoured for distances greater than 50-80 km, since the cost of HVDC converters is less than the costs due to losses in HVAC transmission. Maintenance and installation costs can be higher than for onshore farms; to reduce these costs, drones are now being used to inspect turbines.

The first floating wind farm began operation in 2017 (see Case Study 7.2). Floating wind farms can operate in deep water, and would enable the wind resources off the coasts of the USA and Japan as well as in the Mediterranean to be tapped.

Case Study 7.2 The Hywind offshore wind farm

Besides catching the generally better winds farther offshore, wind turbines mounted on floating platforms can be anchored at sea just over the horizon, out of sight of land. These can still be close to demand centres, as some 40% of the world's population lives within 100 km of a coastline. The world's first floating wind farm, Statoil's Hywind farm, is sited in waters 90 to 120 metres deep 25 kilometres offshore, off Peterhead in Scotland. It consists of five 6 MW wind turbines that can provide power for over 20 thousand homes (see Fig. 7.17). It started production in October 2017 and a capacity factor of over 60% has been achieved. Such a high percentage means that power is more likely to be available when demand is high, which helps to integrate the farm's output into the electricity grid. Coupled with this farm is a 1.3 MWh (1,300 kWh) lithium battery store, called Batwind, which has a maximum output of 1 MW that can be used to help smooth out the variability in the wind production, thereby adding value to the electricity generated.

Fig. 7.17 Hywind floating offshore wind farm (*Source*: Statoil/Oyvind Gravas/Xinhua/Alamy Live News).

The Hywind farm uses a spar buoy platform, developed first for deep water oil extraction. Named after logs that were moored and floated vertically, a spar buoy is a long hollow vertical cylinder with the lower end loaded with ballast, so that the other end is just above water. A spiral fin is attached to the outside of the cylinder to reduce current-induced vibrations, similar to those on tall chimneys—installed for the same reason. The spar's length makes it very resistant to tilting, and the spar buoy is an excellent base for the platform supporting the tower of a wind turbine. The spar buoy design can be used in depths of up to 800 metres, which would open up a huge wind power resource worldwide. In European waters, there is enough wind to meet the total European electricity demand, while off the USA within 200 nautical miles, there is the potential to generate nearly twice the USA demand! More standardization may be possible in the design of platforms as they are independent of subsea soil, which would reduce costs. Installation can also be easier as no foundations are needed.

7.13.3 Hybrid Wind Farms

A hybrid wind farm is one where another energy source—such as a solar farm or an energy store, or both—are incorporated. These can provide a much more reliable power output (e.g. solar and wind often complement each other), and storage enables excess production to be time shifted to when the wind and solar resources are low to maintain output. The world's first hybrid farm was built in China in 2012 and its capacity is 100 MW from wind and 40 MW from solar PV, together with a 36 MW lithium ion battery.

Recently, coupling an electrolyser, to produce hydrogen from sea water, with a wind farm looks an increasingly attractive option. The hydrogen can be compressed and piped online to be stored, or converted to a more easily transportable fuel such as ammonia, and that way provide seasonal energy storage. It can also be used directly in heating homes, or in industry to provide high temperature heat for the production of steel and cement, or in fuel cells to power trains, trucks, or ships (see Chapter 11). Storing hydrogen can avoid curtailment, reduce price volatility, and provide greater energy security. In the 'Hybalance' project in Denmark, excess wind is used to produce hydrogen via electrolysis, with the hydrogen piped onshore and deployed in the transport and industrial sectors in the city of Hobro.

7.14 Developments in Wind Technology

7.14.1 Low-induction rotors

A turbine blade's dimensions are normally chosen to maximise the power output, and this corresponds to choosing an induction factor $a = 1/3$. The thrust on the blades determines the bending moment M at their root, and this in turn determines the maximum blade length R. For large offshore wind turbines, a different design optimization is being trialled in which the blade length and induction factor are varied to optimize the power output subject to maintaining the same bending moment M at the root of the blade.

Using eqns (7.7), (7.10), and (7.11), the thrust between r and $r + dr$ is:

$$dT = \rho 2\left(u_0 - u_1\right)u_1 2\pi r dr$$

Integrating from 0 to R and putting $u_1 = (1-a)u_0$, yields the total thrust as:

$$T = \rho a(1-a)u_0^2 2\pi R^2$$

The bending moment at the root of a blade from the thrust between r and $r + dr$ is:

$$dM = \rho a(1-a)u_0^2 4\pi r^2 dr$$

so the total bending moment is:

$$M = 2TR/3$$

The algebraic expressions for the total thrust, moment, and power from eqn (7.13) in terms of the induction factor a and radius R, are of the form:

$$T \propto a(1-a)R^2; \quad M \propto a(1-a)R^3; \quad P = Tu_1 \propto a(1-a)^2 R^2 \tag{7.41}$$

Expressing R in terms of M and a gives the power output as:

$$P \propto a^{1/3}(1-a)^{4/3} M^{2/3}$$

The value of a that maximizes P holding M constant is $a = 0.2$. The new length R', thrust T', and maximum power P' are related to the original length R, thrust T, and power P by:

$$R'/R = (25/18)^{1/3} = 1.116 \text{ and } (R'/R)^2 = 1.245$$
$$P'/P = (3/5)(6/5)^2 1.245 = 1.076$$
$$T'/T = (3/5)(6/5) 1.245 = 0.896$$

The low induction rotor has a 11.6% longer blade and collects 7.6% more energy in a given time. As the bending moment at the root is the same, we can (to a first approximation) just stretch the blade, maintaining the same cross section and strength; as a result, the mass and cost increase by just 11.6% and the blade component of cost per unit of energy by 3.7%. Since the blades are roughly 20% of the total cost of the turbine plus support structure plus electrical components, the cost of electricity will be lower. The tip speed is higher and the turbine will therefore be noisier, but this is not an issue for offshore turbines. Furthermore, the thrust is 9% lower, and this reduces the downstream wake and interference with other turbines, which will improve the yield of a wind farm.

7.14.2 Multi-rotor designs

For very large turbines, another way that may reduce costs is to mount multiple rotors on a single tower, since the capital cost of several small turbines may be less than that of a single

Fig. 7.18 Vestas's multi-rotor wind turbine (*Source*: Statoil/Oyvind Gravas/Xinhua/Alamy Live News).

large turbine. Consider a turbine with rotors sweeping out a circle of diameter D. The power output is proportional to D^2, but if all parts scale then the size, and hence mass and cost, of the turbine is proportional to D^3. For example, for the same tip-speed the angular velocity drops like $1/D$ and, as the power output is given by the product of torque and angular velocity, the torque rises like D^3, so the mass, and hence cost, of a direct drive generator increases like D^3. In general, this square-cube relation is approximate but provides a useful first estimate. For a multirotor turbine consisting of n turbines, the square-cube relation would predict that the total mass of these turbines would be $(1/n)^{1/2}$ that of a single turbine of the same power. The potential cost savings could be considerable, though there will be additional capital costs in the support structure and yaw mechanism.

An analysis of a four-rotor wind turbine estimated that construction costs would be 15% lower, rather than 50% based on the turbine being the sole cost. It would also be easier to put together and transport, as the components are lighter, and that compared to a single rotor turbine with the same power, the four wakes generated would cause significantly less interference downstream on other turbines in a wind farm: the downstream turbines therefore producing more power and subjected to less stress. Also, if one turbine fails there is still 75% of the output available. Fig. 7.18 shows a prototype 4-rotor turbine.

7.14.3 **Airborne designs**

Wind speeds can be faster and more consistent at higher altitudes than those accessed by ground or sea-based turbines, and there is considerable interest in using **kites** to capture the power in these winds. One technique is for a cable attached to a kite to drive a generator as the cable unwinds. When fully unwound, the kite is feathered so that the pull on the cable is much reduced and the cable is then wound in and the cycle repeated. The kites can be of low mass (and hence cost) and can also be grounded in severe weather conditions. A number of start-up companies are exploring airborne designs.

7.15 Environmental Impact and Public Acceptance

Wind farms are being actively developed because they are sources of renewable energy that produce essentially no global warming or pollution, and also provide energy security. For contrast, the emissions of CO_2 from coal and gas power plants are ~980 and ~480 tCO_2 per GWh, respectively, while only a small amount of emitted CO_2 (~11 tCO_2 GWh^{-1}) is associated with the construction and operation of the wind farms, similar to the amount associated with well-sited hydro or nuclear plants. Also, there are no air pollution-related deaths, and wind farms do not require water. Moreover, generating electricity from wind reduces the amount of fossil fuels used in generation and thus reduces CO_2 emissions.

These are all considerable environmental gains, which should not be forgotten when considering any negative aspects. Before reaching a decision on whether to build a new wind farm, a full **environmental impact assessment** (EIA) is undertaken during which all of the positive and negative impacts of a project are considered. All aspects of the environment that could be adversely affected are identified, and ways to mitigate any negative impacts would then be developed. An example is the concern in the UK over the visual impact of a wind farm, though this is much less of an issue in Germany, China, or Midwest USA. To gain high winds the turbines are often sited on ridges, which means that they can be very visible, and if they are in a region of natural beauty, they can be felt by some to be a blot on the landscape. Clearly, if the site is remote and not visited by climbers or walkers there will be less concern. The issue of visibility clearly favours offshore installations, which can be out of sight, though they must be clear of shipping lanes and not interfere with radar installations. Where a wind farm is visible, experience has suggested that it is more acceptable if there are a few large rather than many small turbines. Making the local population aware of the energy and any job benefits, and involving them in the planning process, can make planning permission easier to obtain.

To avoid noise being a problem for onshore turbines, the tip-speed has to be limited to about 80 m s^{-1}. This limitation arises due to vortex shedding at the blade tips, which gives rise to turbulent eddies that rotate with an angular frequency ω proportional to the tip-speed v_{tip}, and act as dipole sound sources. The intensity of a stationary dipole source depends on ω^4, but when shed from a moving blade tip the resultant sound intensity depends on ω^5, and hence on v_{tip}^5. Offshore tip-speeds can be higher since the greater distance to habitation makes the turbine noise less of an issue. Fish can be affected by underwater noise generated during the construction of the foundations of offshore wind turbines. However, this can be mitigated by surrounding the area by a curtain of air bubbles whose low density attenuates the sound.

Another concern that has been raised is their threat to birds—over a two-year period 183 birds were killed in the Altamont wind farm in California, while in Spain, in 18 wind farms an average of 0.13 birds were killed per turbine in 2003. By comparison, there are estimated to be over 57 million birds killed each year by cars and over 97 million killed by flying into plate glass windows in the USA, while in the UK each year 55 million birds are killed by cats. A study of power stations in Europe and the USA estimated that about 0.35 birds were killed per GWh by wind farms, compared with about 5 per GWh by fossil fuel power stations. So, while care should be taken to reduce bird mortality from turbines by avoiding migratory flight paths or key habitats, the risks should be put in perspective.

The amount of space that wind farms occupy has also been raised as an issue, but it should be noted that the land between turbines can be used for grazing or for growing crops. If we take the minimum area occupied by a turbine, or impacted area, as only D^2, where D is the rotor diameter, then the power density of a wind farm increases from ~2.5 MW km^{-2} to ~100 MW km^{-2}. To get an idea of how much land is required, we estimate the area needed to supply 30% of the UK's electricity.

EXAMPLE 7.6

Estimate the area of land required by wind farms to provide 30% of the UK's electricity demand.

The UK's electricity energy demand in 2015 was ~338 TWh. The number of hours in a year is 8760, so the energy demand would be met by a continuous power P of:

$$P = 338 \times 10^{12}/8760 = 38.58 \times 10^9 \approx 39 \text{ GW}$$

To provide 30%, i.e. 11.7 GW, then at a capacity factor of 0.35, typical of new onshore turbines, wind farms with a rated output (or capacity) of ~33.5 GW would be required. This capacity would be provided by $(33.5 \times 10^3/4) = 8375$ 4-MW turbines.

A typical diameter D for a 4-MW turbine is ~120 m. So, if each turbine occupied an area of $5D \times 8D$, the total area A required would be:

$$A = 8375 \times 600 \times 960 = 48.24 \times 10^8 \text{ m}^2 \approx 70 \times 70 \text{ km}^2$$

The area of the UK is ~240,000 km^2, so ~4900 km^2 is a small fraction (~2.0 %) of the land area. And if most of the turbines were offshore the impact of the wind farms would be minimal.

Noise from wind turbines has been a concern, but improvements in blade design have reduced the noise from modern turbines. Although the magnitude of noise from a wind farm is relatively low (see Table 7.5), the perception of noise is partly subjective. When the wind is blowing strongly the noise from the turbines is masked by that from the wind itself. Only when they are close to built-up areas is noise generally an issue. Offshore turbines can have a higher tip speed as noise is not a problem.

Table 7.5 Noise levels in dB

Noise	Noise level (dB)*
Threshold of pain	140
Pneumatic drill at 7 m	95
Busy general office	60
Wind farm at 350 m	35–45
Rural night-time background	20–40
Threshold of hearing	0

*$I(\text{dB}) = 10 \log_{10}(I/I_0)$, where I_0 is the threshold of hearing (at 1000 Hz $I_0 = 10^{-12}$ W m^{-2}).

Source: Peake *Renewable Energy*.

7.16 **Economics of Wind Power**

The economics of wind power depends on the capital costs to build the wind turbines, the ongoing costs to run the equipment, called operation and maintenance costs (**O&M**), and the revenue from the sale of the electrical energy (kWh) produced over the lifetime of the turbines (20–30 years). There are no fuel costs. With the increased cumulative production of wind turbines, the cost of wind generated electricity has been falling. For onshore wind power, the learning rate was about 19% over the period 1985 to 2014, and saw costs fall from around 57 to 7 cent per kWh as cumulative capacity rose from approximately 1 GW to 350 GW (see Fig 1.14). By 2018, tenders were around 3–5 cent per kWh in several countries. This has meant that onshore wind is now competitive with fossil fuel generation in many parts of the world, a situation that is termed '**grid-parity**'.

Technological improvements that are reducing costs include increasing capacity factors by improving power generation at lower wind speeds through using larger diameter turbine rotors, economies of scale, and improved financing. There is a general tendency towards longer turbine blades and larger hub heights. Wind is now the most economic source in many countries. The average LCOE of onshore wind generation commissioned in 2018 was US $56 per MWh, 35% lower than in 2010, reflecting the reduction in the average cost of a turbine per MW of output, the improvement in the average capacity factor, and the decrease in other costs. The cost is continuing to fall, with average auction and power purchase agreement (PPA) results for 2021 USD $31 per MWh.

The rapidly falling cost of onshore wind means that wind will soon provide electricity at a lower LCOE than existing coal and gas plants—in China this may be for coal by 2024 and in the US it was gas in 2020. But LCOE only partly determines the choice because wind is variable, and additional flexibility is required. This can be provided by lithium-ion batteries, whose costs are also falling fast, and co-located wind and battery-storage plants are expected to compete directly with, for instance, new pithead coal plants in India by 2024. Additional strategies and costs for handling the variability of wind and solar PV generation are discussed in Chapter 10.

Fig. 7.19 Offshore wind farm in the North Sea.

Offshore prices are higher than onshore (by about a factor of two), and in 2018 the average LCOE was 0.13 USD kWh^{-1}, but they are falling fast. In 2017 the first zero subsidy bid, equivalent to a LCOE of ~£60 MWh^{-1}, was awarded to DONG Energy for two wind farms in the North Sea to be commissioned in 2024, where there is already considerable deployment (see Fig. 7.19).

The difference in capital and O&M costs for onshore and offshore wind farms is partly offset by the higher wind speeds offshore and the planned deployment of larger and taller wind turbines of up to 15 MW capacity. In addition, a lower cost of capital from policy initiatives such as contracts for difference (see Chapter 12 Section 12.2.4), give more certainty of revenue and hence a reduced financial risk and lower interest rates. These factors are reflected in the global weighted average auction and power purchase agreement (PPA) results for 2025 of USD 0.071 kWh^{-1}, which is at grid-parity, and support the estimate for the LCOE per MWh to fall to around USD \$50 – \$90 by 2030 and USD \$30 – \$70 by 2050. Improvements in monopile foundations, in transmitting the power from the turbines to shore, and in construction should also help.

A typical wind farm generates the amount of energy used in its manufacture in about 3–10 months. Furthermore, a wind farm's decommissioning costs, a particular concern with nuclear power plants, are roughly covered by its scrap value. Financial payback estimates, however, are complicated and very site-specific, but an important factor affecting the LCOE is the effect of **discounting**. A simple payback estimate does not quantify what return on their investment a utility company can expect. Future revenue has to pay off both the capital and the interest that the capital would have earned during the operation of the wind farm to break even. Confidence in the project, as well as the market interest rates, affects the value of the interest rate charged. Typical values for the interest rate and period of operation are 8% and 30 years. We also have to add the maintenance and operations charge each year, which we can estimate as 4% of the capital. Using these values, the LCOE is given by:

$$\text{LCOE} = 1.47 \times C_{\text{capital}}\,(\text{k\$ kW}^{-1})/CF \quad \text{\$cents kWh}^{-1} \tag{7.42}$$

where C_{capital} is the cost in k\$ per kW of capacity, and CF is the capacity factor. (The details of discounting are explained in Chapter 1 Section 1.5.7.) The average cost per kW in 2018 was 1.50 k\$ and an estimate for a current capacity factor is 0.4, which gives LCOE = 5.5 \$cents kWh^{-1} and is typical of what modern onshore wind farms can provide.

For offshore wind, the average cost of a turbine in 2018 was US \$4350. Estimating M&O as 50% higher than onshore makes M&O 2% of capital. Assuming a favourable discount rate of 5% over a 30-year lifetime and a CF of 0.55 gives a LCOE of US \$77 per MWh; an 8% discount rate would then provide \$98 per MWh, significantly higher.

Electricity from offshore wind will be competitive with fossil fuel generation by the early 2020s without subsidies and from floating wind farms by 2030, with support for grid connections.

7.17 Wind Variability

The output of wind farms is variable, and how this variability is handled depends in particular on the mix of electricity generators available, the interconnectivity with different regions, the amount of energy storage available, and on the ability to vary demand to suit supply. Good interconnectivity is desirable to ensure that any excess can be exported and balancing power can be imported. For example, western Denmark has very strong grid connections with Germany and Norway, and also has the ability to balance wind with Norwegian hydropower.

Connecting wind farms together over large areas helps to smooth out variations in supply. In a study of the global potential (Lu 2009), the correlation between different regions of the contiguous USA was evaluated from a consideration of the wind resource in Montana, Minnesota, and Texas. They found that the wind was essentially uncorrelated between the three regions during the winter months, with r values (coefficient of regression) of less than 0.07. In the summer months, though, there was an appreciable correlation, with r values of 0.28 (Montana–Texas) and 0.37 (Montana–Minnesota). The incorporation of other sources, in particular PV, could help smooth out the overall supply in the summer. Good interconnectors to other grids can help smooth with errors in the wind forecast, though state-of-the-art wind forecasts that are using AI can now predict the timing and amplitude of events to a considerable degree of accuracy 24 hours in advance.

Adding storage to wind turbines or farms enables excess energy to be stored and used when the supply is low, and makes the capacity factor higher and the supply more reliable. The developments in electricity storage and the problems raised by having larger penetration of variable supplies, in particular solar PV and wind, are discussed further in Chapter 10, where the associated costs are also considered.

7.18 **Global Wind Potential**

About 0.5% (900 TW) of solar energy goes into generating wind energy, and about half of this is dissipated in the Earth–atmosphere boundary layer, where there is a loss of kinetic energy through frictional dissipation. The average downward flux of kinetic energy is ~1 MW km^{-2} and this provides a limit on the total global wind power that can be extracted. Over the world's unglaciated land area the flow of kinetic energy is ~112 TW. In the presence of wind turbines, this power is divided into dissipation by boundary layer turbulence and by the turbines. The maximum theoretical amount that can be extracted by the turbines is ~25% (see Exercise 7.18) and, of this, the amount that can be converted into electricity is ~75%, which gives an upper estimate of the onshore wind potential as ~20 TWe (~175,000 TWh). The alteration in global surface energy dissipation from the maximum wind turbine coverage is predicted to have some effect on the climate, but for a total power generation of 2 TWe, a significant fraction of the total global electricity generation of ~3 TWe, the estimated effect is insignificant—peak changes in seasonal temperatures might be ~0.5 °C, but with almost no effect on the global mean temperature.

Higher estimates of the global potential come from considering the areas with wind speeds suitable for wind farms with a typical turbine spacing of ~4–5D × 7–8D. While this spacing allows the wind speed to pick up within the wind farm, the large-scale dynamics of the atmosphere limits the maximum potential from a large region. There are also practical, environmental, and social restrictions on the areas that can be exploited. The resultant potential depends on the restrictions that are applied.

An estimate of the technical potential was made by Lu, McElroy, and Kiviluoma (Lu2009), which was similar to an earlier estimate by Archer and Jacobson in 2005. Lu et al. considered the output from a network of 2.5 MW turbines located in rural areas that are free from ice and unforested and with sufficient wind that the turbines operate at greater than 20% capacity. This corresponds to about 20% of the global land area. For offshore, they considered wind farms within 50 nautical miles of the coast. Their estimate of the total global generation potential was 840,000 TWh y^{-1}.

A more recent assessment by the NREL (Eurek2017) gave a similar total global wind generation potential: 560,000 TWh y^{-1} for onshore wind, and 315,000 TWh y^{-1} for offshore wind within 200 nautical miles of the coastline. These estimates assumed 3.5 MW wind turbines with 90 metre hub heights, and 5 MW km^{-2} deployment density in non-excluded areas. The authors noted that reducing their calculated generation potentials by a factor of five would account for the estimated effects of wind speed depression from very large-scale wind turbine deployment.

The estimates by Lu et al (Lu2009) are shown in Table 7.6 for the eight highest consuming countries plus the UK, together with estimates that the European Environment Agency (EEA) has made for Europe's onshore and offshore technical potential. Also shown are the electricity demands of each country in 2018, and it can be seen that for all of these countries the total onshore and offshore wind potentials far exceed their demand; for Europe and the US the offshore potentials alone are many times their electricity requirements. The restrictions on the siting of wind farms, e.g. not too close to the shore nor in shipping lanes, avoiding wild life reserves, fishing areas, and interference with radar, reduces the technical potential considerably. But around Europe, the remaining potential, estimated by the EEA as 3500 TWh, would still provide the total demand.

Table 7.6 Electricity consumption in 2018, and technical wind potentials for the eight highest electricity-consuming countries plus the UK and Europe

| Country | Electricity* (TWh) | Technical potential (TWh y^{-1}) | |
		Onshore	Offshore
China	6167	39,000	4,600
USA	3971	74,000	14,000
India	1243	2900	1100
Japan	1020	570	2700
Russia	929	120,000	23,000
South Korea	563	130	990
Canada	529	78,000	21,000
Germany	529	3,200	940
Brazil	524	250	160
UK	307	4,400	6,200
Europe	3400	45,000	30,000

Source: *Enerdata.

Areas with great potential are shown on the map in Fig. 7.20 of the global wind speeds. These include northern Europe along the North Sea, in South America near the southern tip of the continent, in Tasmania, in east and north-west Africa, in the Great Lakes and central region of North America, in parts of China, and along the north-east and north-west coasts of North America.

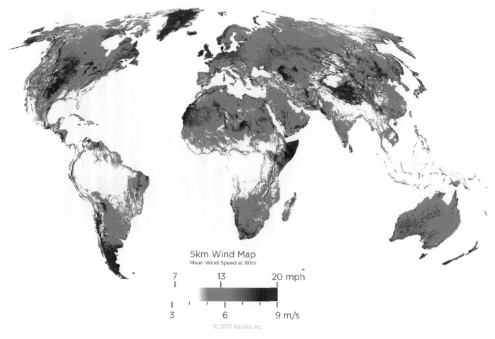

Fig. 7.20 Global wind speed map (Copyright 2015 Vaisala Inc).

7.19 **Outlook for Wind Power**

The global installed capacity of wind turbines was 651 GW in 2019 (see Figure 7.21), which generated 5.3% of the world's electricity. In some countries the percentage of demand met by wind power is significantly higher, notably Denmark 47%, Uruguay 35%, and the UK 21%, while across Europe as a whole 15%. The four countries with the largest wind capacities are China (236 GW), the US (105 GW), Germany (61 GW), and India (37.5 GW). The desire to reduce pollution from coal burning has helped to stimulate the growth of wind power in China and India. Globally, however, significant improvements in transmission grids are needed to accommodate the large capacities required. In Europe and Asia, installation of off-shore wind farms has been growing fast and is attracting investment from the oil giants, who are looking to reduce their emissions and to maintain their viability.

The number of small-scale turbines (<50 kW) is increasing with about a million operating worldwide by the end of 2015: a third in China, a quarter in the USA, and a seventh in the UK. They can be deployed in rural locations off-grid to provide electricity, water pumping, and to displace diesel generators, though competition from PV has affected growth in the US.

The global cumulative wind power capacity has been growing at an annual rate of 14% since 2010 (see Figure 7.21). In order to meet the Paris objective of limiting global warming to 1.5°C, the International Renewable Energy Agency (IRENA2019) has estimated that about 6000 GW of capacity would be needed: 5000 GW onshore and 1000 GW offshore, and would generate about 20,000 TWh a year. This is very close to the amount that could be installed by 2050 given by the Global Wind Energy Council in 2016, which assumed an average growth rate of ~7.5%, with a strong international commitment to meeting climate goals.

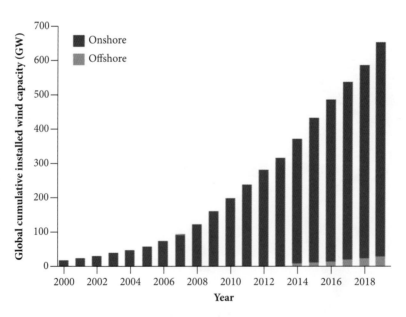

Fig. 7.21 Growth in global wind capacity (GWEC Global Wind Energy Council). Offshore wind capacity in 2019 was 29 GW, and is increasing by around 25% a year since 2015. The four countries with the largest offshore capacities in 2019 were: UK 9.7 GW; Germany 7.5 GW; China 6.8 GW; Denmark 1.7 GW.

This 6000 GW capacity would provide about a third of the predicted global electricity demand and mitigate about 6 GtCO$_2$ per year. In this IRENA scenario, Asia, mainly China, would continue to install the majority of turbines: about 3250 GW by 2050, with North America around 1300 GW, and Europe another 700 GW. To achieve this, the average annual addition in capacity would have to increase from around 50 GW in the period 2014–2018 to around an average of 200 GW over the period 2020–2050, and the required annual investment by around three times. There would be less investment and subsidies required for fossil fuels, and the saving from less pollution and climate change damages alone would more than cover the cost (IRENA2019). And around 20,000 TWh per year would be a very significant help in reducing our dependence on fossil fuels.

However, the economic fall-out of the COVID-19 pandemic may have a significant impact on the funding available for renewables. Fortunately, the economic downturn did not affect the growth of wind power in 2020, when global wind capacity additions totalled 111 GW, of which 6.0 GW was offshore, 90% more than in 2019. Moreover, further investment in wind power would provide a considerable number of jobs that will be needed as the world recovers and provide energy security, as well as helping tackle climate change, which poses a very significant danger if left unchecked.

 SUMMARY

- Global technical onshore potential is ~20 TWe, greater than the world's electricity usage in 2019 of ~3 TWe.

- Power in the wind is proportional to the cube of the wind speed. Sites with wind speeds greater than 6.5 m s^{-1} onshore and 9 m s^{-1} offshore at the height of the turbine are favoured. Power extracted by a turbine is limited to 16/27 of the incident power (the Betz limit) and is given by $P = \frac{1}{2}C_{\mathrm{P}}\rho u_0^3 A$,where C_{P} is the power coefficient, ρ is the density of air, u_0 is the incident wind speed, and A is the area swept out by the turbine blades. C_{P} is typically ~0.5 at the optimum tip-speed ratio $\lambda = v_{\mathrm{tip}}/u_0$, where v_{tip} is the speed of the blade tip.

- Modern wind turbines are horizontal axis wind turbines (HAWTs) and have maximum (rated) outputs of typically 4 MW onshore and 12 MW offshore with diameters $D = 120$–210 m. On wind farms, turbines are spaced at ~(4–5)D × (7–8)D, with tighter spacing if space is at a premium. The capacity factors for new turbines are typically 0.35 onshore and 0.5 offshore.

- The global installed capacity in 2019 was 651 GW, with 23 GW offshore, and the estimated accessible potential by 2050 is ~20,000 TWh.

- Wind is a potentially large source of low-carbon electricity. The cost per kWh from onshore wind farms is already competitive with fossil fuel alternatives in many regions. Onshore wind farms typically require 50-90 square kilometres to generate 1 TWh per year. Offshore sites require about 30-50 square kilometres, and LCOE bids at grid-parity have been accepted for 2024. The installed capacity is about roughly 5 MW km^{-2} onshore or offshore.

FURTHER READING

DTU Wind Energy, Global Wind Atlas 1.0. *globalwindatlas.info/about/method* (DTU).

Eurek, K. et al. (2017) An Improved Global Wind Resource Estimate for Integrated Assessment Models. National Renewable Energy Laboratory.

GWEC (2020). Global wind report.

IRENA (2019). Renewable power generation costs in 2018.

IRENA (2019). Future of wind. International Renewable Energy Agency. (IRENA2019)

Jamieson, P. (2018). *Innovation in wind turbine design*, 2nd edn. Wiley, Chichester.

Lu, X., McElroy, M.B., and Kiviluoma, J. (2009). Global potential for wind-generated electricity. *PNAS* 106 (27) (Lu2009).

Peake, S. (ed.) (2018). *Renewable energy*. Oxford University Press, Oxford. Useful information on wind power.

www.osti.gov/biblio/9460-fatigue-analysis-wind-turbines H. Sutherland, Fatigue analysis of wind turbines (Sand99).

REN21 (2019). Renewables Global Status Report.

cleanpower.org American Clean Power Association.

www.earth-syst-dynam.net/2/1/2011 L.M. Miller, F. Gans, and A. Kleidon, Estimating maximum global land surface wind power extractability and associated climatic consequences.

www.pnas.org/content/101/46/16115.full.pdf D.W. Keith et al., The influence of large-scale wind power on global climate.

yearbook.enerdata.net/electricity-domestic-consumption-data-by-region.html Electricity consumption (Enerdata).

EXERCISES

7.1 Calculate the power in a wind blowing with a speed of 12 m s^{-1} incident on a wind turbine whose blades sweep out an area of diameter 110 m.

7.2 A simple drag machine (Fig. 7.22) consists of two flaps attached to a rotating belt.

The drag force F_D is given by $\frac{1}{2}C_D \rho A u_{rel}^2$ where A is the cross-sectional area of the flaps, C_D is the drag coefficient for the flap, ρ is the density of air, and u_{rel} is the wind speed relative to the flap. Show that the power P_D is given by:

$$P_D = \frac{1}{2}C_D \rho A (u_0 - v)^2 v$$

and that the maximum power P_D(max) is given by:

$$P_D(\max) = \frac{1}{2}(4/27)C_D\rho A u_0^3$$

Deduce that the power coefficient C_P for such a drag machine is equal to $4C_D/27$. The maximum value of C_D is ~1.5 for a cup-shaped flap, so the maximum efficiency of a drag machine is 22%.

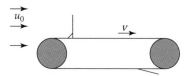

Fig. 7.22 A simple drag machine.

7.3 Consider stream-tubes of air (Fig. 7.23) before and after a turbine, but not across the turbine because the flow is unsteady and not streamlined.

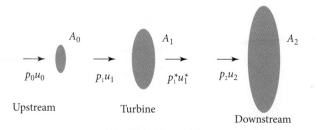

Fig. 7.23 Stream-tubes of air.

Applying Bernoulli's principle (eqn 2.42), and noting that $p_0 = p_2 =$ atmospheric pressure and that from the conservation of mass $u_1 = u_1^*$, show that:

$$(p_1 - p_1^*)/\rho = \frac{1}{2}(u_0^2 - u_2^2)$$

$$F_{\text{thrust}} = \frac{1}{2}\rho(u_0^2 - u_2^2)A_1$$

Hence show that the maximum thrust is given by:

$$F_{\text{thrust}} = \frac{1}{2}\rho u_0^2 A_1 \times 8/9$$

This is similar to a circular disc of area A_1 which has a drag force:

$$F_D = \frac{1}{2}C_D\rho u_0^2 A_1 \text{ and } C_D \sim 1$$

7.4   Using the result for F_{thrust} in Exercise 7.3, calculate the force in tonne-weight on a turbine in a wind speed of 10 m s^{-1} whose blades have a radius of 50 m.

7.5   Deduce the algebraic form of eqn (7.2), $P \propto A\rho u^3$, by using dimensional analysis.

7.6*   Show that the maximum power coefficient for two identical turbines placed one behind the other in line with the wind direction is 16/25.

7.7*   Consider a diffusing duct with a circular cross section mounted on a wind turbine. The cross section at its exit is A_1 and at its entrance A_{ent}, with $A_1 > A_{\text{ent}}$. When the wind turbine is mounted at the exit, show by considering the pressure drop across the turbine that the maximum power extracted is given by $\beta = A_1/A_{\text{ent}}$ times that of a turbine of area A_1 with no duct added, and is independent of where in the duct the turbine is mounted. Comment that the enhancement is not given by the cube of the ratio of the wind speeds in the entrance and exit of the diffuser when there is no wind turbine, i.e. β^3.

7.8   Take the extent of the disturbance of the wind in the direction of the incoming wind, i.e. parallel to the axis of the turbine, as πw, where w is the width of the turbine blade. Show that the ratio of the time for the wind to travel a distance πw to the time for one blade to reach the position of the next blade is greater than $4\pi/(3\lambda)$.

Hence show that the reaction of the blades occurs over a large fraction of the area swept out by the blades.

7.9   Calculate, using eqn (7.25), the width and twist of the optimal turbine blade of length $R = 48$ m for a tip-speed ratio $\lambda = 10$ at radii of 12, 24, 36, and 48 m. Take $\alpha_0 = 5°$, $C_L = 1$, and $n = 3$.

7.10   The two blades of a vertical axis wind turbine (VAWT) are rotating at a speed v in a wind speed u (Fig. 7.23). Show from considering the direction of the wind over the blades that the turbine will rotate as illustrated no matter what direction the wind comes from.

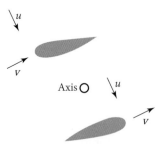

Fig. 7.24 Two blades of a VAWT.

7.11*   In Derivation 7.3 at the end of the chapter, show that the value of a that minimizes

$$dP = \frac{1}{2}\rho(2\pi r dr)4a(1-a)^2 u_0 /(1+a')$$

subject to the constraint:

$$a'(1+a') = a(1-a)/\lambda_r^2$$

is given by the equation:

$$\lambda_r^2 = (4a-1)^2(1-a)/(1-3a)$$

and that $a' = (1-3a)/(4a-1)$. Hence show that for $\lambda_r \geq 2$, $a \approx \dfrac{1}{3}$ and $a' = 2/(9\lambda_r^2)$. (The method of Lagrange multipliers states that the extremum values of a function $f(a,a')$ subject to the constraint $g(a,a') = 0$ are given by the solutions to $\partial f/\partial a + \lambda \partial g/\partial a = 0$ and $\partial f/\partial a' + \lambda \partial g/\partial a' = 0$.)

7.12 Calculate the number of cycles (N) a turbine blade would make in 30 years on a wind turbine with $D = 100$ m, $\lambda = 10$, and a mean wind speed of 12 m s^{-1}. The data for the fatigue strength of two possible turbine materials can both be represented by $\sigma/\sigma_0 = 1 - b\log_{10}(N)$. Material x has $\sigma_0 = 100$ MPa and $b = 0.10$, and material y has $\sigma_0 = 120$ MPa and $b = 0.12$. Which material would you choose?

7.13 The pressure P_t acting on a wind turbine is given by $P_t = \frac{4}{9}\rho u_0^2$ (Exercise 7.3). Estimate the variation in P_t across the height of the circle swept out by the turbine blades arising at a site with a wind shear characterized by $z_0 = 0.1$, $u_s = 10$ m s^{-1} at $z_s = 10$ m, hub height $= 70$ m, and $D = 50$ m.

7.14 A wind farm is situated on a ridge where the wind speed distribution $f(u)$ is given by

$$f(u) = 2a^2 u\,exp(-a^2 u^2)$$

where $a^2 = \pi/(4c^2)$, and c is the average wind speed at the height of the turbines. Each turbine has blades that are 45 m long and a power coefficient $C_P = 0.45$ (independent of wind speed). (a) Calculate the average power extracted by a single turbine when $c = 8$ m s^{-1}. (b) What is the most probable wind speed in terms of the average wind speed?

The density of air is $\rho = 1.2$ kg m^{-3}; $\displaystyle\int_0^\infty u^4 e^{-a^2 u^2}\,du = \dfrac{3\sqrt{\pi}}{8a^5}$

7.15 An approximate expression for the amount of electricity generated by a small wind turbine at a height of 10 m in one year is $E(kWh) = 2.1D^2 u_{av}^3$, where D is the diameter of the turbine and u_{ave} is the mean wind speed at the height of the turbine. Calculate the cost of electricity (neglecting any interest charges) if the turbine cost $70,000 and lasts for 20 years and has $D = 6$ m, (a) for $u_{ave} = 4$ m s^{-1} and 6 m s^{-1}. (b) the change in cost for $u_{ave} = 6$ m s^{-1} if the height of turbine is lowered to 5 m and the surface roughness parameter $\alpha_s = 2$. (c) Repeat (a) but with $D = 7$ m, height 10 m, and all other parameters the same.

7.16 For a wind farm with a spacing of $8D \times 5D$, estimate the difference in power density (MW ha^{-1}) for a farm with 20 turbines each with a rated output of 5 MW, hub height $= 90$ m, and diameter $= 115$ m: (a) assuming a capacity factor of 0.3; (b) using the formula $P_0 \approx 0.15D^2 \langle u(z)\rangle^3$, with $u_s(10) = 9$ m s^{-1} and $z_0 = 0.01$.

7.17* Evaluate eqn (7.38) when the output of a wind turbine $P(u)$ is approximated in the range by $P(u) \propto (u - u_{\text{cut-in}})$, and show that the capacity factor is given by

$$CF \simeq \frac{\eta_{\text{farm}}}{(x_{\text{rated}} - x_{\text{cut-in}})} \int_{x_{\text{cut-in}}}^{x_{\text{rated}}} \exp(-x^2)dx$$

where η_{farm} is the array loss factor, $x = u/c$, and $c = 2\langle u \rangle \pi^{1/2}$. Show that when the approximation $P(u) \propto u^2$ in the range $0 \leq u \leq u_{\text{rated}}$ is made, the capacity factor is then given by

$$CF \approx \frac{\eta_{\text{farm}}}{x_{\text{rated}}^2}\left[1 - \exp\left(-x_{\text{rated}}^2\right)\right]$$

7.18 Data on four offshore wind farms are given in the following table. Make estimates of the power density (MW km^{-2}) and capacity factors for these wind farms, and compare them with the observed values.

Farm	P/area (MW km^{-2})	Area (km^2)	No	P_{rated} (MW)	Dia (m)	u_{mean} (m s^{-1})	u_{rated} (m s^{-1})	Prod (GWh)	Cap (per cent)
Barrow	2.2	10	30	3	90	9	15	195.1	24.7
Egmond	1.3	27	36	3	90	8.5	15	315.2	33.3
Horns 1	3.2	20	80	2	80	9.7	16	565.8	40.4
Kentish	2.4	10	30	3	90	8.7	14	209	26.5

7.19 At night, the Coriolis force causes the wind to veer more with increasing height because of the increase in wind speed with height (see Section 7.11), and in the Northern hemisphere this is to the right. Explain why in a wind farm, the wind speed in the wake of a turbine will pick up speed more quickly if the turbine rotates anti-clockwise, which is opposite to the common clockwise rotating turbine (viewed from upwind). What are the implications for wind farms?

7.20 The momentum balance in the Earth–atmosphere boundary layer in the presence of wind turbines is given by

$$F - kv^2 - M = 0$$

where F is the rate of momentum transfer into the layer from the atmosphere (assumed constant), kv^2 is the force arising from natural turbulent dissipation in the boundary layer, with k the friction coefficient and v the mean wind speed, and M is the rate of momentum extraction by the wind turbines.

The total extracted power P_{tot} ($= Fv$) is divided between dissipation in the boundary layer (kv^3) and power extracted by the wind turbines $P_{\text{wind}}\left(= \frac{2}{3}Mv \text{ (eqn (7.6)}\right)$ in the Betz limit). Show that the maximum value of $P_{\text{wind}} = 4/(3^{5/2})P_{\text{tot}}(M = 0)$ and that the wind speed is then $3^{-1/2}v_0$, where $P_{\text{tot}}(M = 0)$ is the power dissipation in the boundary layer and v_0 is the wind speed when no turbines are present.

7.21 The angular speed ω of a fixed-pitch wind turbine is controlled by setting the generator torque $\tau_{gen} = K\omega^2$, where $K = \frac{1}{2}\rho AR^3\, C_P^{max}/\lambda_{opt}^3$. The angular acceleration $d\omega/dt$ of the turbine is proportional to $(\tau_{wind} - \tau_{gen})$, where $\omega\tau_{wind} = P_{wind} = \frac{1}{2}C_P\rho A u^3$ and $\omega = \lambda u/R$. Show that the turbine will alter speed so that $\lambda = \lambda_{opt}$, provided $C_P > C_P^{(max)}\lambda^3/\lambda_{opt}^3$.

7.22 Show that optimizing the power output of a wind turbine subject to maintaining the same bending moment M at the root of the blade gives an induction factor $a = 0.2$, and that the increased power output, compared with a turbine with the same bending moment but with $a = 1/3$, is 7.6%.

7.23 What are the advantages of a multi-rotor wind turbine design?

7.24 Investigate the development of airborne kites and discuss their potential.

7.25 Discuss the environmental impact of wind power and its effect on the deployment of wind turbines.

7.26 Estimate what the capital cost of an offshore wind turbine in k\$ kW^{-1} must be for the LCOE to be 6 c kWh^{-1} if the capacity factor is 0.5. How can the capital cost be reduced?

7.27 Discuss the relative merits of onshore and offshore wind farms, and in what countries and their territorial waters they might best be deployed.

7.28 How can the variability in the supply of electricity be reduced when wind power is generating a significant fraction of the demand.

7.29 If an average of 200 GW of wind turbine capacity per year was added over the period 2020–2050, what average capacity factor would give 20,000 TWh per year in 2050?

Derivation 7.3 The dependence of the maximum extraction efficiency on the tip-speed ratio λ

A turbine blade is designed for a particular λ and, as explained previously, we expect C_P to fall away on either side of its maximum (see Fig. 7.9). This maximum value is related to the theoretical limit, which at sufficiently high λ is the Betz limit (neglecting the effect of drag). However, at low λ the theoretical maximum for C_P is decreased, as we now explain.

As wind passes through a turbine the wind acquires a swirling motion as a result of the rotation of the turbine blades. This is because angular momentum is necessarily imparted downstream since the air flow through the turbine imparts a torque to the turbine blades (see Fig. 7.25). If the speed of the blade is much greater than the speed of the wind, i.e. a high λ, then the energy associated with this rotational motion of the wind is much smaller than that associated with the momentum change of the wind in the direction of the wind (axial direction). To see this we will consider what happens to the wind when it passes through a small annular area (see Fig. 7.25).

The power extracted by the turbine from the stream-tube of wind defined by the annular area between r and $r + dr$ equals the torque dG arising from the change in angular momentum

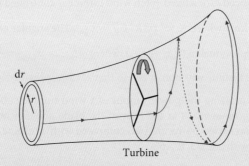

Fig. 7.25 Angular momentum of airflow after passing through a turbine.

of the wind multiplied by the angular speed Ω of the turbine. At a given radius this product equals the force arising from the change in momentum of the wind in the direction of the blade motion multiplied by the speed of the blade.

The power extracted, dP, equals the loss of axial kinetic energy of the wind minus its gain in rotational kinetic energy per second. The mass flow $d\dot{m}$ equals $\rho u_1 2\pi r dr$. If the wind gains an angular velocity ω_1, which will be in the opposite sense to that of the blade, after passing through the turbine, then the rotational kinetic energy gained per second is given by $\frac{1}{2}d\dot{m}r^2\omega_1^2$ and the change in angular momentum per second by $d\dot{m}r^2\omega_1^2$. Hence the torque $dG = d\dot{m}r^2\omega_1$ and dP is given by:

$$dP = dG\Omega = d\dot{m}\omega_1 r^2\Omega = \frac{1}{2}d\dot{m}(u_0^2 - u_2^2) - \frac{1}{2}d\dot{m}r^2\omega_1^2 \tag{7.43}$$

Defining $\omega = \frac{1}{2}\omega_1$ and the **angular induction factor** $a' = \omega/\Omega$, then:

$$dP = dP_B/(1+a') = d\dot{m}\omega_1 r^2\Omega$$

where dP_B is the change in linear kinetic energy per second. As in Derivation 7.1, which neglected wake rotation, dP_B equals the change in linear momentum per second multiplied by the wind speed at the turbine (eqn (7.8)) leading to eqn (7.13), which can be expressed as

$$dP_B = \frac{1}{2}d\dot{m}4a(1-a)u_0^2 \tag{7.44}$$

Substituting gives:

$$dP = \frac{1}{2}d\dot{m}4a(1-a)u_0^2/(1+a') \tag{7.45}$$

Defining the radial tip-speed ratio $\lambda_r = \Omega r/u_0$ and substituting $d\dot{m}\omega_1 r^2\Omega$ for dP gives:

$$a'(1+a') = a(1-a)/\lambda_r^2 \tag{7.46}$$

Substituting $d\dot{m} = \rho u_1 2\pi r dr$ and $u_1 = (1-a)u_0$ gives:

$$dP = \frac{1}{2}\rho(2\pi r dr)4a(1-a)^2 u_0^3/(1+a') \tag{7.47}$$

Maximizing dP by varying a subject to the constraint of eqn (7.46) gives $a' \ll 1$ and $a \cong \frac{1}{3}$ as before, when neglecting wake rotation, for $\lambda_r \geq 2$. The angular induction factor $a' \approx 2/(9\lambda_r^2)$, and if we take a tip-speed ratio of $2\lambda/3$ as representative, then:

$$dP \approx dP_B/[1+1/(2\lambda^2)] \tag{7.48}$$

Hence for $\lambda \geq 4$ the maximum power that can be extracted is very close to the Betz limit, as shown in Fig. 7.9, in the absence of drag.

 For further information and resources visit the online resources
www.oup.com/he/andrews_jelley4e

8 Photovoltaics

✔ List of Topics

- ☐ History of solar cells
- ☐ Solar spectrum
- ☐ p-n junction
- ☐ Solar cells
- ☐ Efficiency of cells
- ☐ Silicon cells
- ☐ Multilayer devices

- ☐ Thin-film cells
- ☐ Developing technologies
- ☐ Solar panels and farms
- ☐ Environmental impact
- ☐ Economics of photovoltaics
- ☐ Global solar potential
- ☐ Outlook for photovoltaics

→ Introduction

The average solar power incident on the Earth is ~1000 W m^{-2} (~100 mW cm^{-2}), around 125,000 TW in total, and is much larger than the global power consumption in 2019 of ~18 TW$_{th}$. About ~11% of the world's final energy demand is currently supplied by **biomass**, which arises from the photosynthesis of plants, but only about 1% of the solar energy absorbed by plants is actually converted into biomass. [By comparison, fossil fuels contribute ~80% of the final energy demand and the burning of fossil fuels releases vast quantities of CO_2 into the atmosphere—the primary cause of global warming.]

In this chapter we describe a much more efficient process (~21%) for converting solar energy directly into electrical power using **photovoltaic (PV) cells**. The photovoltaic effect has been known about for over 160 years, but it was not until some 20 years ago that commercial solar cells started to take off (see Box 8.1). Growth has been particularly fast in the last decade. By 2019 the global installed capacity had increased to 580 GW, providing 720 TWh of electricity, representing 2.7% of the world supply. The cost of producing electricity from utility scale solar PV farms is now lower than that from fossil fuel generators in many regions.

This achievement of better than grid parity has meant that new power plants are increasingly photovoltaic. A considerable expansion of PV systems is also happening in residential systems, particularly in the developing world and in several developed countries. Concentrated solar thermal power (CSP) plants can have a comparable conversion efficiency and include thermal energy storage, but currently their levelized cost of energy (LCOE) is higher than PV and their share of the market is very small; however, their future contribution could be more significant as costs in the last few years have fallen considerably (see Chapter 5 Section 5.3).

Box 8.1 History of solar cells

In 1839 the French physicist **Edmond Becquerel** noticed that uneven illumination by sunlight of electrodes in an acidic solution generated an electric current. But it was not until 1877 that this photovoltaic effect was seen in a solid cell, which was made with the element selenium. A few years later, in 1883, a thin film selenium photocell was produced in the US by Charles Fritz. But it was expensive, and its efficiency was very low, with only about 1% of the energy from the Sun converted into electricity, so it was not a practical way of generating power. The photovoltaic effect was also found in other materials, but the photocells that were fabricated from them had similarly low efficiencies. It was not until the accidental discovery of a silicon photovoltaic device by Russell Ohl in 1940, in the **Bell Laboratories** in the US, that significant progress started to be made: the method of making a silicon rod by slow crystallization had caused just the right kind of different impurities to be at either end of a one-inch long rod for photovoltaic action to be seen when the rod was illuminated by a flashlight.

Improvements in making silicon for **transistors** in 1954 led to a cell that had 6% efficiency, about 10 times higher than that of earlier devices. This cell, invented by Daryl Chapin, Calvin Fuller, and Gerald Pearson in the Bell Laboratories, was considered to be the first practical solar cell, and there was considerable hype over its discovery. However, the cost of these cells was prohibitive except in niche applications, such as in satellites. Research led to a lowering in costs by the 1980s, promoting the application of photovoltaic panels for terrestrial use. Programmes encouraging the installation of panels on rooftops in the 1990s, and the introduction of feed-in tariffs in the 2000s, generated further demand, and Japan and Germany accelerated the industrialization of production. Since 2000, the average annual growth has been a staggering 40%. This growth has resulted in cost reductions, which in turn has helped to increase demand.

China responded to the global growth in demand for solar panels in the late 1990s by starting to make solar panels for the German market. China's own domestic market then took off, motivated by rising pollution from coal-fired power stations, and concern over energy security and climate change. China concentrated on the dominant technology—crystalline silicon—and government support through loans and tax incentives helped companies to build large semi-autonomous factories, which reduced costs dramatically. In the period 2006–2011, which saw global production move away from Japan, Germany, and the US to China, the cost of modules fell by a factor of three. In the next seven years 2011–2018, there was further fall of 3.5 times.

Since its invention, it has taken some 60 years for the efficiency of a silicon solar cell efficiency to increase to over 20%, and for its cost to fall by several hundred times, to the point where the electricity generated by silicon photovoltaic cells is now cost-competitive with that generated by fossil fuels. It has required considerable development and mass production to achieve this, as the processing of silicon to form a solar cell is complex. Silicon cells now account for about 95% of all solar cells. The remainder are based on other photovoltaic materials, such as gallium arsenide and cadmium telluride.

We now look at the solar energy spectrum and how a solar PV cell works. We describe the physical limits to its efficiency and the scope for improvement, and at the different technologies, both established and developing, before describing PV cells, panels, and solar PV farms. We conclude with a discussion of the potential of PV to provide a significant source of low-carbon energy for the world.

8.1 **The Solar Spectrum**

The smooth spectrum shown in Fig. 8.1 is that of a black body at 5800 K. This spectral shape is close to that incident from the Sun on the Earth's atmosphere. The effect of passing through the atmosphere is to reduce the total from 1.37 kW m^{-2} for sunlight incident on the atmosphere, called AM0, to ~1.0 kW m^{-2} for that passing through a typical thickness of the Earth's atmosphere taken to be 1.5 times its height, called AM1.5. AM1.5 corresponds to sunlight incident at an angle of 48° to the vertical. The effect of absorption by water vapour, carbon dioxide, and methane is nearly all in the infrared region, corresponding to photon energies below ~1.75 eV. The energy of the photons in the visible part of the solar spectrum ranges from ~1.75 eV (0.7 μm) to ~3.25 eV (0.38 μm).

Solar energy reaches the Earth's surface by direct radiation (focusable by mirrors) and diffuse radiation (unfocusable). The diffuse percentage is strongly dependent on how clear the sky is, and a typical yearly average is about 30%. The total amount of radiation varies considerably with cloudiness, season, and location, from a yearly total on a horizontal surface of ~2300 kWh m^{-2} in the Tropics to ~800 kWh m^{-2} in the Arctic Circle (latitude 66.5°). The average flux (watts per m^2) on a cloudy day is typically ~10% in the UK and ~50% in the Tropics of the flux on a sunny day. Note that the sky would be black (except for stars and planets) in the absence of diffuse radiation, and is blue because short wavelengths are scattered more than long ones by molecules and particles of size $<< \lambda_{visible}$.

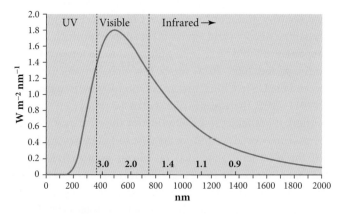

Fig. 8.1 Black-body spectrum at T = 5800 K (see Section 2.3.3). The total intensity is normalized to that of the AM0 solar spectrum of 1.37 kW m^{-2}. The wavelengths corresponding to 3.0, 2.0, 1.4, 1.1, and 0.9 eV are indicated.

EXAMPLE 8.1

Direct sunlight of average intensity 200 W m^{-2} is incident normally on a solar cell. The area of the cell is 0.1 m^2. What is the total incident energy in one day in kWh and in MJ? How is this total energy altered if the sunlight falls at angle of 30° to the normal to the surface of the cell? The incident power P is the intensity I multiplied by the area A of the cell, so:

$$P = I \times A = 200(0.1) = 20 \text{ W} = 0.02 \text{ kW}$$

The incident energy E is the power multiplied by the time t. Hence,

$$E = P \times t = 0.02(\text{kW}) \times 24(\text{h}) = 0.48 \text{ kWh} = 1.73 \text{ MJ}$$

If the sunlight falls at 30° to the normal to the cell, then the incident power is reduced by a factor cos 30° ≈ 0.87, since A cos 30° is the projected area of the cell normal to the beam. The incident energy then becomes:

$$E = 0.42 \text{ kWh or } 1.50 \text{ MJ}$$

The variation of the inclination of sunlight with latitude and season is the main cause of the variation in the Sun's intensity, or irradiance (W m^{-2}), on a horizontal surface with location. The average irradiance on a surface over a period of time is called the *insolation*, often measured in kWh m^{-2} y^{-1}.

In 1905 Albert Einstein realized that light does not only behave as a wave, but also as particles (photons), whose energies are proportional to the frequency of the light. For this insight he received the 1921 Nobel Prize in Physics. The energy of a photon $E_\gamma = h\nu = hc/\lambda$, where $hc = 1240$ eV and λ is the wavelength of the light; so, for example, the photons in a beam of light with a wavelength of 510 nm (green) have an energy $E_\gamma = 1240/510 = 2.43$ eV.

When sunlight falls on a solar photocell, some electrons in the cell material absorb energy from the light and produce a potential difference that can power devices. How this is achieved is by taking advantage of certain properties of semiconductor materials.

8.2 Solar Photocells

Photovoltaic solar cells are made from semiconductor materials, which are characterized by having a low electrical conductivity. This comes about because the outer atomic electrons in the semiconductor fully occupy a band of energy levels, so there are no empty levels available close by for electrons to be promoted into when an electric field is applied (unlike in a good conductor like copper). Conduction can only arise by electrons that have been thermally excited to the next band. This will only occur to a significant extent when the gap, called the **band gap**, between the filled (**valence**) band and empty (**conduction**) band is relatively small, ~1 eV. Such materials are called **semiconductors**.

A particularly important semiconductor is **silicon**, which is used in about 95% of solar cells. As an isolated atom, silicon has four electrons in a half-filled shell, but in a crystal lattice, these electrons are shared (covalent bonds) between four silicon atoms, and a full valence band is formed with a 1.1 eV gap to the next band. Besides silicon, a number of other semiconductor materials are used in photovoltaic cells, including GaAs, CdTe, CuInGaSe$_2$ (CIGS), perovskites, and certain organic compounds.

The presence of *impurity* atoms can give rise to electron states within the band gap, and this contamination must be reduced before the material acts as a pure semiconductor, called an *intrinsic semiconductor*. The development of the technique of **zone-refining**, in which sections of a rod of semiconductor are heated sequentially so the molten region moves from one end to the other, thereby concentrating the impurities at one end of the bar, played a key part in the development of semiconductor devices. Silicon can be produced with less than 1 in 10^9 impurities.

The addition, called **doping**, of certain impurity atoms to intrinsic material can significantly alter the conduction properties of semiconductors. If we include atoms with five outer electrons (pentavalent atoms) within a silicon crystal, then each of these atoms will have one electron which is only weakly bound and can easily be excited into the conduction band. Such atoms are called **donors**, and the doped silicon is called **n-type**. Donor atoms with only three outer electrons (trivalent atoms) can gain an electron quite easily through thermal excitation of an electron from the top of the valence band, which leaves a positively charged vacancy called a hole. Such atoms are known as **acceptors**, and the doped silicon is then called **p-type**. A particularly important device, and the basis of the photovoltaic solar cell, can be made by forming a junction between p- and n-type material, called a **p-n junction**, which gives rise, as we will see, to an internal electric field.

When a solar cell, consisting of a wafer of relatively thick p-type silicon with a thin n-type region on top, is illuminated with light, photons with energies greater than the band gap excite electrons in the p-region from the valence band to the conduction band. There they are free to move around within the wafer, and when these electrons come within the internal electric field, they are swept into the n-type region. The accumulation of electrons there produces a potential difference that, as in a battery, can drive a current through a device.

8.3 **p-n Junction**

Fig. 8.2 shows separate pieces of p- and n-type material and a single piece of semiconductor doped to form a p-n junction. We can see thermally excited electrons in the conduction band of the n-type and positive (i.e. absence of electrons) holes in the valence band of the p-type.

In the piece with both p- and n-type material adjacent, Fig. 8.2 (b), electrons have diffused across the junction from the conduction band on the n-side, as a result of the concentration gradient, and these have filled the vacancies (holes) in the top of the valence band on the p-side. An electric field is set up across the junction and causes a drift of electrons which balances the electron diffusion due to the concentration gradient. (There are corresponding currents of holes.) As a result, the n-side becomes positively charged while the p-side

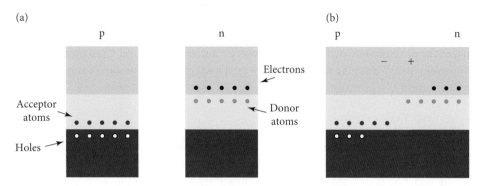

Fig. 8.2 (a) p- and n-type material (b) p-n junction.

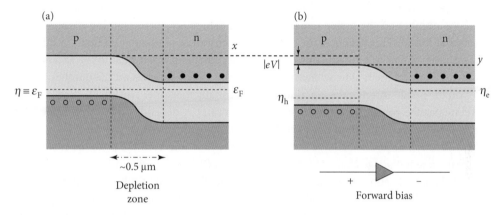

Fig. 8.3 (a) Energy levels and depletion zone in a p-n junction (b) Effect of positive bias on a p-n junction with standard circuit symbol for a diode shown.

becomes negatively charged. The energy of the electrons on the n-side is therefore lowered compared with electrons on the p-side, and the junction region is normally drawn as shown in Fig. 8.3 (a).

The region about the junction where there are no charges is called the **depletion zone**. The electrons (holes) in the conduction (valence) band on the p- (n-) side of the junction are called the minority carriers, and on the n- (p-) side the majority carriers are electrons (holes).

The drift and diffusion currents are driven by opposite gradients in the concentration and electrostatic potentials. The potential relating to the concentration is called the chemical potential. When in equilibrium the sum of these potentials, called the **electrochemical potential** η, is a constant and is given by $\eta = q\phi + \mu$, where q and μ are the charge and chemical potential of the electron or hole, and ϕ is the electrostatic potential. The value of the electrochemical potential η, in a semiconductor, also called the Fermi level ε_F, is the same on the p-side as on the n-side of the junction, as shown in Fig. 8.3 (a). For a current to flow across the junction there must be a difference in the electrochemical potentials on either side of the junction.

Fig. 8.3 (b) shows the effect of biasing the p-region positively by V relative to the n-region. The junction is then **forward-biased**. (The circuit symbol for a diode is shown below the junction.) This lowers the conduction band on the p-side relative to that on the n-side of the junction by an amount $|e|V$, where e is the charge of the electron. This alters the balance between the diffusion and drift currents across the depletion zone and gives rise to a net electron (hole) current from the n- (p-) to the p- (n-) side, called the **forward current**. The total forward current I_F is given by:

$$I_F = I_S[\exp(V/V_T)-1] \tag{8.1}$$

where I_S is called the **saturation current** and $V_T = kT/|e| \approx 0.026$ V at room temperature.

The Boltzmann factor $\exp(V/V_T)$ quantifies the imbalance in diffusion and drift currents when the junction is forward biased by V. When V is zero, the number of electrons in levels thermally excited above the top of the band gap on the p-side is equal to the number on the n-side above that energy (level x in Fig. 8.3 (a)). When forward biased by V, the number on the n-side above that energy (level y in Fig. 8.3 (b)) is greater due to the Boltzmann factor, while the number on the p-side remain the same. (There are corresponding changes in the number of holes.) This produces an electron concentration gradient, and hence a diffusion current, from the n- to on the p-side of the junction that is proportional to the excess number of electrons on the n-side; while the drift current, which depends on the number of electrons on the p-side entering the junction, stays essentially the same. The total forward current therefore has the voltage dependence shown in eqn (8.1). I_S depends on the area of the junction and on the recombination rate of electrons and holes in the cell.

The direction of positive charge flow is across the junction from the p- to the n-side, and the ideal current–voltage characteristics given by eqn (8.1) are shown in Fig. 8.4 (a). We can see that the p-n junction acts as a diode, allowing current to flow easily only when forward-biased. Fig. 8.4 (b) shows the actual characteristics for a silicon p-n junction: until there is a forward bias of ~0.6 V, conduction is less than that for an ideal junction. This is caused by the trapping and generation of electrons and holes within the depletion zone. But the idealized equation for the forward current is a very useful approximation, and we will now see how such a device can act as a photocell.

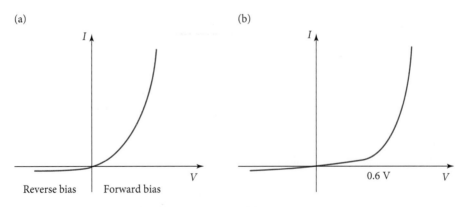

Fig. 8.4 (a) IV characteristics of ideal p-n junction (b) Actual IV characteristics for silicon diode.

8.4 Silicon Photocell

When sunlight falls on a silicon cell some of the photons can create electron–hole pairs through the **photovoltaic effect**, in which a photon is absorbed by an electron, and provides the energy needed to promote it from the valence to the conduction band. The minimum energy that the photon must have equals the band gap E_{gap}. For silicon, the minimum energy is 1.1 eV and corresponds to a wavelength of ~1.1 μm, using the Einstein relation $E_\gamma = h\nu = hc/\lambda = 1.24/\lambda$ eV, with λ in μm. Fig. 8.5 shows the standard silicon solar cell structure that is use in the majority of current solar panels. Photons incident on the cell junction produce electron–hole pairs within the p layer.

The top metal electrode is in narrow strips to let the light fall on the junction. An anti-reflection coating increases the transmission of light into the cell. The surface of the cell is textured (not shown) to increase light absorption (see Box 8.3). To reduce the series resistance, the n-layer is highly doped (labelled n^+). Minimizing surface recombination at the rear contact is achieved by placing a highly doped p-layer (p^+) just in front of the contact. This sets up an internal electric field that reduces the electron concentration in this region, and hence reduces the chance of recombination.

Electrons promoted to the conduction band in the p-layer diffuse and, when within the built-in field across the depletion region, are swept to the n-side. This produces a reverse current I_L (since the electrons flow across the junction from the p- to the n-side), and the net photocell current I_C is given by:

$$I_C = I_L - I_S[\exp(V/V_T) - 1] = I_L - I_S[\exp(I_C R/V_T) - 1] \tag{8.2}$$

where the forward bias V across the junction equals $I_C R$. This forward bias produces a forward current I_F given by the second term, eqn (8.1), flowing across the junction from the p-side to the n-side, which is in the opposite direction to the light-induced current I_L.

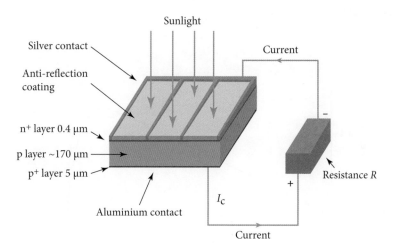

Fig. 8.5 A back-surface field (BSF) silicon solar photocell.

When the resistance R is infinite then $I_C = 0$, and the open circuit voltage V_{OC} (by which the p-side is more positive than the n-side of the junction) is given by:

$$V_{OC} = V_T \ln(1 + I_L/I_S) \approx V_T \ln(I_L/I_S) \tag{8.3}$$

since $I_L \gg I_S$. (V_{OC} is less than E_{gap}, since the maximum voltage must be less than the band gap.) Note that $V_T = kT/|e| \approx 0.026$ V at room temperature.

When the circuit is open then $V = 0$ and the short circuit current $I_{SC} = I_L$. For a finite resistance R, the photocell current I_C generates power P_C given by:

$$P_C = I_C V = I_C^2 R \tag{8.4}$$

Consider I_L fixed. As R increases, V and therefore I_F increase. Initially, the increase in I_F is very small and $I_C \approx I_L$, as shown in Fig. 8.6, and so $P_C = I_C V$ increases. But there comes a point ($V = V_m$) where any further increase in V, and hence I_F, causes a sufficient drop in $I_C = I_L - I_F$ that offsets the rise in P_C from the increase in V. The forward current therefore limits both the voltage and current at maximum power.

The forward current of electrons recombines with holes on the p-side; hence the forward current is also called the 'recombination' current. There is a minimum recombination current when the only recombination mechanism is radiative, and this gives rise to the **Shockley–Queisser limit** for the efficiency of a single junction solar cell (see Section 8.5 and Derivation 8.1 at the end of the chapter). In practice, the recombination is predominantly caused by impurities.

For the solar cell of area 22 cm^2 shown in Fig. 8.6, $I_{SC} = 0.91$ A and $V_{OC} = 0.70$ V. At the maximum power point $P_C = P_m = 0.52$ W, $V = V_m = 0.60$ V, and $I_C = I_m = 0.87$ A. Since voltage equals current times resistance, the resistance R across the cell required to give the maximum output is given by $V_m = I_m R$, so:

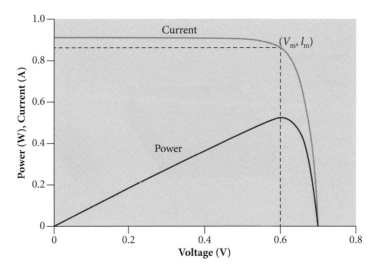

Fig. 8.6 Characteristics of a silicon solar cell (Sandia National Laboratories) with a PERL (passive emitter, rear localized contacts) architecture to reduce recombination at the contacts.

$$R = V_m/I_m = 0.60/0.87 = 690 \text{ m}\Omega$$

The illumination was 1000 W m^{-2}, so this cell of area 22 cm^2 has an efficiency of 23.6%.

We notice that I_m and V_m are both close to I_{SC} and V_{OC}, respectively. **The fill factor (FF)** is defined as the ratio:

$$FF = P_m/(I_{SC}V_{OC}); \text{i.e.} P_m = FF \times I_{SC}V_{OC} \tag{8.5}$$

For the cell shown in Fig. 8.6, the fill factor *FF* equals 0.82. *FF* is a measure of how close the *IV* characteristic is to a rectangle. It is useful for quality control, with good silicon solar cells having *FF* > 0.8.

We now look at why single junction solar cells have efficiencies of typically only ~20–30%.

8.5 Efficiency of a Solar Cell

The **conversion efficiency** is defined as the ratio of the maximum power output to the incident solar power, which for AM1.5 solar radiation (Fig. 8.1) is 1000 W m^{-2}. That this is not 100% is due to several causes: one is that not all the photons have enough energy (>1.1 eV for a silicon cell) to produce electron–hole pairs, as can be seen in Fig. 8.1.

The photons with energies less than 1.1 eV carry 23% of the incident solar energy. Only 1.1 eV of the energy of any higher-energy photons is available to produce power, with the rest lost as heat. The result is that only 47% of the incident solar energy could contribute to the power, with 30% lost as heat.

But there is a further significant loss from the voltage factor, which is the ratio $|e|V_m/E_{gap} \approx 0.6/1.1 \approx 0.55$. This is the ratio of the energy given to an electron at maximum power generation to the minimum energy required to produce an electron. The loss mainly arises from the generation of entropy when light is absorbed from the solid angle Ω_S subtended by the Sun, and emitted effectively into π steradians, when electrons and holes in the forward current recombine. This increases the number of photon states by a factor $W = \pi/\Omega_S$ corresponding to an increase in entropy $\Delta s = k \ln W$ (eqn 2.18), which gives a reduction of $(kT_C/|e|) \ln[\pi/\Omega_S]$ (270 mV in the open circuit voltage, and hence in V_m. There is also a reduction from non-radiative recombination (~200 mV).

The entropy loss can be reduced by concentrating the light from the Sun. The maximum concentration possible is $\pi/\Omega_S \approx 46,000$, which would increase the maximum theoretical efficiency from 30% (see Derivation 8.1 at the end of the chapter) to 39% when $E_g = 1.3$ eV. In practice, the increase in the generated current limits the concentration to about 1000, otherwise the loss through the series resistance of the solar cell becomes too large. The cell must also be kept cool: for silicon, the dependence on temperature is a ~1% loss in absolute efficiency for a 7°C rise in temperature, while for Group III-V semiconductors such as GaAs the corresponding temperature rise for a ~1% loss is 20°C.

The final significant loss comes about because not all the electron–hole pairs make it across the junction: about 10% recombine.

A potentially large loss (~40%) from reflection from the front surface of the silicon can be reduced by a large amount using *quarter-wavelength* layers of material to act as an anti-reflection coating. The reflectance ρ between two media with refractive indexes n_1 and n_2 is:

$$\rho = (n_1 - n_2)^2 / (n_1 + n_2)^2 \tag{8.6}$$

Silicon, since it is partly conducting, has a complex refractive index which is frequency dependent and averages about 3.5. Substituting this value into eqn (8.6) gives $\rho \approx 40\%$. If we add an odd number of quarter-wavelength thick layers with a refractive index n_1 that is intermediary between that of silicon (n_2) and air ($n_0 = 1$) to the silicon, then ρ can be reduced considerably. Fig. 8.7 shows a ray of light incident almost normally on such an arrangement.

The single quarter-wavelength thickness makes the reflected components a and b out of phase. The reflectance at each surface must be equal, so $(n_1)^2 = n_0 n_2$. The effect over the range of solar wavelengths is to reduce ρ to ~6%. Multiple reflection coatings can reduce it still further, to ~1%.

There are also small losses from contacts on the front surface. There is very small loss from those photons which are not absorbed, as a result of the optimization of the silicon thickness plus the addition of reflecting layers on the back of the cell (total ~3%). The overall efficiency η_C after multiplying all these factors together is

$$\eta_C = (0.47)(0.55)(0.9)(0.96) \approx 22\% \tag{8.7}$$

We can see that the efficiency is dependent on the band gap E_g: decreasing E_g increases the photocurrent, since more light can produce electron–hole pairs, but decreases the maximum output voltage, as $|e|V_{OC} < E_g$. For band gaps E_g in the range $0.5 < E_g < 1.75$ eV, under AM1.5 illumination, the short-circuit current J_{SC} is approximately equal to $(800 - 340E_g)$ A m^{-2}; i.e. the short circuit current I_{SC} is:

$$I_{SC} \approx A_{cell}\left(800 - 340E_g\right) \text{A} \tag{8.8}$$

where A_{cell} is the area of cell. V_m is related to E_g, and I_m to I_{SC}, by

$$|e|V_m \approx E_g - 0.5\,\text{eV} \quad \text{and} \quad I_m \approx 0.98 I_{SC} \tag{8.9}$$

The power P is given by $P = I_m V_m$, so there is an optimum value of E_g, which is the value that maximizes the power P. Differentiating the product $I_m V_m$ with respect to E_g and setting the derivative equal to zero (see Exercise 8.14) gives the optimum band gap as $E_{gap}^{opt} \sim 1.4$ eV, and the semiconductors GaAs and CdTe have band gaps close to this optimum value.

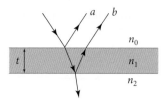

Fig. 8.7 Quarter-wavelength anti-reflection coating (the angle of incidence is exaggerated for clarity).

The maximum efficiency is only slightly reduced for the band gap of 1.1 eV in silicon, so silicon is near optimal as a photovoltaic material.

We now look at the construction of silicon crystalline cells before looking at thin-film solar cells and at devices under development.

8.6 Commercial Solar Cells

By far the majority of solar cells in production today are silicon solar cells, which command 95% of the market. They can produce electricity at or below grid parity in many regions of the world, and look set to dominate the market for single junction solar cells. As we will see in Section 8.7, larger efficiencies can be obtained by building **multilayer solar cells,** in which each layer captures a different part of the solar spectrum. Multilayer cells are currently too expensive to generate electricity cheaply, but a number of prototypes under development look very promising.

The thickness of a silicon sheet needs to be ~300 μm in order to absorb most of incident sunlight, since the light absorptivity of silicon is low. This is a consequence of the nature of the band gap, being indirect rather than direct: when indirect, the excitation of an electron from the valence to the conduction band has to be accompanied by a change in the vibrational state of the lattice. To use a thinner silicon layer (~170 μm) in a solar cell requires light-trapping techniques that make the light pass through the silicon layer several times (see Box 8.3). However, the use of highly light-absorbing direct band gap semiconductor material reduces the amount of material required considerably to a thickness of order 1 μm.

The first thin film cell to appear on the market in the late 1970s was based on **amorphous silicon**, a-Si, and these cells were used to power pocket calculators. At the time silicon cells were very expensive, and it was thought that thin-film cells could be fabricated much more cheaply than silicon cells, as they used much less material. Most types, though, have lost out in the rapid fall in the cost of silicon cells, but some are still used, and some under development (see Section 8.8) may prove very effective. The main type of silicon cell currently in use is the standard BSF cell described above in Section 8.4. The processes involved in their manufacture are described in Box 8.2.

Box 8.2 Manufacture of silicon cells

For good performance, pure silicon is required as the starting material, and until 1997 the silicon used in solar cells was obtained from the waste of the electronic industry. But with increasing global demand this supply became inadequate. Moreover, the purity required for solar cells is not as stringent as that for electronic components, so dedicated plants tailored to solar cell requirements were built. The principal method used is to reduce quartz (silicon dioxide) with carbon in an electric furnace. The silicon is chemically purified and then melted, and a very small quantity of boron added to make what is called p-type silicon.

To produce the wafers for the cells, a monocrystalline p-type silicon rod about 200 mm in diameter and 2–3 metres long is produced by pulling a small silicon crystal out of a crucible of molten silicon. The method was discovered by accident by the Polish chemist Jan Czochralski in 1916 when he accidently dipped his pen into a crucible of molten tin, rather than into an ink pot, and pulled out a single crystal thread of tin! Alternatively, multi-crystalline ingots of silicon can be made by crystallizing the molten silicon; this is a cheaper process, but produces slightly less efficient cells, as there are more defects.

The p-type silicon rod (or ingot) is then cut into thin wafers, about 170 microns (2 sheets of paper) thick, using diamond impregnated wires. The front surfaces of the wafers are textured to reduce reflection. Phosphorous is then diffused a very short distance into the wafer to give a very thin surface region of what is known as n-type silicon. Then all but the top thin n-type layer is etched away, the top surface is textured to enhance light absorption, and an anti-reflection coating is applied. Finally, contacts are added by screen printing silver paste to the top surface to make fine line electrodes, which minimize shading by only covering a small fraction of the area, and by applying a thin film of aluminium to the back surface. The aluminium forms a p^+ layer that sets up a back field which reduces recombination. The surfaces are also treated to optimize charge collection (called passivation). The challenge in manufacturing cells has been throughout in adapting techniques for obtaining the highest efficiency in a research laboratory into low-cost methods suitable for mass-production.

Although the first solar cell was n-type, i.e. the silicon wafer was predominantly n-type with a thin p-type layer, it turned out that p-type cells were less damaged by radiation, and as space applications were then a priority, p-type cell technology was developed and commercialized. However, p-type silicon suffers from light induced degradation caused by boron-oxygen defects, and n-type cells can be more efficient. These cells are taking over an increasing share of the market, and a schematic of one of their designs is shown in Fig. 8.8.

In this cell, electrons promoted by incident light from the valence to the conduction band in the predominant n-type region create a potential difference that can drive a current through an external device. As in a p-type cell, electron flow through the junction from the n- to the p-region is opposed by the internal electric field. Both contacts are on the back surface, which eliminates the shading loss caused by front surface electrodes, and the electrodes can be close

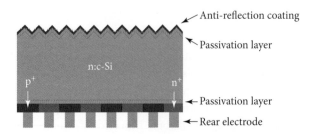

Fig. 8.8 An IBC-SHJ back contact n-type silicon solar cell.

together to lower the series resistance, which reduces the loss in operation. An interdigitated back contact (IBC) is made by doping a thin amorphous silicon (a-Si) layer alternately n^+ and p^+ to make back field and junction regions, respectively. As the n- and p-regions in the n-p junction are formed in different semiconductor materials, a-Si and c-Si, the junction is called a silicon heterojunction (SHJ) and the architecture IBC-SHJ. The front surface is shown textured to give light trapping (see Box 8.3) and has an anti-reflection coating.

A prototype IBC-SHJ cell with an area of 180 cm^2 has achieved an efficiency of 26.6%, suggesting that commercial module efficiencies of 25% will be possible in the near future.

About a third of silicon cells manufactured in 2017 were made from crystalline (c-Si) rather than multi-crystalline (m-c-Si) silicon and the share of c-Si is growing, primarily because their efficiency is higher. We now describe a few important thin film cells and then multilayer cells. At the end of this section we consider some of the technologies that are under development.

8.6.1 Thin-film Cells

There are a number of semiconductor materials that have good solar light absorption and have been made into cells with good efficiency: in particular, GaAs, CdTe and CuInGaSe$_2$ (CIGS), and to a lesser extent amorphous hydrogenated silicon (a-Si:H). All of these films only need to be about ~1 μm thick, so much less material is required ($\lesssim$1%) than for silicon cells. Materials are only part of the cost, and the challenge in thin-film technology is to develop techniques for fast deposition of films while maintaining film quality. Cells with a large area (~1 m^2) generally have more imperfections and so have a lower efficiency than small-area (~1 cm^2) devices.

GaAs cells

GaAs has a band gap of 1.4 eV, close to optimal and a GaAs cell has the highest single junction efficiency of 29.1%. It can withstand high temperatures since the band gap is sufficiently large to keep thermal excitation small. This enables light concentrators to be used with GaAs cells (see Section 8.8.5). These focus the solar radiation onto the active cell area and can increase the flux by up to 1000 times. The cell architecture is similar to that of the BSF cell shown in Fig. 8.c, but the thickness of the absorbing region is only 3 μm as GaAs has a direct band gap. However, fabricating the cells is relatively expensive and only small area GaAs cells are available. Where higher performance is required, multijunction cells (see Section 8.7) are used, e.g. with concentrators and in space applications.

EXAMPLE 8.2

A 4 cm^2 GaAs solar cell has a saturation current I_S of 4×10^{-15} mA. Under normal illumination of AM1.5 solar radiation the short circuit current I_{SC} is 127 mA. What is the conversion efficiency under normal and under ×1000 illumination?

The open circuit voltage V_{OC} under normal illumination is given by eqn (8.3) as:

$$V_{OC} = V_T \ln(I_{SC}/I_S) \approx 0.026 \ln[127/(4 \times 10^{-15})] = 0.988\,\text{V}$$

We could find the maximum output power P_m by finding I and P for different V, using eqns (8.2) and (8.4), noting that $I_{SC} = I_L$. However, good approximations (see Exercise 8.6) for V_m and I_m are given by:

$$V_m = V_{OC}(1 + x_{OC} \ln x_{OC}) \text{ and } I_m = I_{SC}(1 - x_{OC}) \tag{8.10}$$

where $x_{OC} = kT/(|e|V_{OC}) = V_T/V_{OC}$, $V_T \equiv kT/|e| \approx 0.026$ V at room temperature. So, the fill factor FF is given by

$$FF = (1 - x_{OC})(1 + x_{OC} \ln x_{OC}) \tag{8.11}$$

For normal illumination $x_{OC} = 0.0263$. Substituting in eqn (8.10) gives:

$$V_m = 0.89 \text{V} \quad \text{and} \quad I_m = 124 \text{mA}$$

So, $P_m = 0.89(124) = 110$ mW

Under ×1000 illumination the open circuit voltage increases as the short circuit current is 1000 times larger, i.e.

$$V_{OC} = 0.026 \ln[127\ 000/(4 \times 10^{-15})] = 1.168 \text{V}$$

The value of $x_{OC} = 0.0223$ and the maximum output power is when $V_m = 1.07$ V and $I_m = 124$ A, corresponding to:

$$P_m = 1.07(124) = 133 \text{W}$$

Under AM1.5 illumination the solar intensity is 100 mW cm^{-2}, so as the area of the cell is 4 cm^2 a solar power of 400 mW falls on the cell. The conversion efficiency is the output power P_m over the incident solar power, i.e. 110/400 = 27.5%.

Under ×1000 illumination the incident solar power is 400 W and the output power P_m is then 133 W, giving a conversion efficiency of 33.3%. Increasing the illumination gives an improvement of a factor 1.21 in conversion efficiency. The fill factor remains approximately constant, changing from 0.881 to 0.895.

CdTe and CIGS solar cells

CdTe and CIGS have both been made into large-area (~1 m^2) solar cells and accounted for about 4% and 1%, respectively, of global production in 2019; their band gaps are, respectively, ~1.5 eV and, dependent on CIGS composition, ~1.1–1.7 eV. Both are examples of heterojunction devices as the p- and n-regions are in different semiconductors. In CdTe cells, a thin (~0.1 μm) n-doped CdS layer is on top of an ~5 μm thick p-doped CdTe layer, while in CIGS cells, a thin n-doped CDS layer is on top of an ~1 μm thick p-doped copper indium gallium selenide layer. Both cells are fabricated on thin glass sheets with a film of a transparent conducting oxide to provide a good electrical contact to the thin CdS layer. A back metallic film makes contact with the p-doped region. Anti-reflection coatings reduce reflections. Both can give efficiencies greater than 20%.

However, given the scarcity of their constituent materials (Te, Ga, and In), very large-scale production of these cells is unlikely. Glass substrates are fragile and there may be specialized applications with the cells on polymer films. These will give flexibility, low weight, easy mounting, as well as use on buildings and for mobile power units.

Amorphous silicon solar cells

The first amorphous silicon (a-Si) solar cell was made in 1976. a-Si is produced by electrically decomposing silane, SiH_4, together with a small amount of boron dopant. The hydrogen provides additional electrons that combine with dangling bonds in the a-Si, producing an intrinsic semiconductor, a-Si:H. The material is a silicon–hydrogen alloy with 5–20 (atomic) per cent of H, and the resultant band gap is ~1.7 eV. The disordered arrangement of atoms, together with the hydrogen, also gives a high optical absorption, which allows the device to be only ~1 μm thick.

The cell is built upon a glass substrate, with a transparent indium tin oxide layer providing the electrical contact to the p-doped region of a hydrogenated amorphous SiC layer. The SiC has a larger band gap than a-Si and so allows a larger fraction of the solar radiation through to the intrinsic hydrogenated a-Si layer (a-Si:H). Below the a-Si:H there is an n-doped region with a contact layer.

In a crystalline silicon cell, electrons and holes produced by light within a few diffusion lengths of the junction contribute to the photocurrent. But in the p- and n-regions of the a-Si the diffusion lengths are quite small, and most of the contribution is from electron–hole pairs produced in the a-Si:H region. The field between the doped regions causes the holes to flow to the p-region and the electrons to the n-region. The layers form a single p-i-n junction, where i stands for the intrinsic a-Si:H layer.

While the technology is well established and robust cells can be made on flexible substrates, its limitation is its relatively low efficiency of ~10%. Part of the problem is that there is some degradation under illumination, but more critical is the large voltage loss arising from its amorphous nature. Its share in the market has been dropping, and it is likely to be overtaken by higher efficiency thin-film cells.

The thickness of silicon solar cells has reduced from around 300 um to around 170 μm thick layers now. This has been made possible by using **light trapping**. This method, described in Box 8.3, can lead to an increase in path length of ~$4n^2$, where n is the refractive index of the material. For silicon n is ~3.5, so the increase is ~50. The limit to thickness now lies in the fabrication.

Improving the efficiency of the cell could give a more cost-effective cell and this is possible by making a multilayer device consisting of several p-n junctions on top of one another.

Box 8.3 Light trapping in thin films

Silicon has a low absorption, so light trapping is required if thinner layers are to be used. There is also less recombination in these thin layers, which increases the open circuit voltage and hence the efficiency. One way of enhancing the light absorption is to texture the top surface of the silicon and make the back surface reflective, as illustrated in Fig. 8.9.

The texturing of the top surface can cause the light to make two reflections. This reduces the overall amount reflected, but more important is the increase in path length for light within the silicon. Silicon has a high refractive index of $n \sim 3.5$, so if silicon is in a medium with a refractive index of unity then only light incident within a cone of half-angle $\theta_c = \sin^{-1}(1/n)$ will escape. Otherwise the light is totally internally reflected, as shown in Fig. 8.9 (a).

Light passing through the top surface of the silicon undergoes refraction, and as the surface is irregular it acts as a diffuser. For an ideal diffusing surface the intensity per unit solid angle $B(\theta)$ is related to the incident intensity I_0 by:

$$B(\theta) = (I_0/\pi)\cos\theta \; \text{(Lambert's law)} \tag{8.12}$$

After each reflection the fraction f of light that escapes is given by:

$$f = \int_0^{\theta_c} (1/\pi)\cos\theta \, 2\pi\sin\theta \, d\theta = \sin^2(\theta_c) = 1/n^2 \tag{8.13}$$

which is the fraction of light that falls within the critical angle θ_c. Fig. 8.9 (b) shows the amounts lost and intensities remaining after each reflection (NB The actual light is scattered in all directions.) The mean distance D that light travels between each reflection off the back surface is given by:

$$D = \int_0^{\pi/2} (1/\pi)\cos\theta (2W/\cos\theta) 2\pi\sin\theta \, d\theta = 4W \tag{8.14}$$

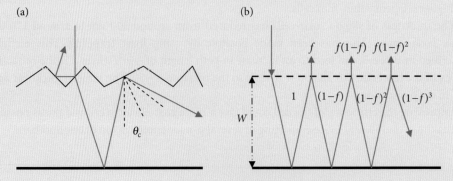

(a)

(b)

Fig. 8.9 (a) Textured top and reflective back (b) Schematic of multiple reflections.

where W is the depth of the silicon layer and $(2W/\cos\theta)$ is the distance travelled by light reflected at an angle θ.

The mean path P travelled is obtained by summing the fraction of light undergoing 1, 2, 3 … reflections, each multiplied by $4W$, which from Fig. 8.9 (b) is given by:

$$P = 4W + (1-f)4W + (1-f)^2 4W + \ldots = 4W/f = 4n^2W \tag{8.15}$$

where the identity $(1-x)^{-1} \equiv 1 + x + x^2 + \ldots \ (x < 1)$ has been used to give the sum of the geometric series. Since n for silicon is ~3.5, this path length is ~50 times greater than that with no texturing of the front surface and no back reflective surface. One way the front surface can be textured is by using an acid etch.

8.7 Multilayer Thin-film Cells

A multilayer device can utilize different regions of the solar spectrum. Consider a two-layer cell with a wide band gap material as the upper layer, and a narrow band gap as the lower layer. High-energy photons are absorbed in the upper layer, while lower-energy light is transmitted by the upper layer and absorbed in the lower layer. An example of such a device under development has a silicon lower layer and a **perovskite** upper layer (see Section 8.8.1). Perovskites are crystals with the same structure as the mineral calcium titanium oxide, originally discovered in the Ural Mountains and named after the Russian mineralogist Lev Perovski. The construction of such a two-layer solar cell is shown in Fig. 8.10. (These are also called **tandem** cells.)

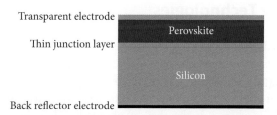

Fig. 8.10 Two-layer or tandem perovskite-silicon solar cell.

The layers are in series, so the current passing through both layers is the same and is limited by the layer producing the smallest current. By using perovskite with E_g ~1.7 eV on top of silicon E_g ~1.1 eV, their short circuit currents are about equal.

A four-layer device with band gaps of 1.8, 1.4, 1.0, and 0.7 eV has a theoretical efficiency greater than 50%, and 46% has been achieved. These are also called multijunction cells.

EXAMPLE 8.3

A two-layer Perovskite-Silicon solar cell has band gaps of 1.7 eV and 1.1 eV. Under standard illumination (1000 W m^{-2}), the photocurrent density from photons with energies between 1.1 and 1.7 eV is 19 mA cm^{-2}, while that from photons with energies greater than 1.7 eV is

21.3 mA cm^{-2}. Compare the output power and efficiency for this two-layer solar cell with that of a simple silicon solar cell ($E_g = 1.1$ eV). Assume the open circuit voltage of each layer, and of the silicon cell, is given by $V_g - 0.4$ V. Take the fill factor to be 0.8.

The two layers are in series, so the voltage across the cell will be the sum of the voltages across each layer. The short circuit current density through the layers is that from the lower layer, 19 mA cm^{-2}, as it is the smaller. The open circuit voltage is the sum of V_{OC} from each layer, so:

$$V_{OC} = (1.1 - 0.4) + (1.7 - 0.4) = 2.0\,\text{V}$$

The maximum power P_m is given by eqn (8.5):

$$P_m = FF \times I_{SC} V_{OC} = 0.8(19)2.0 = 30.4\,\text{mW cm}^{-2}$$

Since 1000 W m$^{-2} \equiv 100$ mW cm^{-2}, the efficiency is 30.4%. The simple silicon solar cell has a short circuit current given by:

$$I_{SC} = 19 + 21.3 = 40.3\,\text{mA cm}^{-2}$$

The open circuit voltage is $(1.1 - 0.4) = 0.7$ V. The output power is therefore:

$$P_m = 0.8(0.7)40.3 = 22.6\,\text{mW cm}^{-2}$$

and the efficiency is 22.6%. We can see that the power output, and hence the efficiency, is significantly higher in the two-layer cell.

8.8 Developing Technologies

An interesting method for concentrating and converting sunlight that is attracting research is to load a sheet of transparent material (glass or plastic) with a suitable dye and a wavelength shifting material that causes the sunlight to be absorbed and re-emitted at a longer wavelength. A significant fraction of the emitted light is then channelled by total internal reflection to the edges of the glass where it is absorbed by photocells. No light is lost by reabsorption because of the wavelength shift. The challenge is to improve the efficiency of the device, which is called a **luminescent solar concentrator**.

A technique that is under development that could be applied to several types of cells is to use both sides of a cell. Such **bifacial cells** would require transparent electrodes on both the front and back of the cell. By mounting these cells about a metre above a reflective surface such as light-coloured stones or white-painted concrete, the output could be increased by up to 30%, and potentially produce cheaper electricity. A modification that is already employed is to cut panels in half. These **half-cut panels** have slightly improved performance as the current and series resistance are both halved, which halves the resistance power loss ($I^2 R$) and also keeps the panel cooler. They are also more shade tolerant.

One of the most significant developments in recent years is that of the perovskite solar cell, where the efficiency has increased from 3.8% in 2009 to 25% in 2020.

8.8.1 Perovskite Solar Cells

Perovskite cells emerged from research on dye sensitized solar cells (DSSC). Light is absorbed in a dye coated on a porous TiO_2 layer producing electron-hole pairs. The electrons are conducted by the TiO_2 to one electrode and the holes are filled by electrons from negatively charged ions in the electrolyte that come from the opposite electrode. When the dye was replaced by a perovskite in 2009, the cell had an efficiency of 3.8%. But by 2013 Henry Snaith et al. at Oxford University had shown that efficiencies of over 15% were possible using just the perovskite as both a light absorber and charge transporter. The perovskite semiconductor was found to conduct electrons and holes, produced when the cell absorbs light, over sufficiently long distances that it can operate like a p-i-n type of solar cell, as in an a-Si cell (see Section 8.6). A schematic of the structure of a perovskite cell is shown in Fig. 8.11.

The n-type material sets up a depletion region at the boundary with the perovskite, as does the p-type material, and there is an internal electric field within the perovskite. The n-type collects electrons, as in the p-n junction solar cell illustrated in Fig. 8.5, the p-type holes. When light is incident on the cell it creates electrons and holes within the perovskite absorber, and these diffuse and are swept as in a p-i-n type of solar cell, to the n-type and p-type layers. There is good charge collection with relatively little recombination (other than radiative), so the ratio of open-circuit voltage to band gap is high, which is necessary for high efficiencies to be obtained.

Excellent cells have been made with the organometal trihalide perovskite ABX_3, methylammonium lead trihalide perovskite with $A \equiv (CH_3NH_3)$, $B \equiv Pb$, and $X \equiv I$. It has a direct band gap of ~1.6 eV and needs to be only 330 nm thick. It can be made by solution-based processing, which requires very little energy and which should enable production costs to be low. The band gap can be changed over the range 1.2–2.3 eV by altering the composition of the perovskite. However, the perovskite material is sensitive to air and water and researchers have fine-tuned the cell's composition to improve its stability. Encapsulation has also helped and cells can now run for thousands of hours at elevated temperatures and high humidity. The use of lead has caused safety concerns, though the quantities used are very small and, provided the encapsulation is sound, the lead would be contained.

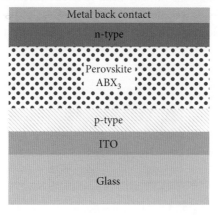

Fig. 8.11 Schematic of a perovskite solar cell. (ITO is indium tin oxide, a transparent conducting material.)

Perovskites can also be printed onto flexible as well as solid materials. Also, changing the composition of the perovskite alters its colour, which could be attractive when used in building integrated photovoltaic (BIPV) systems. Development is being directed at making a tandem cell with a perovskite cell on top of a silicon cell. It is estimated that adding the wider band gap perovskite could increase the overall efficiency up to 30%–35%, and in January 2020 an efficiency of 29.15% was achieved with a small 1 cm^2 cell. Scaling up is possible and these cells should also be cheap to manufacture, so should not increase the module plus balance of system (Section 8.11) cost significantly. Provided they also prove to be stable and non-polluting, they could be a very significant advance in solar cell technology.

8.8.2 Organic Photocells

In the search for cheap solar cells there has been considerable research and development on organic thin-film solar cells, which might be produced at low cost by being fabricated on a flexible plastic substrate. If printing technology could be used, then vast areas of these organic photovoltaic cells (OPVs) might be produced quickly and cheaply (like the printing of newspapers). These cells can have a coloured appearance or be semi-transparent, which is attractive in building integrated photovoltaic applications. Currently their competitiveness has been affected by their relatively low efficiencies but recently an 18% OPV has been reported.

OPVs are made of p-type donor and n-type acceptor organic semiconductors. These materials are typically sandwiched between a transparent anode and a metal cathode. This photoactive layer can be created from a solution containing the two components to form an interlacing network of regions, in which the donor and acceptor materials are connected, respectively, to the hole (anode) and electron (cathode) transporting electrodes (see Fig. 8.12). When solar photons are absorbed, they generate excitons, which are coulombically bound electron-hole pairs. These diffuse typically ~10 nm and are separated by the electric field at the boundary of the donor and acceptor materials; there the electrons and holes travel through the acceptor and donor regions to the electrodes. A thickness of ~100 nm is needed to absorb the light, which is why the photoactive layer is composed of interlacing regions.

These bulk heterojunction cells initially used fullerene derivatives as acceptors and conjugated polymers as donors. But these acceptors have poor light absorption and are difficult to tune and the highest efficiency achieved was only 8%. Recently, though, 18% efficiency has

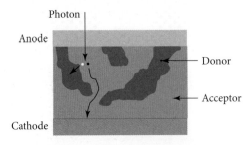

Fig. 8.12 Bulk heterojunction organic solar cell.

been reported using a fused ring compound as acceptor, together with a copolymer donor. Such efficiencies make OPVs possible candidate top cells in tandem solar cells, as well as in applications where flexible light weight cells are needed. Modules weigh around 0.5 kg compared with about 11 kg for silicon panels per square metre. With economies of scale these may offer applications in the developing world and elsewhere for lighter and temporary structures, as well as for mobile deployment.

8.8.3 **Thermo-photovoltaic Cells (TPVs)**

Thermophotovoltaic cells (TPVs) enable the direct conversion of heat into electricity, and were conceived in the late 1950s and early 1960s. The heat source can be the combustion of fossil-fuels, biofuels, or renewable fuels, such as hydrogen, nuclear or solar power, or waste heat. A heat source raises the temperature of a surface to a temperature typically greater than ~1000 °C. This surface emits mainly infrared radiation which is converted to electricity in a photovoltaic cell. Although the emitter temperature is much lower than that of the Sun, its proximity to the PV cell means that the electrical output power density can be much higher than that from a solar cell. While the intensity of solar radiation is highest at a photon energy of 2.6 eV, for an emitter at 1200 °C the peak is at 0.5 eV. Narrow band gap materials are therefore needed, and gallium antimonide GaSb (Eg = 0.73 eV) or indium gallium arsenide InGaAs (E_g = 0.75 eV) cells are commonly used. Thermophotovoltaic systems are quiet and require little maintenance, so are suitable for remote locations. They can also provide both heat and power.

The technology has the potential of high efficiency by using a selective emitter that concentrates the radiation into a band corresponding to photon energies just above the band gap. However, this would reduce the power density considerably, and generally the aim is to select all photons above the band gap, sometimes with the help of a filter. It has proved difficult to find effective materials that withstand high temperatures, and the relatively complex construction and the losses in the several energy conversions have meant the technology has not yet proved cost-effective.

One possible improvement is to use a band-pass filter made out of material with a refractive index n, for then the intensity can be enhanced by n^2. The photovoltaic cell does not have to be in contact, but must be within less than a micron of the filter to benefit from this enhancement. This method could increase both efficiency and power density, but such tolerances will be challenging to achieve.

An effective strategy reported recently is to suppress the below band gap photons by using the band-gap selectivity of the photocell. A highly reflective gold mirror is evaporated on the cell's back surface to reflect below band gap photons back onto the hot emitter where they are absorbed and re-emitted with the range of energies characteristic of the emitter's thermal radiation. The mirror also increases the light collection of the cell, boosting its efficiency. This recycling of photons gave an efficiency at an emitter temperature of 1207 °C of 29.1% for the ratio of power out of the cell to incident light intensity on the cell. Efficiencies of greater than 50% are predicted to be possible. Such an efficient TPV device could be useful for remote power generation, combined heat and power, and waste heat conversion to electricity.

8.8.4 **Thermo-radiative Photovoltaics**

An alternative method of converting heat to electricity, which in principle could be used for power generation at night, is to arrange for the thermal radiation emitted by a photovoltaic cell to be greater than any thermal radiation absorbed by the cell. This is opposite to what happens in a solar cell and the current generated in such a thermo-radiative cell is in the opposite sense. For a cell at an outdoor temperature of ~300 K pointing at the night sky, a selective emitter tuned to the atmospheric window lying between 8 μm and 13 μm, and which reflects all other wavelengths, must be attached to the surface of the cell.

The cell must have a bandgap of less than 0.1 eV for a sufficient fraction of the 300 K thermal radiation to be effective. The selective emitter couples the cell at a temperature of ~ 300 K to outer space at a temperature of ~3 K. The whole unit would need to be encapsulated to restrict conduction and convection. It has been estimated that potentially more than 10 W m^{-2} could be generated under typical sky conditions, but the technical challenges are very significant. (In principle, a surface coating with the above selective emitting properties (sometimes called a super-cool material) could be used to provide **radiative cooling** for a building, and there are a number of start-ups developing this technology.)

8.8.5 **Concentrated Photovoltaics (CPV)**

The principle underlying concentrated PV (CPV) is that by concentrating sunlight onto PV cells, the area of cells needed per watt output is sufficiently reduced compared with a conventional PV system so that the total cost of the cells (either of special cells for high concentration or of silicon cells for low concentration, and of the optics and trackers for the CPV system) would be less than that of the PV panels. There would also be a slight improvement in conversion efficiency as the illumination is greater (see Example 8.4).

Around 2009, the cost of conventional silicon PV panels was such that the economics of CPV systems looked very competitive. However, the recent very sharp fall in silicon PV prices has put considerable pressure on CPV and the industry has contracted significantly; only 370 MWp had been installed by 2017. As GaAs multi-junction cells can be used, CPV can operate at higher temperatures than conventional silicon panels and with efficiencies approaching 40%; also tracking can give higher late afternoon output when demand is typically high. However, it cannot utilize diffuse radiation; it is most cost-effective in regions with a DNI > ~2000 kWh m^{-2} y^{-1}, but even then its LCOE is higher than that of PV: €0.10 kWh^{-1} to €0.15 kWh^{-1}, and at locations with a DNI of 2500 kWh m^{-2} y^{-1} €0.08 kWh^{-1} to €0.12 kWh^{-1}, c.f. PV (2019) €0.04 kWh^{-1}. (DNI is the direct normal irradiance.)

The challenge will be in gaining sufficient global market penetration to obtain the economies of scale and from learning given the current lack of 'bankability' for CPV. Fabrication of lightweight compact optics, and very small cells, which dissipate heat more easily, using techniques developed for LED manufacture, may help. Improving 'bankability' will require demonstrating cost reductions and system reliability.

8.9 Solar Modules, Panels, and Solar Farms

Solar cells are typically some 240 square centimetres in area and, since they are very fragile, around 60 cells are mounted together between glass sheets in an aluminium frame to form a solar module. The cells are connected together in series, giving an operating voltage of about 36 volts. One or more modules make up a solar panel. A output of a solar panel is determined by the intensity of the sunlight, and is reduced when it is cloudy to between about a tenth to a third of what it is on a clear sunny day. In the Tropics the maximum intensity is close to 1 kilowatt per square metre; typical efficiencies are 21%, so a panel, which has a cell area of 1.44 square metres, would then produce 300 Watts ($1000 \times 0.21 \times 1.44$). This output under 1000 W m^{-2} illumination is termed the Watt-peak (Wp) output, and the cost of modules is often given in terms of $ per Wp or € per Wp.

The amount of power that a solar panel will produce over a year depends on how much sunlight the location receives. A 1-kWp array of panels gives an output of 1 kW when the solar intensity is 1 kW m^{-2}. The yearly output (without losses) from a 1-kWp array is therefore numerically the same as the annual amount of solar energy per square metre, which is made up of both direct and diffuse radiation. The total falling on a horizontal surface is the global horizontal irradiance (*GHI*), but solar panels are typically tilted away from horizontal by an angle close to the latitude of their location to better collect the direct radiation. The global tilted irradiance (*GTI*) is therefore higher than the *GHI*; for example, by about 15% in Seattle (47.6 N) and Los Angeles (34.1 N).

The actual output is given by the performance ratio (*PR*) of the PV system. This is the ratio of actual to theoretical annual energy production, which is the annual amount of incident solar energy multiplied by the PV module's efficiency, so the *PR* is a measure of all the system losses. The *PR* has improved from ~70% before 2000 to 80–90% by 2019. So, the actual yearly output is given by *PR*×*GTI,* and can often usefully be approximated by *GHI* in mid latitudes, since the gain from tilting roughly compensates the system losses.

The capacity factor of a PV system is the ratio of the annual output to the output if operating all the time at its rated capacity, which for a 1-kWp array would be 8760 kWh. In Germany, for example, $GHI \approx 1100$ kWh m^{-2} y^{-1}, corresponding to a capacity factor of 12.5%. This is for the DC output of the panels, but **inverters** are used to convert the output to AC. For solar farms, the DC power rating are typically higher than their AC ratings and the inverter loading ratio (ILR) might be 1.3. Oversizing the array of panels is increasingly less costly, and will generate more energy and revenue, particularly in winter months and in late afternoons. A smaller inverter is cheaper, and the output is only clipped occasionally, since the solar intensity is often rarely high enough to give peak DC capacity. The capacity factor on the grid is then increased by close to the ILR, which would give an AC CF in Germany of around 16.5% with ILR = 1.3. The global average of the capacity factor is ~0.18 (2018).

Coupling the solar farm with storage avoids the need to clip the output and the excess can be stored for use when the Sun is down, which is often when the demand is high; furthermore, the ILR and AC CF can be increased. For example, a solar farm with 100 MW (AC) output, coupled to a 50 MW/4hr storage system might have an ILR of 1.9 to maximise the revenue to cost ratio. Such farms could provide fast dispatchable power to help maintain grid stability.

8.9.1 **Solar Farms**

Photovoltaic solar farms comprise a large array of solar panels used to generate electricity for a grid. For example, in the south west of the US is the California Valley 250 MW capacity farm, which has a capacity factor of 25%. Its panels, made up of 20% efficient silicon solar cells, are mounted on a single-axis tracking system, which increases the output relative to a fixed orientation by about 20%. It covers an area of 796 ha and generates 0.55 TWh per year, corresponding to an area requirement of ~15 sq km per TWh. Most of the global growth in photovoltaic capacity is through large solar farms, such as the one in Qinghai province in China illustrated in Fig. 8.13, which can have capacities of over 500 MW. (NB Tracking is not worthwhile when adding panels is cheaper than adding trackers.)

Fig. 8.14 illustrates the optimum orientation for a fixed solar panel when located at a latitude of φ. The panel points south, and its normal is tilted at an angle of φ to the vertical. When the Sun is at an equinox (21 March or 21 September) the Earth's axis is perpendicular to the direction of light from the Sun, and the Sun rises and sets as shown in Fig. 8.14.

Fig. 8.13 Photovoltaic panels in the Qinghai Golmud Solar Park in China (copyright iStock.com/zhudifeng).

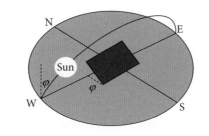

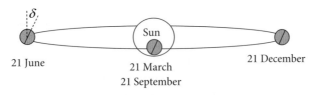

Fig. 8.14 Optimum orientation of a fixed solar panel.

The tilt (declination) δ of the Earth's axis to the plane of the Earth's motion around the Sun, which is illustrated in Fig. 8.14 varies from 0° to ± 23.5°, lowering the path across the sky in winter and raising it in summer. For a panel oriented towards the equator, and with its slope angle equal to its latitude, the intensity I on the panel is:

$$I = S\cos\delta\cos H \tag{8.16}$$

where S is the solar intensity on a surface perpendicular to the direction of the Sun's rays (the global tilted irradiance, *DTI*), H is the angle that the Sun has moved in its plane of motion since noon, and δ is the declination. The angle H is given by $H = 15°(t - 12) = \pi(t - 12)/12$ radians, where t is the time in hours.

EXAMPLE 8.4

Fixed silicon solar panels with an efficiency of 20% are located at a latitude of +35°. The active area of panels is 10,000 m². Estimate the maximum daily output in kWh of the solar farm in September, when the number of hours of sunshine each day is 7. Take the intensity of the Sun as 1 kW m^{-2}.

The optimum orientation is for the solar panels to be pointing south and tilted by the angle of latitude, in which case eqn (8.16) holds. In September $\delta \sim 0$, so the total solar radiation R in one day is given by:

$$R = A \times \int_a^b S\cos[\pi(t-12)/12]\mathrm{d}t$$

where $A = 10^4$ m² is the panel area, and a and b are the times the Sun rises and sets. In this case, $a = 8.5$, $b = 15.5$, and $S = 1$ kW m^{-2}. Integrating yields

$$R = A(12S/\pi)[2\sin(3.5\pi/12)] = 10^4(12/\pi)1.587 = 6.06 \times 10^4 \, \text{kWh}$$

The efficiency of the panels is 20% so the output = 12.1 MWh.
(If the panels tracked the Sun, then $R = 7 \times 10^4 \, \text{kWh}$.)

8.9.2 Distributed Generation with Photovoltaics

A steady expansion of solar photovoltaics is also happening in residential systems, where the cost of electricity from domestic rooftop installations is now often comparable to electricity rates from the grid; as in the US, Australia, and in Germany. These systems can give their owners the possibility of selling excess generation back to the grid, called **net metering**, or storing it in batteries for evening use. Almost 40% of global capacity is distributed in such systems, rather than in farms. There has been concern over net metering as it pushes the burden of maintaining the grid onto those without solar panels; however, this is offset by home ownership of solar panels reducing the need for utilities to invest in infrastructure and lowering peak demands when electricity prices are highest.

In the developing parts of Asia and sub-Saharan Africa, solar panels are making a difference not just in supplying electricity to grids but also in distributed generation through rooftop solar installations; and furthermore, in powering mini-grids in regions where grids are non-existent or poor in quality. Almost one billion people (13% of the world population) are still without access to electricity, mainly in sub-Saharan Africa (600 million) and in India (200 million). In sub-Saharan Africa, the distances are often so great that building a grid would be very expensive, and even where a village is on a grid, the cost of connection for a household can often be unaffordable. The rapidly falling cost of solar panels has meant wider access to affordable clean electricity (see Fig. 8.15). Many homes are now solar powered, with modern

Fig. 8.15 Solar panel recharging a mobile phone in Malawi (copyright Joerg Boethling/Alamy Stock Photo).

energy services provided by increasingly efficient and cheap LED lighting and appliances. As battery costs fall, cooking by electricity will be increasingly available, and diesel generators for supply at night will no longer be needed.

Payment for these systems is being made easier for those with little capital by pay-as-you-go (PAYG) schemes coupled with mobile banking. But ensuring finance for these initiatives is not straightforward, and electricity is not yet reaching the very poor. Where both off-grid and grid expansions are taking place, these need to be complimentary, and government involvement is important, as, for example, in Nigeria. While the cities and most villages in India are connected to a grid, which provides their schools and public institutions with electricity, there can be many homes in a village that cannot afford or do not want to be connected, as the supply is often unreliable. Solar powered homes and **mini-grids** can give a more reliable supply in such situations. In Bangladesh, several million households have solar home systems; many on mini-grids that enable electricity trading, using **blockchain** to ensure secure exchange of information, which helps to balance supply and demand.

EXAMPLE 8.5

A solar panel is made up of 4 modules in parallel each with 40 silicon cells in series of area 0.013 m^2, open circuit voltage 0.7 V, and fill factor 0.8. The short circuit current density of a panel under AM1.5 illumination is 400 A m^{-2}. In the south of the UK there is about 850 kWh m^{-2} y^{-1} of solar radiation. If an area of 15 m^2 is available on a house in that region, estimate the amount of energy a year that could be provided by solar panels.

AM1.5 illumination gives 1 kW m^{-2}. An annual amount E_S of 850 kWh m^{-2} is equivalent to an average illumination $\mathcal{P}_{inc}$ given by:

$$\mathcal{P}_{inc} T = E_S$$

where T is one year (i.e. 8760 h). Substituting gives $\mathcal{P}_{inc} = 97 \text{ W m}^{-2}$. The average short circuit current density J_{SC} is proportional to the intensity of solar radiation, so

$$J_{SC} = (97/1000)400 = 38.8 \, A \, m^{-2}$$

A solar panel has an area of 2.08 m^2 and $V_{OC} = 40 \times 0.7 = 28$ V. Its short circuit current $I_{SC} = J_{SC} \times (0.013 \times 4) = 2.02$ A.

With 15 m^2 available, 7 solar panels could be mounted, and the total average short circuit current $I_{SC} = 2.02 \times 7 = 14.1$ A. The average output power P_{out} is given by eqn (8.5) as:

$$P_{out} = FF \times I_{SC} V_{OC} = 0.8(14.1 \times 28) = 316 \, W = 0.316 \, kW$$

The amount of energy E produced in a year is given by the product of the power by the time, so:

$$E = 0.316(8760) = 2768 \, kWh \approx 2.77 \, MWh$$

Solar panels are often combined with **battery storage**. This allows operation of equipment at night, but the battery can also provide a load that is quite close to optimal. The voltage across a battery remains reasonably constant while being charged. The power provided by the panel is the product of the battery voltage and the current produced by the solar panel. This power can be quite close to the maximum power point over a wide range of current and hence solar intensity. For example, a 12 V battery requires a charging voltage between 12 and 15 V. This voltage can be provided by a 24-cell silicon module with an insolation varying between values of 0.2- and 1-kW m^{-2} (See Exercise 8.14).

Solar panels can provide power in remote locations; for example, for telecommunications equipment and lighting, and also for small electronic devices. Where AC power is required, an inverter is used that converts the DC output to AC. Resistive loads give a voltage proportional to the current and so are not well matched to a solar panel supply. For such loads, a DC–DC converter is used, where the input DC voltage is close to the optimal voltage for the panel.

8.9.3 Hybrid Solar Farms

Solar farms are increasingly being built with energy storage from lithium-ion batteries. A recent very large-scale project was approved in late 2019 to provide 6–7% of Los Angeles's electricity demand with a 400 MW solar farm and a 1200 MWh Li-ion battery. This will be built on 2,650 acres in the north-western Mojave Desert in eastern Kern County, California. It will be able to provide electricity for more than 150,000 homes at a cost of less than 4c per kWh, and for 4 hours each night, cheaper than that from fossil fuel generation. It demonstrates the enormous economies that have come from manufacturing solar cells on a large scale, and there are similar projects around the world. For instance, in the UK, a 350 MW solar farm at Cleve Hill, Kent, with 700 MWh storage to follow, and in Australia a 720 MW Solar Farm in New South Wales, which will be combined with a 200 MW/400 MWh lithium-ion battery.

Solar farms do not have to be built on land—floating installations are also being developed: China has built a 40 MW solar photovoltaic farm on a lake. This has the advantage of not impacting on land use, and also of maintaining efficiency as the panels do not overheat. Coastal locations could also be used. There are also a number of projects looking at putting solar PV farms on hydropower reservoirs. Thailand is heavily dependent on fossil fuels and its good sunshine makes solar PV attractive. However, land is at a premium for food production, so the government is looking to locate solar farms on 15 hydropower plant reservoirs with a combined capacity of 2.7 GW. To see whether their generation would be competitive, a prototype 45 MW floating solar farm is planned for the reservoir at the Sirindhorn hydropower plant. The co-location enables a constant output, saves money on the electrical distribution, reduces the evaporation of water, and its production during the day will enable a larger hydroelectric output at night. In Africa, a study found that a 1% coverage of existing hydropower reservoirs with solar PV would increase output by ~50%.

Large solar farms can be co-located with wind farms and with electrolysers. The electrolysers can use excess generation to produce hydrogen from water (see Chapter 10 Box 10.2). Hydrogen can be used as a seasonal energy store or directly to provide heat in homes and industry, or in fuel cells to provide power. Other ways to handle the variability of solar PV are discussed in Chapter 10.

8.10 **Environmental Impact of Photovoltaics (PV)**

Solar photovoltaic power in operation produces no pollutants, no greenhouse gases, and is a safe way of generating electricity. There are no moving parts, which reduces maintenance and also results in no noise pollution, and no water is required (except some for cleaning). However, for silicon solar cells, the mining of quartz can put miners at risk of the lung disease silicosis, and in the production of pure silicon there can be environmental impacts without due care. Hydrochloric acid is used, which is poisonous, corrosive and very reactive, and one of the more toxic by-products is silicon tetrachloride, but this can be recycled safely back into the production process. With effective safeguards and regulations, the risks in manufacturing solar panels can be kept very small and acceptable. In Europe, it takes between one and two and a half years, dependent on location, to generate the same amount of energy that was used in making the panels. This is called the **energy payback time** (EPBT).

Production mainly uses fossil fuel energy at the moment; but since panels last for 30 years or more, the carbon footprint of their electricity generation is only about 10% of that from gas turbine power plants, and this percentage will reduce as more power is generated from renewables. However, most panels are produced now in China, where electricity is still mainly from coal-fired power stations, and these have about twice the carbon footprint of those built in Europe. When the panels are recycled, about 90% of the materials used can be recovered for use in new modules, and the energy requirement is then approximately a third of that needed when starting from raw materials. But as in the production of the panels, care must be taken in the disposal of chemicals and of any waste to avoid risks to health and any environmental impact.

In regions with good sunshine, an area of about 7.5–20 square kilometres of land, roughly corresponding to GHI ~2500–1000 kWh m^{-2} y^{-1}, is required to generate one TWh per year, which would power around 300,000 European homes [for comparison, about 70 square kilometres of land would be needed for a wind farm to produce this amount of electricity]. A photovoltaic farm would cover a good fraction of this area, unlike a wind farm with space between turbines, but sheep can graze under and between the panels. However, when located on agricultural land, the clearing of the land of vegetation with herbicides prior to installing the panels can degrade the soil and displace wildlife, and the layout of a PV farm can lead to fragmentation of the landscape and create barriers to animal movement and availability of food. The use of chemicals when cleaning the panels can also cause contamination of rivers, and the large area of panels can affect the microclimate and the reflected light disorientate insects and birds. Moreover, a clear view by the panels of the Sun is often maintained with herbicides and dust suppressants.

These adverse effects and their impact on agricultural production can be mitigated by locating the solar PV panels on marginal lands, such as brownfield sites, or in deserts in regions where there is high solar isolation, little cloud cover, and small biodiversity. Solar farms can

Fig. 8.16 PV panels on the side of a building.

also be mounted on water (as noted in Section 8.9), but should avoid affecting aquatic bio-diversity. The visual impact of photovoltaic panels can also be lessened by integrating them into buildings (**building-integrated PV** or BIPV) (see Fig. 8.16), which can save money as smaller quantities of conventional building materials are required, or by applying them to the existing structure (**building-applied PV** or BAPV). Modules are now available which look like roof tiles. The use of the urban environment avoids the impacts on biodiversity that location in the country can cause.

8.11 **Economics of Photovoltaics**

The most significant development in solar cells over the period 2006–2018 has been the dramatic fall in their cost by a factor of about 11. The cost is taken as that of a module whose area in full sunlight (1 kilowatt per square metre) would generate one watt of power, called one **Watt-peak** (Wp). The cost has fallen from about \$3.5 to \$0.3 per Watt-peak, so a module that produces 300 watts in full sun would have cost around \$90 in 2018. About 95% of modules are based on silicon.

While efficiencies have increased from 15% to 21% in the last decade, costs have mainly fallen through using thinner wafers, applying less silver, by developing high throughput processes, and by economies of scale. The learning rate, which is the percentage reduction in costs for every doubling in global capacity, has been 24%, as can be seen in Figure 8.17. This is in line with Swanson's law, which predicts a percentage drop of 20%. (The bump around 2006 was due to a shortage of silicon.) The cost of a solar panel system depends not only on the cost of the modules but also on the cost of structural and electrical components, and the installation. These balance-of-system costs are now roughly two thirds for large photovoltaic farms, where costs have been reduced by using robots in assembly; but are more for residential systems, where savings can be made by integrating the panels into buildings. Costs are

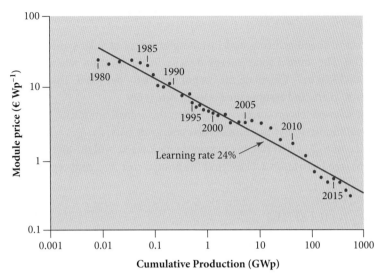

Fig. 8.17 Learning curve for solar modules 1980–2018 (Fraunhofer Institute for Solar Energy Systems ISE).

calculated assuming a lifetime for the panels of typically 20–30 years. However, if the efficiency of solar panels only decreases at 0.5% per annum, the output will still exceed >75% of their rated capacity after 50 years, which would decrease the cost of PV electricity still further.

The cost of electricity depends in particular on that of the solar panel system and on the solar intensity. In the US, the utility scale cost of electricity dropped by a factor of eight during 2009-2018: costs fell from about 36c to 4.5c per kWh, and it is now cheaper than electricity produced from coal, and about the same as that from gas; the costs from these fossil fuel plants are in the range 4-14c per kWh. However, the output of photovoltaic farms is variable and there are costs associated with managing this variability. As with wind farms, these depend on the mixture of electricity generators, the interconnectivity with different regions, the amount of energy storage available, and on the ability to vary demand to suit supply. How the variability of solar power is handled and what are the associated costs are discussed further in Chapter 10.

As the amount generated depends on the amount of sunlight, the cost of electricity will generally be more expensive in less sunny regions and countries; for example, in Germany the mean *GHI* is 1100 kWh m^{-2} y^{-1}, while for the contiguous US the mid-range *GHI* is 1925 kWh m^{-2} y^{-1}—these values are reflected in the differing cost of electricity, which was in 2018 around 4.5 eurocents per kWh and 3 USD cents per kWh, respectively (€1≈ $1.1).

Subsidies are starting to no longer be needed to promote photovoltaics since their costs have fallen so much, and we are now seeing prices set through auctions. The improvements in technology and the competition that these auctions encourage has seen prices fall in 2018 to as low as 2-3 USD cents per kWh in Egypt, India, Saudi Arabia, UAE (Dubai), and US (Texas). Prices are also helped by a low cost of borrowing money, and by favourable support policies. It looks quite possible that by 2030 prices of 2-4 USD cents per kWh will be widespread across the globe.

EXAMPLE 8.6

The capital cost of manufacturing and installing an array of solar panels that will produce 1 kWp is $1200. The annual solar energy density in the location where the panels will be installed is 2000 kWh m^{-2}. Calculate the cost of electricity per kWh. Take the lifetime of the solar panels as 30 years, and assume the discount rate is 6%.

As the area of solar panels produces 1 kWp, then the annual amount E_{elec} of electricity produced will equal (numerically) the annual solar energy density, i.e. $E_{\text{elec}} = 2000$ kWh.

The capital cost C_{capital} is $1200, the discount rate R is 6%, and the lifetime N is 30 years, so using eqn (1.6) (see Derivation 1.1) we find the annual cost A_{cost} that repays the capital from the formula:

$$C_{\text{capital}} = A_{\text{cost}}[1-(1+R)^{-N}]/R$$

So:

$$A_{\text{cost}} = R \times C_{\text{capital}}/[1-(1+R)^{-N}] = 0.06(1200)/[1-(1.06)^{-30}] = \$87$$

The cost of electricity C_{elec} is given by:

$$C_{\text{elec}} = A_{\text{cost}}/E_{\text{elec}} = 87/2000 = 0.044 = 4.4 \text{¢ kWh}^{-1}$$

8.12 Global Solar Photovoltaic Potential

Figure 8.18 shows a world map of the annual amount of solar energy on a square metre of flat ground. Photovoltaics could make a significant contribution to meeting electricity demand in nearly all populated regions.

As can be seen, the average difference in sunshine in countries in the Tropics to those in Central Europe is about a factor of two. In the southern half of India and in the north of Nigeria the levels of sunlight are such that about 2100 kWh could be produced annually for every 1 kWp of panels. The levels in the south-west of the United States and the southern half of Spain would yield 1800 kWh, in Japan 1500 kWh, and in Germany 1100 kWh.

To estimate the potential contribution that photovoltaics can make to a country's energy demand, we need to identify suitable areas. For solar photovoltaic farms, urban areas, forest, ice-covered regions, protected areas, and mountainous terrain have to be excluded. Only a small percentage of agricultural land is suitable, with progressively larger percentages of grassland, barren areas, and deserts. A recent estimate of the potential electricity production from farms covering these areas is 600,000 TWh per year, which is some six times the world's final energy consumption. Furthermore, the amount predicted from panels on rooftops could be considerable: in the US, the National Renewable Energy Laboratory have estimated 40% of US current electricity demand could be met.

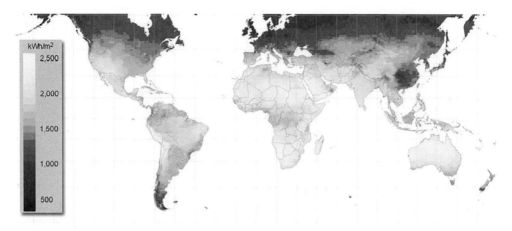

Fig. 8.18 Average annual sunshine in kWh per square metre − the amount is low in south central China where it is mainly cloudy and overcast (© 2019 The World Bank, Source: Global Solar Atlas 2.0, Solar resource data: Solargis).

8.13 **Outlook for Photovoltaics (PV)**

With the cost of producing electricity from utility scale photovoltaic farms now lower than that from fossil fuel plants in many regions of the world, new generators are increasingly photovoltaic. In 2018, these totalled 100 GW—about the same capacity as added by all non-renewable generators. China dominates, having about a third of the global capacity, and both China and India have ambitious plans for expanding photovoltaic generation. But both countries need to upgrade their grids to accommodate their increasing renewable generation. The sharp fall in the cost of solar power is already helping to displace coal generation, with some proposed plants cancelled. Nonetheless, China and India are still building coal plants, which can be easier to bring online to meet local increased demand than a solar farm, unless the farm has storage or other backup supplies (see case studies in Chapter 3). Other countries are also seeing an expansion in photovoltaics (see Fig. 8.19), and as the cost of battery storage falls, the investment in solar panels coupled with storage is picking up rapidly.

In 2019 solar photovoltaics provided 2.7% of the world's electricity demand from a capacity of 580 GW. While this is still small, the global production of solar panels was 115 GW in 2019, and capacity has been doubling about every three years. The International Renewable Energy Agency (IRENA) in 2019 projected that global capacity could be 8500 GW by 2050 producing around 15,000 TWh a year (a capacity factor of 0.2) in their scenario to limit global warming to 1.5 $^\circ$C. This would require an average annual additional capacity of ∼250 GW, and an 68% increase in annual investment in solar PV compared to that in 2018.

More ambitious estimates were made by the Fraunhofer Institute in 2015 of up to 15,000 GW capacity generating 20,000 TWh a year by 2050 from a capacity factor of 0.15, and given a breakthrough in deployment following massive investment, 40,000 TWh a year. Such enormous growth would see the cost of solar photovoltaics fall even more, and would be a way of

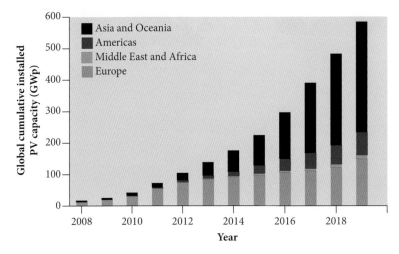

Fig. 8.19 Global installed photovoltaic capacity. In 2019, China had 59%, Japan 18%, and India 10% of the Asia and Oceania market, and the USA 79% of the Americas market (IRENA, International Renewable Energy Agency, Renewable Capacity Statistics 2020).

providing affordable electricity to millions in China, India, North and South America, Africa, and elsewhere.

However, the COVID-19 pandemic may have a significant impact on the expansion of solar PV. Fortunately, the growth in 2020 was 30% more than in 2019, with the total global PV capacity additions in 2020 of 127 GW. Moreover, investment in PV is vital to help tackle climate change, which poses a very significant danger if left unchecked, and would provide a considerable number of jobs that will be needed as the world transitions off fossil fuels and recovers from COVID-19, as well as energy security in the areas where deployed.

 SUMMARY

- Solar radiation is a huge resource, offering more than 5000 times current world power consumption. The solar intensity on a clear sunny day is ~1000 W m^{-2}.

- Silicon solar cells have, under AM1.5 illumination, $V_{OC} \sim 0.7$ V, $J_{SC} \sim 400$ A m^{-2}, and $FF \sim 0.8$. Output power density $\mathcal{P}_m \sim 210$ W m^{-2}, equivalent to a conversion efficiency of ~21%.

- Area of a solar PV farm required to generate 1 TWh per year is about 7.5–20 sq. km.

- The cost of generating electricity by photocells has dropped very significantly over the last 10 years and has now reached grid parity in many sunny regions with a LCOE ~$40 MWh^{-1}.

- Solar PV as a source of low-carbon electricity has great potential and with favourable policies and massive investment might provide annually 40,000 TWh of global electricity by 2050, approximately up to 40% of global demand.

FURTHER READING

Battaglia, C., Cuevas, A., and De Wolf, S. *High-efficiency crystalline silicon solar cells: status and perspectives*, Energy and Environmental Science 5 (2016).

Fraunhofer ISE (2019). *Photovoltaics report.*

Goodall, C. (2016). *The switch*. Profile Books, London. Stimulating argument for the case that photovoltaic cells are now cheap enough for the world to switch from fossil fuels to solar PV.

IRENA (2019). *The Future of Solar PV.*

Jaffe, R. and Taylor, W. *The Physics of Energy*, Cambridge University Press, Cambridge (2018). Clear advanced discussion of the solid-state physics of solar cells.

Peake, S. (ed.) (2018). *Renewable energy*, 4th edn. Oxford University Press, Oxford. Good qualitative discussion of solar photovoltaic technologies.

REN21 (2019). *Renewables Global Status Report.*

Shockley, W. and Queisser, H.J. (1961). *Detailed Balance Limit of Efficiency of p-n Junction Solar Cells*. J. Appl. Phys. 32, 510 (1961); *http://dx.doi.org/10.1063/1.1736034.*

Svaram, V. *Taming the Sun; innovations to harness solar energy and power the planet* (The MIT Press, 2018).

Wurfel, P. (2005). *Physics of solar cells*. Wiley, New York. More advanced textbook on the physical principles of solar cells.

en.wikipedia.org/wiki/Solar_cell Good summary of solar cells.

EXERCISES

8.1 Blue light of wavelength 475 nm falls on a silicon photocell whose band gap is 1.1 eV. What is the maximum fraction of the light's energy that can be converted into electrical power?

8.2 Sunlight of intensity 600 W m^{-2} is incident on a building at 60° to the vertical. What is the solar intensity, or insolation, on (a) a horizontal surface, (b) a vertical surface?

8.3 When it is heated to temperatures of about 1700 K, approximately 20% of the radiation emitted by the rare-earth oxide ytterbium oxide, Yb_2O_3, is in a narrow band around 1000 nm. Would this source of radiation be suitable for a Si or a GaAs photocell? (Band gaps: Si 1.1 eV, GaAs 1.4 eV.)

8.4 The intensity I_{AM0} of solar radiation incident on the Earth's atmosphere (AM0) is given to a good approximation by

$$I_{AM0} = \sigma T^4 \Omega_S / \pi \text{ Wm}^{-2}$$

where Ω_S is the solid angle subtended by the Sun and σ is the Stefan–Boltzmann constant. Find T, given that the intensity I_{AM0} is 1367 W m^{-2} and $\Omega_S = 6.8 \times 10^{-5}$ sr.

8.5 Plot the current through an ideal p-n junction which has a saturation current of 10^{-11} A, for bias voltages of -0.4 to $+1.0$ V in 0.2 V steps.

8.6* The current–voltage relation for an ideal diode is given by

$$I = I_L - I_S[\exp(V/V_T) - 1]$$

where $V_T \equiv kT/|e| \approx 0.026$ V at room temperature. By differentiating the power P, given by $P = VI$, with respect to V and equating the derivative to zero, show that the maximum power P_m occurs when

$$(1 + V/V_T)\exp(V/V_T) = I_L/I_S + 1$$

By approximating $(1+V/V_T)$ by V_{OC}/V_T where $V_{OC} = V_T \ln(I_L/I_S)$, and $(I_L/I_S + 1)$ by I_L/I_S, show that V_m and I_m are given by

$$V_m = V_{OC}(1 + x_{OC} \ln x_{OC}) \text{ and } I_m = I_{SC}(1 - x_{OC})$$

where $x_{OC} = V_T/V_{OC}$. Note that $I_{SC} = I_L$.

8.7 Explain how a silicon solar cell works.

8.8 What are the advantages of an n-type silicon solar cell over a p-type cell?

8.9 What limits the efficiency of silicon solar cell to ~30%, and why is a high value of $|e|V_{OC}/E_g$ in a solar cell important for high efficiency?

8.10 Using eqns (8.8) and (8.9) for I_{SC} and eV_{OC}, show that the optimum value of E_g that maximizes

$$P = FF \times I_{SC}V_{OC} \text{ is } E_g =\sim 1.4 \text{ eV}$$

8.11* In the Shockley-Queisser limit, the maximum open circuit voltage is given by eqn (8.19) (see Derivation 8.1 at the end of the chapter):

$$eV_{OC} = kT_{Cell} \ln(Q_{Sun}/Q_{cell}) - kT_{Cell} \ln(\pi/\Omega_S)$$

where the ratio Q_{Sun}/Q_{cell} is the ratio of the integrals of the Planck distribution above the band gap E_g of radiation at T_{Sun} to that at T_{cell}, and equals

$\int_{E_g}^{\infty} dE \, E^2/(e^{\frac{E}{kT_{Sun}}} - 1) / \int_{E_g}^{\infty} dE \, E^2/(e^{\frac{E}{kT_{cell}}} - 1)$. The integrals are given to a good approxima-

tion by $F(x) = (kT)^3(x^2 + 2x + 2)e^{-x}$, where $x = E_g/kT$.

Calculate V_{OC} for E_g = 1.1, 1.2, 1.3, 1.4 and 1.5 eV, given π/Ω_S = 46,000. Take T_{Sun} = 6000 K and T_{cell} = 300 K.

The photocurrent density $J_L = P_{inc}\left(\dfrac{15}{V_g\pi^4}\right)x_S e^{-x_S}(x_S^2 + 2x_S + 2)$, where P_{inc} is the inci-

dent light intensity, $x_S = E_g/kT_{Sun}$, and $E_g = eV_g$. Show the efficiency is a maximum for $E_g \approx 1.35$ eV, and find its value assuming the fill factor is given by eqn (8.10).

Show that eV_{OC} equals E_g when $T_{cell} = 0$ and the sunlight is fully concentrated such that $\pi/\Omega_S = 1$.

8.12 (a) A silicon photocell has an area of 4 cm^2 and is illuminated normally with AM1.5 solar radiation. The short circuit current is 160 mA and the saturation current is 4×10^{-9} mA. Calculate the maximum power output and the corresponding load resistor. (b)* What is the output power when the load resistor is 10% higher than the optimum value?

8.13 A household uses 4000 kWh of electricity in a year. Estimate what area of solar panels would be required to produce 1000 kWh of electricity per year. The insolation in the region is 800 kWh m^{-2} y^{-1}.

8.14 A 24-cell silicon solar panel has a saturation current density $J_S = 10^{-9}$ A m^{-2}. Show that this panel could be used to charge a 12 V battery by calculating the peak power voltages V_m for insolation values of 0.2, 0.4, 0.6, 0.8, and 1.0 kW m^{-2}. An insolation of 1 kW m^{-2} gives a short circuit current density of 400 A m^{-2}.

8.15 In a region where the solar insolation is 1800 kWh m^{-2} y^{-1}, estimate the area of solar panels that would be required to produce an average output of 100 MW of electricity.

8.16 A reasonable approximation to the dependence of the short circuit current density J_{SC} on the band gap E_g for $0.5 < E_g < 1.8$ eV under AM1.5 illumination is

$$J_{SC} = (800 - 340E_g) \text{ A m}^{-2} \text{ with } E_g \text{ in eV}$$

A three-layer multijunction solar cell has an upper layer with $E_g = 1.8$ eV. (a) Determine the optimal band gaps for the lower two layers. (b) Calculate the output power under AM1.5 solar illumination, assuming a fill factor of 0.8 and an open circuit voltage for each layer given by $V_{OC} = V_g - 0.4$ V. (c) What is the conversion efficiency?

8.17* A thin-film silicon solar cell has a thickness W. The upper surface is polished flat and has an anti-reflection coating. On the back surface there is a perfectly diffusing reflective coating. Show that light will have an effective path length within the silicon of $(4n^2 + 1)W$, where n is the refractive index for silicon.

8.18 A CIGS photocell of area 10 cm^2 and band gap 1.5 eV is illuminated with laser light of wavelength 800 nm. The photocell has a saturation current of 10^{-10} mA. The light power is 150 W. Use eqns (8.3), (8.10), and (8.11) for V_{OC}, I_m, V_m, and FF to estimate the conversion efficiency of the photocell.

8.19 What are the advantages and disadvantages compared with silicon cells of (a) organic solar cells (b) concentrated photovoltaic (CPV) systems?

8.20* A solar cell with an open circuit voltage of 0.4 V utilizes quantum dot photon absorbers. These absorbers emit two electrons when the energy of the photon $E_\gamma > 1.6$ eV and only one when $0.8 < E_\gamma < 1.6$ eV. Compare the conversion efficiency under AM1.5 illumination with that of a p-n junction solar cell with a band gap of 0.8 eV and an open circuit voltage of 0.4 V. Assume that both cells have the same fill factor.

8.21* In a TPV system, the central cylindrical emitter is surrounded by two concentric quartz cylinders. The quartz cylinders transmit radiation with wavelength $\lambda < \lambda_{max}$. Wavelengths with $\lambda > \lambda_{max}$ are absorbed and re-emitted in all directions. Surrounding the quartz cylinders are the photocells. Show that only a third of the radiant energy with $\lambda > \lambda_{max}$ from the central emitter is transmitted by the quartz cylinders to the photocells.

8.22 A solar farm consisting of fixed silicon photovoltaic panels is located at a latitude of 25°, where the average number of hours of direct sunlight is 10 per day. The panels have an efficiency of 18% and a total active area of 3×10^4 m^2. Estimate the maximum average power output and corresponding capacity factor of the farm.

8.23 Why is the inverter loading ratio for solar panels typically greater than one?

8.24 What improvement could be expected to be gained by tracking the PV panels in the solar farm describe in Exercise 8.22, and at what cost would it be worthwhile?

8.25 What would the capital cost and installation charges need to be for 1 kWp of solar panels for the cost of electricity to be 5¢ per kWh? The insolation is 2000 kWh m^{-2} y^{-1}. Take the lifetime of the panels to be 30 years and the discount rate to be 7%. Neglect any maintenance charges.

8.26 The capital cost of manufacturing and installing a 1 kWp array of solar panels is €1200. The annual solar energy density in the location where the panels will be installed is 1800 kWh m^{-2}. (a) Calculate the cost of electricity per kWh. Take the lifetime of the solar panels as 30 years and assume the discount rate is 5%. (b) What will be the cost of electricity if (i) there is an annual maintenance charge of €20, (ii) in addition there is a battery storage system that costs €150 per kWh, half the electricity generated is used at night, and the system lasts 11,000 cycles?

8.27 An organic photocell with an efficiency of 16% is printed on flexible sheet. Estimate the area required to produce an average annual electrical output of 100 GW in a sunny region and compare with the area printed in a year of a country, state, or city newspaper available in that region.

8.28 Why are perovskites potentially such important photovoltaic materials?

8.29 Describe the relative merits of PV solar farms and distributed PV generation.

8.30 Why has the 'learning' effect been so significant for PV deployment in the last decade?

8.31 Might PV-generated electricity be a viable way to provide clean cooking in the developing world within a few years?

8.32 What are the advantages of hybrid solar farms?

8.33 Can solar PV provide all of the world's energy demand within a few decades?

Derivation 8.1 The Shockley Queisser limit for the efficiency of a solar cell

The first effective solar cell was made in Bell Labs in 1954 and had an efficiency of 6%. Solar panels were soon being considered for space applications, and their maximum possible efficiency was of great interest. Semiempirical limits had been obtained but no absolute one, like the Carnot efficiency for any thermal engine, had been found. In 1961, Shockley and Queisser derived a very important fundamental upper limit based on the *principle of detailed balance*.

Consider a cell, consisting of a single p-n junction with a band gap E_g, enclosed in a hemispherical cavity; see Fig. 8.20 (a). Both the cell and the cavity are at room temperature

(a)   (b)

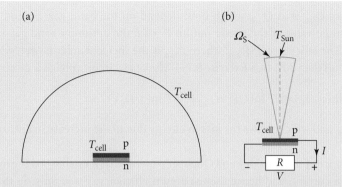

Fig. 8.20 (a) p-n cell in hemispherical cavity (b) p-n cell illuminated by sunlight.

($T = T_{\text{Cell}} = 300$ K). The photons in the thermal radiation from the cavity with energies greater than E_g can create electron-hole pairs in the cell, and these pairs can recombine and emit photons. By the principle of detailed balance, when the cell is in equilibrium the rates of these inverse processes are equal; i.e. the rate of generation of pairs equals the rate of their radiative recombination.

Shockley and Queisser assumed that this radiative recombination was the only mechanism by which electrons and holes recombined; i.e. that the material was ideal with, for instance, no electron traps. The cell of area A absorbs photons with energies greater than the band gap which generates electron-hole pairs. The generation rate G_0 is calculated by assuming the cavity is a blackbody, and integrating, over the energies above the band gap and over the hemisphere, the Planck distribution for $T = 300$ K, which yields $G_0 = \pi A Q_{\text{cell}}$. The recombination rate R_0 equals G_0 and is also proportional to the number of electron-hole pairs in the cell, $n_0 p_0$. So,

$$R_0 = \pi A Q_{\text{cell}} = k n_0 p_0 \tag{8.17}$$

where k is a constant.

When the solar cell is illuminated by sunlight, taken to be from a blackbody at $T = 6000$ K, and connected to a load, a voltage V is created across the cell; see Fig. 8.20 (b). The effect of the voltage is to increase the number of electron-hole pairs in the cell by the Boltzmann factor $\exp(eV/kT)$, where $T = T_{\text{cell}}$ and $e = |e|$ (see Section 8.3). Using eqn (8.17), the recombination rate R_V when the voltage is V is then given by:

$$R_V = k n_0 p_0 \exp(eV/kT) = \pi A Q_{\text{cell}} \exp(eV/kT) \tag{8.18}$$

The rate at which sunlight creates electron-hole pairs is calculated by integrating, over the energies above the band gap and over the solid angle Ω_S subtended by the Sun, the Planck distribution for $T = 6000$ K, which yields $\Omega_S A Q_{\text{Sun}}$. The photocurrent I is then given by:

$$I = I_L - I_S[\exp(eV/kT) - 1] \tag{8.2}$$

where the reverse current $I_L = e \Omega_S A Q_{\text{Sun}}$ and the saturation current $I_S = e \pi A Q_{\text{cell}}$

By evaluating Q_{Sun} and Q_{cell}, which both depend on the band gap E_g, and varying V to give the maximum value of the power output $P = VI$, the optimum band gap is found to be $E_g \approx 1.35$ eV. The efficiency, P/P_{Sun}, where P_{Sun} is the incident solar power, whose intensity is 1.37 kW m^{-2}, is then 30% (see Exercise 8.11). The open circuit voltage is given by eqn (8.3)

$$eV_{OC} = kT \ln\left(I_L/I_S\right) = kT \ln\left(Q_{Sun}/Q_{cell}\right) - kT \ln(\pi/\Omega_S) \tag{8.19}$$

where $T = T_{cell}$, and which shows the loss in energy due to the increase in entropy $k \ln(\pi/\Omega_S)$ (see eqn 2.18).

The importance of the Shockley and Queisser paper is that it established the minimum possible saturation current, and hence the maximum efficiency, based only on the principle of detailed balance and a few simple assumptions.

For further information and resources visit the online resources
www.oup.com/he/andrews_jelley4e

9 Nuclear Power

Introduction

Nuclear power is associated in some people's minds with nuclear weapons and nuclear waste. However, it is also an abundant source of low-carbon energy that could help in combating global warming. There are two forms of nuclear energy: one from controlling fission, the reaction used in the first 'atomic' bombs; the other from controlling fusion, the energy source in stars. Fusion power is now only at the prototype stage, but it holds the promise of almost unlimited power.

Commercial nuclear power plants are fission reactors, most of which use uranium for fuel, an element that occurs in many parts of the world, with Canada and Australia currently the main producers. Compared with the amounts of coal, oil, or gas required to fuel a conventional power station, remarkably small amounts of uranium are needed for a nuclear reactor: roughly 1 tonne of uranium will deliver an amount of energy equivalent to 25,000 tonnes of coal!

Uranium (U) is roughly as common as tin or zinc. It is present in many rocks and in the sea. The average concentration in the Earth's crust is 2.8 ppm (ppm is parts per million by mass). Granite contains about 4 ppm U, while the sea has ~0.003 ppm, which corresponds to 4600 Mt. At \$130 per kg U or less, the known reserves are 4.7 Mt, equivalent to 85 years of operation at the present consumption of ~2500 TWh y^{-1} (IAEA), with an estimated additional 35 Mt recoverable that would come from unconventional sources such as phosphate rocks. High-grade ore containing ~2% U is the cheapest to mine.

In this chapter we describe nuclear power from fission, and the progress on obtaining power from fusion. We will see that both come from converting part of the mass of nuclei into energy. In fission, we find that the process is initiated by neutrons that then yield more neutrons, giving rise to the possibility of a **chain reaction** that has to be controlled safely in a nuclear reactor.

We explain how a chain reaction is brought about, how it is controlled, what the power output from a reactor is, and what the fission waste products are. This is followed by a discussion of its economics, safety, environmental impact, and its contribution to global low-carbon electricity generation now and in the future. The chapter concludes with a description of current research on fusion power.

9.1 Binding Energy and Stability of Nuclei

In order to understand why energy is released in the fission of heavy nuclei or in the fusion of light nuclei, we need to consider the relationship between the mass and the stability of nuclei. A nucleus consists of protons and neutrons (collectively referred to as nucleons) bound together by a short-range attractive force. Its mass is less than the sum of the masses of its constituent nucleons, and the size of the difference ΔM gives the **total binding energy** B_E through Einstein's relation $B_E = \Delta M c^2$. B_E is the energy that would be required to pull apart the nucleus into its constituent nucleons, and determines whether a nucleus is stable or unstable.

The nuclear force between nucleons is short-range and attractive, unless the separation between the nucleons is very small ($\lesssim 1$ fm), when it becomes repulsive. Nucleons are therefore on average the same distance apart and interact primarily with their nearest neighbours. So we expect the total binding energy of a nucleus to be approximately proportional to the number of nucleons A, or the binding energy per nucleon, $B_E/A \equiv b(A)$, to be approximately constant; while it is roughly constant above $A \sim 12$, it has a maximum near iron (Fe; $A \sim 60$), see Fig. 9.1.

The reason for the maximum in the binding energy is that nuclei would be most tightly bound if the number of neutrons N equalled the number of protons Z, were it not for the electrostatic repulsion between the protons, which increases with increasing Z. This causes heavy nuclei such as uranium to have more neutrons than protons, and produces the fall in $b(A)$ seen above $A \sim 60$. Below $A \sim 60$ the effect of the increase in the relative number of nucleons on the surface of the nucleus, which are less tightly bound as they have fewer neighbouring nucleons than those within, outweighs the reduced electrostatic repulsion from fewer protons, and so $b(A)$ is lower.

We can now see why the fission of a heavy nucleus releases energy. When uranium splits into two lighter nuclei, these are very neutron rich and typically two or three neutrons are emitted promptly. The resulting nuclei are typically excited and decay by γ– and β– decays to stable nuclei, emitting energetic photons, and electrons and anti-neutrinos, respectively. Some of the β-decays lead to excited states of nuclei, which are unstable to neutron emission. This process is called **beta-delayed neutron emission** because the neutron is emitted only after a β-decay, and happens about 0.65% of the time in the neutron-induced fission of uranium.

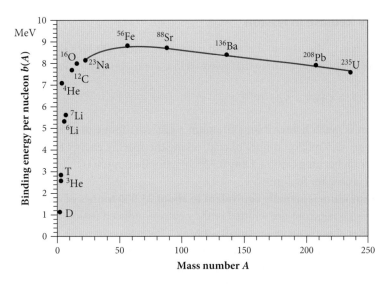

Fig. 9.1 Binding energy per nucleon $b(A)$ as a function of A.

The total mass of the final products is less than that of uranium, since the binding energy of the lighter nuclei is greater, so by Einstein's relation energy is released. On average, ~200 MeV of energy is deposited in the surrounding material by the fission products (~85%) and other emitted radiation (~15%) (see Exercise 9.1).

9.2 Neutron-induced Fission of Uranium

Although energy is released by the fission of uranium, the natural occurrence of this process (called **spontaneous fission**) is very rare. This is because there is a barrier to decay as uranium requires about 6 MeV initially to split, and decay can only take place through a process called **quantum tunnelling** with a half-life of $\sim 10^{16}$ y for ^{238}U, which has a comparatively shorter half-life by α-decay of 4.5×10^{9} y.

The probability of fission of uranium is increased enormously when a uranium nucleus captures a **slow neutron**, i.e. one with a kinetic energy less than 1 eV. For a ^{235}U nucleus this capture produces an excited ^{236}U* nucleus, $n + {}^{235}U \rightarrow {}^{236}U^{*}$, whose energy is above the height of the fission barrier and can therefore fission promptly. However, for ^{238}U the excited ^{239}U* nucleus is ~1 MeV below the top of the barrier.

This difference in excitation arises because a neutron in ^{236}U is slightly more strongly bound than in ^{239}U. In ^{236}U, which has 92 protons, all the neutrons are paired off, while in ^{239}U there is one unpaired neutron and this is less tightly bound than a paired-off neutron. It therefore takes more energy to separate a neutron from ^{236}U than from ^{239}U, and so the excited state formed when a slow neutron is captured by ^{235}U is more highly excited than when captured by ^{238}U.

9.2.1 **Energy Release in Fission**

The energy release in the fission of uranium is approximately 200 MeV $\equiv 3.2 \times 10^{-11}$ J, which is about *50 million* times more than that released in a chemical combustion reaction such as:

$$C + O_2 \rightarrow CO_2 + 4.1\,eV \equiv 6.6 \times 10^{-19}\,J$$

A carbon atom is much lighter than a uranium atom, so 1 tonne of ^{235}U is equivalent as an energy source to ~3.5 million tonnes of coal, taking the specific energy from burning coal as ~70% of that from burning carbon. Only 0.72% of natural uranium is ^{235}U, 99.28% being ^{238}U, so if only ^{235}U is used for fission:

1 tonne of uranium is actually equivalent to about 25,000 tonnes of coal.

On average ~2.4 neutrons are emitted in the neutron-induced fission of ^{235}U, with a broad range of energies about a mean energy of ~2 MeV. The release of more than one neutron in the neutron-induced fission of ^{235}U opens up the possibility of a **chain reaction**, which will occur if on average at least one of the neutrons released induces fission of another nucleus. Typically, the fission is asymmetric, and as noted above not all of the neutrons are emitted promptly. About 0.65% are β-**delayed neutrons** with a mean delay time of 13 seconds, and we later show how this delay enables reactors to be controlled safely (see Section 9.4).

For a nuclear power station, we need to control the chain reaction so that a steady release of energy occurs—a condition called **criticality**—and we can then harness this energy to produce electricity. We first consider the conditions for a chain reaction to occur in uranium by looking at the relative probabilities for different neutron reactions on ^{235}U and ^{238}U.

9.3 **Chain Reactions**

Whether a chain reaction actually occurs depends on the probability of neutron-induced fission relative to neutron loss. The dominant cause of loss is neutron capture, which is followed by γ emission. The main reactions that neutrons with energies from a fraction of an eV (thermal neutrons) to several MeV undergo with uranium are scattering (both elastic and inelastic), capture, and induced fission.

Natural uranium consists of 99.28% ^{238}U and 0.72% ^{235}U. The average number of neutrons emitted per fission, called v, is ~2.4, and their energies range between ~0 and ~10 MeV with a mean energy of ~2 MeV. In natural uranium, these neutrons are most likely to scatter off ^{238}U, and it is only when they have energies less than ~1 eV that neutron-induced fission of ^{235}U is more likely than capture by ^{238}U. Neutrons with energies below 1 eV are called *slow neutrons*, and those above 100 eV *fast neutrons*.

For neutrons with energies between ~1 and ~100 eV, sharp peaks (called **resonances**) exist in the probability of neutron capture by ^{238}U, which correspond to excited states in ^{239}U being formed. At these peaks the probability of capture is close to 1, since the probability for capture is much larger than that for scattering. In this region, the change in the energy of a neutron after each elastic scattering off ^{238}U is sufficiently small that the chance that its energy falls under a peak in the ^{238}U capture probability is high, with the result that very few neutrons reach energies less than ~1 eV and induce fission of ^{235}U.

Only a small percentage of fast neutrons induce fission of ^{238}U (as the probability of inelastic scattering to below the fission threshold is high), or of ^{235}U (as most fast neutrons capture on ^{238}U), so there is no chain reaction in natural uranium.

We can conclude that one way to produce a self-sustaining chain reaction in uranium, is for the isotopic abundance of ^{235}U to be increased significantly from its naturally occurring percentage of 0.72%, by a process called **enrichment**, so that the percentage of induced fission by fast neutrons of ^{235}U is much larger. Only a few such reactors, called **fast reactors** (see Section 9.5) have been built, and they operate with fuel enriched to ~10–30%, with the chain reaction sustained by fast neutrons with energies greater than ~1 keV.

Modern enrichment plants use a gas centrifuge process. The main advantage of the gas centrifuge process compared to the older gaseous diffusion process is that it requires only about 2% of the energy for an equivalent enrichment. Since the facilities used to enrich uranium for use in nuclear reactors could also be used to provide highly enriched weapons-grade ^{235}U, the development of nuclear power in any country has considerable political and security implications.

Alternatively, the probability of capture by ^{238}U could be reduced by increasing the average change in a neutron's energy after elastic scattering off nuclei in the core, which can be achieved by effective **moderation**, so that the fission of ^{235}U is induced by **slow neutrons**. Nearly all commercial reactors use moderation, and most also use slightly enriched fuel to achieve criticality.

9.3.1 Neutron Moderation

If nuclei with a low atomic number, called **moderators**, are added to the reactor core, then the change in kinetic energy of a neutron following an elastic collision with a moderator nucleus can be sufficiently large that the chance of capture by ^{238}U in one of the resonance peaks is significantly reduced for a sufficiently high moderator-to-fuel ratio.

When a neutron elastically scatters off a nucleus of mass M it will lose some kinetic energy ΔE, the amount depending on the angle of scatter and on the mass M: the larger the mass M, the smaller the loss ΔE. Rather than using ΔE, which depends on E, the moderating power of a material with mass number A is given by the logarithmic decrement ξ, which for small ΔE equals the average relative energy loss $\Delta E/E$. For $A > 1$, ξ can be approximated by the formula:

$$\xi \approx 6/(3A + 2) \tag{9.1}$$

For $A = 1$, $\xi = 1$.

We can now see why neutrons slowing down in natural uranium by scattering off the uranium nuclei have only a small chance of not being captured by one of the ^{238}U resonances. Each peak in the capture probability occurring between ~1 and 100 eV extends over a small energy range, which is given by the width of the resonance. The peaks correspond to the formation of excited states in ^{239}U that are unstable, and the widths (Γ) are related to the mean lifetimes (τ) of the states through $\Gamma = h/(2\pi\tau)$, which is an example of the uncertainty principle. If the peak cross section (a measure of the probability) is at an energy E_0 then the cross section is reduced by a factor of 2 at $E_0 \pm \Gamma/2$ and by a factor of 5 at $E_0 \pm \Gamma$.

One of the strong capture resonances is at a neutron energy of 6.7 eV, with a total width of 0.024 eV. The average neutron energy loss ΔE for a neutron with energy E when scattering from ^{238}U is given by ξE, so, for a 6.7 eV neutron, $\Delta E \approx 0.056$ eV by eqn (9.1). There is therefore a good chance that, while losing energy through elastic scattering from U nuclei, a neutron has an energy within Γ of the peak energy. As the peak cross section is over 100 times the elastic scattering cross section, the neutron is likely to be captured. We therefore need to choose a material with low-mass nuclei as a moderator, which will have a larger $\Delta E/E$, to reduce the chance of resonant capture. To be an efficient moderator, both ξ and the probability of scattering should be large and that of absorption small.

Once below the resonance region the probability of induced fission is large, and a chain reaction can be maintained with neutrons with energies ~0.05 eV; neutrons with these energies are called **thermal neutrons** since they are at the same temperature as the uranium fuel. The use of moderators is the basis of commercial thermal reactors.

9.4 Thermal Reactors

Commercial nuclear reactors operate using thermal neutrons (obtained by the use of a moderator), and most use fuel enriched in ^{235}U to a few per cent. This enables a chain reaction to occur. As the time between a neutron from one fission inducing another fission is typically only ~0.1 ms, the chain reaction would be uncontrollable if it were not for the influence of the small fraction of **beta-delayed neutrons** that are also released. The mean delay time of these neutrons is ~13 s, which allows sufficient time for the chain reaction to be controlled mechanically using **control rods**. Control rods contain nuclei with very high neutron absorption cross sections (e.g. ^{10}B, ^{113}Cd), and the total number of neutrons inside the reactor can be controlled by continually adjusting how far the rods are inserted into the reactor core.

We now look at the layout of the most common thermal reactor: a pressurized water reactor (PWR).

9.4.1 Pressurized Water Reactor (PWR)

The PWR shown in Fig. 9.2 is the most widespread commercial reactor; out of a total of 448 nuclear reactors in 2016, 292 were PWRs. The PWR was initially developed for submarines, since, unlike internal combustion engines, nuclear-powered submarines do not need oxygen and can therefore remain underwater for much longer than diesel-electric powered submarines. The heat from the reactor produces steam to drive a turbine, and the relatively compact core proved a cost-effective design that could be scaled up to ~1 GW. The first prototype was operated in 1953. The fissile material (the fuel) is in the form of fuel rods, which allows for easy refuelling.

In a PWR (see Fig. 9.2), water is circulated past the fuel rods in the primary loop at a high pressure of ~15 MPa to keep it in the liquid phase at a temperature of ~315 °C. It is passed

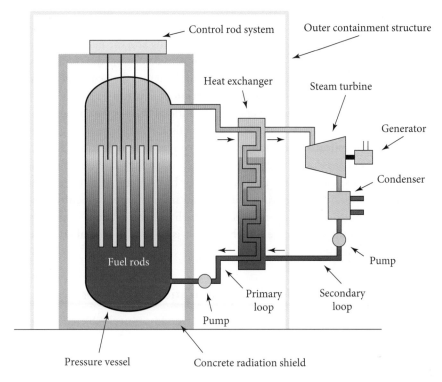

Fig. 9.2 Main components in a PWR.

through a heat exchanger, where the water in the secondary loop, which is at a lower pressure of ~5 MPa, is heated to produce steam to drive the turbine, after which the steam is condensed and the water is returned to the heat exchanger. The high neutron flux in the core activates the cooling water and makes it radioactive. This radioactivity is kept within the primary loop and inside the containment vessel. The thermal efficiency of a PWR is typically ~33%.

The water acts as a moderator as well as a coolant. It also absorbs neutrons. Should the pressure drop in the primary loop and the water start to boil, the creation of bubbles (voids) decreases the moderation and also the absorption. The effect on the moderation is more significant, the chain reaction stops, and the reactor is no longer critical. The moderation is also decreased if the core temperature rises, as this broadens the ^{238}U resonances. Hence there is a *negative temperature coefficient of reactivity*, which tends to stabilize the power output, since an increase in power causes the temperature to rise and the reactivity to fall, and vice versa.

Over a long period, the high neutron flux causes embrittlement of the reactor vessel as the metal becomes less ductile, and this limits the lifetime of the reactor. Corrosion in the steam generating tubes must also be monitored. The remote possibility of a loss-of-coolant accident (LOCA), in which the water in the primary loop is lost, requires additional emergency cooling to be available. While the core would no longer be critical, the heat generated from the decay of the fission products in the fuel rods could cause the core to melt. This possibility has been termed the **China syndrome**, referring to the (mistaken) idea that the molten core would bore through the Earth all the way from the USA to China!

9.4.2 **Criticality**

Criticality is the state in which a reactor is operating steadily and the neutron flux in the core is constant. The first aspect of the design of a reactor is to determine the conditions required of the fuel and moderator for a chain reaction to occur. Whether the neutrons emitted following neutron-induced fission (called the **next generation of neutrons**) actually lead to a chain reaction in a piece of fissile material depends on the ratio of the number of neutrons producing fission in one generation to the number producing fission in the previous generation. This ratio is called the **multiplication constant** k, and a chain reaction is possible if $k \geq 1$.

The multiplication constant k is determined by the probability of neutron-induced fission relative to neutron loss, via other neutron reactions or by escaping from the reactor; the size of the core determines the escape probability. For an infinite core, no neutrons can escape and the multiplication constant is then called k_∞, which is greater than k. A chain reaction is possible if $k_\infty > 1$, and the size of core for which $k = 1$ is called the **critical size**. We will first consider the factors affecting k_∞.

9.4.3 **Four-factors Formula for the Multiplication Constant k_∞**

In a core of uranium that is sufficiently large that neutron loss is negligible, there are four factors that determine k_∞:

fast fission factor ε

resonance escape probability p

thermal utilization factor f

number of neutrons produced per thermal neutron absorbed η

Of the average initial number of fast neutrons emitted per fission, v, the loss through capture is slightly offset by the gain through fast neutron-induced fission of ^{238}U. For ^{235}U, we have $v = 2.425$. The ratio of the number below the ^{238}U threshold of ~1 MeV to the initial number is called the **fast fission factor** ε. For low enrichment, the fast fission factor ε is close to unity.

Below the ^{238}U threshold, a fraction escape resonance capture by ^{238}U and manage to slow down to thermal energies. The **resonance escape probability** p depends strongly on the choice of material for the moderator. When fuel and moderator are separated, the value of p can be ~0.9.

The fraction f of thermal neutrons that are absorbed by U, rather than by the moderator or the fuel can, is called the **thermal utilization factor** f, and depends on their relative absorption probabilities (see Derivation 9.1).

The **thermal fission factor** η is the ratio of the number of fission neutrons produced to the number of thermal neutrons absorbed. This is less than the average number v of neutrons emitted, since some of the neutrons are captured by ^{238}U and by ^{235}U without emission of neutrons (see Derivation 9.1). For the thermal neutrons only absorbed by ^{235}U, the number of fast neutrons emitted is η_0.

Finally, combining all the above factors yields the **four-factors formula**:

$$k_\infty = \varepsilon p f \eta \tag{9.2}$$

Derivation 9.1 Neutron reaction rate, mean free path, and macroscopic cross section

The probability that a neutron reaction occurs can be described in terms of a **cross section** σ. This can be visualized as the effective cross-sectional area within which a target nucleus and an incident neutron will interact and give rise to a particular reaction. Its units are **barns** (b) $\equiv 10^{-28}$ m^2. For comparison, the cross-sectional area of a uranium nucleus is ~2 b. The value of σ may be much larger than the cross-sectional area of a nucleus (see Table 9.1). This is a consequence of the wave-like properties of a neutron. The cross section for any interaction is the total cross section σ_t, and that for absorption equals the sum of the capture and neutron-induced fission cross sections, i.e. $\sigma_a = \sigma_c + \sigma_f$.

Consider neutrons moving at a speed v through uranium where the number of ^{235}U nuclei per unit volume is n_f. If the cross section for neutron-induced fission is σ_f, then in a small-time t a neutron travels a distance vt and passes through a volume $\sigma_f vt$ within which the chance of the neutron causing a fission is $n_f \sigma_f vt$. So, the probability of a reaction per unit time is $n_f \sigma_f v$ and per unit distance $n_f \sigma_f$.

For n neutrons per unit volume, the reaction rate R is therefore given by:

$$R = n n_f v \sigma_f = \Phi n_f \sigma_f \tag{9.3}$$

where $\Phi = nv$ is the **neutron flux**. Since the probability of an absorption on ^{235}U per unit distance is $n_f \sigma_a$, it follows that the **mean free path** before an absorption λ is the inverse of this value, i.e. $\lambda = 1/n_f \sigma_a$ (see exercise 9.5).

The product of the number density and the cross section σ_f:

$$\Sigma_f = n_f \sigma_f \tag{9.4}$$

is called the **macroscopic cross section** for neutron-induced fission of ^{235}U. Its units are m^{-1}. The macroscopic cross sections determine the relative probabilities for a neutron giving rise to a particular reaction with different nuclei. For example, the ratio of $\Sigma_f^{235}/\Sigma_c^{238}$ gives the relative probability that a neutron induces fission of ^{235}U rather than is captured by ^{238}U. The thermal utilization factor f is given by the ratio of the macroscopic cross section for absorption in the fuel (F) to that for absorption in the core (C):

$$f = \Sigma_a(F)/\Sigma_a(C) \tag{9.5}$$

The thermal fission factor is given by:

$$\eta = \frac{v\Sigma_f(F)}{\Sigma_a(F)} \tag{9.6}$$

which, from eqn 9.4 and Table 9.1 for uranium fuel enriched to x% in ^{235}U, equals:

$$\eta \cong \frac{5.28x}{1+2.53x} \tag{9.7}$$

Table 9.1 Nuclear and material properties*

Material	Density (kg m^{-3})	n (10^{28} m^{-3})	σ_f (b)	σ_s (b)	σ_a (b)	η_0**	v
C (Graphite)	1 760	8.80	–	4.9	0.0045		
^{233}U	18 500	4.78	529	12.2	575	2.296	2.493
^{235}U	18 700	4.78	583	16.0	681	2.075	2.425
^{238}U	18 900	4.78	–	9.4	2.68		
^{232}Th	11 720	3.04		11.8	7.37		

*σ are averaged over thermal energies; **η_0 is the value of η when thermal neutrons are only absorbed by the fissile isotope*

As an example, in Derivation 9.2 we calculate k_∞ for a reactor core containing a mixture of enriched uranium and carbon in the form of graphite, for which the nuclear cross sections and densities are given in Table 9.1. The uranium is enriched to 1.7% in ^{235}U, and the ratio of $n_s/n_f = 500$, where n_s is the number density of the graphite moderator, and n_f is the number density of fuel nuclei. For such a core $p = 0.736$, $f = 0.863$, $\eta = 1.691$, and for this low enrichment and high moderator to fuel ratio, $\varepsilon \approx 1$, so:

$$k_\infty = \varepsilon p f \eta = 1.074$$

Derivation 9.2 The multiplication constant k_∞ of a reactor core

To illustrate the four-factors formula for k_∞, we will consider a reactor core containing a mixture of uranium, enriched to 1.7% in ^{235}U, and graphite, with a ratio of $n_s/n_f = 500$, where n_s is the number density of the graphite moderator.

An approximate expression for the resonance escape probability p for a homogeneous reactor core, where elastic scattering is predominantly by the moderator and loss is mainly by absorption by ^{238}U in the uranium fuel, is given by:

$$p \cong \exp[-2.4[n_f/(n_s\sigma_s)]^{1/2}/\xi$$

where σ_s is in barns. The logarithmic decrement of the mixture is essentially the same as that of graphite as $n_s \gg n_f$. Using the values for the cross section given in Table 9.1 and the logarithmic decrement ξ given by eqn (9.1) gives the resonance escape probability as

$$p = \exp\{-2.4[n_f/(n_s\sigma_s)]^{1/2}/\xi\} = \exp\{-2.4[1/500 \times 4.9)]^{1/2}/0.158\} = 0.736$$

The thermal utilization factor f for this graphite-moderated enriched-uranium mixture is given by eqn (9.5) as:

$$f = \Sigma_a(\text{fuel})/[\Sigma_a(\text{fuel}) + \Sigma_a(\text{graphite})]$$

So, applying eqn (9.5) the thermal utilization factor f is given by:

$$f = (0.017\sigma_a^{235} + 0.983\sigma_a^{238})/[0.017\sigma_a^{235} + 0.983\sigma_a^{238}) + (n_s/n_f)\sigma_a(\text{graphite})]$$

Substituting in values from Table 9.1 and using $n_s/n_f = 500$ gives $f = 0.863$.

The ratio η of the number of fission neutrons produced to thermal neutrons absorbed is, from eqn (9.6),

$$\eta = v\Sigma_f(\text{fuel})/\Sigma_a(\text{fuel})$$

where $\Sigma_f(\text{fuel}) = (0.017\sigma_f^{235})n_f$ and $v \approx 2.425$. Substituting, we obtain $\eta = 1.691$.

For this low enrichment and high n_s/n_f ratio, $\varepsilon \approx 1$, so $k_\infty = \varepsilon p f \eta = 1.074$.

This means that a chain reaction is possible and will be critical, i.e. $k = 1$, if the size of the reactor is such that the fractional loss, $(k_\infty - 1)/k_\infty = (1.074 - 1)/1.074 = 0.069$. This loss is the probability for neutrons to escape from the core.

9.4.4 Reactor Core Design

The neutron loss can be reduced by increasing the core size. This is because neutrons created in the core diffuse away a finite distance from their point of origin, since they typically undergo many elastic scatterings before reacting. The fraction of neutrons that are sufficiently close to the outer surface of the core to have a good chance of escaping therefore reduces with increasing core size as the surface-to-volume ratio decreases. When the core is just critical, i.e. $k = 1$, the probability that neutrons do not diffuse out of the core equals $1/k_\infty$. The critical size of the core can be decreased by surrounding the core with a reflector—a material that reflects neutrons (such as graphite).

Making the reactor core heterogeneous by putting the fuel in the form of rods increases the resonance escape probability p significantly. Part of this increase arises because some of the fast fission neutrons, which have a good chance of escaping the fuel rods, are thermalized by the moderator before interacting with another fuel rod, and hence escape capture. However, the fuel rods are generally sufficiently close that this is not so significant.

The greater effect arises because the flux of neutrons with energies in the resonance region is significantly reduced within a fuel rod because of the very strong absorption of these neutrons within the surface layer (<0.01 cm) of the rod. This means that most of the fuel is shielded from these neutrons, so the chance that a neutron is captured is considerably lower than if the fuel were uniformly mixed throughout the moderator (homogeneous reactor); hence the escape probability is higher and the value for p can be ~0.9.

A flux reduction also occurs within the fuel rods for thermal neutrons due to the large probability of absorption. However, since the mean free path for thermal neutron absorption in natural uranium is about 2.5 cm, the effect on the thermal utilization factor f is much less than on the resonance escape probability p, with the result that the product fp is increased.

Fuel in the form of rods also increases the fast fission factor ε. The probability that the emitted fast neutrons induce fission in ^{238}U is larger because, typically, the fast neutrons travel through pure fuel a distance of the order of the radius of the fuel rod before being moderated.

Besides these nuclear physics reasons for the fuel to be in the form of rods, there are excellent engineering reasons as well. The use of fuel rods allows good cooling, and hence good heat transfer, as well as easy refuelling.

9.4.5 Reactor Control

It is essential to maintain the multiplication factor k close to unity, in order for the neutron flux to be almost constant. In particular, as the fuel is '**burnt**', k will decrease, so the neutron absorption must be lowered. This is achieved by using electromagnetically operated control rods, which control the neutron flux in the core of the reactor. These contain nuclei with a high cross section for neutron absorption, such as ^{10}B and ^{113}Cd, and can be lowered into the reactor core to reduce the flux or raised out of the core to increase the flux. The **reactivity** ρ, defined by $\rho \equiv (k-1)/k$, is generally used when describing the time dependence of the neutron population.

In a chain reaction there is a short period of time between the release of a neutron from a fission and this neutron initiating another fission and thereby producing more neutrons. This period, called the **generation time** τ_g, for neutrons released promptly following fission, is given by $\sim \lambda_a/u$, where λ_a is the absorption mean free path and u is the neutron speed. A typical value of τ_g for prompt thermal neutrons is $\sim 10^{-4}$ s. If the number of neutrons per unit volume in the reactor core is n, then the build-up of neutrons will be governed by the rate equation:

$$dn/dt = (k-1)n/\tau_g \tag{9.8}$$

The -1 term arises because one neutron is absorbed in the chain reaction to produce the next generation, which has k neutrons, and this occurs on average every τ_g seconds. The population of neutrons therefore grows exponentially with a time constant:

$$\tau = \tau_g/(k-1) \tag{9.9}$$

Hence, if k increases to 1.001, the neutron population increases by a factor of 20 in 0.3 seconds, which would make mechanical control exceedingly difficult. Fortunately, this is not what actually happens in practice because a small percentage of neutrons, $\beta \sim 0.65\%$ for ^{235}U, are emitted following the β-decay of neutron-rich fission fragments, with a mean delay time $\tau_d \sim 13$ s ($\tau = t_{1/2}/\ln 2$).

For k near unity, it turns out (see Exercise 9.12*) that τ_g effectively depends on the delayed neutrons alone and equals the average neutron lifetime τ_a, given by:

$$\tau_a = \beta(\tau_d + \tau_p) + (1-\beta)\tau_p = \beta\tau_d + \tau_p \tag{9.10}$$

where τ_p is the generation lifetime for prompt neutrons (~ 0.1 ms). The time constant for $k = 1.001$ then becomes ~ 85 s, and mechanical control of the reactor is therefore possible.

When a reactor is started up, k is deliberately made sufficiently greater than unity that the flux increases to its operating value Φ_o within a reasonable space of time.

9.4.6 **Reactor Stability**

It is very important that the reactivity should decrease if the temperature of the core rises, to ensure that the core temperature will be stable to small fluctuations in reactivity. In a thermal reactor this comes about through a broadening of the ^{238}U capture resonances with increasing temperature. The ^{238}U nuclei are vibrating and have a mean square speed that is proportional to the temperature of the uranium atoms. The neutron energy required to form an excited state of ^{239}U depends on the relative velocity of the neutron and ^{238}U nucleus and so has a spread Δ that increases with increasing temperature. This has the effect of increasing the neutron absorption rate. Increasing the core temperature therefore decreases the resonance escape probability and hence decreases k.

9.4.7 **Reactor Fissile Inventory**

In a reactor the total energy absorbed as heat per fission is close to 200 MeV. Typically, over 95% of the uranium in the core is ^{238}U, so nearly all the neutrons that do not induce fission are captured by ^{238}U and produce ^{239}U, which β-decays, first to ^{239}Np and then to the long-lived fissile nucleus ^{239}Pu. A nucleus that can be converted to a fissile nucleus by neutron capture is called **fertile**. In a few per cent enriched uranium PWR, about 30–60% of the output comes from the neutron-induced fission of ^{239}Pu produced through conversion of the ^{238}U. (See Box 9.1.)

Box 9.1 Conversion and breeding

Conversion: The **conversion ratio** C is defined as the ratio of the amount of fissile material produced from fertile nuclei to the amount consumed. There is a contribution from fast neutrons resonantly captured by ^{238}U, C_f, as they slow down, and another from the capture of thermalized neutrons, C_{th}. As there are η_0 fast neutrons per thermal neutron absorbed by ^{235}U, and a factor ε increase through fast neutron-induced fission, the fast neutron contribution is $C_f = \eta_0 \varepsilon (1 - p)$, where p is the resonance escape probability. There are η fast neutrons per thermal neutron absorbed by ^{235}U and ^{238}U, so the thermal neutron contribution $C_{th} = (\eta_0/\eta) - 1$. The initial conversion ratio C is therefore given by:

$$C = \eta_0 \varepsilon (1 - p) + (\eta_0/\eta - 1)$$

N fissile nuclei will therefore produce CN fissile nuclei from fertile nuclei, and these in turn will produce $C^2 N$ fissile nuclei, and so on. The total increase is by a factor of $(C + C^2 + C^3 + \cdots) = C/(1-C)$ when $C < 1$. So if x is the initial percentage of fissile nuclei, conversion makes the percentage available P_a equal to $x + xC/(1-C)$, i.e.

$$P_a = x/(1-C) \tag{9.11}$$

Breeding: When $C > 1$, C is then called the **breeding ratio** and fissile material continues to be produced while fertile material remains, greatly increasing the amount of fuel. For breeding to be possible, $\eta_0 > 2$, since at least one neutron is required to keep the chain reaction going and one to produce a fissile nucleus from a fertile one, and the **breeding gain** is $(C-1)$. In practice, however, owing to neutron leakage and neutron absorption by elements in the core other than the fuel, η_0 must be greater than 2.2. For the fissile elements ^{233}U, ^{235}U, and ^{239}Pu, only ^{233}U can be used to breed with thermal neutrons. ^{233}U can be bred from ^{232}Th. The process is:

$$n + {}^{232}Th \rightarrow {}^{233}Th \rightarrow {}^{233}Pa \rightarrow {}^{233}U$$

where the last two reactions occur via β-decay. Thorium is about three times as abundant as uranium. India has large reserves of thorium and has a nuclear power programme designed to utilize them. They are estimated at ~800,000 tonnes, which is a huge energy source, since 1 tonne of thorium corresponds to 2.6 GWy.

The thermal neutron contribution to conversion in a core fuelled by ^{232}Th enriched with ^{233}U is higher than in a U fuelled core with the same enrichment of ^{235}U, as the capture probability of ^{232}Th is higher than of ^{238}U, and breeding is theoretically possible with just thermal neutrons. However, the few operating breeder reactors have been fast reactors (see Section 9.5) that have taken advantage of the larger η_0 values for fast than for thermal neutrons, and ^{233}U, ^{235}U, and ^{239}Pu can all be used. Fissile material continues to be produced while fertile material remains. This increases the percentage of uranium used to produce power up to a practical limit of around 50%, which is about a 50-fold increase in the potential amount of energy available compared with conventional nuclear fission reactors, i.e. over 1000 TWey c.f. about 25 TWey from the conventional reserves of 4.7 Mt of natural uranium. A measure of how good a reactor is at breeding is the time it takes for the amount of fissile material in the reactor to double (see Exercise 9.16).

As shown in Box 9.1, the percentage of fissile material available with conversion is increased from an initial percentage x to a percentage $x/(1-C)$, where C is given by:

$$C = C_f + C_{th} = \eta_0 \varepsilon (1-p) + (\eta_0/\eta - 1) \tag{9.12}$$

Conversion is particularly important for reactors using natural uranium ($x = 0.72$) as a fuel, for then $C_f \approx 0.25$ and from eqn (9.7) $\eta \approx 1.35$, while $\eta_0 = 2.075$, so $C_{th} = (2.075/1.35 - 1) = 0.54$, giving a conversion ratio $C = 0.79$. The percentage of fissile nuclei available with conversion is therefore $0.72/(1-C) = 3.4\%$, rather than 0.72% with no conversion (neglecting changes in C caused by changes in the fuel composition and assuming that ^{239}Pu acts like ^{235}U).

For a reactor using fuel enriched to a few per cent, the conversion ratio is smaller; for example, for 4% enrichment from eqn (9.7) $\eta = 1.90$, and assuming C_f is the same as for natural uranium, then $C = 0.34$ and 6.1% is the maximum amount of fissile material that could be consumed. The time interval between refuellings of a reactor is determined by the **burn-up** of the fuel, which is measured in GWday per tonne of fuel; modern reactors are designed to give 60 GWday per tonne burn-up.

In a reactor, the heat per fission of 200 MeV translates to an energy output of 0.95 GWday per kg of ^{235}U, or about 1 GWday per kg of fissile nuclei. To produce 1 GW for a year requires 384 kg of fissile nuclei. However, this is thermal power, not electrical power; the thermal efficiency of a typical PWR is ~33%, so about 1150 kg of fissile nuclei per GWey is actually needed.

9.4.8 Power Output of a Thermal Reactor

In a reactor fuelled with uranium, thermal neutrons travel a mean distance λ before an absorption on ^{235}U, which has an 86% probability of inducing a fission, where λ depends inversely on the number density of ^{235}U (see Derivation 9.1). If the neutron number density is n and their average speed is v, then the distance travelled by neutrons per second per unit volume is nv, and so the number of fissions per second per unit volume R is given by:

$$R = 0.86\ \Phi/\lambda \tag{9.13}$$

where $\Phi = nv$ is the neutron flux, which has units of neutrons per square metre per second The higher the flux, the higher the power output. For a core with an initial mass M_{235} of ^{235}U, the starting power P can be expressed in terms of the mass M_{235} (in tonnes) and flux Φ (in units of $10^{18}\ \mathrm{m}^{-2}\ \mathrm{s}^{-1}$) as

$$P = 4.75 M_{235}\Phi\ \mathrm{GW} \tag{9.14}$$

EXAMPLE 9.1

A thermal reactor has an initial core mass of ^{235}U of 4.7 tonnes and a neutron flux of $2 \times 10^{17}\ \mathrm{m}^{-2}\ \mathrm{s}^{-1}$. The fuel is 5% enriched initially, for which from eqn (9.7) $\eta = 1.93$, and is replaced when the percentage of fissile material is reduced to 1.35%. Calculate the initial power output, and estimate the conversion ratio and the fuel burn-up in GWday per tonne of fuel.

From eqn (9.14) the initial power output P is:

$$P = 4.75 \times 4.7 \times 0.2 = 4.47\ \mathrm{GW}$$

The conversion factor is given by $C = 0.25 + (2.075/1.93 - 1) = 0.325$, assuming $C_f = 0.25$. The total percentage of fissile nuclei consumed is $(5 - 1.35)/(1 - 0.325) = 5.4\%$, i.e. 54 kg per tonne. At 0.95 GWday per kg, the burn-up is 51 GWday per tonne.

9.4.9 Fission Products

During the operation of a reactor there is a build-up of fission products within the fuel rods. Some are very long-lived actinides arising through successive neutron capture reactions on uranium. When the amount of fissile material in a fuel rod is insufficient to maintain criticality, the rod is removed, and the remaining fissile material is extracted chemically (called **fuel reprocessing**). It is reutilized in new fuel, and the waste products are separated for storage. The presence of these actinides means that the waste must be stored safely for many thousands of years (see Section 9.10).

Some of the fission products are volatile fission fragments, and it is these that could be most easily released in the event of a major accident. For example, ~6% of fissions of ^{235}U produce ^{135}I, which has a 6.7 h half-life, and the activity from its decay in a 1 GWe reactor is ~ 5.6×10^6 TBq (see Exercise 9.14). This activity is very large, and in the case of a core melt-down some 20% of the ^{135}I in the fuel rods could be released into the reactor containment building and subsequently leak out into the atmosphere. The radioactive gas could then be dispersed by wind. Close to the reactor building the direct dose from the radioactive nuclides within the reactor building is dominant, while farther away the internal thyroid dose from iodine isotopes is generally the largest. These considerations are clearly important in determining where to site a reactor and how much radiation shielding is required.

9.4.10 Radiation Shielding

Shielding is required around the reactor core to reduce the radiation dose to safe levels. (See the discussion of radiation in Box 9.2.) Generally, it is only necessary to shield against γ-rays and neutrons, since α and β particles have very short ranges in matter. Most of the prompt γ-rays emitted in fission are attenuated by material in the core, and it is the fast neutrons that are the most difficult to shield against. For neutrons, the probability of absorption is much higher at low energies ($E_n \lesssim 1$ keV) than at high energies ($E_n \geq 100$ keV), so good moderation is required of the fast fission neutrons. Water is a very effective neutron shield, and boron can be added to the water to improve the absorption of low-energy neutrons. Concrete is also effective due to its large hydrogen atom density (1/4 that of water). It can be cast into the required shape and is strong, so it provides most of the shielding of a reactor.

Box 9.2 Radiation

Radiation affects tissues as it causes ionization, which breaks molecules apart and gives rise to free radicals, which can damage cells. The scale of the effect depends on the amount of energy deposited per unit mass of tissue, the dose D, and the type of radiation. Charged particles, such as α particles, cause relatively more damage than γ-rays or electrons depositing the same energy, since their energy loss per unit distance travelled (the **linear energy transfer**, LET) is higher. Likewise, neutrons transfer their energy in matter to nuclei, so their LET is also high. This difference in LET is taken into account by using a weighting factor w to give the equivalent radiation dose, $H = wD$. In SI units, H is measured in sieverts (Sv) and the dose D in grays (Gy), with 1 Gy $\equiv$ 1 J kg^{-1}. An older unit for H still in use is the rem $\equiv 10^{-2}$ Sv. The weighting factor for X-rays, γ-rays, and electrons is 1, for α particles, fission fragments, and heavy nuclei it is 20, and for neutrons its maximum value is 20.

The dose received at a distance r in air from a radioactive source is proportional to $1/r^2$ and to the decay rate or **activity** of the source. The unit of activity is the becquerel (Bq), which is one disintegration per second. An older unit still used is the curie (Ci), where $1\,\text{Ci} \equiv 3.7 \times 10^{10}\,\text{Bq}$.

At a distance of r (m) from a source with activity A (MBq) emitting γ-rays of energy E (MeV) an approximate expression for the equivalent dose rate D_{rate} is:

$$D_{\text{rate}}(\mu\text{Svh}^{-1}) \approx A(\text{MBq}) \times E(\text{MeV})/6r^2(\text{m}^2)$$
$$\approx 2F(10^6\,\text{m}^{-2}\text{s}^{-1}) \times E(\text{MeV}) \tag{9.15}$$

when E is in the range ~0.1 to several MeV, and the flux $F = A/4\pi r^2$.

The equivalent dose gives a measure of the biological effect when the whole body is irradiated uniformly. If only certain organs receive a dose, the effect on the person is less, and this is taken into account in the effective dose. An example is the effective dose following radioactive iodine inhalation. Iodine concentrates in the thyroid, and if the thyroid receives an equivalent dose of 25 μSv then the effective dose is about 1 μSv.

Environmental radiation

Naturally occurring radiation exists in the environment that comes from radioactive minerals and from cosmic rays. Uranium and thorium and their decay products, in particular radon, and the potassium isotope ^{40}K contribute most. Typical values for the annual effective dose in the UK from various sources are in μSv: cosmic rays 260; food 300; environmental* 1650; medical 370; miscellaneous 0.4; fall-out 5; occupational 8. (* higher in granite areas).

The cosmic ray contribution increases with altitude and in an airliner at 10,000 m, it is ~150 times larger than at sea level and approximately 15 times the average annual dose rate. The amount of radon in the air is higher in granite areas such as Cornwall, UK, where the total annual dose is about 6900 μSv.

For people working with radioactive materials, such as in the nuclear power industry, there is the possibility of increased radiation exposure. The principal long-term risk from exposure to significant amounts of radiation is an increased risk of cancer. The limits set on the annual radiation dose and on the amounts of any radioactive isotopes that might be ingested or inhaled are such that receiving these amounts would not cause the worker a significant risk in comparison with other occupational risks. The current whole-body annual dose limit for radiation workers in the UK is 20 mSv.

9.5 Fast Reactors

As mentioned previously, a chain reaction can be maintained predominantly by fast neutrons ($E_n > 1$ keV) if the fuel is sufficiently enriched. However, making a fast reactor is technically more challenging than a thermal reactor because it has a much higher energy density in the reactor core, and there are very few fast reactors in operation. Criticality is maintained using control rods, and the time constant for control is determined, as in thermal reactors, by the mean delay time of the β-delayed neutrons.

The global reserves of uranium are such that breeders are not essential in the short term. However, when the supply of uranium eventually becomes more limited, fast reactors are likely to become particularly attractive for conserving uranium stocks, since the conversion of fissile material is generally much higher than in a conventional fission reactor. For conversion, the core is surrounded by fertile material and the emitted neutrons convert it to fissile material. For example, when fuelled by ^{235}U some of the fission neutrons are absorbed by ^{238}U and produce ^{239}Pu, which is fissile, via the reactions:

$$n + {}^{238}U \rightarrow {}^{239}U \rightarrow {}^{239}Np \rightarrow {}^{239}Pu \qquad (9.16)$$

where the last two reactions occur via β-decay.

Conversion can be sufficient to yield more fissile material than is used in the core. When this happens, the reactor is called a **breeder reactor** and the utilization of uranium can be increased from ~2% to ~50%, an increase by a factor of ~25 in the size of the nuclear energy reserve. The details of breeding are described in Box 9.1. The proven conventional fossil and uranium (for thermal reactors) fuel reserves are ~4×10^{22} J and ~3×10^{21} J, respectively. Breeder reactors could therefore provide an enormous amount of energy, without any significant greenhouse gas emissions.

In a fast reactor, to ensure that fission is maintained predominantly by fast neutrons, no moderating material (or very little) is used and the fuel is enriched to 10–30%. As a result, the core is very compact. For a reasonable power output, the energy density in the core must be high, and this requires a coolant with excellent heat transfer properties. Also, the coolant must not moderate the neutrons, thus favouring coolants that do not contain low atomic number nuclei. Sodium has been used in many designs. It has a high boiling point of 882 °C at 1 atmosphere, so the thermodynamic efficiency is high without requiring high pressures and a large pressure vessel. However, sodium is chemically very reactive and is also activated through neutron absorption, producing ^{24}Na, which has a 15 h half-life, and so the coolant loop must be heavily shielded. As a result, the design of a reliable, economic sodium-cooled fast reactor has proved to be a difficult challenge.

9.6 **Thermal Reactor Designs**

The first reactor was built in a squash court in Chicago in 1942 under the direction of the Italian physicist Enrico Fermi, as part of the **Manhattan Project**, which developed the first fission bomb. It was a graphite-moderated pile containing natural uranium fuel in the form of rods embedded in graphite blocks. The development of graphite-moderated reactors using natural uranium fuel was the approach adopted in the UK for the first generation of nuclear power stations, owing to the difficulty of producing uranium enriched in ^{235}U. These reactors were gas-cooled, using high-pressure carbon dioxide as the primary fluid, and used uranium oxide fuel in a magnesium alloy container, hence their name: **Magnox reactors**. They operated at a relatively low temperature of 400 °C, which gave a low thermal efficiency of ~30%. In Canada, a heavy water moderated and cooled reactor (CANDU) was successfully developed. The use of non-enriched fuel, however, makes the core large, and the **pressurized water reactor** (PWR) developed in the USA, which uses enriched fuel and a more compact core, has been much more widely adopted, and has been described above in section 9.4.1.

The early prototype reactors, called generation I, were made during the 1950s and 1960s. Most of the reactors operating today are generation II reactors and were built during the 1970s and 1980s. The 448 reactors operating in 2017 included PWRs (292), boiling water reactors BWRs (75); heavy-water reactors HWRs (49), light water and gas-cooled graphite reactors, LWGR (15) and GCRs (14), and fast neutron reactors FNR (3). Fig. 9.3 shows the Daya Bay power plant in China, which has two 944 MWe Gen II+ PWRs built in the late 1980s.

Boiling water reactors (BWRs) use water as both the coolant and moderator and use enriched fuel, as do PWRs. The difference is that there is only a single loop circulating water through the core where it is heated and boils, producing steam for a turbine generator. The exiting steam is condensed and returned to the core to complete the loop. The pressure is less than in a PWR, about half, and no heat exchanger is required, but the power density is less, and all the components of the steam turbine and condenser must be shielded.

Fig.9.3 Daya Bay nuclear PWR power plant in China (Skyscan Photolibrary/Alamy).

9.6.1 **Advanced Nuclear Power Reactors**

Many light-water reactors (LWRs) in the USA were one-off designs, which led to duplication of effort and greater costs, and the Chernobyl accident in 1986 emphasized the importance of safety. These issues prompted the development in the early 1990s of standardized advanced light-water reactor designs which are simpler, require fewer components and less piping, are easier to build, and include additional passive emergency cooling systems. (These are sometimes referred to as generation III or III+ designs.) Passive cooling is an example of a **passive safety** system and depends on gravity or temperature differences, rather than pumps, and as a result it is expected to be much more reliable and closer to a fail-safe system. A few of these generation III reactors have now been built. The advantage of standardization in design is seen in France, where, over nearly two decades, 34 0.9 GWe and 20 1.3 GWe Gen II nuclear plants have been built. They supply ~75% of France's electricity.

Most of the advanced LWRs typically have a 60-year design life, a burn-up greater than 50 GWd per tonne, and are able to alter their output to half of their capacity or lower; i.e. they will be load following. Most French reactors can already be operated in this way. India is developing an advanced heavy-water reactor as part of its programme to utilize thorium. It is designed to be self-sustaining with ^{233}U being bred from ^{232}Th, and is expected to use 65% of the energy of the fuel, with two-thirds from thorium via ^{233}U. There is low-fissile Pu production, and it has good proliferation resistance. There is also cooperative international research on six reactor technologies, all of which would give improved performance; these are called Gen IV reactors (GIF). Prototypes of the generation IV reactors are planned for operation in 2020–2030.

9.6.2 **Small Modular Nuclear Reactors (SMR)**

In an attempt to reduce capital costs, there has been increasing interest in the development of small modular nuclear reactors (<300 MWe). These could provide both process heat for industry and electricity, and off-grid power in developing countries. It is hoped that by producing a large number of modules, costs would decrease both through economies of scale and through the learning effect. Due to their small size, there would be less radioactive material in the core, and they could be partly constructed in a specially equipped factory, ensuring good quality control. There could also be passive safety features.

Several designs of SMR designs are under consideration, including thermal-neutron reactors, fast neutron reactors (FNRs) and molten salt reactors (MSRs). In an MSR, the uranium fuel is dissolved in a sodium fluoride salt coolant which circulates through a graphite moderator. Fission products are removed continuously and the actinides are fully recycled, which reduces the high-level waste. The operating temperature is high enough for hydrogen production, and there is passive cooling of the core, which improves safety.

SMRs offer an attractive means of providing power in remote locations, or as a back-up to other forms of power. For example, since 1974 the Bilibino Nuclear Power Plant in Siberia has operated the world's smallest nuclear reactor, consisting of four light-water graphite reactors each with a capacity of 12 MWe. At Tuoli, in China, a fast breeder reactor with a capacity of 25 MWe has been supplying the electricity grid since 2011. Rajasthan Atomic Power Station, in India, operates a 100 MWe pressurized heavy-water reactor—the fourth smallest commercial SMR in the world.

The development of SMRs is now attracting private investment in the West, due to their lower capital requirement compared with large nuclear plants and their low carbon footprint. However, gaining regulatory approval is impeding their development, since the regulatory authorities are geared to large nuclear plants and lack the expertise needed for small scale reactors. Another concern is that the manufacture of SMRs could increase the proliferation of nuclear weapons.

Nonetheless, in 2020, approval was given to NuScale (based in Portland Oregon) to build a 60 MW SMR comprising 12 power modules, due to be operational in 2029. It occupies only 1% the volume of a conventional PWR, and though it generates less power per tonne of nuclear fuel, it gains in terms of flexibility, e.g. for a desalination plant or for use in developing countries. The projected cost of electricity is $55 per MWh, and while this is considerably less than recent cost estimates for plants in the West (see Section 9.8), it is higher than the present average cost from solar PV and onshore wind farms in the USA (see Chapter 1 Table 1.2).

Also, Rolls-Royce is proposing to build up to 16 SMRs in the UK over the next decade, each generating 440 MW of electricity. At a cost of £2bn each they would be better value for money than the 3 GW nuclear plant at Hinkley Point C, costing around £22bn.

9.7 Safety of Nuclear Power

There are about 440 nuclear plants worldwide (2020), most of them LWRs, operating at ~80% of maximum annual output. Operating experience has led to improved capacity as well as improved safety. However, public concern remains over the use of nuclear power, due in part to four serious accidents in commercial plants since the first nuclear power plants in the 1950s.

In 1952, there was a serious accident in a reactor at **Windscale** in the UK and another in 1979 in a reactor at **Three Mile Island** in Pennsylvania in the USA—both resulted in only a limited release of radioactivity. However, in 1986 there was an uncontrolled reactor power increase in a reactor at **Chernobyl** in the Ukraine, and a huge release of radioactivity. Fortunately, its effect on health was much less than initially feared, but public confidence in nuclear power was severely dented. The fourth major accident was in 2011 following a massive tsunami at the **Fukushima Daiichi** reactor site in Japan which further eroded confidence.

The Fukushima Daiichi accident led to reviews of nuclear reactors around the world, and while these have reaffirmed that nuclear plants can be operated safely, the resulting recommendations on improving the defence of plants against natural disasters will increase the capital costs of new reactors and affect the operation of some existing plants. Since any serious nuclear accident has global consequences, internationally approved designs and inspections would clearly separate the regulation from the management of nuclear plants and help build public confidence. Regulators need to be strong and independent, and a safety culture must be promoted.

There is another aspect of reactor safety, and that is the ability of a reactor to withstand a terrorist attack. This has received particular attention following the 9/11 attacks. Fortunately, the strength of the construction in nuclear power plants would provide considerable protection from the effects of the crash of a hijacked aircraft or of a truck bomb. The spent fuel, which is initially stored on site, is in reinforced concrete pools and so is also protected. Even so, the security of nuclear power plants has been further increased as a result of 9/11.

9.8 Economics of Nuclear Power

Nuclear reactor plants require large capital investment, and the financial risks associated with capital expenditure can be more easily absorbed within a large state-protected framework (or borne by the consumers rather than by the suppliers, with a consequent reduction in the cost of capital), and projects with a long-term payback can be funded as part of the national infrastructure for economic security. A competitive electricity supply market, however, tends to favour less capital-intensive energy sources and ones with shorter construction times. Long-term prices need to be guaranteed to ensure investment in a capital-intensive industry. As a result, small modular reactors are being considered as construction and financing should be easier.

The trend in new reactors has been for increased power output and higher safety levels. In the West there has been an increased cost for such first-of-a-kind (FOAK) reactors, but elsewhere, where there is state backing, costs have been more competitive. In the West, costs per kW of output have increased from around $3000 per kW in 2002 to around $8000 (or higher) per kW in 2016. In 2009 an estimate (MIT2009) of the capital cost was $4000 per kW, and at that price nuclear could be competitive at 6.6 cents kWh^{-1} with coal or gas if the interest rate on the capital cost were the same. A simple estimate for the LCOE from a plant with no fuel costs and an interest rate of 8% (see eqn (1.1)) is,

$$LCOE = 1.47 \times C_{capital} \, (\text{k\$ per kW}) \, \text{\$cent per kWh}$$

which gives 5.9 cent per kWh, in good agreement, as the cost of uranium is less than 1 cent per kWh. But at $8000 per kW, the LCOE at about 12 cents per kWh is uncompetitive.

Moreover, the cost of borrowing capital is higher for nuclear than for coal or gas, due to nuclear's poor track record for construction, coupled with the regulatory risk for new designs, since it can take much longer to obtain approval, and delays are expensive. Furthermore, there is a technical risk with any first-of-a-kind project. In the West the increased cost of borrowing and amount of capital is making nuclear power uncompetitive. However, where capital and interest rates are lower, then nuclear is more competitive.

Another factor that can increase the chance of delays and hence costs is the experience of the construction workers. Relatively few reactors have been built during the last decade, and in the West skilled labour has been lost, which makes delays more likely. An expansion would help drive down costs through the learning-curve effect (see Chapter 1 Section 1.5.8).

9.9 Environmental Impact of Nuclear Power

The principal environmental advantage of nuclear power is the very small amount of associated CO_2 emissions, which makes it a very good energy source in the light of global warming. The main environmental considerations are over the siting of nuclear reactors; in particular, the seismology (as the Fukushima accident has demonstrated), meteorology, geology, risk of flooding, and population distribution in the vicinity of the reactor. Meteorology is important, since, in the event of an accident, the effect of the prevailing weather conditions on the dispersal of radioactivity has to be assessed. In addition, the land requirements, the effect of thermal discharges to the environment, and the storage and disposal of waste have to be considered.

9.10 Nuclear Waste Disposal

The waste generated in a nuclear reactor is classified into **high-**, **intermediate-**, and **low-level waste**. High-level waste is the main concern and comprises ~25 tonnes of spent fuel per year for a 1 GWe PWR, which represents ~95% of the activity produced. It contains both trans-uranic and fission products: the transuranic waste contains long-lived isotopes of plutonium, americium, neptunium, and curium, which are all **actinide elements** and are the dominant hazard after 1000 years; the fission products, mainly ^{90}Sr and ^{137}Ce, account for most of the heat and penetrating radiation.

The spent fuel is first stored for several years on site, to allow the intense short-lived activity to decay. On site, it is kept in storage pools to remove the heat and provide shielding of the radiation. The spent fuel can then be stored or reprocessed to recover the uranium and plutonium and the remainder immobilized by vitrification, in which the waste is incorporated into borosilicate glass. Another method is to incorporate the waste in natural stable mineral lattices. The spent fuel or vitrified waste can then be placed in a corrosion-resistant can and stored in an underground repository. Reprocessing may not be required, since fuel cost is not a major factor in reactor economics. It could also be eventually disposed of as new tech-nologies emerge in the future, such as **accelerator-induced transmutation** of the long-lived nuclides to shorter-lived ones.

Caverns in dry, stable rock formations, deep boreholes, or salt deposits have been proposed for these repositories. The main concern is whether the waste could leak into the ground-water and contaminate drinking water. The challenge is to identify sites that are geologically stable for thousands of years. The amount of waste generated during the early years of nuclear reactors, called legacy waste, is far greater than that generated by the current generation of reactors and also needs to be stored safely. As yet, however, there is no country with a perman-ent deep underground site for storing high-level waste, though in Finland such a depository is under construction (2020).

9.11 Outlook for Nuclear Power

The desire for both a secure energy supply, as there are diversified sources of uranium, and a reduction in global carbon emissions, is a strong argument for nations to use nuclear power as a source of low-carbon electricity. It is a very compact fuel source with a high-energy den-sity that can be easily stockpiled and can provide steady base-load power. More standardiza-tion and simplicity of design would improve the competitiveness of nuclear power, and the inclusion of both passive, active safety systems will reduce the likelihood of accidents. Public trust is essential, safe operation and waste disposal are paramount in any new reactor building programme, and political and security issues arising from possible nuclear proliferation and terrorist attacks are a concern that needs to be addressed.

The production of electricity from nuclear power rose steadily until 2004, but since then it has been around ~2500 TWh per year, with a capacity of 392 GW in 2017. Its share of the total world production has been decreasing since ~1996, when it was 17.5%, to about 10% in 2019. The banking crisis of 2008-2010 made the financing of large capital projects difficult and slowed growth of many of them. Also, the Fukushima Daiichi nuclear power plant accident in 2011 caused renewed concern over the safety of reactors and reduced public confidence in nuclear power, and resulted in a few countries phasing out their nuclear programmes, including Germany and Italy.

The lack of effective carbon pricing, the falling price of renewables, and in the US shale gas, are making nuclear less attractive in several Western Countries, but in Asia and Russia nuclear is more competitive and some expansion is planned. In 2019, the World Energy Organization estimated that nuclear generating capacity could approximately double, and its share remain roughly 10% of electricity generation in 2060. However, the IEA in their report on nuclear power in 2019 concluded that without action nuclear power in advanced economies could fade significantly by 2040, making meeting the Paris goals harder to achieve. Although about 50 reactors are under construction, about a quarter of the existing capacity of ~440 reactors is expected to shut down by 2025, and with the falling price of renewables nuclear is increasingly uncompetitive. The IEA recommended as the most cost-competitive action extending operation of existing nuclear plants as long as it could be done safely, supporting new build, and encouraging new lower cost technologies (IEA2019).

Nuclear can provide valuable low-carbon base-load dispatchable generation and energy security and could help in reducing global emissions, but with its relatively high price coupled with safety and waste disposal concerns, the extent of its contribution is somewhat uncertain. Its potential contribution by 2050 is estimated as 5000 TWh y^{-1}.

9.12 Energy from Fusion

The binding energy increases with mass number for light nuclei (Fig. 9.1), which means that the fusion of two light nuclei to form a heavier nucleus results in the release of energy. This fusion is the source of energy in stars. In many stars, including our Sun, the energy results from a series of nuclear reactions that convert hydrogen into helium (the p–p chain), in which the initial step is the fusion of two hydrogen nuclei (protons) to form deuterium. This process is a weak interaction, as one of the protons changes to a neutron with the emission of a positron and a neutrino, and is the rate-determining step in the 'burning' of hydrogen.

In order for two protons to react they must have sufficient energy to overcome their mutual electrostatic (Coulomb) repulsion. According to classical mechanics, protons require more kinetic energy than the Coulomb barrier (~700 keV) in order to fuse. However, quantum-mechanical tunnelling enables the reaction to take place at the much lower proton energies that occur at the centre of the Sun (~1 keV). The protons have a Maxwell energy distribution, $N(E) \propto E \exp[-E/(kT)]$, so the strong energy dependence means that the small high-energy tail of the distribution contributes most to the fusion reaction rate in the Sun.

The energy release in the fusion of hydrogen to form helium in the p–p chain is about 25 MeV. This is some 10^6 times larger than in a typical chemical reaction involving electrons (as in burning fossil fuels). This is a similar yield to that in fission, and again there is no release of greenhouse gases. However, the fusion of hydrogen is totally impractical for a fusion reactor, since the reaction rate is too low, as it only involves the weak interaction. There are, though, other fusion reactions that occur via the strong interaction with a similar energy yield and with a reaction rate that can be made fast enough for a practical reactor. In particular, the fusion of deuterium and tritium to form helium and a neutron has the highest rate and releases 17.6 MeV. This has the potential to provide the world's energy needs for thousands of years.

9.12.1 **D–T Fuel Resources**

The fuels for D–T fusion are deuterium, and lithium from which tritium is produced. Deuterium is very plentiful as there are about 30 g m^{-3} of seawater (1 in 6500 hydrogen atoms are deuterium). The Earth's reserves of lithium are estimated to be ~13.5 million tonnes. Less easily available are the amounts in the sea where there are ~230 billion tonnes, at a concentration of Li$^+$ of about 175 mg m^{-3}. The energy density of the fuel for a fusion reactor is enormous: 50 mg of D plus 75 mg of tritium has the same energy content as one tonne (~1200 litres) of oil, and only 100 kg of D and 150 kg of tritium, obtained from ~300 kg of ^{6}Li, which corresponds to 4 tonnes of natural lithium, are required to produce ~1 GW of electricity for one year (Example 9.2). The cost of deuterium is ~€1 per g and lithium is ~€20 per kg, so the contribution of the cost of the fuel to the cost of electricity is negligible at 0.001c per kWh. The reserves of lithium in the Earth alone, ~13.5 million tonnes, would provide the world with ~3 TWy annually for ~1000 years!

EXAMPLE 9.2

What is the energy released when 100 kg of deuterium and 150 kg of tritium are consumed in 1 year in a fusion reactor? If the reactor's thermal plant is 35% efficient, find the average continuous electrical power output over the year.

100 kg of deuterium and 150 kg of tritium each correspond to 5×10^4 moles. The energy release E is therefore:

$$E = 5 \times 10^4 \times 6 \times 10^{23} \times 17.6 \times 10^6 = 5.28 \times 10^{35}\,\text{eV}$$
$$= 5.28 \times 1.6 \times 10^{16}\,\text{J} = 8.45 \times 10^{16}\,\text{J}$$

An amount of energy E in a thermal plant of efficiency ε can provide a continuous electrical power output P for a time t given by:

$$\varepsilon E = Pt, \text{ or } P = \frac{\varepsilon E}{t} = 0.35 \times 8.45 \times \frac{10^{16}}{3.15 \times 10^7} = 0.94\,\text{GW}$$

9.13 **D–T Fusion**

For the fusion of deuterium and tritium to occur at a sufficient rate, the nuclei must have energies ~10 keV, corresponding to a temperature ~10^8 °C. At these temperatures deuterium and tritium are highly ionized and form a gas of charged particles called a **plasma**. No solid material can survive at such temperatures, so the plasma must be kept away from any containing walls. In the Sun, the hydrogen plasma is so massive that it is contained by the huge gravitational force of attraction that balances the outward force resulting from the plasma pressure. In a terrestrial reactor the plasma must be contained by some other means. One method is to contain the plasma using magnetic fields, called **magnetic confinement**. Another is to compress the plasma by means of huge implosive forces, called **inertial confinement**. We will concentrate on magnetic confinement, as it is more developed. We first look at the characteristics of D–T fusion.

The fusion reaction of deuterium (D ≡ ^{2}H) and tritium (T ≡ ^{3}H) forms ^{4}He (α particle) plus a neutron:

$$D + T \rightarrow\ ^4He + n + 17.6\ MeV \qquad (9.17)$$

(The nucleus of a deuterium atom is called a deuteron, and that of a tritium atom a triton.) The 17.6 MeV of energy released is shared by the α particle (3.5 MeV) and the neutron (14.1 MeV). Deuterium is stable, but tritium is unstable with a half-life of 12.3 years; however, tritium can be produced using the emitted neutrons. The energetic neutrons, which take ~80% of the fusion reaction energy, can be stopped in lithium external to the plasma containment vessel. The energy deposited would be used as a source of heat for a thermal power station, and the neutrons would produce (breed) tritium in the lithium via the following two reactions:

$$n +\ ^6Li \rightarrow T + \alpha + 4.78\ MeV \qquad (9.18)$$

$$n +\ ^7Li \rightarrow T + \alpha + n - 2.87\ MeV \qquad (9.19)$$

The charged α particles from the D + T reaction provide a source of heat to maintain the plasma at a temperature of 10^8 °C, which must be kept away from the vessel walls. There is a loss of energy by the plasma through radiation, particle diffusion, and heat conduction to the walls. For steady-state operation, this loss would be balanced by the heat generation in the bulk of the plasma from the α particles. There would also be some additional heating and fuel injection used to maintain optimum running conditions. The ejected particles would be diverted away from the main body of the plasma, and thereby exhaust the helium ('ash') and heat from the plasma.

The 14.1 MeV neutrons from the D + T reaction interact with the walls of the plasma chamber and other components. Their interaction can produce radioactive nuclei, i.e. activate the materials. A low-activation stainless steel has been developed in which Cr, W, and Ti replace Mo, Ni, and Nb, which allows the steel to be recycled after 50–100 years. Ceramic and fibre-composite materials are also being examined. The level of waste is significantly less than that from a fission reactor and there would not be a large nuclear waste disposal problem. Also, the tritium does not pose a serious radiation risk.

For a D–T fusion reactor the fusion power P_{fusion} produced by the plasma is divided between the α particles, P_{α} (20%), and the neutrons, P_{n} (80%). The α particles are stopped within the plasma, so P_{α} heats the plasma. The neutrons escape the plasma and pass through the containment walls. Outside they are stopped in the lithium blanket, where tritium is produced, and thereby heat the fluid for the power turbine.

External power P_{ext} may be required in addition to P_{α} to compensate for losses from the plasma and to optimize the plasma conditions. The power loss P_{loss} arises from particles and heat that diffuse from the centre of the plasma to the walls of the container. In addition, there is radiation (bremsstrahlung and synchrotron), together with line radiation from impurities emitted by the containment walls. The power loss P_{loss} is related to the **energy containment time** τ_{E} and the total plasma energy W by:

$$P_{\text{loss}} = W / \tau_{\text{E}} \tag{9.20}$$

where τ_{E} is the time for plasma energy to be lost to the walls when the plasma is in its operating state but with no energy input. (NB τ_{E} is not the same as the plasma duration (burn) time: P_{α} and P_{ext} heat the plasma during a burn, so the duration time is longer.)

A quality factor Q is defined by:

$$Q = P_{\text{fusion}} / P_{\text{ext}} \tag{9.21}$$

Break-even is defined as $Q = 1$, and **ignition** as $Q = \infty$. Ignition corresponds to $P_{\alpha} = P_{\text{loss}}$ and $P_{\text{ext}} = 0$. We show in Derivation 9.3, when deriving the **Lawson criterion**, that for a 50–50 mixture of D and T the requirement for **ignition** is:

$$n\tau_{\text{E}}T_{\text{keV}} \geq 3 \times 10^{21}\,\text{m}^{-3}\text{s keV} \tag{9.22}$$

where T_{keV} is the plasma temperature expressed in keV, where $T_{\text{keV}} \equiv 10^{-3}\,kT/|e|$ so 10^{8} K ~10 keV. For $T_{\text{keV}} \sim 10$ keV and $n \sim 10^{20}$ m^{-3} the Lawson condition requires a confinement time $\tau_{\text{E}} \geq 3$s.

Derivation 9.3 Lawson criterion

For a D–T reactor the reaction rate per unit volume R, or reactivity, is given by eqn (9.3):

$$R = n_{\text{D}}n_{\text{T}} \langle v\sigma \rangle$$

where $\langle v\sigma \rangle$ is the average value of the product of the relative velocity of the D and T nuclei and the fusion cross section.

Each reaction produces energy E_{fusion}, where:

$$D + T \rightarrow \alpha + n + E_{\text{fusion}} \quad \text{and} \quad E_{\text{fusion}} = E_{\alpha} + E_{n} = 17.6\,\text{MeV}.$$

The power released is the reaction rate multiplied by the energy released per reaction, i.e.

$$P_{\text{fusion}} = RV \times E_{\text{fusion}} \tag{9.23}$$

where V is the volume of the plasma.

For a 50–50 mixture of D and T, $n_D = n_T = \frac{1}{2}n$. Plasma neutrality means that the electron and ion densities are equal, so that $n_e = n$. Hence the total kinetic energy per unit volume W/V is given by:

$$W/V = 2n\left(\frac{3}{2}kT\right) = 3nkT \tag{9.24}$$

assuming the electron and ions are in thermal equilibrium (i.e. at the same temperature). Using eqn (9.3) with $n_D = n_T = \frac{1}{2}n$, P_{fusion} (eqn (9.23)) becomes:

$$P_{fusion} = \frac{1}{4}n^2\langle v\sigma\rangle E_{fusion} V \tag{9.25}$$

Ignition corresponds to $P_\alpha = P_{loss}$, where $P_\alpha = \left(\dfrac{E_{alpha}}{E_{fusion}}\right)P_{fusion}$ and $P_{loss} = W/\tau_E$, eqn (9.20), i.e.

$$\frac{1}{4}n^2\langle v\sigma\rangle E_{alpha} V = 3nkTV/\tau_E$$

In the operating temperature region of 10–20 keV the value of $\langle v\sigma\rangle$ is approximately proportional to T^2, i.e.:

$$\langle v\sigma\rangle \sim 1.2\times10^{-24}\, T_{keV}^2\ \mathrm{m^{-3}\ s^{-1}} \tag{9.26}$$

where $T_{keV} \equiv 10^{-3}kT/|e|$ is in keV, so this condition can then be expressed as one on the triple product $n\tau_E T_{keV}$. Noting that $E_{alpha} = 10^6|e|E_{alpha}(MeV)$ and that $E_{alpha}(MeV) = 3.5$, the requirement for **ignition** is:

$$n\tau_E T_{keV} \geq 3\times10^{21}\ \mathrm{m^{-3}\ s\ keV}, \quad \text{equivalent to } p\tau_E \geq 5\ \mathrm{atms}$$

where the relation $p = nkT$, eq (9.29), has been used. This is known as the **Lawson criterion**. For $T_{keV} \sim 10$ keV and $n \sim 10^{20}\ \mathrm{m^{-3}}$, $\tau_E \geq 3$ s.

9.14 Plasmas

Any time-independent external electric field is shielded from the interior of a plasma by a boundary layer of charge (plasma sheath) that is typically a fraction of millimetre thick in a fusion reactor. However, magnetic fields can penetrate plasmas and can be used to contain the charged particles.

The typical distance that charged particles are separated in a plasma is of order

$$r_s \sim n^{-1/3} \tag{9.27}$$

The relative kinetic energy of a pair of charged particles is of order kT, so the separation r_c, when their electrostatic potential energy $U_c = e^2/(4\pi\varepsilon_0 r_c)$ equals their relative kinetic energy, is given by:

$$r_c \sim e^2/(4\pi\varepsilon_0 kT) \tag{9.28}$$

If their typical distance apart is much greater than r_c, i.e. $r_s/r_c \gg 1$, then binary Coulomb interactions are rare. It is then a good approximation to treat such plasmas as **weakly coupled**.

The pressure p exerted by the plasma is given by:

$$p = nkT \tag{9.29}$$

EXAMPLE 9.3

A plasma has a temperature of 10 keV and a number density $n = 10^{20}$ m^{-3}, typical for fusion. Calculate the plasma pressure and the ratio r_s/r_c.
Using eqn (9.29) for p, and noting that $kT/|e| = 10^4$,

$$p = 10^{20} \times 10^4 \times 1.6 \times 10^{-19} = 1.6 \times 10^5 \text{ Pa} = 1.6 \text{ bar}$$

The particle distance r_s and electrostatic distance r_c are given by eqns (9.27) and (9.28), so:

$$r_s/r_c = (4\pi\varepsilon_0 kT)n^{-\frac{1}{3}}/e^2 = 4\pi \times 8.85 \times 10^{-8} \times 10^{\frac{20}{3}}/(1.6 \times 10^{-19}) = 1.5 \times 10^6$$

Hence $r_s/r_c \gg 1$, so the plasma is weakly coupled.

9.14.1 Plasmas in a Toroidal Magnetic Field

In a magnetized weakly coupled plasma, the charged particles spiral quite freely along the magnetic field lines with a radius of gyration ρ, which is the Larmor radius given by:

$$\rho = mv_\perp/qB \tag{9.30}$$

where $v_\perp$ is the component of velocity perpendicular to $\boldsymbol{B}$. Since in a plasma at a temperature T the mean kinetic energy $\frac{1}{2}mv^2$ equals $\frac{3}{2}kT$, only one component of velocity is parallel to $\boldsymbol{B}$, so $mv_\perp^2 = 2kT$. For a temperature of 10 keV and a B field of 5 tesla (T), typical values in a fusion plasma are:

$$Deuterons: \rho = 4.1 \text{ mm} \quad \omega_c = 239 \times 10^6 \text{ rads}^{-1} = 38 \text{ MHz}$$
$$Electrons: \rho = 67 \text{ μm} \quad \omega_c = 879 \times 10^9 \text{ rads}^{-1} = 140 \text{ GHz}$$

where $\omega_c = qB/m$ is the frequency (cyclotron frequency) at which the particles gyrate about the magnetic field lines.

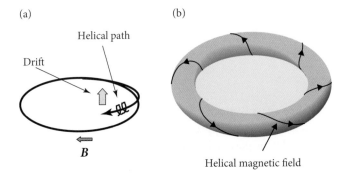

(a)

(b)

Helical path

Drift

B

Helical magnetic field

Fig. 9.4 (a) Drift in a toroidal magnetic field space (b) Helical field from adding a poloidal field.

In a toroidal magnetic field the field lines are continuous, which suggests that particles spiralling round the field lines on helical paths would be contained within the field. However, the curvature and non-uniformity of such a field cause the spiralling motion of a charged particle to drift away from the toroidal magnetic field lines, as shown in Fig. 9.4 (a). This is because when a charged particle experiences a force F in a magnetic field B it drifts at a velocity v in the direction $F \times B$ such that the Lorentz force qvB balances the force F.

In a toroidal magnetic field the field varies as $1/r$ in the radial direction. Hence the magnetic energy of the charged particle arising from the interaction of its magnetic moment μ, associated with its spiralling motion, with the magnetic field B, varies with position, and the particle therefore experiences a force F in the direction of the varying field. It also experiences a further radial force as the particle is on average accelerating radially. The result is that ions and electrons will drift in the direction $F \times B$, i.e. vertically (Fig. 9.4 (a)), but in opposite directions, which will set up a vertical E field and thus a radial $E \times B$ outward drift, leading to loss of confinement.

This drift can be counteracted by adding a poloidal component to the magnetic field producing a helical field, as illustrated in Fig. 9.4 (b). The helical field lines connect the top and bottom of the torus, with the result that drifts cancel, no charge separation arises, and there is no $E \times B$ drift outwards. The charged particles are therefore confined by such a field.

The helical toroidal magnetic field can be produced by a series of twisting magnetic coils in a configuration called a **stellarator**, or by a **tokamak**. Fusion research with tokamaks is the most advanced, and we now describe the main features of a tokamak.

9.15 **Tokamaks**

In a tokamak (a Russian acronym for 'toroidal chamber with magnetic coils'), the toroidal magnetic field is produced by a series of toroidal field-coils (Fig. 9.5). For continuous operation in a fusion reactor, superconducting coils would be used to reduce the power

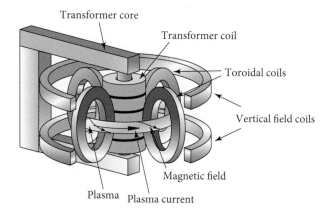

Transformer core

Transformer coil

Toroidal coils

Vertical field coils

Magnetic field

Plasma   Plasma current

Fig. 9.5 Principle components of a tokamak and the confining helical magnetic field.

requirements for generating the large fields of several Tesla. The vertical field-coils are used to position and shape the plasma. Negative feedback is used to maintain the plasma in position.

To create the poloidal field a current is induced in the plasma; the resulting field is helical, as shown in Figs. 9.4 (b) and 9.5. The current is induced in the plasma by a large time-varying magnetic field that passes through the centre of the torus. The plasma acts like the secondary circuit in a transformer. If the flux linkage is Φ, then:

$$V = d\Phi/dt \quad \text{and} \quad I = V/R$$

where R is the resistance of the plasma. In the JET (**Joint European Torus**) apparatus at Culham in the UK, a voltage of 10 V was required to establish the current, since initially R was relatively high. Once the plasma was fully ionized by ohmic heating then a voltage of 0.3 V was sufficient to drive a 3.5 MA current. JET had a flux swing of 34 V s and was able to sustain a current for about a minute ($[\Phi_{max} - \Phi_{min}] = -\int V \, dt$).

The need to apply an external field reduces the efficiency, or availability, of a fusion reactor. Fortunately, a radial pressure gradient within the plasma causes a toroidal current to flow. As it is produced by the plasma itself, it is called a **bootstrap** current and reduces the amount of external current drive required to maintain the toroidal current.

In the early tokamak experiments, the temperature of the plasma was raised by ohmic heating. However, the resistivity of the plasma decreases with increasing temperature (as $T^{-3/2}$), owing to the decrease in the separation of colliding particles (see eqn (9.28)) and the time spent colliding. The ohmic heating from the current required to give the necessary poloidal field raised the temperature to only a few 10^7 °C. At this temperature the power losses due to radiation and transport balanced the power input by ohmic heating. We therefore need to raise the temperature of the plasma by other means.

One method of raising the temperature is to inject a beam of neutral deuterium atoms into the plasma. Charged deuterium ions are accelerated and then neutralized by passing them through a neutral deuterium gas. In the gas, charge exchange yields high-energy (~80 keV) neutral deuterium atoms that can cross magnetic field lines. Once inside the plasma they transfer energy to the plasma particles and also become ionized. (D or T would be used in a fusion reactor.)

Another technique to heat the plasma is to use high (radio) frequency electromagnetic waves: a technique called **rf heating**. The oscillating electric field in the rf wave resonates with the cyclotron motion of the plasma particles, either ions or electrons, depending on the frequency. The frequencies are 20–50 MHz for ion cyclotron resonance heating (ICRH), or 70–140 GHz for electron cyclotron resonance heating (ECRH). Both the neutral beam and rf can also be used to drive a current within the plasma. Using high-speed pellets of frozen D, a high central plasma density can also be obtained.

9.15.1 Plasma Containment in a Tokamak

When a current is induced in the plasma in a tokamak, the current flows in a ring. A short section of the current can be approximated by a cylindrical section. The current gives rise to a very large poloidal magnetic field, as shown in Fig. 9.6, which exerts an inward force on the plasma current, the **pinch effect**. For equilibrium the plasma current must be large enough to generate sufficient magnetic force to balance the outward pressure of the plasma.

We can estimate the magnitude of this current I by finding the magnetic field B at the edge of the plasma, which has a radius R. The magnetic field B due to the plasma current is $\mu_0 I/(2\pi R)$. Approximating the current to be all at a radius R, the force on a length l of plasma is IBl. The force acts inwards over an area $2\pi Rl$, so the magnetic 'pressure' (i.e. force per unit area) p_B is:

$$p_B \sim \mu_0 I^2 l/[(2\pi R)^2 l] = \mu_0 I^2/(2\pi R)^2 = B^2/\mu_0$$

The pressure of the plasma p in the cylindrical section is nkT or $NkTl/(\pi R^2 l)$, where N is the number of charged particles per unit length and l is the length of the plasma section. The magnitude of current required to balance p is therefore given by $I^2 \sim 4\pi NkT/\mu_0$. The exact relationship (Bennett pinch formula) for the case of a cylindrical plasma is:

$$I^2 = 16\pi NkT/\mu_0 \tag{9.31}$$

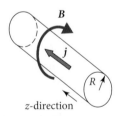

Fig. 9.6 z-pinch configuration.

EXAMPLE 9.4

Estimate the current required to contain a tokamak plasma at a temperature of 10^8 K, a number density $n = 10^{20}$, and a cross-sectional area of 1 m^2.

The linear number density is $N = 10^{20}$. For equilibrium we require a current:

$$I = (16\pi 10^{20} \times 1.38 \times 10^{-23} \times 10^8 / 4\pi 10^{-7})^{1/2} = 2.35 \text{ MA}$$

This is the order of magnitude of the current required in a tokamak.

9.15.2 Energy Confinement Time

The energy confinement time τ_E is determined by the time taken for particles and energy (both kinetic and radiation) to be transported from the hot plasma to the outside wall. A typical deuteron in a plasma at a temperature of 10 keV has a velocity of ~1000 km s^{-1}, so in a pulse of a few seconds duration the deuteron travels a great distance. Plasma particles spiralling along a magnetic field will scatter and transfer energy to other particles at a rate that is affected in particular by turbulence. The process is not well understood, and much research effort is being directed towards modelling turbulence.

A significant advance in improving confinement came with the discovery of an H-mode (high confinement mode) in the ASDEX tokamak in Germany. It was associated with the use of a divertor in which the magnetic field is altered to keep the main interactions of the plasma with any material away from the central plasma region.

By elongating the field vertically in a tokamak it is possible to form a cross-over, called a **separatrix**, in the magnetic field configuration near the wall. This configuration diverts the charged particles that have diffused out away from the main central region onto **divertor plates**. The charged particles are incident over a larger area which helps reduce the power density. In addition, gas is added near the plates to increase the radiative loss and, through charge exchange, neutralize the energetic charged particles. Both the radiation and the neutral particles spread the heat load.

The divertor needs to be made from a low-Z material so that any evaporated atoms are fully ionized in the hot plasma; carbon, with its very high sublimation temperature of 3825 °C, and beryllium have been used.

9.16 Outlook for Controlled Fusion

Large tokamak development

Fig. 9.7 shows the progress in obtaining better confinement since 1968, when the results from tokamak T3 were announced. The increase of $n\tau_E T$ in the following 40 years has been four orders of magnitude, and the successful culmination of experiments at the Joint European Torus (JET) in 1997, when fusion with a 50–50 D–T plasma produced 16 MW of power corresponding to a $Q \sim 0.6$ and obtained $\tau_E \sim 1$ s, made the prospect of a commercial fusion reactor

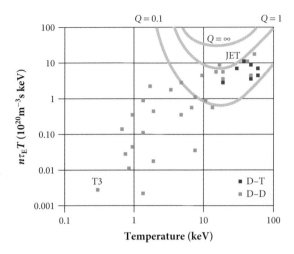

Fig. 9.7 Progress on increasing $n\tau_E T$ since the Russian tokamak T3 in 1968 (*Source:* Tok).

by 2050 much more feasible. The H-mode provides a fusion plasma that can be confined and from which the output heat and particles can be extracted. However, considerable research is still required on the transport problem and on the materials to be used, particularly for the divertor, and on avoiding excessive heat loads from instabilities.

These problems will be addressed in the **International Thermonuclear Experimental Reactor (ITER)**, which will be a tokamak designed to give $Q \sim 10$ and provide data upon which a demonstration reactor (DEMO) can be built. ITER will produce significant power under quasi-stationary conditions, investigate α-particle heating and particle wall interactions, and test the tritium breeder lithium blanket design. An international collaboration of China, the EU, the UK, Switzerland, India, Japan, South Korea, Russia, and the USA is involved. ITER is planned to have an inductive drive that will last for 8 minutes. It will also have auxiliary current drives that could extend operation up to 30 minutes. The fusion power output is planned to be 500 MW. In particular, ITER will provide information on the lifetime of the torus wall components. It is currently under construction in Cadarache, near Aix-en-Provence in southern France, but due to delays, the first plasma is now expected around 2025 and $Q > 1$ in the early 2030s. The tokamak will be nearly 30 m tall and weigh 23 000 tonnes, with a plasma volume of 830 m^3. ITER will use superconducting magnets to reduce energy consumption. By comparison, the largest currently operating tokamaks, JET in the UK and JT-60 in Japan, both have plasma volumes of 100 m^3.

Small and alternative fusion reactors

Conventional wisdom is that larger tokamaks, like JET and ITER, are better at preventing the hot plasma from heating the walls of the container than smaller ones. Recently, however, over 20 research groups around the world have begun to explore the possibility of building a small fusion reactor that they hope might become a practical reality within the next 10 years or so.

Advances in high temperature superconducting magnets that give higher magnetic fields, super-fast electronics and real-time computer modelling, have raised hopes that the plasma can be controlled more easily than in previous decades, attracting significant financial support from commercial as well as national funding agencies. One such design is SPARC a compact tokamak proposed by Commonwealth Fusion Systems and MIT. Tokamak Energy have opted for a spherical device, which requires weaker magnetic fields than a toroidal shape, but presents greater problems in removing heat from the centre of the plasma. Lockheed Martin have built a series of cylindrically symmetric reactors, the latest being 18m long and 7m in diameter, of comparable size to a fission reactor in a nuclear submarine.

Other projects are also aiming to reduce the cost and scale of tokamaks. For example, in 2020, first plasma was achieved in a more compact spherical tokamak design called MAST at the Culham laboratory in the UK. A **stellarator**, the Wendelstein-7X, started up in Germany in 2015, and its reported performance in 2019 is encouraging. In a stellarator, the helical field is provided by external coils, so no induced current is required, which eliminates any instabilities arising from the current and allows the reactor to run for longer.

Research is also ongoing into **inertial confinement fusion**. In this method, small spheres containing deuterium and tritium are highly compressed to raise the temperature and pressure sufficiently high that the fuel reacts before it can fly apart; i.e. it is confined by its own inertia. At the National Ignition Facility in California, intense laser beams have been used to heat the D–T spheres, but progress is hindered by instabilities arising during compression. In particular, the Rayleigh–Taylor instability (which arises when two fluids in contact have density and pressure gradients of opposite signs) occurs in the implosion of the D-T spheres and much research still needs to be done on understanding and controlling such instabilities. Also, hybrid approaches are being explored: in General Fusion's design a cylindrical magnetically confined plasma is heated and compressed by shock waves.

Whether any of these smaller projects actually leads to a commercial fusion reactor in the next few decades or so remains to be seen, but it has at least injected fresh thinking into the field. Controlled fusion offers an enormous source of power with essentially no greenhouse emissions, and though a commercial reactor may still be several decades from fruition, the long-term benefits to future society could be very considerable and justify continued investment in fusion research.

 SUMMARY

- Nuclear power is a very compact source of low-carbon electricity, with 1 tonne of uranium equivalent in energy to 25 000 tonnes of coal.

- The known reserves of uranium are 4.7 Mt, equivalent to 85 years of operation at present consumption (IAEA), with an estimated additional 35 Mt recoverable. The predominant reactors are PWRs and BWRs, in which water is both the coolant and the moderator. The neutrons are moderated to thermal energies to allow the chain reaction to proceed with fuel only a few per cent enriched in ^{235}U.

- Nuclear power provides good energy security and very low carbon emissions, which is particularly important in combating global warming. It is currently providing ~2500 TWh per year, and its potential by 2050 is ~5000 TWh per year. But concerns about cost, safety, waste disposal, and nuclear proliferation make its future role uncertain.

- Fusion has a very high-energy density: 100 kg of deuterium and 150 kg of tritium would produce 1 GWe for one year. Tritium is produced by neutron reactions in lithium.

- Resources of lithium and deuterium are sufficient to provide the world with 10 TWy of fusion energy annually for ~1000 years.

- For a self-sustaining fusion plasma, the particle density n, the temperature T, and the containment time τ_E must satisfy the Lawson criterion:

$$n\tau_E T_{keV} \geq 3 \times 10^{21} \text{ m}^{-3} \text{ s keV}$$

where $T_{keV} \equiv 10^{-3} kT/|e|$. For $T_{keV} \sim 10$ keV and $n \sim 10^{20}$ m^{-3}, $\tau_E \gtrsim 3$s.

- The increase of $n\tau_E T$ over the last 40 years has been four orders of magnitude, with JET obtaining a ratio of fusion to input power of about 0.6 and a τ_E of ~1 s in 1997 with a D + T plasma.

- ITER, which is a tokamak design, is under construction in the south of France. Data from this facility will be used to design a demonstration reactor (DEMO). It is hoped that the first commercial reactor will be operational by 2050.

- The contribution of controlled fusion as a means of providing an almost unlimited supply of carbon-free energy, both for improving the standard of living worldwide and for tackling the consequences of climate change, could be very important.

 FURTHER READING

Lilley, J. (2001). *Nuclear physics: principles and applications*. Wiley, New York. Clear discussion of the physics of nuclear reactors.

Nature Physics 12 (2016). Informative series of articles about fusion power.

World Nuclear Association: World nuclear performance report 2018.

www-fusion-magnetique.cea.fr/gb/ Useful quantitative discussion of fusion.

www.gen-4.org/ GenIV International Forum on next-generation nuclear energy (GIF).

www.iaea.org/ International Atomic Energy Agency (IAEA).

www.iea.org/reports/nuclear-power-in-a-clean-energy-system (IEA2019).

scipub.euro-fusion.org/wp-content/uploads/2014/11/JETR99013.pdf J. Wesson, The science of JET. Interesting description of the development of JET.

www.iter.org/ ITER tokamak fusion reactor.

web.mit.edu/nuclearpower/ (MIT2009).

9.1 From Fig. 9.1, estimate the energy released when the fission of ^{235}U yields two stable nuclei with mass numbers 141 and 92 together with the emission of two neutrons [About 10 MeV of this energy is not absorbed in the reactor core as it is taken away by neutrinos.]

9.2 How many tonnes of uranium fuel enriched to 3% in ^{235}U is equivalent to 100 000 barrels of oil?

9.3 Estimate the amount of natural uranium required annually to provide 10% of a primary global power consumption of 18 TW.

9.4 The global production of electricity by nuclear power in 2017 was about 2500 TWh. Estimate the annual consumption of uranium. At this rate, how many years would the conventional reserves of 4.7 Mt of uranium last: the reserve-to-production (R/P) ratio?

9.5* Show that the neutron mean free path λ in uranium before an absorption on ^{235}U is given by $\lambda = 1/\Sigma_a$, where $\Sigma_a = n_f \sigma_a$, n_f is the number of ^{235}U nuclei per unit volume, and σ_a is the neutron absorption cross section.

9.6* Deduce that the critical size a_c of a reactor is of order $\delta/\sqrt{(k_\infty - 1)}$, where δ is the mean distance travelled by a neutron before it is absorbed, from equating the fraction of neutrons within δ of the surface of the core $\sim (\delta/a_c)^2$ to the fractional loss of neutrons $(k_\infty - 1)/k_\infty$. Deduce that $\delta \sim 1/\sqrt{\Sigma_a \Sigma_s}$.

9.7* Consider a spherical reactor core containing a homogeneous mixture of graphite and uranium enriched to 2.0% in ^{235}U and with a ratio of graphite to uranium n_s/n_f of 600. (a) Find p, f, η, and k_∞ assuming $\varepsilon = 1$. (b) Estimate the critical radius a_c.

9.8* For a core of thorium enriched to 2% with ^{233}U and moderated with graphite with a ratio of $n_s/n_f = 400$ and $k_\infty = 1.05$, show that the conversion factor $C = 1.07$. Take $\varepsilon = 1$. What is the implication?

9.9* After the shutdown of a nuclear reactor, show that the number of ^{135}Xe nuclei n_p and of ^{135}I nuclei n_I satisfy the differential equations:

$$dn_p/dt = \lambda_I n_I - \lambda_p n_p \quad dn_I/dt = -\lambda_I n_I$$

Show that the number of ^{135}Xe nuclei grows to a maximum ~11 h after shutdown. Neglect the equilibrium number of ^{135}Xe compared with the equilibrium number N_0 of ^{135}I present in the core prior to shutdown.

If the maximum amount of ^{135}Xe that can be compensated for by removing control rods is $0.2N_0$, how long is the reactor out of action after the shutdown? (For ^{135}I, $t_{1/2} = 6.7$ h; for ^{135}Xe, $t_{1/2} = 9.2$ h.)

9.10* (a) Consider a nuclear fission reactor in which it is assumed that neutrons are produced directly by fission and also indirectly by the β-decay of a single fission product. The concentration n of neutrons and the fission product C satisfy the differential equations:

$$dn/dt = [k(1-\beta) - 1]n/\tau + \lambda C \quad dC/dt = \beta n/\tau - \lambda C$$

Explain the physical meaning of each of the four parameters k, β, τ, λ in these equations, and show that $k = 1$ in the steady state.

(b) Consider a reactor transient in which the 'reactivity' $\rho = (k - 1)/k$ is slightly positive. By considering solutions of the above equations of the form $n \propto \exp(mt)$ and $C \propto \exp(mt)$, show that m has two possible values, given by

$$m_1 \approx -\left(\frac{\beta - \rho}{\tau}\right), \quad m_2 \approx \left(\frac{\lambda\rho}{\beta - \rho}\right)$$

Assume that $(\beta - \rho)/\tau \gg \lambda$ and $[(\beta - \rho)/\tau]^2 \gg 4\lambda\rho/\tau$. Typically, $\lambda = 8 \times 10^{-2}\ \text{s}^{-1}$, $\rho = 2 \times 10^{-3}$, $\beta = 6 \times 10^{-3}$, and $\tau = 10^{-4}$ s.

9.11 The value of $\beta\tau_d$ for the delayed neutrons from the fissile nucleus ^{239}Pu is 0.0324. Calculate the time constant for control of a ^{239}Pu fuelled reactor when $k = 1.0005$. Assume $\tau_p = 0.1$ ms.

9.12 A PWR nuclear reactor is generating 1 GWe of electrical power. The reactor core contains 100 tonnes of uranium and the neutron flux density is 4×10^{17} neutrons $\text{s}^{-1}\ \text{m}^{-2}$. Estimate the enrichment in ^{235}U of the fuel.

9.13 Estimate the ^{235}U enrichment required to give a burn-up of 60 GWday per tonne, if the fuel is replaced when the percentage of fissile material is reduced to 1.75%. Assume $C_f = 0.25$.

9.14 The fission yield b of the nuclide ^{88}Kr ($t_{1/2} = 2.79$ h) is 0.0364. (a) Estimate the activity in the core from ^{88}Kr one day after a 2 GWe reactor is shut down. Assume production of ^{88}Kr ceases on shut down. (The fission rate F is related to the thermal power output P in GW, assuming 200 MeV per fission, by $F = 3.13 \times 10^{19}P$ fissions per second. The decay rate R of an unstable nuclide, after $t \gg t_{1/2}$, equals the production rate bF, assuming no contribution from a fission product decay.) (b) ~6% of fissions of ^{235}U produce ^{135}I; show that the activity from its decay in a 1 GWe reactor is $\sim 6 \times 10^6$ TBq.

9.15 A beam of 3 MeV γ-rays of flux $F = 10^9\ \text{m}^{-2}\ \text{s}^{-1}$ is incident on a 1 m-thick water shield. The linear attenuation coefficient μ of water for 3 MeV γ-rays is 3.96 m^{-1}. Calculate (a) the effective γ-ray flux F_{eff}, (b) the effective dose rate $D_{\text{rate}}(\mu\text{Sv h}^{-1})$, where $F_{\text{eff}} = F \times B(\mu x)\exp(-\mu x)$ and the build-up factor $B(\mu x)$ allows for the increase in flux primarily from lower-energy γ-rays arising from Compton scattering, and for the $x = 1$ m-thick shield is 3.55.

9.16* For a breeder reactor operating at constant power output P_0, show that if m kilograms of fissile material are consumed per day per gigawatt of output, then the time t_D to double the mass of fissile material equals $M_i/[(C - 1)mP_0]$, where M_i is the initial mass and C is the breeding ratio.

A 2 GW breeder reactor contains ^{235}U surrounded by ^{232}Th. For an initial amount of ^{235}U of 1000 kg and a breeding gain of 0.05 to breed ^{233}U, calculate the doubling time t_D.

9.17 (a) Estimate the volume of spent fuel generated annually by a 1 GWe PWR (without reprocessing). The fuel is uranium dioxide enriched to 3% in ^{235}U and the density of the fuel is 4×10^3 kg m^{-3}. (b) Reprocessing reduces the volume of waste by a factor of about 4. Estimate and comment on the volume of waste generated annually with reprocessing by PWRs with a total output of 0.9×10^6 GWh of electricity (~25% of the annual consumption in the USA).

9.18 What options have been considered for dealing with nuclear waste, and to what extent do you think that they are effective?

9.19 Devise a demonstration (practical or computer) that illustrates how moderators increase the resonance escape probability in a reactor.

9.20 Discuss the relative merits of using carbon, heavy water (D_2O), or water as a moderator, given that the ratio of the average energy E_f of a neutron after impact to the energy before impact is $E_f/E_i = (A^2 + 1)/(A + 1)^2$, where A is the atomic mass number of the target nucleus.

9.21 Describe the principle features of a PWR and explain how a BWR differs.

9.22 Summarize the main economic considerations that affect the choice of nuclear power compared with an alternative source of power.

9.23 Discuss *critically* the statement that the environmental risks associated with nuclear power are far less than those associated with global warming.

9.24 Does public opinion preclude the expansion of nuclear power?

9.25 Explain what is meant by 'passive safety features'.

9.26 Is it better to build a single nuclear power plant or several smaller nuclear plants with the same total output?

9.27 Can nuclear power make a significant impact on averting global warming?

9.28 Calculate the amount of tritium required to fuel a 5 GWe power station for three years.

9.29 Using the conservation of momentum, explain why in the fusion of D + T the α particles receive 3.5 MeV and the neutrons 14.1 MeV.

9.30 Calculate the number density of a plasma at a temperature of 10^8 °C and a pressure of 2 bar.

9.31 A plasma has a pressure of 2 bar and a number density $n = 10^{20}$ m^{-3}. Calculate the plasma temperature, and the ratio r_s/r_c.

9.32 A current of 3 MA flows in a tokamak plasma which has $n = 2 \times 10^{20}$. Estimate the temperature of the plasma.

9.33 Estimate the distance travelled by a triton in a 15 keV plasma contained for 5 seconds.

9.34 A plasma containing a 50–50 mixture of D and T has a pressure of $1.6 \, 10^5$ N m^{-2}, an energy confinement time τ_E of 3.5 s, and a temperature of 15 keV. Calculate the triple product $n\tau_E T$ and compare with the Lawson criterion.

9.35 Assuming that heat loss is mainly by via particle diffusion and heat conduction, estimate the confinement time in JET with a cross-sectional area of its toroid of 8 m^2, given that the tokamak TFR had a confinement time of 20 ms with a cross-sectional area of 0.13 m^2.

9.36 Describe the main features and principles of operation of a tokamak.

9.37 Investigate the progress being made on small fusion reactors.

9.38 Discuss *critically* the statement: 'Even though fusion power plants that can contribute significantly to base-load power will not be available until after 2050, and therefore will not contribute to reducing our carbon dioxide emissions before then, it is still very important to invest in their development'. Source: DOE2007

For further information and resources visit the online resources
www.oup.com/he/andrews_jelley4e

10 Electricity and Energy Storage

Introduction

In the earlier chapters we considered the ways in which various forms of primary energy are exploited to provide electricity and heat. We now describe how electricity is generated, transmitted, and distributed. We also investigate the integration of the variable supplies of electricity from wind and solar PV, and describe various forms of energy storage, in particular batteries.

10.1 Generators of Electricity

A generator is a machine for converting mechanical energy to electrical energy. In general, a rotating shaft causes the magnetic field passing through a coil to change, which (by Faraday's law of electromagnetic induction) generates a voltage that can drive a current. The basic principle of an electrical generator is illustrated in Fig. 10.1. Consider a planar loop of conducting wire of area A, rotating at a constant angular velocity ω in a uniform magnetic field B. Suppose the loop consists of N turns and subtends an angle $\theta = \omega t$ to the magnetic field at some instant t. The magnetic flux intersecting the loop is given by:

$$\phi = NBA \cos \theta = NBA \cos \omega t \tag{10.1}$$

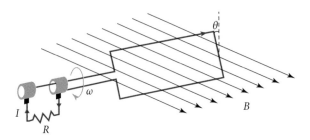

Fig. 10.1 Loop rotating in a uniform magnetic field.

By Faraday's law of electromagnetic induction, the electromotive force ε (equal to the potential drop across R in Fig. 10.1), called the back emf, is given by the rate of change of magnetic flux. Thus:

$$\varepsilon = -d\phi/dt = NBA\omega \sin \omega t \qquad (10.2)$$

If the electrical circuit is completed by connecting the ends of the wire across a resistance R, then the alternating electric current flow through the resistance is given by:

$$I = \varepsilon/R = I_0 \sin \omega t \qquad (10.3)$$

where:

$$I_0 = NBA\omega/R \qquad (10.4)$$

In order to make electrical contact between the moving and stationary parts, slip rings and graphite brushes are inserted between the rotating conductor and the external circuit. It should be noted that the current induced in the wire is an effect that exists only while the conductor rotates in the magnetic field; hence $I \to 0$ as $\omega \to 0$.

EXAMPLE 10.1

A coil with 1000 turns of cross-sectional area 1m^2 rotates at 50 Hz in a uniform magnetic field of 0.5 T. What is the maximum current flowing through a load of 1000 Ω?
From eqn (10.4), the maximum current is given by:

$$I_0 = \frac{NBA\omega}{R} = \frac{10^3 \times 0.5 \times 1 \times 2\pi \times 50}{10^3} = 157 \text{ A}$$

In a typical large generator, the magnetic field is produced by passing a direct current through coils mounted on a central rotating shaft connected to a turbine, called the **rotor**. There is a small loss of energy due to resistance heating of the coils, but it is more economic than using permanent magnets. The rotor is surrounded by a set of stationary coils wound around an iron core, called the **stator**. The rotating magnetic field due to the rotor intersects the stationary coils in the stator and induces a current. The frequency of generation is determined by the angular velocity of the rotor. It is chosen to be high enough to avoid flickering of electric lights: 60 Hz in North America, South America, and parts of Japan, and 50 Hz in Europe.

The configuration of the windings in the stator is very complicated. The general principle is to maximize the emf (electromotive force ≡ voltage generated), but it is also necessary to minimize the flow of **eddy currents**, which are self-circulating currents that produce unwanted components of magnetic field and cause losses. Eddy currents are reduced by increasing the resistance of the paths through which eddy currents flow, i.e. by laminating the core using thin sections of steel alloy, thereby forcing the current to flow mainly through the laminations. The evolution of the design of rotors and stators has been something of a 'black art', and details of particular machines tend to be commercially sensitive.

Electricity is usually generated as **three-phase current** rather than single-phase current; this is achieved by employing three independent sets of windings, 120° apart, around the generator (Fig. 10.2). The idea of using three phases originated with **Nikola Tesla**, a pioneer in the early years of electricity generation and transmission. Unlike single-phase power, three-phase power never drops to zero and the power delivered to a resistive load is constant in time (Exercise 10.5). Another advantage of three-phase current is that it needs only about 75% of the material to conduct the same quantity of power as that in single-phase.

Heat is dissipated in the stator, which needs to be cooled in order to maintain it at a constant temperature; at higher temperatures the resistance of the wiring increases and the life of the insulation is shortened. For medium-sized machines, forced cooling using a rotor mounted fan is sufficient, but, for large machines stators are cooled using deionized water or hydrogen (which is a better heat conductor than air).

Mechanical stability is also an important consideration, particularly in large machines. A plant outage to repair a large generator is a time-consuming and costly business, and usually means that less efficient generating plant has to be used. It is therefore normal practice to monitor mechanical vibrations for early signs of metal fatigue and cracks before a major incident occurs.

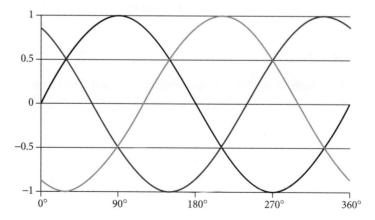

Fig. 10.2 Three-phase current.

Apart from electricity generation by conventional turbines (i.e. steam, gas, or water turbines) described earlier in the book, special-purpose generators are needed for exploiting certain kinds of renewable energy. For solar PV, the dc output from the panels is first converted into ac for transmission by units called *inverters* (see Section 10.4).

The case of wind turbine generators is described in Box 10.1.

Box 10.1 Wind turbine generators

Generators work by moving magnetic field lines through conductors and thereby inducing a current. The magnetic field can be provided by current flowing in coils, or by permanent magnets. The declining cost of rare-earth magnets (in particular NdFeB magnets) has led to permanent magnet generators being built with capacities of several MW. NdFeB has a very high *BH* product (where *B* and *H* are the magnetic field density and the magnetic field strength, respectively), which means that only a small amount of magnetic material is required to produce a strong magnetic field. The magnets are attached to a cylindrical drum (the rotor) connected to the turbine shaft. Surrounding the rotor are conducting coils attached to the inside of a fixed cylinder (the stator). As the magnets rotate, an alternating current is induced in these coils.

In a variable-speed turbine, the alternating current is first rectified to direct current and then converted back to alternating current at a fixed frequency for connection to the grid. A torque τ is required to turn the rotor, since the induced current produces a magnetic field that opposes the motion of the magnets (Lenz's law). The opposing tangential force per unit area (i.e. the shear stress), S, depends on the surface current density K (amperes per metre) and on the strength of the magnetic field B. (NB For a single conductor of length L carrying a current I normal to a field B, the force is BIL.) The magnitude of the current density is limited by dissipation, and the maximum B field is limited by magnetic saturation; as a result, S is typically between 25 kPa and 100 kPa.

The size of generator required depends on the torque τ produced by the wind turbine, which is given by the product of S, the surface area of the rotor and the radius of the rotor, i.e.:

$$\tau = 2\pi r l S r \quad or \quad \tau/V_R = 2S$$

where V_R is the volume of the rotor. From Chapter 7 Section 7.5, the power absorbed from the wind is given by:

$$P = \frac{1}{2} C_p \rho A u_0^3 = \tau \Omega$$

where C_p is the power coefficient, ρ is the air density, u_0 is the wind speed upstream of the turbine, Ω is the angular velocity, and A the swept area of the turbine. Ω is related to the radius R of the turbine, the tip-speed ratio λ, and the wind speed u_0 by:

$$\Omega = \lambda u_0 / R$$

Hence:

$$V_R = C_P \rho A u_0^3 / (4\Omega S) = \pi R^3 C_P \rho u_0^2 / (4\lambda S)$$

For a 2 MW turbine operating with $C_P = 0.45$, a wind speed of $12\ \text{ms}^{-1}$, $R = 35$ m, and a tip-speed ratio $\lambda = 8$, the volume of the rotor with $S = 80$ kPa is $\sim 4.1\ \text{m}^3$ ($\rho \sim 1.2\ \text{kg m}^{-3}$), i.e. ~ 2 m in diameter and 1.3 m long. We can see that the generators become very large for outputs of order MW, but the simplicity of having no gearbox reduces maintenance costs (and noise) and improves reliability.

10.2 High Voltage Power Transmission

To transmit large quantities of power over large distances the following issues need to be addressed:

What is the optimum voltage for long-distance transmission?

How can the voltage be increased and decreased?

Is it better to transmit AC or DC?

To answer the first question, we consider the heating due to the resistance of a long-distance transmission line, known as **ohmic** (or **Joule**) **heating**. In principle, a superconducting cable would be the perfect solution, but no material has yet been found that is superconducting at ambient temperatures.

Consider a wire of resistivity ρ, cross-sectional area A, and length L. The total resistance of the wire is:

$$R = \rho L / A \tag{10.5}$$

Suppose the wire conducts a current I at a voltage V. The loss of power due to the resistance of the wire is:

$$\Delta P = RI^2 = \rho L I^2 / A$$

Putting $I = P/V$, we have:

$$\Delta P / P = \rho P L / A V^2 \tag{10.6}$$

So, for a given power P and a given investment in cabling cost, which limits the cross-sectional area of cable A for a certain length L, it follows from eqn (10.6) that the operating voltage should be as high as possible to minimize the fractional loss of power $\Delta P/P$. In practice, there is also an upper limit to the current density I/A that can be conducted; otherwise the wire would get too hot. The total loss of power for a national grid due to long-distance high voltage transmission and local distribution is typically about 5–10%.

EXAMPLE 10.2

A power plant transmits 100 MW of power along a transmission line of length 50 km with a resistance of 0.01 Ω km^{-1}. Calculate the percentage loss of power if the line is at (a) 10 kV; (b) 400 kV.

The resistance of the complete line is $R = 0.01 \times 50 = 0.5\ \Omega$. For a line at 10 kV, the current needed to transmit a power of 100 MW is $I = P/V = (10^8)/(10^4) = 10^4$ A. The fractional power loss is $\Delta P/P = (RI^2)/P = (0.5 \times 10^8)/10^8 = 50\%$. For a line at 400 kV, the current is $I = P/V = (10^8)/(4 \times 10^5) = 250$ A and the fractional power loss is $(RI^2)/P = (0.5 \times 250^2)/(10^8) = 0.03\%$.

The maximum voltage for transmission is determined by the electrical breakdown strength of air. The maximum electric field occurs at the surface of the cable, and if it gets too large the surrounding air becomes ionized, forming a **corona discharge** that conducts electric current to ground. The electrical breakdown strength of dry air is around 3×10^6 Vm^{-1}, but is lower in damp conditions.

Overhead transmission cables consist of twisted strands of conducting wire. The outer strands are usually made of aluminium (due to its low resistivity and low cost) and the inner strands of steel, for mechanical strength. The electric field at the outer surface varies inversely with the radius of the cable. For voltages over 110 kV, cables are usually configured in bundles, which reduces corona losses. Two-cable bundles are used for 220 kV lines and three- or four-cable bundles for 400 kV lines.

High voltage overhead lines operate between about 110 kV and 1200 kV. In Europe, the typical voltage for long-distance transmission is around 400 kV. Underground lines can be up to three times the cost of overhead lines with the same load capacity and are used where overhead lines are unacceptable, e.g. built-up areas and underwater crossings, and also increasingly to avoid problems with gaining planning permission. For AC transmission, increasing or decreasing the voltage is straightforward using transformers. However, in the case of DC transmission, electronic devices are required which add to the overall cost.

10.3 High Voltage AC (HVAC) Transmission

Power is normally transmitted when AC as three-phase, each phase being conducted at a different height above the ground. The tops of the transmission towers are connected by a cable at earth potential, which helps to protect the transmission cables from lightning strikes. The height of the cables is set to ensure that the electric field at ground level is too low to endanger life.

The electricity in power stations is typically generated at around 18–20 kV. The power is fed into a substation, where a series of **step-up transformers** increases the voltage to that of the transmission line. Conversely, a **step-down transformer** is used to reduce the voltage from the transmission line to that of the local distribution network. A transformer basically consists of two coils of wire wrapped around a common iron core. In the case of a step-up transformer, the number of turns N_2 in the secondary coil is greater than the number N_1 in the primary coil, and vice versa for a step-down transformer. The iron core has a high magnetic permeability in order to ensure that the bulk of the magnetic flux is concentrated in the iron core. The ratio of input to output voltages is given by $V_2 = (N_2/N_1)V_1$. Large power transformers (over 50 MVA) are typically over 99% efficient. The main energy losses are due to the resistance of the windings, and cooling systems are used to maintain transformers at a steady temperature.

However, high voltage AC transmission has significant drawbacks. The inductance and capacitance of an AC transmission line generate losses which do not arise in DC transmission. In addition, capacitors and other components are required to control the reactive power flow and maintain the stability of the system voltage. The net effect is to limit the transmission distance and the transmission capacity of a high voltage AC transmission line. Also, alternating current is predominantly conducted in the outer layer of a cable—the **skin effect**—which increases the effective resistance of the line. Furthermore, it is not possible to make a direct connection between two AC systems operating at different frequencies, and even connections at the same frequency can lead to system instability.

10.4 High Voltage Direct Current (HVDC) Transmission

The long-distance transmission of electricity began with direct current, around the end of the nineteenth century, but this was soon displaced by alternating-current systems, which were much easier to step up and step down to different voltages (using transformers) and because alternating three-phase synchronous generators were technically superior to direct-current generators.

However, for very-long-distance transmission overland, and for transmission undersea and under cities, HVDC transmission systems can now be more economic than HVAC systems, despite the additional cost of converters at each end of the transmission line. The breakeven point is around 600 km for overland and about 50 km for below ground or sea. There are numerous undersea HVDC links in Europe under the Baltic, the North Sea, and the Mediterranean. The lack of reactive losses means that the efficiency of HVDC power transmission can be around 97% over a 1000 km line.

China has invested heavily in HVDC transmission to integrate wind, solar and hydropower resources lying at great distances from population centres. In 2019, the world's largest UHVDC line started transmitting 12,000 MW of wind and solar power from the Northwest to the East of China at 1100 kV over 3000 km, which will supply 66 billion kWh annually, meeting the power demand of 50 million homes. There are also UHVDC projects in India and Brazil and many HVDC ones around the world.

Direct current is produced by rectifying alternating current. The original devices for AC to DC were gas discharge valves called **mercury arc rectifiers**, but in the mid-1970s they were displaced by solid-state devices called **thyristors**. Inverters for DC to AC basically switch the direction of the DC input rapidly, while rectifiers only conduct when the AC input is in one direction; in both cases, the outputs can be shaped, into an AC sine wave or a constant DC voltage, respectively, by using capacitors and inductors. High voltage DC can be produced by first generating relatively low voltage AC and then stepping up the voltage with a transformer before rectification; following transmission the HVDC is inverted, and the high voltage AC is stepped down before distribution.

Most of the present HVDC systems use converters that employ thyristors. In these, the conversion of DC to AC is controlled by the line voltage of the AC and the devices are called line-commutated converters (LCC). They have a high current and voltage capability and are best suited for long distance point to point transmission. Converters that use integrated gate bipolar transistors (IGBTs) enable the direction of the DC current to be easily switched, and are now increasingly competitive; in these voltage-source converters (VSC), conversion is independent of the external circuit. The power carrying capability of LCC is presently higher than for VSC but the control afforded by VSC using IGBTs makes it ideal for multi-terminal DC networks, allowing for current reversal between nodes. Improvements in HVDC circuit breakers, using hybrid breakers, are also making DC grid fault protection easier.

10.5 National Electricity Grids

In the 19th century electricity was generated close to where it was needed, but economies of scale in the 20th century led to centralized power plants, long-distance power lines, and local sub-stations. In most countries in the world electricity is now supplied by such a grid. National electricity grids are complex networks managed by a central control unit, which decides how much power each plant is allowed to generate, taking account of the cost of the electricity generation, plant availability, load distribution, and the need to minimize the energy losses along the transmission lines, subject to various constraints such as the maximum current rating for each line.

This system is designed for the supply to meet the demand, with the minimum demand, called the baseload, met by the cheapest generators. Up until recently, these were typically coal-fired (or nuclear or hydropower plants, where available) and ran for most of the time. They were supplemented by other power stations, usually combined cycle gas turbine plants, to meet the daily variations in load, and by fast-acting gas turbine or diesel generators for surges in demand or for power station outages. Interconnecting transmission lines between the power stations and the sub-stations meant that the electricity supply could be maintained even if a line or power station went down. The grids enabled electricity to be delivered to far flung communities, as well as accessing remote sources of electricity.

10.6 Integrating Renewable Energy Supplies

Solar and wind farms are now providing an increasing proportion of electricity on many grids. This is changing the requirements on power plants. Typically, a mix which varies throughout the day using renewable and conventional power plants generates electricity most economically, instead of large conventional generators. Wind and solar farms' operating costs, called

marginal costs, are the cheapest, because they have no fuel costs (as well as having no emissions), and are called on first. To ensure that the greatest fraction of generation from wind and solar farms can be accommodated, additional power plants that can respond quickly to changes in supply and demand are best; and preferably these should also run economically at a small fraction of their maximum load. Generally, coal and nuclear plants do not ramp up or down quickly, and gas-fired and renewable plants are better. Depending on location, hydro-power, biomass, geothermal, and concentrated solar power (with thermal storage) plants can all be used as flexible generators.

Fossil fuel power plants can store their fuel and provide electricity on demand. Unlike these generators that can supply at will (called **dispatchable** or **firm** supplies), wind and solar farms give a variable output that is dependent on the weather. While they can complement each other, with wind speeds typically higher in the winter than in the summer, and sunshine greater in summer than in winter, there can still be significant gaps in availability. However, contrary to what some people imagine, grids with significant wind and solar generation are able to provide power when required. Variations in the output of solar and wind farms are generally well anticipated, through good weather forecasting that uses artificial intelligence (AI) to obtain the best results. When the renewable supply is up to 30% of demand, these variations can be readily met with the fast-reacting power plants already installed on the grid to meet changes in demand. Coping with a large 1000 MW power station unexpectedly tripping (caused by an equipment fault or an overload) can be far more challenging than a sudden drop in wind or solar power. Stand-by reserve plants have to come on-line quickly, and wind and solar farms, if not run at full capacity, can provide additional valuable back-up when the weather is windy and sunny, by ramping up their outputs rapidly.

10.6.1 Electricity Mainly from Renewables

In order to provide clean, secure, and affordable power, and drastically reduce carbon emissions by mid-century to avoid dangerous climate change, we must power grids predominantly by renewables. The percentage of electricity from renewables can be raised on a grid by installing overcapacity, increasing their geographical spread, and by interconnections with other grids. Increasing their generating capacity partly compensates for when weather conditions are poor, and connecting solar and wind farms over a wide area provides a smoother and more reliable output. In Europe, Denmark helps balance supply and demand by trading electricity with Norway, Sweden, or Germany: exporting it when their own wind generation is high and importing it when wind generation is low.

However, building an intercontinental renewable grid is not straightforward. There was a proposal (DESERTEC) to transmit solar power generated in North Africa to Europe, but it floundered due to political instability, and because there were objections arising from conflicting demands from different regions and countries on the proposed grid. Moreover, the sharp fall in the cost of solar panels has made the advantage of more sunshine less significant in some regions: it can be more economic to increase the size of a solar array to compensate for less sunlight, than to pay for the long-distance transmission. Local generation also gives security of supply, by not having to rely on fossil fuel imports.

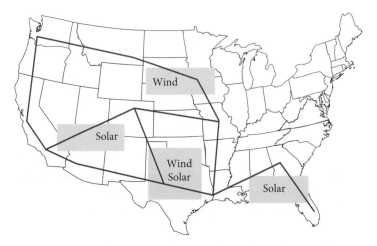

Fig. 10.3 Proposed HVDC grid for renewables in the USA (NREL 2018).

However, balancing demand and supply can be helped significantly by an extensive grid, and Fig. 10.3 shows a HVDC grid proposed by the NREL for transmitting across the US solar power generated in the south-west and south, where the solar irradiance is high, and wind power from the plains. The grid would enable back-up power plants to be shared, and by enabling excess generation to be used where demand is high, considerable savings can be made: for this 'macro grid' an estimated $48 billion in savings compared to $19 billion to construct. The time difference of 5 hours between the West and East coasts means that both solar and wind power could contribute to the peak evening demand on the East coast. HVDC can also help with the planning difficulties that have hindered linking the three largely independent grids (Western, ERCOT (Texas), and Eastern) in the past by using underground cabling, which can give protection from extreme weather, along existing rights of way (roads and railways), and by converting overhead HVAC lines to HVDC, for which the voltage and current can be higher (see Exercise 10.8) enabling greater power transmission.

With voltage source converters (VSC), HVDC can be part of a multi-terminal network linking renewable energy across a huge area, and allowing transmission to asynchronous power grids. Several cross border HVDC links are installed in Europe where, for instance Germany has in operation (and planned) connections with offshore wind farms, and where the compactness of VSC is advantageous. China now has several HVDC projects for linking wind farms with its grid, and the 'Plains and Eastern clean HVDC' project in the US should help with establishing a wider national HVDC grid.

10.7 Demand Response and Smart Grids

The need for reserve plants can be reduced by altering the demand to match the supply, and this is called **demand response**. This can be the cheaper option, as the fast-acting power plants used to meet the peak load are the most expensive to run. On a grid with a high input from solar PV, the supply is maximum in the middle of the day, while the total load tends to rise in the morning then flatten off before rising again in the evening [the supply can extend a little later by having the panels west facing]. The load, met generally by conventional dispatchable generation to supplement the solar PV, called the net load, therefore falls during the day and rises steeply in the evening. It is called a **Duck Curve**, because its profile resembles that of a duck. In regions of high solar irradiance there is often a large air-conditioning load in the evening and the need for a large (and expensive) standby capacity can be reduced by shifting this load. This **load shifting** can be achieved, for example, by using solar PV electricity to make ice during the day that is then used to provide cooling in the evening.

More precise adjusting of the load to match the supply can be done using a smart grid that allows two-way communication between the grid operator and the user. This allows just the right amount of demand to be taken off-line or added. Many operations exist where interrupting or reducing the electricity supply for a short time is possible: operations with thermal inertia—such as keeping something, e.g. iron or bitumen, molten, or food in a refrigerator cold; or when heating or cooling a building—or where a stockpile of items is produced first, before the items are assembled into products. Likewise, demand can be increased by turning on an electric furnace, or a large electrolyser, or (to help cope with droughts from climate change) a desalination plant. We are only at the beginning of the **smart grid** revolution with its digital technology, which will enable significant changes in load to be made; this will ease the transition to renewable energy and bring lower costs for customers.

Smart meters, smart sockets, and smart devices have been available to savvy consumers for some years to time their electricity consumption to coincide with periods of the day when electricity prices are lower. A limited version of this principle has been available for a few decades in some countries, such as France, with a predominantly nuclear generation capacity, where cheaper electricity tariffs at night-time have been exploited to drive washing machines, dishwashers, etc., and for heat storage. A smart grid takes the idea a stage further by controlling the operation of tens of millions of smart domestic and industrial devices (each fitted with sensors and communication systems) from a central grid control. The potential economic benefits to countries with smart grids are considerable, significantly reducing peak demand and improving system stability, maximizing the exploitation of renewable energy supplies (notably wind and solar) and thereby mitigating global warming as well as providing cheap electricity for an electric-car society.

Encouraging customers to alter their demand can be done through price differentials. A simple scheme in Italy has the capital for power stations and the cost of distribution recovered through fixed charges that depend on the maximum power used, and production costs through a price per kWh. By restricting the power demand, which makes the electricity cheaper for the consumer, the use of appliances, such as a kettle, washing machine, and an oven, have to be spaced out during the day; if used all at once the power supply trips. This reduces the peak load for which the cost of generation is highest. Cheaper prices for off-peak (e.g. night time use) is another way. But for better adjustment a smart grid and smart meters are required. Then customers can see details about their consumption, and opt to only use certain devices when electricity prices are low or have an override high price button.

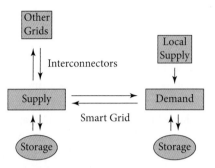

Fig. 10.4 Schematic showing the balance of supply and demand on a modern grid.

10.7.1 **Grid Electricity Storage**

High percentages of renewable generation are greatly helped by having electrical storage available. A mix of solar and wind farms with the capacity to more often meet the demand in an evening will tend to over-produce during the day, and cause electricity prices to fall. Without storage, this surplus must be exported if possible, or lost through curtailing the supply. Short-term storage can shift some production from the afternoon to the evening, so a smaller capacity can meet the daily demand. As the cost of batteries falls sharply, this form of storage is becoming increasingly available, and is also starting to displace fast-reacting fossil fuel plants.

The storage can be both by the generator or by the consumer, and how storage, interconnectors and a smart grid can help balance supply and demand is illustrated schematically in Fig. 10.4.

The local supply shown in Fig. 10.4 could be from rooftop solar panels that might be part of a mini-grid. This would be an example of a **behind-the-meter** system, as its electricity does not pass through the meter registering the consumption of electricity from the utility supplier. Trading between users on the mini-grid can be handled using **blockchain** technology. Local storage can be provided by batteries but also with heat stores (e.g. hot water tanks or generating ice for cooling).

10.8 **Economics of Renewable Electricity**

The cost of changing to renewable generation comes not only from the huge investment in wind and solar farms, but from the cost of strengthening the grid to transport electricity from remote sites. In Germany, the rate of growth of variable renewable generation was slowed while transmission lines were upgraded; and in China rooftop solar panel installations have been encouraged, which will lessen the need for long distance transmission and reduce the load on the grid. While renewable capacity grew rapidly in China in the last decade, **wind curtailment** (i.e. the percentage of electricity available from wind that was not used) reached 17% in 2016. Directed policies, opening up electricity trading between provinces, and new transmission lines have reduced curtailment by 2019 to 4% for wind and 2% for solar.

In some countries, local generation with solar panels can be the best source of electricity, particularly where no grid exists, as in much of sub-Saharan Africa; where grid electricity is very expensive as in parts of Australia; or where the grid has been poor, as in India. Residential systems can give their owners the possibility of selling excess generation back to the grid, called **net metering**. But there has been concern over net metering as it pushes the burden of maintaining the grid onto those without solar panels.

While the cost of producing electricity from wind and solar farms is now very competitive with fossil fuel generation, the variability of solar and wind adds additional costs to operating a grid. This is because of the expense of back-up generators for when the supply is low, which adds to the consumer's electricity price. But with greater percentages of renewable generation, fewer large conventional plants are required; and as wind and solar farms are becoming cheaper, this helps offset the cost of the additional reserve plants that must run occasionally to balance supply and demand. Electricity markets typically operate on the expected demand a day ahead, and the amount of reserve depends on the forecast error; if the market operates on a shorter timescale the amount required decreases.

With more renewable generation, intraday markets are increasingly important; for example, the Nord Pool enables intraday trade between 15 Baltic and Nordic countries, and intraday markets are now used in many European and other countries. In 2020 India introduced a near real-time intraday market, which includes Myanmar, Sri Lanka, Nepal, Bhutan and Bangladesh, to help with the integration of the output from wind and solar farms, and improve the utilization of power plants throughout the region. It is part of an ambitious 'One Sun One World One Grid' plan being promoted by India. China has also in the last two years introduced electricity markets designed to enable renewable energy and energy storage to help reduce curtailment and to encourage more trading.

Increasing the capacity of the variable renewable plants helps to provide sufficient output when the renewable source is low, but it can have a distorting effect on an electricity market when it is high. This is because the cost to provide an additional unit of kWh or MWh, called the **marginal cost**, of the renewable electricity is then effectively zero, since the resource is free. When a market is based on marginal costs, the price of electricity in a deregulated market will drop when there is a large amount of renewable capacity, so the revenue to cover the cost of new capacity will fall.

The shortfall in revenue from the low marginal cost of renewables is called the **missing-money problem**. It is already causing some existing conventional power plants to close before time, and too few new ones to be built to sustain the required capacity to meet demand. Moving back to a more regulated market or to more public ownership could help, as might having a **capacity market** which encourages continued investment in generating capacity. Participants are paid a per MW rate for the capacity they offer which must be available to provide power for balancing supply and demand when necessary.

While electricity power markets, introduced in many countries to promote competition in the 1980s and 1990s, are having to adapt to support short-term generation, the fraction of renewable generation on a grid can be very high with demand response, distributed generation, interconnectors, and short-term storage. For example, Denmark has a substantial renewable capacity, which is mostly onshore wind, about equal to its power demand. It has coped with such a variable supply partly by being integrated with the grids of neighbouring countries, but also by its integration of electricity and heat supply through its extensive CHP network. Half of its electricity supply comes from small CHP plants that supply local district heating networks which include large tanks of water as heat stores. This allows the CHP plants to alter their proportion of electricity to heat output to help match variations in the output of the wind farms. In 2019, the share of electricity generation by wind and solar in Denmark was 50%, while in the first half of 2020 selected regional percentages were: World 10%; Germany 42%; UK 33%; Australia 17%; USA; 12%; China; 10%, and India 10%.

A study for Europe in 2018 from Chalmers University of Technology (Sweden) of the costs of increasing the amount of variable renewable energy on the grid from 20% to 80%, found prices per kWh rising from around 5 to 8.5 eurocents, with most of the electricity generated by wind power; with low cost for both solar and batteries, the range went down to about 4 to 6 eurocents. Variations in supply would be principally handled by importing and exporting electricity, together with flexible generators (gas-powered). The increase in cost would be far less than that from the damage caused by continuing to burn fossil fuels.

Emissions can be further lowered when there is access to biomass, hydrogen-fuelled turbine plants, concentrated solar power plants with storage, or flexible nuclear power or hydropower plants, which can reduce the dependence on natural gas-fired generators to balance the grid. Obtaining a high percentage of renewables in some regions would be helped by having long-term (several months) storage to meet the large inter-seasonal mismatch between supply and demand that can occur. Using surplus renewable generation to produce hydrogen, which is called green hydrogen as the electricity is from renewables, looks promising as an energy store, but also as a fuel for heavy transport powered by fuel cells, and as source of clean heat for industry.

Some short-term mismatch may soon be met, as costs fall, by over-capacity solar farms with a day's battery storage. In the USA, a report by the Rocky Mountains Institute in 2018 concluded that advances in renewable energy and distributed energy resources, including batteries, over the last decade meant that these can now provide as reliable a level of supply as new gas plants, at a comparable or lower cost.

Solutions that enable a large fraction of renewable generation to power a grid will depend on the resources available in a region, but one key area where further cost reduction would help significantly is in energy storage.

10.9 Energy Storage

Energy storage reduces generation costs during periods of peak demand and enables the grid controllers to cope with sudden changes in electricity demand or unexpected losses in generation capacity until alternative generating units can be brought into action. Electricity grids generally need the flexibility of being able to store energy over time intervals varying from a few seconds to several hours. There are many ways that energy can be stored and converted into electrical energy when needed. As we have seen, energy storage will also be an important factor in the case of renewable energy sources of generation such as solar and wind, due to their inherent variability; i.e. it is important to be able to store the energy and to supply it when it is needed.

We now describe some of the important forms of energy storage and first consider batteries. These are electrochemical devices for storing energy in a form that can be readily converted into electrical energy. The chemicals are stored within the device from manufacture, unlike a fuel cell in which the chemicals are renewed continuously (see Chapter 11 Section 11.4.3). Batteries are either non-rechargeable (primary) or rechargeable (secondary), but only rechargeable batteries are of interest for energy storage, in particular Li-ion batteries used in many electric vehicles and increasingly for large scale storage.

10.10 **Basic Principles of Batteries**

A battery contains **galvanic** (also called **voltaic**) **cells**, each of which has two or more half-cells: one for **oxidation** and one for **reduction**. Each half-cell contains an electrolyte solution and an electrode. A simple example is the copper–zinc galvanic or **Daniell cell** shown in Fig. 10.5. When the zinc and copper electrodes are joined by a conductor, electrons flow via the external circuit from the zinc electrode to the copper electrode, where they combine with copper ions in solution, which are deposited on the copper electrode. The following half-reactions occur at the electrodes:

$$Zn \rightarrow Zn^{++} + 2e^- \text{ oxidation } \quad Cu^{++} + 2e^- \rightarrow Cu \text{ reduction}$$

Zinc loses electrons and is oxidized more readily than copper, with the result that zinc goes into solution, and copper ions accept electrons at the copper electrode and are reduced. The zinc ions in solution are neutralized by sulphate ions flowing through the porous barrier, which completes the flow of charge around the circuit. The barrier stops copper ions diffusing and depositing on the zinc electrode when no current is flowing. Zinc is the anode or negative electrode, and copper is the cathode or positive electrode.

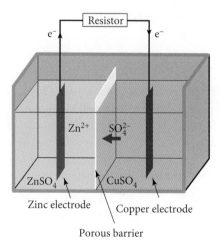

Fig. 10.5 Daniell cell.

The energy required to remove an electron from zinc is less than that for copper. The available energy (as explained in Chapter 2 Section 2.7.2) is minus the change in the Gibbs free energy $-\Delta G$ in the overall reaction, called a **redox reaction**:

$$Zn + Cu^{++} = Zn^{++} + Cu \tag{10.7}$$

The value of ΔG depends on the concentrations; if these are one molar (1 M) at 25°C then ΔG° = −213 kJ mol^{-1}. In reaction (10.7) a charge $Q = nN_A e$ is transferred, where $n = 2$ (since the ions are doubly ionized), N_A is the Avogadro constant, and e is the electronic charge, so the potential difference V° between the electrodes is related to the change in the Gibbs free energy by the formula:

$$\Delta G^{\circ} = QV^{\circ} = nN_A e V^{\circ} = -nFV^{\circ} \tag{10.8}$$

where F, the Faraday constant, is equal to 96,485 Coulombs per mole. Substituting the value of ΔG° gives what is called the **standard potential** of the Daniell cell as $V^{\circ} = 1.10$ V.

10.10.1 The Standard Potential and the Nernst Equation

The standard potential of a metal or compound is taken to be the voltage between the substance and a hydrogen electrode, which consists of hydrogen gas flowing over a platinum electrode that catalyzes the reaction $H_2 \rightarrow 2H^+ + 2e^-$. Some standard potentials for half reactions used in batteries are given in Table 10.1:

The dependence of the voltage of a battery on the concentration of the solutions is determined by how the Gibbs free energy depends on concentration. The result (see Chapter 2 Section 2.7.3) is:

$$\Delta G = \Delta G^{\circ} + RT \ln Q \tag{10.9}$$

where Q, the reaction quotient, is the ratio of the concentrations of the products to that of the reactants. Substituting eqn (10.8) gives the **Nernst equation** relating the standard potential to the concentration of the ions in the battery:

$$V = V^{\circ} - \frac{RT}{nF} \ln Q \tag{10.10}$$

For the Daniell cell, the overall reaction is given by eqn (10.7), so at $T = 25°C$ (298 K) and noting $\log x = \log e \times \ln x$, the potential difference V will be given by:

Table 10.1 The standard potential V° for some half-reactions used in batteries

$Li^+ + e^- \rightarrow Li$ −3.04 V	$Na^+ + e^- \rightarrow Na$ −2.71 V
$V^{+++} + e^- \rightarrow V^{++}$ −0.26 V	$2H^+ + VO_2^+ + e^- \rightarrow VO^{++} + H_2O$ +1.00 V
$Zn^{++} + 2e^- \rightarrow Zn$ −0.76 V	$Cu^{++} + 2e^- \rightarrow Cu$ +0.34 V

$$V = V^\circ - \frac{0.059}{n} \log \frac{[Zn^{++}]}{[Cu^{++}]}$$

where $[Zn^{++}]$ and $[Cu^{++}]$ are the concentrations of the Zn^{++} and Cu^{++} ions, respectively. The activities of the metals (the electrodes) do not change and can be set to unity, as it is only the difference in concentrations (see eqn (10.10)) that gives rise to a change in potential. If the concentrations start at one molar, then after a current I has flowed for a time t, we have:

$$\frac{[Zn^{++}]}{[Cu^{++}]} = \frac{1 + It/(nF)}{1 - It/(nF)}$$

The logarithmic dependence on concentration means that the potential falls only very slowly until nearly all the Cu^{++} ions have deposited on the copper electrode, dropping only 14 mV after 0.5 mole, 38 mV after 0.9 mole, and totally discharging to $V = 0$ volts after 1 mole of charge has flowed.

10.11 Pb, Na-S, NiMH, and Liquid Metal Batteries

For energy storage the overall chemical reaction in the battery needs to be reversible, i.e. the direction of the reaction is reversed if an opposing emf is applied that is larger than the cell potential. The battery can then be recharged. The first practical rechargeable battery was the **lead–acid battery**, invented in 1859 by the French physicist, **Gaston Planté**.

10.11.1 Pb: Lead–acid Batteries

The battery consists of a series of lead anodes and lead oxide cathodes immersed in sulphuric acid (H_2SO_4), with each cell connected in series (Fig. 10.6). Its energy density is comparatively quite low, but it can provide large currents and is still used in cars today. A lead–acid battery can be recycled several hundred times, depending on the discharge rate and depth of discharge. Those designed to produce large currents (as for starting cars) have more and thinner plates than those designed for discharge to lower levels.

On discharge, the lead anode oxidizes to lead sulphate by the half-reaction:

$$Pb + H_2SO_4 \rightarrow PbSO_4 + 2H^+ + 2e^-$$

and the lead oxide cathode reduces to lead sulphate by the half-reaction:

$$PbO_2 + 2H^+ + 2e^- + H_2SO_4 \rightarrow PbSO_4 + 2H_2O$$

The overall reaction is:

$$Pb + PbO_2 + 2H_2SO_4 \rightarrow 2PbSO_4 + 2H_2O \quad V^\circ =\sim 2.0 \text{ V}$$

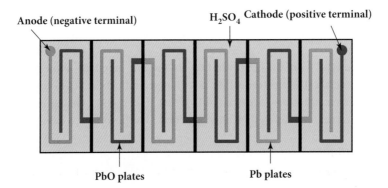

Fig. 10.6 Lead–acid battery with six cells: output voltage ~12 V.

The dens<insert two spaces>ity and concentration of the sulphuric acid decrease during discharge. The voltage also therefore drops during discharge and, as in the Daniell cell, the fall is slow initially with a rapid drop close to full discharge (see Exercise 10.20). The specific energy (energy per unit mass) is ~20–30 Wh kg^{-1}.

In a conventional lead acid battery, the charging rate and lifetime are limited by the dissolution rate and build-up of lead sulphate on the negative electrode. Adding carbon to the negative electrode improves the charging rate, and lowers the accumulation of lead sulphate giving improved cycle life. While still having a low-energy density, these lead–carbon batteries are proving cost-effective and can provide cheap energy storage when size and weight are not important. In Europe and the USA over 95% of lead–acid batteries are recycled.

10.11.2 Na-S: Sodium–sulphur Batteries

The Na–S battery, first developed in the 1960s, uses a solid electrolyte (beta alumina) that selectively conducts Na$^+$ ions, a liquid metal (sodium) negative electrode, and molten sulphur in porous carbon as the positive electrode. The battery operates at a temperature of ~300 °C, and during discharge the sodium ions pass through the beta alumina and combine with sulphur to form sodium polysulphide. The overall reaction can be represented by:

$$2Na + 3S \rightarrow Na_2S_3 \quad V° = \sim 1.9 \text{ V}$$

The materials are cheap and have low densities. The theoretical specific energy is high, ~720 Wh kg^{-1} (see Exercise 10.21), and values of ~200 Wh kg^{-1} have been obtained. Their response time is about 1 ms. The cycle lifetime is long and the charge efficiency (the ratio of energy from discharge to energy required to charge) is high (~85%), but economies of scale, the high temperature, and the corrosive nature of the materials make this battery most suitable for large-scale energy storage, such as for the grid. Typical lifetimes are 15 years or 4500 cycles. Examples are a 34 MW/245 MWh unit used for stabilization on a wind farm in Japan and a massive 108 MW/648MWh unit in Abu Dhabi providing 6 hours of storage to be used for load balancing.

A related molten-salt battery operating at ~300 °C that uses a liquid sodium anode, separated by beta alumina from a molten sodium tetrachloroaluminate ($NaAlCl_4$) electrolyte, with a nickel chloride cathode, called a ZEBRA battery, has been developed that also has a high specific energy of ~120 Wh kg^{-1} and has been used to power vehicles.

10.11.3 NiMH: Nickel Metal Hydride (NiMH) Battery

These rechargeable batteries first became available commercially in the late 1980s and have a considerably higher energy density than lead–acid batteries. The positive electrode is a hydrogen-absorbing alloy, and the negative electrode is nickel oxyhydroxide, NiOOH. The discharge half reactions are:

$$M + e^- + H_2O \rightarrow MH + OH^- \quad \text{and} \quad Ni(OH)_2 + OH^- \rightarrow NiOOH + H_2O + e^-$$

where M stands for an inter-metallic compound, which typically includes a mixture of rare earths, nickel, and manganese. The electrolyte is generally potassium hydroxide. Their specific energy is ~80 Wh kg^{-1}. The chemicals used in the NiMH batteries are less reactive than those in lithium-ion batteries. While they are heavier than Li-ion batteries, they currently last longer in very hot conditions. The batteries are used in some hybrid electric cars, but have been mainly superseded in pure electric cars by lithium-ion batteries.

10.11.4 Liquid Metal Batteries

More recently at MIT in 2006, Donald Sadoway revived research into liquid metal batteries, where both electrodes and the electrolyte are liquids that are separated vertically just through the difference in their densities. On discharging, positive anode metal ions diffuse through the molten salt electrolyte into the lower liquid cathode layer where they alloy with the cathode metal. On charging the process is reversed. Advantages are that many candidate materials are inexpensive, and the liquid state gives fast discharge-charge cycles during which volume changes do not cause mechanical stresses. They have potentially long lifetimes, and have a simple construction as they lack membranes and separators. However, they have relatively low voltages and energy densities, and are sensitive to movement, so are most suitable for grid storage, as they are potentially easily scaled, fast reacting, and cheap.

The high operating temperature of 700°C in MIT's original cell, which used a magnesium anode, $MgCl_2$ electrolyte, and antimony cathode, caused degradation, but promising results have been obtained recently with different liquid metal alloy electrodes that operate at significantly lower temperatures. The MIT start-up Ambri is commercializing a battery based on a calcium alloy anode and an antimony cathode operating at 500°C, while a research group at UT Austin reported in 2020 a cell operating at 20°C based on a Na-K anode and a gallium-based alloy cathode. However, the fast-falling cost of the established Li-ion battery is hurting investment and development of new battery technologies.

10.12 **Lithium-ion Batteries**

A very significant advance in battery technology was made in the 1980s and 1990s with the development of the lithium-ion battery. Lithium has the highest electronegativity (see Table 10.1) and a low density, so it is ideal for providing high-energy density. However, it is chemically very reactive, but a way of controlling its reactivity was found when it was discovered that lithium ions could be moved into or out of graphite without breaking up its structure— a reversible process called **intercalation**. Lithium can also be intercalated into certain lithium oxides that have a layered structure, such as $LiCoO_2$, $LiMn_2O_4$, and $LiNiMnCoO_2$ (NMC). Cells need to be built with safeguards to avoid operation outside safe limits of voltage and temperature, and there have been a few lithium battery fires, notably one on a Boeing 787 Dreamliner jet at Boston in 2013.

The lithium-ion battery contains a negative electrode (anode) of usually graphite and a positive electrode (cathode) of lithium oxides. The cell voltage, ~3.7 V, is the difference in free energy between Li^+ ions in the crystal structures of the two electrode materials. (The use of a silicon rather than a graphite anode offers increased capacity.) The operation of the battery involves the transfer of lithium ions from one electrode to the other—'the rocking chair' effect—with the intercalation of the lithium controlling its reactivity. The electrodes are separated by a Li^+-conducting electrolyte, a lithium salt in an organic solvent. The majority of installations in 2020 used Lithium Nickel Manganese Cobalt Oxide (NMC) and Lithium Iron Phosphate (LFP) chemistries. The best lithium-ion batteries have specific energies of ~250 Wh kg^{-1} (2020).

A schematic of a lithium-ion battery is shown in Fig. 10.7. On charging, lithium ions move out of the lithium cobalt oxide cathode, through an electrolyte, and combine with graphite in the anode to form lithium graphite. When a device is attached, a current of electrons flows through it from the anode to the cathode, because electrons are more strongly attracted to the lithium cobalt oxide, than to the lithium graphite. The current is driven by a voltage of

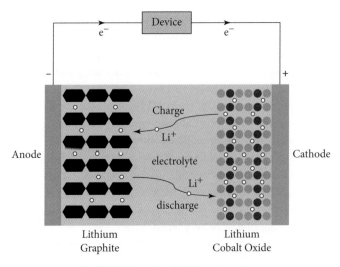

Fig. 10.7 Schematic of a lithium-ion battery.

about 3.7 volts and powers the device. During discharge, lithium ions are released into the electrolyte at the anode, and electrons into the external circuit. While at the cathode, the electrons from the external circuit combine with lithium ions, which originate from the anode, and cobalt oxide to form lithium cobalt oxide.

Research is ongoing to increase the charging rate of Li-ion batteries, particularly for EV applications where charging a battery in less than a tenth of an hour would be ideal. Currently, charge rates are such that battery swaps, as used with power tools, or charging for around 30 minutes with a fast charger to be 80% full are typical.

10.12.1 Economics of Lithium-ion Batteries

Lithium-ion batteries are now dominating the battery market, and as global production rises, their costs have been falling fast in the last 10 years (see Fig. 10.8)—some 18% each time the amount manufactured worldwide doubles. They are predicted to cost about $100 per kWh in 2024, and by around then electric vehicles will be cost-competitive with conventional cars. They are being used increasingly for large-scale electricity storage; e.g. a 100 MW/185 MWh unit collocated with the Hornsdale 315 MW wind farm in South Australia, where it increases wind power production by reducing curtailment.

The batteries can be used for frequency regulation as well as for support for renewables to increase reliability and reduce curtailment. But single use generally underutilizes batteries as they can carry out many other tasks, each of which has a value that can improve the economics. Demand-charge management is one where batteries are charged during periods when a building's demand is low, and discharged when it is high to reduce the peak demand charges; this typically uses them 5–50% of the time. Time-of-use arbitrage is another where a battery is charged when electricity prices are low and discharged when they are high, and batteries can be used as a back-up if the power fails. In combination with solar and wind farms, they are increasingly competitive with gas-peaking plants.

The use of Li-ion batteries as an energy store adds to the cost of electricity. For a battery with a 10-year lifetime, i.e. ~3500 cycles, the additional cost (neglecting any interest) for a

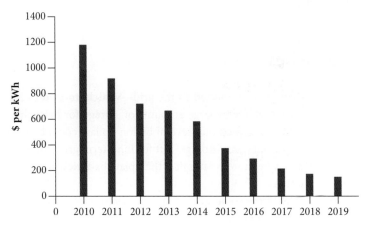

Fig. 10.8 Reduction in lithium-ion battery price 2010–2019 (BloombergNEF).

battery pack costing $150 per kWh is ($150/3500) = ~4 cents per kWh, assuming all the capacity of the battery can be used. For a long battery life, though, operating between 80% and 20% of full capacity is often recommended, which will increase the additional cost to ~7 cents per kWh. This indicates that for batteries with a ~3500 cycle lifetime we would like battery pack costs of less than $50–100 per kWh to be competitive with gas power plants.

10.12.2 Lithium–air Batteries

Lithium–air batteries have a very high theoretical energy density of ~3500 Wh kg^{-1}, but there are still many unresolved problems in realizing this potential. One design uses a porous carbon cathode and a lithium metal anode with a polymer electrolyte. During discharge, the lithium ions migrate to the porous cathode where they combine with oxygen from the air to form lithium oxide, while on charging lithium migrates to the anode and oxygen is given off from the cathode. The electrochemical reactions are:

$$\text{Anode: } 2Li \leftrightarrow 2Li^+ + 2e^- \qquad \text{Cathode: } 2Li^+ + O_2^+ + 2e^- \leftrightarrow Li_2O_2$$

$$\text{Total: } 2Li + O_2 \leftrightarrow Li_2O_2 \quad \text{voltage } \sim 3V$$

The low-density cathode and anode of a lithium–air battery means that an energy density of about five times that of a lithium-ion battery, i.e. ~1000 Wh kg^{-1}, might be achieved. This would give a comparable energy density to that of an internal combustion engine (ICE), the higher efficiency (~90%) and higher power-to-weight ratio (~6 kW kg^{-1}) of an electric motor—compared with a thermal efficiency of ~25% and power-to-weight ratio of ~1.0 kW kg^{-1} for an ICE—offsetting the higher energy density (~13,000 Wh kg^{-1}) of the ICE fuel (petrol (gasoline) or diesel) (see Exercise 10.22).

But currently lithium-air batteries have generally low-energy efficiency, short cycle life, and poor rate capability, and there is considerable R&D to be undertaken. The use of nanotechnology to increase the surface area of the electrodes and to produce materials with the right properties may help in achieving a reliable, very high-energy density Li–air battery that can be cycled (discharged and charged) many times. Such a battery could have a major impact in decarbonizing transport.

10.13 Flow Batteries

Batteries in electric cars could be connected to the grid—**vehicle-to-grid** (V2G), when the cars are parked, and could provide very large amounts of storage for balancing supply and demand. A study for California showed that several billion dollars could be saved if electric vehicles were used in place of stationary storage. But for long-term grid storage (several days or longer), flow batteries may be more economical than plate batteries, and provide a breakthrough in large-scale storage.

Flow batteries store their electrical energy within their electrolytes rather than within their electrodes, so their capacity is limited only by the volume of their electrolyte containers; the power and capacity of the battery are therefore decoupled. Generally called redox flow

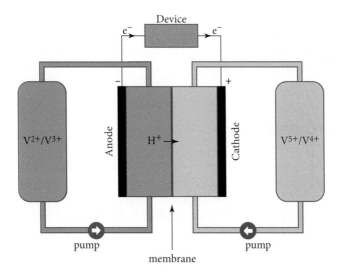

Fig. 10.9 Schematic of a vanadium flow battery.

batteries (RFB), they can have a high efficiency over a very large number of charge and discharge cycles.

The most developed and already commercialized example is the vanadium redox flow battery, which has two separated uncharged electrolytes, each containing vanadium ions in different charge states in an acid. Originally the acid was sulphuric, but a mixture of sulphuric and hydrochloric has been found to give improved vanadium solubility, and hence greater capacity, and a wider operating temperature range. A schematic of a flow battery is shown in Fig. 10.9.

The electrolytes are separated by a semipermeable ion-exchange membrane (IEM) which allows protons (H^+) ions to pass through to maintain charge neutrality, and the catalyzed electrodes have a porous structure that can be formed from carbon-fibre paper, carbon felt, or carbon nanotubes. Such electrodes give a large surface area and good permeability. The half reactions are:

$$Anode: V^{++} \rightarrow V^{+++} + e^- \quad Cathode: 2H^+ + VO_2^+ + e^- \rightarrow VO^{++} + H_2O$$

The arrangement is similar to that of a Daniell cell in that the electrolytes are separated.

However, although quite abundant, vanadium is currently expensive, and several flow batteries with different chemistries are now under development. As the energy storage requirements on the grid are many millions of kWh, millions of tonnes of material will be needed, so reducing the material cost will improve cost-competitiveness. Possible candidates are a sulphur flow battery from Form Energy and an iron flow battery from ESS, both supported by Breakthrough Energy Ventures, a billion-dollar fund chaired by Bill Gates to fund new energy technologies.

Typical energy densities are 10–50 Wh/kg, which is small compared with what can be achieved with plate batteries, but they should scale-up more cost-effectively and their capacities can be ~10 MWh. Flow batteries are therefore good candidates for grid-storage. There are already a number of commercial flow batteries installed; for example, a 4 MW/6 MWh Vanadium RFB at the Subaru Wind Farm in Japan for energy storage and wind power stabilization, and a 2 MW/6 MWh zinc–iron RFB is to be used to provide ancillary services for Ontario's electricity operator (OIESO).

10.14 **Supercapacitors**

While not a battery, a supercapacitor is an electrochemical device using electrodes and electrolytes that complement a battery in its energy storage capabilities. Electrochemical double-layer capacitors (EDLCs) are made from porous carbon electrodes immersed in an electrolyte and separated by a thin porous membrane (see Fig. 10.10). When a voltage is applied across the electrodes, positive ions are attracted to the negative electrode and diffuse through the separator, which provides electronic insulation, and form an electric double layer of charge separated by about an ion diameter. A similar double layer forms on the other electrode. The carbon electrodes form a porous structure that has a very large surface area ($\sim$3000 m^2 g^{-1}), which combined with the small separation of the charge layers gives a very large capacitance. The maximum voltage is limited by the breakdown voltage of the electrolyte, typically 1–3 V.

The two electrodes, separator, and electrolyte form a thin strip which is rolled up into a cylindrical or rectangular shape. The capacitance C can be considered to be that of two parallel plate capacitors in series, each of area A and plate separation d, immersed in an electrolyte with dielectric constant ε, i.e.

$$C = \varepsilon\varepsilon_0 A/2d$$

where $\varepsilon_0 = 8.85 \times 10^{-12}$ F m^{-1}. We can estimate C by taking $d = 0.5$ nm, $\varepsilon = 2$, and $A = 3000$ m^2 g^{-1}, which gives $C \approx 50$ F g^{-1}. A typical voltage is $V = 2.5$ volts, so the energy stored is given by $E_S = CV^2/2 = 150$ Jg^{-1} = 0.15 MJ kg^{-1} = 40 Wh kg^{-1}. (More accurate estimates require evaluating the effect of image charges.) Supercapacitors using graphene-based electrodes have a higher conductivity than activated carbon-based ones, which could enable them to be used for higher-frequency applications. Supercapacitors have self-discharge times that are shorter than batteries, typically weeks rather than months.

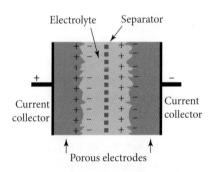

Fig. 10.10 Schematic diagram of a supercapacitor.

The maximum power P_m that a supercapacitor can deliver is when the load resistance R_L is equal to its equivalent series resistance R_s, and is given by $V^2/4R_s$. For $V = 2.5$ volts, values of $R_s \approx 0.1$ mΩ kg^{-1} are typical, giving $P_m \approx 10$ kW kg^{-1}.

Since no charge is transferred between the electrodes and the electrolyte, supercapacitors can be cycled a very large number of times ($\sim 10^6$) and can release their stored energy quickly (~ 1 s). They are therefore ideal in an electric vehicle for storing the energy generated when braking electromagnetically and for providing boosts in power. They are also used to provide short injections of power to optimize the angle of wind turbine blades when the wind changes.

10.15 Specific Energy and the Ragone Plot

In a battery, the charge capacity Q_c of an electrode is given by $Q_c = nF/(3600 \times M)$ Ah kg^{-1}, where F is Faraday's constant, n is the charge state of the ion, and M is the molecular weight. For example, for a LiMn$_2$O$_4$ electrode in a lithium-ion battery, $n = 1$ and $M = 0.181$ kg mol^{-1}, so $Q_c = 148$ Ah kg^{-1}. The energy density E_s can be estimated by $E_s = V_p Q_c$, where V_p is the potential of the battery when half depleted. For a lithium-ion battery $V_p \sim 3.7$ V so $E_s \approx 550$ Wh kg^{-1}. This is the theoretical upper limit, since the practical limit must account for the mass of the other components of the battery and also the fact that not all the lithium can be utilized. As of 2020, the best lithium-ion batteries have an energy storage of ~ 250 Wh kg^{-1}.

When a battery is discharged very slowly, the voltage is close to that given by the standard reduction potentials. However, when discharged quickly there are significant voltage drops associated with the higher current through the internal resistance, with the voltage required to drive the reactions (activation potential), and with concentration gradients within the battery, which affect the voltage through the Nernst relation (eqn (10.10)). These voltage drops mean that the amount of energy extracted from a battery by the time it reaches a particular voltage is less than when it is discharged slowly, and the effective battery capacity is reduced. There is also a reduction in battery capacity at high discharge rates, caused by the diffusion rate of ions being insufficient to match the current drawn.

The charge or discharge rate of a battery expressed in terms of the fraction of its charge capacity that the battery gains or loses in 1 hour is called the **C-rate**; e.g. a charge rate of 0.5 C would mean that the battery gained 0.5 of its charge capacity in 1 hour. The lifetime of a battery is shortened if its C-rates are too high; for lithium-ion batteries for consumer electronics a C-rate of less than ~ 0.5 C is recommended (2015).

The performance of batteries and supercapacitors can be compared on a Ragone plot (see Fig. 10.11). This is a graph of the energy stored per unit mass (specific energy) versus deliverable power per unit mass (specific power).

For a car, the specific energy is related to the range of the car, and the specific power to the acceleration. Also plotted are the corresponding values for an internal combustion (IC) engine and for fuel cells. While the weight of a battery remains the same during discharge, an IC engine's fuel is consumed, so the specific energy value is for the initial mass of fuel, approximately a quarter kilogramme of fuel per kilogramme of engine. Taking the efficiency of an IC engine as 30% then gives about 750 Wh kg^{-1} of fuel plus engine. For pure electric cars, the weight of the battery dominates as it is roughly five times that of the electric motor.

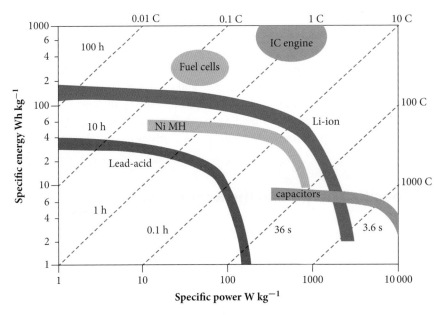

Fig. 10.11 Ragone plot of specific energy versus specific power.

10.16 **Energy Storage in Hydrogen**

Instead of storing electricity in batteries, electricity can be used to generate hydrogen, through the electrolysis of water (see Box 10.2) that is then stored for subsequent use as a fuel to provide heat through combustion in industry or in homes, or electricity via a fuel cell, or as a feedstock for making other fuels. The hydrogen acts as a store of energy and the lower heating value (LHV) of hydrogen is very high at 121 MJ kg^{-1}, but its volumetric energy density at ambient pressure and temperature is very low at 0.01 MJ litre^{-1} and how it is stored or transported depends very much on its application.

10.16.1 **Hydrogen for Transport**

While there is as yet no way of storing hydrogen very compactly, it can now be compressed up to 70 MPa ($\equiv$ 700 bar) in lightweight carbon-fibre reinforced composite containers—the energy needed is about 10% of its LHV. One type (Faurecia) has an external volume of 69 litres and weight of 37.5 kg, when storing 2.75 kg of H$_2$ at 700 bar, corresponding to energy densities of 8.9 MJ kg^{-1}, 4.8 MJ litre^{-1}, and 7% by weight. These can be compared with those of petroleum-based fuels of 45 MJ kg^{-1}, 35 MJ litre^{-1}, and 38% by weight.

Compressed hydrogen is too bulky a store for cars, but it is used for heavy goods vehicles and trains. On these, there is the space to store the cylinders (see Exercise 10.27), and their gravimetric energy density of 8.9 MJ kg^{-1} is equivalent to 2472 Wh kg^{-1}, some 10 times higher

Box 10.2 Electrolysis of water

Electrolysis is the decomposition of an ionic compound by passing an electric current through a liquid containing the compound. The process is important in the extraction of certain metals such as aluminium or sodium from their ores or oxides. The passage of a current provides the energy required to make a chemical reaction occur in the non-spontaneous direction, i.e. in the direction that results in an increase in the Gibbs free energy. The electrolysis of water is an important application in the synthesis of low-carbon fuels and as a way of storing energy through the production of hydrogen.

In pure water the fraction that is ionized is very small (H^+ as H_3O^+ and OH^- concentrations are 10^{-7} M at STP), so the water acts as an insulator and requires the addition of an electrolyte to be easily electrolysed. The electrolyte must be more difficult to oxidize or reduce than water for the products of electrolysis to be hydrogen and oxygen. Suitable electrolytes are sulphuric acid or potassium hydroxide, as they are essentially completely ionized when in aqueous solution, the sulphate ion SO_4^{2-} is very hard to oxidize, and K^+ has a lower electrode potential than H^+ and so will not be reduced instead. Three systems: alkaline water (AWE), polymer electrolyte membrane (PEM), and solid oxide electrolysis (SOEC), illustrate the main features.

The overall reaction in the electrolysis of water is $2H_2O \rightarrow 2H_2 + O_2$. The change in the Gibbs free energy under standard conditions (i.e. 1 molar at 25°C) corresponds to a standard potential of 1.23 V, and is the voltage required for the electrolysis of water to occur under these conditions. The **efficiency** of modern electrolyzers can be as high as 80%.

Alkaline Water Electrolysis (AWE)

AWE is the most established and uses generally a solution of KOH as the electrolyte, with the electrodes separated by a diaphragm that is permeable to OH^- ions. These ions react at the anode to give oxygen and water releasing electrons that travel through the external circuit. At the cathode the electrons react with water molecules, generating hydrogen and OH^- ions. The inexpensive nickel-based electrodes catalyze the reactions and the operating temperature range is 40-90 °C. Its disadvantages are that the diaphragm causes high resistive losses, limits the maximum current, and allows some mixing of O_2 and H_2 that restricts the maximum operating pressure.

Polymer Electrolyte Membrane Electrolysis (PEM)

The anode and cathode are separated by a solid polymer electrolyte membrane which conducts protons (H^+), keeps separate the product gases, and allows operation at high pressures, which reduces resistive losses. However, expensive platinum group metals are needed to catalyze the reactions. The operating temperature range is 20–100 °C.

Solid Oxide Electrolyzer (SOEC)

In this high temperature electrolyser, which is under development, the electrodes are separate by an O^{2-} ion conducting ceramic electrolyte. These ions release electrons at the anode and generate oxygen, while at the cathode the electrons react with water, producing hydrogen and O^{2-} ions. The operating temperature is higher at 700–1000 °C, which enhances the O^{2-} conductivity and improves the efficiency of electrolysis. With increasing temperature, the change in the Gibbs free energy (ΔG_e) required to electrolyse water, decreases while the overall energy needed (ΔH_e) only increases very slightly. Since $\Delta H_e = T\Delta S_e + \Delta G_e$ in a reversible process and the amount of electrical work (ΔG_e) decreases with temperature, a larger amount of heat $(T\Delta S_e)$ can supply the energy required. Higher temperatures also improve the reaction rate and reduce the energy losses during electrolysis.

High-temperature electrolysis may well be a good way to produce hydrogen if low-carbon sources of heat and electricity, such as solar or nuclear plants, are used, and the reliability of the technology is established.

than lithium-ion batteries. While the efficiency of PEM fuel cells (see Chapter 11 Section 11.4.4) is only up to ~60% compared with ~90% for electric motors, the weight savings on fuel are considerable. Refuelling can be in less than 10 mins, which is quicker than currently possible with batteries, and the materials in the hydrogen gas containers are environmentally friendly. As a result, the market for fuel cell powered trucks is growing fast.

Hydrogen can also be stored as hydrides but so far the efficiency and ease of production and release of the hydrogen, which often requires heat and catalysts, has been poor. There are many material-based hydrogen storage systems, but for transport applications compressed hydrogen is currently the most attractive.

10.16.2 Large-scale Hydrogen Storage and Distribution

Large amounts of hydrogen can be stored in salt caverns and could provide valuable seasonal energy storage. Hydrogen can be mixed with natural gas and distributed and stored in gas distribution networks. It can also be transported as pure hydrogen in polypropylene piping. For transport by road, rail or ship, hydrogen can be liquified which is a well-developed technology. Its energy density, both gravimetric and volume, are only a little larger than gas storage at 700 bar, and it suffers from only a 70% efficiency on liquification and from loss through boil-off.

Attaching hydrogen to an organic liquid, such as toluene, has been developed and may prove cost-effective. Another possibility actively being pursued is to form ammonia from hydrogen and then transport liquified ammonia, a method that is used internationally already (ammonia is a key feedstock for fertilizers). These power-to-gas technologies (see Chapter 11 Section 11.2.1) are currently expensive and sometimes not all that efficient.

10.17 **Pumped Storage**

Pumped storage plants (PSP) are the most cost-effective form of large energy storage available. In 2019 there was an estimated 158 GW of PSP generating capacity installed, providing about 94% of global electricity storage. For countries with suitable high-level terrain, **pumped storage** can be an attractive option. It can be situated by a river where water is pumped up to a reservoir, as in the Koepchenwerk pumped-storage plant in Germany (see Fig. 10.12). The same machine is used for pumping water to the upper reservoir and for generating electricity, i.e. it acts as a reversible pump-turbine. Pumped storage is essentially a form of hydropower, and has been briefly described in Chapter 6 Section 6.6. The operating principles are illustrated in Fig. 10.13. The environmental impact of pumped storage is generally confined to mountainous areas (since large heads are required) and is usually hidden from view. Its global technical seasonal storage potential is estimated as 17.3 PWh at <50 USD MWh^{-1}.

A major advantage of pumped storage is that it can respond continuously to fluctuations in demand and also to a sudden surge in demand, e.g. due to the loss of a generator elsewhere on the grid. The Dinorwig pumped storage plant in North Wales has a working volume of $6 \times 10^6 \, \text{m}^3$, a head of 600 m, and a total storage capacity of 7.8 GWh. There are six generating units, each of which can deliver 317 MW in 16 seconds from rest, or 10 seconds if they are already spinning in air.

Fig. 10.12 The Koepchenwerk pumped-storage plant on the Ruhr river in Germany (imageBROKER/Alamy Stock Photo).

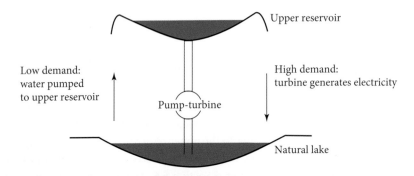

Fig. 10.13 Pumped storage.

An estimate of the energy density of pumped storage can be obtained from the expression (eqn (6.1)) for the power $P = \eta \rho g h Q$ of a dam with a head h and an efficiency η. Since $P = E/t$ and $Q = V/t$, where V is the volume of water that flows for a time t, then the energy density $U \equiv E/V$ is given by $U = \eta \rho g h_{ave}$, where h_{ave} is the average head. For the Dinorwig pumped storage facility, $h_{ave} \approx 500$ m and $\eta \approx 0.75$, so $U \approx 1$ kWh m^{-3} of water.

EXAMPLE 10.3   Pumped storage

It is proposed to build a pumped storage plant with a head of 500 m. How large a working volume is needed if the plant is required to generate 100 MW for 3 h a day? Assume the efficiency of generation is 90% and the density of water is 10^3 kg m^{-3}.

From eqn (6.1) the volume throughput required to generate 100 MW is given by:

$$Q = \frac{P}{\eta \rho g h} = \frac{10^8}{0.9 \times 10^3 \times 10 \times 5 \times 10^2} = 22.2 \text{ m}^3 \text{ s}^{-1}$$

The working volume needed to deliver 100 MW for 3 h is:

$$V = Qt \approx 22.2 \times 3 \times 60 \times 60 \approx 240000 \text{ m}^3$$

10.17.1 Gravitational Storage

A recent innovation is a variation on pumped storage that could be used widely. The idea, proposed by the company Energy Vault, is to raise cheap concrete blocks to store energy, rather than water, using a six-arm crane. Several thousand 35 tonne blocks would be stacked to form a tower about the height of a 35-storey building. When power is required, the blocks are attached to the crane and dropped. This drives the crane motor in reverse, which generates electricity. The Tata Power company in India have ordered such a tower system that can store 35 MWh with a 4 MW peak output and a fast response time.

10.18 **Compressed Air Energy Storage (CAES)**

Another method of storing energy using cheap off-peak electricity is to pump air at high pressure into a large chamber until needed. Storing compressed air in large air-tight pressure vessels is uneconomic, but underground caverns or disused mines are feasible provided they are leak-tight. Aquifers (see Chapter 5 Section 5.7.1) and salt caverns are particularly suitable, salt being self-sealing under pressure, though there is public concern over the risk of an explosion. But it has been uncompetitive, and there are only two large scale plants operational, one in Huntdorf in Germany and the other in Alabama in the USA. There are plans, though, to build more efficient plants, with one in the Netherlands for use with wind farms. And in the UK, it has been suggested that using porous sandstone under the North Sea for storing compressed air could provide valuable inter-seasonal energy storage.

The Huntdorf compressed air storage facility uses an old salt cavern with a capacity of 300,000 m^3 and air compressed to around 70 bar. The plant can generate 290 MW for 2 h. A compressed air energy storage in McIntosh, Alabama, generates 226 MW. The energy required to compress n moles of air isothermally (approximated when the heat of compression is dissipated with heat exchangers) from V_i to V_f is given by:

$$E = -\int_{V_i}^{V_f} p\,\mathrm{d}V = -\int_{V_i}^{V_f} \frac{nRT}{V}\,\mathrm{d}V = nRT\ln\left(\frac{V_i}{V_f}\right) = p_f V_f \ln\left(\frac{V_i}{V_f}\right) \tag{10.11}$$

assuming air acts like an ideal gas. When electricity is required, the compressed gas is expanded approximately isothermally by mixing the air with natural gas and burned in a combustor, the heat of combustion replacing the dissipated heat of compression, after which it is expanded through a gas turbine and generates electricity.

The McIntosh CAES plant uses the ejected heat from the gas turbine generator to pre-heat the air from the cavern using a heat exchanger, which improves its efficiency. In the Huntorf plant for an output of 1 kWh, 0.8 kWh of electricity for compressing the air and 1.6 kWh of gas for combustion are used; for the McIntosh plant the corresponding values are 0.69 kWh for compression and 1.17 kWh of gas.

We can take the efficiency of the process to be the ratio of the amount of output work to the amount of input work. The output work is the electricity generated less the work equivalent of the heat required to maintain the output temperature of the air, while the input work is the electricity required to compress the air. The amount of electricity produced by a gas turbine from 1 kWh of gas input is ~0.5 kWh. With this definition of efficiency, the Huntorf and McIntosh plants' efficiencies are ~25% and ~60%, respectively. The advantages are quick start-up (about 10 minutes) and long storage capability. However, there are still CO_2 emissions from the use of the gas.

The option of expanding air alone (~adiabatically) is not feasible because the air temperature at discharge would be very low, causing material and seal degradation. To improve efficiency, the heat of compression could be stored, e.g. in a spray of water, and then used for the expansion. This has been under investigation using carbon fibre reinforced cylinders to store the air, but making an above-ground facility cost-effective is proving very difficult.

An adiabatic CAES scheme under development and evaluation is the ADELE-ING project in which the heat generated in compression will be stored at ~600 °C, with the compressed air stored in a cavern. A round-trip efficiency of about 70% was confirmed in 2017. The economic case for large grid storage is likely to be stronger when the penetration of renewables is above ~40%.

EXAMPLE 10.4   Compressed air energy storage

A compressed air energy storage chamber has a capacity of 300,000 m^3 and compresses air from 1 bar to 70 bar. Assuming that the compression process is at constant temperature, estimate:

(a) the energy required to compress the air;

(b) the power output if the air is discharged isothermally over a period of 2 h, assuming the output of the gas turbine generator when burning 1 kWh of gas with the compressed air produces 1.2 kWh of electricity for every 1 kWh used in compression;

(c) the carbon intensity given renewable electricity is used for compression and that 2.4 $kgCO_2 \, kg^{-1}$ of natural gas are emitted.

From eqn (10.11), the energy required is:

$$E = 3 \times 10^5 \times 70 \times 10^5 \times \ln(70) \approx 8.92 \times 10^6 \, MJ = 2.48 \times 10^6 \, kWh.$$

The average power output over a 2 h period is $P = \dfrac{1.2 \times 2.48 \times 10^6}{2} kW = 1.49 \, GW.$

The energy of natural gas is ~43 $MJ \, kg^{-1}$ and 2.4 $kgCO_2 \, kg^{-1}$ are emitted. So, 1 kWh ≡ 3.6 MJ of gas produces 0.2 $kgCO_2$, and the carbon intensity $= (0.2/1.2) = 0.17 \, kgCO_2 \, kWh^{-1}$.

10.19 Thermal Storage

Thermal storage has been used for a long time in buildings where solar heat absorbed by material within the building during the day is emitted during the night. The idea can be extended to seasonal thermal stores where heat is pumped into an underground store, e.g. into the soil itself or a water tank when cooling a house in the summer, and then extracted for heating during the winter. Another scheme makes a large amount of ice that is stored underground when energy is cheap and the temperature is low. The store is compact, since ice has a large latent heat of fusion. The ice is then used to cool the building during the summer.

Phase-change materials have also been used to both regulate and store heat. One method is to fill a cavity wall with a wax-impregnated insulation chosen such that the wax melts at a suitable room temperature, e.g. 22°C. In the morning, the Sun's radiation heats up the wall to 22°C and the wax then melts, with the wall remaining at 22°C. The amount of wax is designed so that not all of it melts by sunset, after which the wax will re-solidify, giving up its latent heat to the house.

As discussed in Chapter 5 Section 5.3, thermal storage is also used in concentrated solar thermal power (CSP) plants where the solar heat is stored in a suitable material (e.g. molten salt) during the day and the stored heat is then used overnight to generate steam for a steam turbine generator. The thermal energy is typically stored as sensible heat, though latent heat stores, which are more compact, are being investigated.

An interesting method, suggested by researchers at MIT, is based on a very old technology. Firebricks, baked from a clay that can withstand high temperatures, were used first by the Hittites some 3000 years ago in their iron smelting kilns. An insulated stack of firebricks could be used as a thermal store, heated to about 850 °C by electrical resistance heating. For higher temperatures, silicon carbide firebricks might work well. Using excess production from wind and solar farms would be a cheap source of electricity and avoid curtailing the farms' output. When required, air could be blown over the bricks to provide high temperature heat for industrial furnaces at a competitive price, as the materials and air blowers are inexpensive.

Electricity can be stored and recovered by heating a well-insulated material and then using when required the stored heat to run a generator. Efficiencies of around 50% are possible and its attraction is its simple technology, cheap materials, and potentially large energy storage capacity that can offset its relatively low efficiency. Stiesdal and Siemens Gamesa are developing hot rock stores at 600 °C and 750 °C, respectively, to enable round-the-clock wind and solar energy.

10.19.1 Liquid-air Thermal Energy Storage

An interesting method of energy storage is to liquefy air and store it in well-insulated cryogenic containers at atmospheric pressure. When energy is required, the liquid air is first compressed to high pressure, then vaporized and heated to the ambient air temperature. The high-pressure superheated air (since the gas is above its critical point) passes through a turbine generator, in the same way as superheated steam is used in a thermal power station (see Chapter 3 Section 3.11). When waste heat is used to heat the air, then about 70% of the energy used to liquefy the air can be recovered. The challenge with this method is to achieve a sufficient rate of liquefaction of air. Highview Power is building a store using this technology, called Cryobattery, with a capacity of 50 MW/250 MWh in Greater Manchester, England. It is planned to be in commercial operation in 2023.

10.19.2 Reversible Thermal Energy Storage

In principle, it is possible to store electrical energy by using an engine to pump heat from one reservoir to another and to then recover it by running the system in reverse. The maximum amount of heat that can be ejected at T_2, when work W pumps heat from T_1 to T_2, is $|Q| = W \times T_2/(T_2 - T_1)$, while the maximum work from a heat engine with $|Q|$ flowing in at temperature T_2 and with heat ejected at temperature T_1 is $|Q| \times (T_2 - T_1)/T_2 = W$. The difficulty with such a process is in keeping the overall efficiency high, i.e. keeping the process as reversible as possible, since there are inefficiencies in the conversion of electrical energy to shaft work, shaft work to compressive energy, and compressive energy to thermal energy (arising from the temperature difference required for adequate heat flow), and back again.

Assuming that the thermodynamic losses are dominated by the heat transfer losses, we can model the pumped heat store by embedding an ideal engine operating between temperatures higher and lower than the temperatures of the reservoirs (see Chapter 3 Section 3.13 for a description of the power cycle). If the heat flows out and into the hot reservoir are equal then the roundtrip efficiency is given by (see Exercise 10.33):

$$\eta_r = \frac{(1-\tau)\left(2-\dfrac{\tau_c}{\tau}\right)}{2-\dfrac{\tau_c}{\tau}-\tau_c} \tag{10.12}$$

where $\tau_c = T_0/T_1$, and $\tau = T_0^-/T_1^-$, where T_0^- and T_1^- are the temperatures of the reservoirs of the embedded Carnot power cycle engine. For $\tau = \tau_{max} = \sqrt{(T_0/T_1)}$, the power cycle is running at maximum power (see Chapter 3 Section 3.13) with an efficiency $\eta_{max} = 1 - \tau_{max}$, and then $\eta_r = (2 - \tau_{max})/(2 + \tau_{max})$.

As an example, consider a pumped heat energy store operating between 175 K and 700 K. Then $\tau_c = 0.25$, $\eta_{max} = 1 - \tau_{max} = 0.5$, and $\eta_r = 0.6$. Running the embedded engine at less than maximum power such that $\eta = 1 - \tau = 0.65$, then $\tau = 0.35$ and $\eta_r = 0.81$; i.e. a roundtrip efficiency of 81%. Malta Inc, a Google spin-off company, is developing this technology using stores of molten salt at 565°C and antifreeze at –65°C, with a projected efficiency of around 60% for the pilot plant and higher for larger systems. In the UK, a pilot system that compresses and expands argon gas between gravel stores at 500°C and –160°C is under development at the University of Newcastle.

However, it is not only the efficiency that matters but also the size and cost of the store. For a thermal store using gravel as its storage medium (specific heat capacity 800 J C^{-1} kg^{-1}, density 2400 kg m^{-3}), the cost of the storage material is very low, and the energy density for a temperature difference of 200°C, assuming an efficiency of 40%, is considerably higher at ~40 kWh m^{-3} than that of pumped storage, which is typically ~1 kWh m^{-3} (see Section 10.17). Making such a thermal store cost-effective, though, has yet to be demonstrated (2020).

10.20 Flywheels

Another means of storing mechanical energy is in the form of rotational kinetic energy. The idea is not new:

- Grid controllers use the 'spinning reserve' of rotors to make minor adjustments to power supply and frequency.
- Flywheel-powered buses were used in Switzerland in the 1950s.
- The flywheel in a car provides kinetic energy to keep the engine turning between piston strokes.

In recent years, flywheels for energy storage have been developed for hybrid cars (e.g. the Honda Accord) and for niche markets, e.g. providing power for testing switchgear equipment, which would otherwise cause large disturbances to the local distribution network due to sudden drops in current.

Conventional flywheels for energy storage are metallic with mechanical bearings and rotate at up to around 4000 rpm. By using strong and light materials such as plastics, epoxies, and carbon composites, together with magnetic bearings in vacuum to minimize friction, up to 100 000 rpm can be achieved.

The kinetic energy of a flywheel with a moment of inertia I and angular velocity ω is given by $E = \frac{1}{2}I\omega^2$, where the moment of inertia is of the form $I = kmr^2$; m is the total mass of the flywheel, and r is the outer radius. Hence:

$$E = \frac{1}{2}kmr^2\omega^2 \tag{10.13}$$

$k = 1$ for a thin ring and $k = \frac{1}{2}$ for a uniform disc. Since the kinetic energy varies as the square of the angular velocity but linearly with the mass, it is more effective to rotate flywheels faster than to make them heavier. For dynamic equilibrium, the centrifugal force is balanced by the tensile stress; the maximum angular velocity is determined by the maximum tensile stress σ_{max} that the material can withstand without breaking (see Exercise 10.34).

Typically, the energy storage capacity of flywheels made from epoxies or plastics is about 0.5 MJ kg^{-1}, i.e. about 10 times that of steel. For a 100-tonne flywheel, the storage capacity is about 15 MWh. The storage capacities of modern flywheels are comparable with batteries, but flywheels can be energized and de-energized much more rapidly than batteries. The typical efficiency of a flywheel is about 80%. The major problems are safety and cost. Flywheel explosions due to material failure or bearing failure can be catastrophic, so strong containment vessels are essential, adding to the capital cost. The cost of flywheels is typically an order of magnitude greater than that of batteries, but they are particularly suited where high power is repeatedly required for a short time.

EXAMPLE 10.5   Flywheel energy storage

A 1-tonne flywheel is a uniform circular disc of radius 1 m and rotates at 4000 rpm. Calculate the kinetic energy of the flywheel.

Putting $k = \frac{1}{2}$ in eqn (10.13), we have:

$$E = \frac{1}{2} \times \frac{1}{2} \times 10^3 \times 1^2 \times \left(\frac{2\pi \times 4 \times 10^3}{60}\right)^2 \approx 43.8 \text{ MJ}$$

10.21 Superconducting Magnetic Energy Storage

Superconducting magnetic energy storage (SMES) is the storage of energy in a magnetic field due to the flow of direct current in a superconducting material. The energy stored in the magnetic field is released by discharging the current in the coil. Since superconductors have no resistance, the current and the associated magnetic field do not decay with time once a direct current has been induced to flow in a superconducting coil.

Essentially, a SMES system consists of three components: a superconducting coil, a cooling system, and a power conditioning system (which converts AC to DC and vice versa). The overall efficiency of SMES is typically 95% after allowing for losses in AC/DC conversion and for the cryogenic cooling of the superconducting material.

The advantages of SMES are that there are no moving parts, power is available almost immediately, and a very high-power output can be delivered for short periods. The disadvantages are that superconductors operate only at low temperatures, their capital cost is high, and the energy content is fairly small. The magnetic energy stored per unit volume is given by:

$$E = \frac{B^2}{2\mu_0} \approx 4 \times 10^5 \ B^2 \ \mathrm{Jm}^{-3}$$

Thus, a magnetic field of 4 T has an energy density of about 6.4 MJ m^{-3}.

The main applications of SMES are for improving power quality for electricity supply utilities, e.g. smoothing fluctuations due to intermittent loads on transmission lines, and for manufacturing processes where an ultra-smooth power supply is important.

SUMMARY

- Electricity is transmitted over long distances by HVAC or HVDC via a grid.

- Increasing amounts of wind and solar PV are requiring new ways of operating grids that will involve demand response, smart grids, and more back-up supplies.

- For off-grid and for high penetrations of wind and solar PV on the grid, energy storage of electricity will become increasingly important.

- The storage of green hydrogen produced by the electrolysis of water using renewably generated electricity, can contribute to storing energy and decarbonizing heat and transport (see Chapter 11).

- Pumped storage still dominates grid storage (~94%), but Li-ion and flow batteries are increasingly being used.

- The cost of Li-ion batteries is expected to be below $100 per kWh by 2024, which will help the transition from fossil fuel to electric vehicles and the handling of the variability in output of wind and solar farms.

- Supercapacitors are useful energy stores that complement batteries.

FURTHER READING

Bloomberg New Energy Finance (BNEF2019) Analysis of clean technologies.

www.energy.gov US Department of Energy website, appraising current programmes of research on batteries and fuel cells.

www.lazard.com Levelized cost of energy and storage.

www.nrel.gov/analysis/seams.html US national grid study (NREL2018).

www.rmi.org Rocky Mountain Institute on sustainability and low-carbon technologies.

? EXERCISES

10.1 How many turns are needed for a coil of resistance 1 Ω, cross-sectional area 0.1 m^2, and rotating at 50 Hz in a uniform magnetic field of 0.5 T to generate a current of 1000 A?

10.2* Propose a circuit to generate three-phase electricity.

10.3 Prove that the power dissipated through a resistance by a three-phase current is independent of time.

10.4 A coil with 500 turns of cross-sectional area 2 m^2 rotates at 60 Hz in a uniform magnetic field of 0.3 T. What is the maximum current flowing through a load of 10,000 Ω?

10.5* Derive the algebraic form for the torque on a loop of current-carrying wire in a magnetic field.

10.6 A power plant transmits 2000 MW of power along a transmission line of length 10 km with a resistance of 0.02 Ω km^{-1}. Calculate the percentage loss of power if the line is at (a) 110 kV, (b) 400 kV.

10.7 Compare the advantages and disadvantages of HVDC compared with HVAC transmission.

10.8 Estimate the increase in power transmission for the same current from converting a HVAC line to HVDC using the same towers. Can the same cable transmit a higher current when transmission is DC rather than AC?

10.9 It is desired to increase the reliability of part of a grid by installing extra components. Assuming the components have a probability of failure of 0.01, what is the total probability of failure if two components are connected (a) in series, or (b) in parallel?

10.10 It is desired to reduce the voltage of a transmission line from 400 kV to 12 kV. Calculate the ratio of the number of turns from the primary to the secondary side of a transformer.

10.11 How are renewables changing the requirements on power plants on a national grid?

10.12 How can a variable demand for clean electricity at an affordable price be met mainly by renewable power plants?

10.13 Explain how demand response can help with the integration of renewables into a grid.

10.14 Explain the missing money problem arising from the low marginal cost of renewables.

10.15 Discuss for a power plant supplying electricity to a grid, whether it is the capacity or the product of the capacity factor and the capacity of the power plant that is most important.

10.16 Discuss the implications of the 'Duck Curve'.

10.17 Explain the advantages of distributed generation.

10.18 What problems can countries face in integrating more renewable energy supplies?

10.19* The equilibrium constant for the water–gas shift reaction (which produces hydrogen)

$$CO(g) + H_2O(g) \rightleftarrows CO_2(g) + H_2(g)$$

is approximately given by:

$$K = \exp\left(-4.33 + \frac{4577.8}{T}\right)$$

Calculate the equilibrium concentrations of the gases at $T = 750$ K if only CO and H_2O are present initially, at concentrations of 0.2 mol litre^{-1}.

10.20* An approximation for the discharge reaction in a lead–acid battery is

$$Pb(s) + PbO_2(s) + 2H^+(aq) + 2HSO_4^-(aq) \rightarrow 2PbSO_4(s) + 2H_2O(liq)$$

The standard Gibbs free energy values in kJmol^{-1} are: Pb(s) 0.0; PbO$_2$(s) –217.3; H$^+$(aq) 0.0; HSO$_4^-$(aq) –755.9; PbSO$_4$(s) –813.0; H$_2$O(liq) –237.1.

Show that the standard voltage for this cell is 1.92 V.

The standard state of a liquid (liq) or solid (s) is the pure substance, so the activities (concentrations) of Pb(s), PbO$_2$(s), and PbSO$_4$(s) are constant and equal to 1. Assuming that the activity (concentration) of H$_2$O is unaffected by the dissolved sulphuric acid, calculate, using the Nernst equation (eqn (10.10)), the cell voltage when the molality of the sulphuric acid is 5, 3, 1, 0.5, 0.1, and 0.02 molal.

10.21 Calculate the theoretical specific energy of a sodium–sulphur battery, whose overall reaction can be represented by $2Na + 3S \rightarrow Na_2S_3 + {\sim}1.9$ V.

10.22 Calculate the specific energy of a battery that would be comparable to that of an internal combustion engine running on (a) petrol, or (b) bioethanol. State your assumptions.

10.23 Write an article of 1000 words for a popular science magazine on the prospects for fuel cells. Do not assume that the readers have any prior knowledge of the subject.

10.24 Explain the principle of operation and advantages of a flow battery.

10.25 Estimate the additional cost per kWh of storing electricity when the cost of lithium-ion batteries is $50 per kWh, and discuss the implications.

10.26 Discuss the advantages of converting electricity from wind and solar farms into hydrogen.

10.27 A 40-tonne hydrogen fuel cell powered truck has a frontal area of 10 m^2 and a drag coefficient of 0.6. Its rolling resistance is 0.5% of its weight. Assuming a constant speed of 90 km per hour and 50% efficiency, estimate (a) the energy per kilometre, and (b) the volume and weight of the hydrogen storage required for a 400-kilometre journey. Take the density of air as 1.2 kg m^{-3}.

10.28 A pumped storage plant has a head of 600 m and a working volume of 500,000 m³. How much power can be generated if the plant is required to operate at maximum output for 2 hours per day? Assume the efficiency of generation is 85% and the density of water is 10^3 kg m^{-3}.

10.29 A wind farm produces an average output of 250 MW. A pumped storage facility is planned to provide up to 300 MW for 5 days. The drop from the upper reservoir, whose depth is 30 m, to the power generator is 500 m. Estimate the area required for the wind farm and the area of the reservoir. Assume an efficiency of 90% for the pumped storage.

10.30 Explain the principle of adiabatic compressed air energy storage.

10.31 A solar heated thermal store is made from 30 kg of zinc, which melts at 420°C. What is the maximum amount of energy (in MJ and in kWh) available if the maximum temperature that the zinc is heated to is 600°C and the lowest useful temperature of the store is 150°C? (Specific heat of zinc = 0.39 kJ kg^{-1} °C^{-1}; latent heat of zinc = 118 kJ kg^{-1}.)

10.32 The 50 MWe Andasol concentrated solar thermal power plant uses two tanks of molten salt as its thermal store. Each tank contains 28 500 tonnes of a 60%/40% mixture of sodium and potassium nitrate which has a specific heat of 1.47 kJ kg^{-1} °C^{-1}. The tanks operate between 393 and 293°C. What is the amount of energy stored in each tank in MWh (thermal)? How long can the plant operate at night if the efficiency of the 50 MWe plant is 16%?

10.33* A thermal energy store is modelled by an endoreversible engine operating in two heat cycles, as shown in Fig. 10.14.

In the power cycle: $\dot{Q}_1^- = C\left(T_1 - T_1^-\right)$ and $\dot{Q}_0^- = C\left(T_0^- - T_0\right)$

In the pumped heat cycle: $\dot{Q}_0^+ = C\left(T_0 - T_0^+\right)$ and $\dot{Q}_1^+ = C\left(T_1^+ - T_1\right)$

The efficiency η of the Carnot power cycle engine is given by:

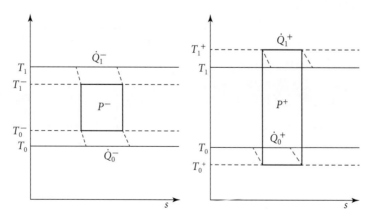

Fig.10.14 (a) Power cycle (b) Pumped heat cycle.

$$\eta = 1 - T_0^- / T_1^- \equiv 1 - \tau$$

The roundtrip efficiency η_r is the ratio of the output to input efficiencies:

$$\eta_r = (P^- / \dot{Q}_1^-) / (P^+ / \dot{Q}_1^+), \text{ where } P^- = \dot{Q}_1^- - \dot{Q}_0^- \text{ and } P^+ = \dot{Q}_1^+ - \dot{Q}_0^+.$$

If the heat flows out and into the hot reservoir are equal in magnitude, i.e. $\dot{Q}_1^- = \dot{Q}_1^+$, show that η_r is given by eqn (10.21):

$$\eta_r = \frac{\left(1 - \tau\right)\left(2 - \dfrac{\tau_c}{\tau}\right)}{2 - \dfrac{\tau_c}{\tau} - \tau_c}$$

where $\tau_c = T_0 / T_1$.

10.34 Consider a ring of inner radius a and outer radius $a + t$ with a square cross section ($t \ll a$) that is rotating at an angular velocity of ω. The material has a maximum tensile stress σ_m and a density ρ. Deduce that ω_m, the maximum value of ω, is given by $\omega_m^2 = \sigma_m / (a^2 \rho)$, and that the maximum kinetic energy per unit mass equals $\frac{1}{2}\sigma_m / \rho$.

10.35 A uniform cylindrical flywheel of radius 10 cm and mass 10 kg rotates at 100,000 rpm. Calculate the kinetic energy of the flywheel.

10.36 Can behavioural changes avoid the need for energy storage in a low-carbon society?

For further information and resources visit the online resources
www.oup.com/he/andrews_jelley4e

11 Energy Demand in Buildings, Industry, and Transport

→ Introduction

Global final energy consumption has doubled since the mid-1970s and over the period 2010–2018 grew by 1.5% a year. **Buildings** (residential and commercial), **industry**, and **transport** account for ~89% of final energy use (about equally) and ~90% of energy-related CO_2 emissions, which in 2016 were ~36 $GtCO_2$. To limit global warming to 1.5 °C, these emissions need to be reduced to almost zero by 2050. The scale of the challenge to achieve this goal requires drastic action in the coming decades, in both **switching to low-carbon** sources of energy and in **reducing energy demand** and thereby the generation capacity required.

Electricity accounts for about a quarter of the world's final energy consumption. Due to the intrinsic inefficiency of fossil-fuel power plants, this produces a third of global energy related CO_2. Another third of CO_2 emissions are released from the burning of coal, gas, and oil used for heat in industry and for heating buildings and from industrial process emissions, and a tenth from other processes. The remaining quarter of CO_2 emissions comes from oil-derived fuels used in engines to power transport.

In particular, heat-intensive industries like **steel** are responsible for a great deal of pollution and carbon dioxide emissions, which would be avoided by switching to renewable energy. However, using electricity from renewables to produce high-temperature heat (e.g. industrial oven) is more expensive than burning gas, or other fossil fuels, and the development of cheaper **power-to-gas** alternatives is very important.

About 20% of the world's energy consumption is used for space heating and cooling **buildings**, and for heating water. The heat is required at moderate temperatures, and a very efficient low-carbon technology already exists that can provide this using renewable electricity: the **heat pump** (see Chapter 5 Section 5.2).

The challenge in the case of the **transport sector** is that internal combustion engines are inefficient, and it is not practical to capture the CO_2 emitted by road vehicles and ships. **Biofuels** and **power-to-fuel** can play a part—particularly in aviation, shipping and trucking—but one of the most effective ways of reducing these emissions is to avoid combustion altogether by **electrifying transport**.

Providing the enormous renewable electricity generating capacity needed to decarbonize the industrial, building, and transportation sectors will be made easier by reducing energy demand, and the potential savings are very significant.

We first look at the demand in buildings and how it might be significantly reduced.

11.1 Energy Demand in Buildings

In 2016, the building sector accounted for around 120 EJ (~30%) of final energy use and 9.7 $GtCO_2$ (~28%) of energy-related emissions. Improving the insulation and glazing of buildings has helped to restrict the increase in energy demand to 20% in the period 2000–2017 caused by the growth in global population, larger living spaces as incomes have risen, and an increase in air-conditioning—cooling now accounts for about 10% of global electricity demand, with large increases in China and India. Some developing countries are dependent on **biomass**, where it plays an important role in meeting demand in buildings, and where pollution and inefficiency are the biggest concern (see Chapter 4 Section 4.5).

Most energy is consumed in the use, rather than in the construction, of buildings, so CO_2 emissions can be reduced by improving the energy efficiency (both thermal and electrical), decarbonizing the energy supply—electricity as well as heat—since about half of all electricity currently consumed is in buildings, and by reducing demand by improving the insulation and design of buildings.

Changes in **lifestyle and behaviour** can make a considerable difference to energy consumption in buildings, e.g. only heating and lighting that part of the building being occupied, lowering indoor temperatures, and relying less on mechanical systems. Most of the technologies involved are well developed, but there are considerable barriers to implementing them. These include high initial costs, lack of public awareness of their potential, entrenched practices, split incentives, and the lack of a **carbon price**. Moreover, new skills need to be acquired by the building trade for the benefits of the low-carbon technologies to be fully realized. There is also a need for higher appliance standards, building codes, and energy efficiency requirements; for instance, of the ~135 million air conditioners installed each year, many are less than half as efficient as the best available.

There is enormous scope for making much greater use of renewable energy for domestic buildings. Between 2009 and 2019, the installation cost of a 6 kW **residential solar thermal system** in the US dropped from $51,000 to $17,760. However, there can be a significant difference in the solar energy supply between summer and winter, so some alternative form of heating is generally necessary. (NB Long-term heat storage is an expensive option.) In the UK, the cost of **solar thermal** for hot-water heating (see Chapter 5 Section 5.1) for a typical household is £3,000–£6,000, including the solar collectors, control panel, pipework, hot water tank, and installation. For comparison, the cost of a new gas boiler is around £1,500–£3,000, and the average annual cost of gas is £550, so the lifetime cost of solar thermal is competitive, and does not produce any CO_2 emissions!

Moving from using fossil fuels for heating to heat pumps (see Chapter 5 Section 5.2) will increase the need for low carbon electricity, and reducing demand is essential to enable sufficient supply to be built. More integrated urban design can enable district heating to be viable. While combined heat and power (CHP) makes for higher efficiency, it does not provide a low-carbon solution to using fossil fuel plants. Biomass boilers can be very effective if a sustainable supply such as forest residues is available.

Smart controls that can vary demand will also help reduce the need for extra generating capacity. The priorities will differ from country to country and depend on the climate, economic development, and population. The demand E can be usefully broken down as follows:

$$E = h \times \frac{p}{h} \times \frac{area}{p} \times \frac{E}{area} \tag{11.1}$$

where h is the number of households, p/h is the number of people per household, $area/p$ is the area per person, and $E/area$ is the energy demand per unit area. The demand for cooling in hot climates is expected to triple by 2050, so adapting the building envelope (walls, windows, roof, doors, and floor) with better insulation and reflective surfaces and using very efficient air conditioning is important. Better use of natural lighting and LEDs would also help (see Box 11.1).

Box 11.1 Light-emitting diodes (LEDs)

A light-emitting diode (LED) is a semiconductor that emits photons when a current flows through the forward-biased diode, and the wavelength, and hence the colour of the light emitted, depends on the band gap (see Chapter 8 Section 8.2). The first LED was invented by Nick Holnyak in 1963 and used the semiconductor GaAs, which emits in the red. Blue GaN LEDs were invented in 1992 by Akasaki, Amano, and Nakamura (Nobel Prize 2014). By using phosphors that absorb some of the blue and emit green and red light, the emitted light appears white. White LEDs can also be made by combining blue, red, and green LEDs, and warm-white LEDs with an output similar to that of a blackbody of temperature 2700–3000 K (called colour temperature) are now available.

The light output (lumens) of LEDs has increased dramatically whereas the cost per lumen has dropped enormously over the past 5 decades (Haitz's law), as shown in Fig. 11.1.

High power white LEDs are far more efficient than incandescent lamps and fluorescent tubes (120 lm W^{-1}, 15 lm W^{-1}, 60 lm W^{-1}, respectively) and last about 25,000 hours. Despite their higher capital cost, LEDs are more economic overall than traditional lighting devices, and are transforming the way we illuminate buildings and numerous other applications. Lighting accounts for ~15% of global electricity consumption, and

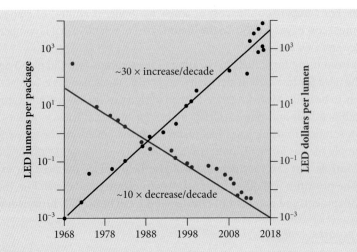

Fig. 11.1 Light output versus cost of LEDs over last 5 decades (LED2017) (c.f. 60 W bulb: 500 lumens, ~1 dollar; 5 W LED : 500 lumens, ~4 dollars).

the widespread adoption of LEDs could reduce this percentage considerably. Estimating a 10% saving globally, then the 2018 demand would have been significantly reduced by ~2300 TWh from 23,000 TWh, a saving in the energy-related CO_2 emissions of ~1.2 GtCO_2 out of a total of ~37 GtCO_2.

The burning of **biomass** or of biomass-derived products accounts for about 8% of the primary energy demand: ~46 EJ in 2016, much of which is used for **cooking**, with stoves that are very inefficient and polluting (see Chapter 4 Section 4.2.2). There is enormous scope to develop more efficient and healthier stoves, which would also reduce the time spent in gathering firewood in some parts of Africa.

At present, space heating and cooling, plus water heating accounts for roughly half of global energy consumption in **buildings**. Of primary importance is improving the building envelope, but it is also important that electrical resistance heating and incandescent light bulbs are phased out. All new buildings should have low-energy consumption, use passive heating and cooling techniques, and low-carbon technology. Also, in the European Union, Russia, and America, over 60% of the current buildings will still be in use in 2050, so it is very important to retrofit such buildings with energy-saving technology.

We now consider how buildings can be made more energy efficient and the challenges of retrofitting old buildings. We also look at how making the best use of natural resources in a hot climate can significantly reduce energy demand.

11.1.1 Energy-efficient Buildings

Maintaining buildings at a comfortable temperature is one of the biggest uses of energy, for heating in winter and cooling in summer, and **passive design** is a very effective way of reducing the amount of energy required. A passive design is one that uses the local environmental

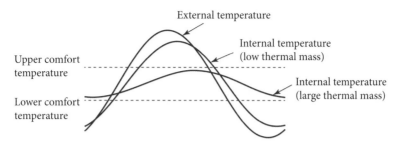

Fig. 11.2 Effect of thermal mass over a 24-hour period.

conditions and appropriate materials to help maintain a comfortable temperature, by optimizing the effects of shading, orientation, natural ventilation, insulation, and thermal mass in the basic design of the building such as to minimize the energy requirement. The idea originated in North America during the oil embargos of the 1970s, to build houses with minimal energy consumption. The concept was much developed in Germany in the 1990s, where an Institute for Housing and the Environment was founded in Darmstadt, which established the **Passivhaus standard**.

Fig. 11.2 shows the effect of **thermal mass** over a 24-hour period. It compares the internal temperatures of a building with low thermal mass (e.g. timber cladding and stud walls) and one with large thermal mass (e.g. brick or concrete blocks). The building with low thermal mass closely follows the external temperature, whereas that with large thermal mass is less responsive and exhibits a significant time lag in peak temperature.

Maintaining a large office or school classroom with large windows at a reasonable temperature on hot summer days can be helped by means of a thick concrete floor, which acts as a large heat sink (and cools down overnight). Thermal mass is also important in winter, when the interior walls, floors, and furnishings absorb solar radiation through windows, and later radiate the heat into the room when the temperature falls.

Thermal mass is more significant in those parts of the world where there is a large difference between the daytime and night-time temperatures. Traditional houses in hot, dry climates tend to be built with very thick walls. During the day, solar heat is absorbed by the outer layer of the walls without allowing much heat transfer to the interior of the building. At night, when the outside temperature is much lower, the heat is radiated back to the environment and transferred by conduction to the interior of the building.

For cooling the occupants of a building, evaporative cooling is effective and requires moderately dry (<60% humidity) air. The air is cooled by passing it through moist pads where it provides the latent heat to evaporate the water. Relatively cool water can be obtained by using a roof pond, particularly in a hot desert climate: the pond is exposed at night to lose heat by radiation and is insulated during the day.

11.1.2 **Thermal Insulation**

Insulation of the building envelope (the walls, windows, roof, doors, and floor) is crucial in most buildings to achieve good performance, and we now explain how **thermal insulation** is quantified.

We express the rate of heat transfer through a material in the form:

$$\dot{Q} = UA\Delta T = \frac{1}{R}A\Delta T \tag{11.2}$$

where A is the area, U is the thermal conductance or U-value ($\mathrm{Wm^{-2}\,K^{-1}}$), R is the thermal resistivity or R-value ($\mathrm{m^2\,K\,W^{-1}}$), and ΔT (K) is the temperature difference across the material (e.g. through glass or stationary air). It follows from eqn (11.2) that U and R are related to the thermal conductivity k and thickness d (see eqn (2.4)) by the formula:

$$U = \frac{1}{R} = \frac{k}{d} \tag{11.3}$$

When there are several different layers of insulation, the total thermal resistivity is equal to the sum of the resistivities of the individual elements. For a single glazed window, in addition to the resistivity of the glass, the resistivity of the layers of air that conduct heat by convection must be included. This is illustrated in Example 11.1.

EXAMPLE 11.1

The U-value of a window is given by $1/U = 1/h_i + 1/h_o + d/k$, where $(1/h_i + 1/h_o)$ is the thermal resistivity of the inner and outer surface air layers ($\sim 0.16\ \mathrm{W^{-1}\,m^2\,K}$) and d/k is the thermal resistivity of the window. For a double-glazed unit, with two panes of glass of thickness d_g separated by an air gap d_a, the thermal resistivity of the window is given by $0.16 + 2d_g/k_g + d_a/k_a$, where k_g and k_a are the respective thermal conductivities (see Exercise 11.3). Calculate the U-value for the glass component (i.e. neglecting the frame) of (a) a single-glazed and (b) a double-glazed window with dimensions $d_g/2 = d_a = 5$ mm, $k_g = 1.5\ \mathrm{W\,m^{-1}\,K^{-1}}$, and $k_a = 2.1 \times 10^{-2}\ \mathrm{W\,m^{-1}\,K^{-1}}$.

$1/U = 0.16 + 0.005/1.5 = 0.163\ \mathrm{W^{-1}m^2K}$ so $U = 6.1\ \mathrm{W\,m^{-2}K^{-1}}$

$1/U = 0.16 + 0.01/1.5 + 0.1/0.021 = 0.643\ \mathrm{W^{-1}\,m^2\,K}$ so $U = 1.6\ \mathrm{W\,m^{-2}\,K^{-1}}$

In the UK, the installation of central heating raised the average temperature in older houses from 12 °C to 18 °C, with about 60% of total domestic energy being used for space heating. Domestic heating bills and carbon emissions can be significantly reduced by decreasing the U-value of the building elements: installing thicker roof and floor insulation, cavity wall insulation, and double- or triple-glazing. However, the gaps in the building envelope allow a significant amount of warm air to escape and be replaced by cooler air from outside, and thus provide ventilation. This **ventilation is an important loss of heat**.

11.1.3 Ventilation

The specific heat of air at atmospheric pressure is close to 1000 J $\mathrm{kg^{-1}\,K^{-1}}$ and its density 1.2 kg $\mathrm{m^{-3}}$. If the number of air changes per hour is N, the volume of the house V, and the area of floor A, then the heat loss per square metre of floor will equal $0.33N \times V/A$ W m^{-2} K^{-1}. The ratio V/A equals the spacing of the floors in the building (height of the rooms), and is typically ~2.5 m; so, the heat loss from ventilation is given by:

$$H_V = 0.33N \times V/A \sim 0.8N \text{ Wm}^{-2}\text{K}^{-1} \tag{11.4}$$

A minimum number of air changes per hour recommended for well-being is $N = 0.35$ (although this can be higher depending on the number of occupants). To reduce the heat loss further, mechanical ventilation with heat recovery (MVHR) can be used. The outgoing warm but stale air is passed through a heat exchanger, where it transfers most of its heat to the incoming cool fresh air from outside, thus conserving energy. The efficiency of transfer can be 75%, in which case the heat loss from ventilation is reduced to ~0.2 N W m^{-2} K^{-1}; e.g. for $N \sim 0.4$, $H_V \sim 0.08$ W m^{-2} K^{-1}.

We now see the effect of insulation and ventilation for three different kinds of homes.

11.1.4 Heat Loss in Domestic Houses

Typical percentage heat losses from an old domestic house without good insulation, i.e. with solid walls two bricks in thickness, single glazed windows, and no roof insulation, are shown in Fig. 11.3.

Table 11.1 compares the heat losses from an old detached house, a new detached house and one which meets the Passivhaus Standard (which has excellent thermal insulation, airtightness, and uses MVHR, see Box 11.2). All 3 houses have the same footprint of 7 m × 7 m and have two floors with ceiling heights of 2.5 m. The table also gives estimated U-values for the various building elements and the percentage loss in each case.

The element thermal bridging refers to connections between the inner and outer building surfaces, made when constructing the house, that have good thermal conductance. These cause heat loss, and the estimated U-value is that for the whole surface area of the building envelope. For the Passivhaus, it can be seen that both MVHR and low thermal bridging are as important as good insulation. To work out the total loss over a whole year we need to know

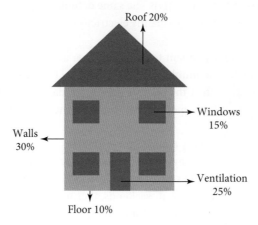

Fig. 11.3 Typical heat losses from an old domestic house.

Table 11.1 Typical U-values ($Wm^{-2}K^{-1}$) for building elements and percentage heat loss (HHLC)

Building element	Passive House		Recent building		Old building	
Walls	0.15	25%	0.4	21%	1.5	30%
Roof	0.15	11%	0.3	7%	2.3	19%
Floor	0.12	8%	0.24	5%	0.8	7%
Window	0.8	22%	2.0	17%	4.8	15%
Door (unglazed)	1.5	8%	3	5%	3	2%
Air changes/hour	0.4	12%	1	36%	1.5	21%
Thermal bridging	0.04	14%	0.08	9%	0.15	6%
Total loss W $°C^{-1}$	69		223		589	

the average temperature difference between the inside and outside of the house. This depends on location and is given by the number of degree-days.

11.1.5 Degree-days and the Performance Line

A useful measure of the space heating requirement of buildings in any particular location is given by the **number of heating degree-days**. It is assumed that buildings are heated only when the ambient temperature falls below a nominal base temperature. For buildings in the UK with good insulation and a high internal heat gain, the base temperature may be as low as 10 °C, but is otherwise taken to be 15.5 °C (60 °F). [NB The unit of temperature is normally chosen to be appropriate to the country concerned.] The internal heat gain includes the heat emitted by the occupants (~80W per person), the appliances, and any solar gain. The number of degree-days in a given period is defined as:

No. of degree-days = [average ambient temperature − base temperature inside the house]
× [fraction of the day occupied by the period]

For periods when the average ambient temperature exceeds the base temperature, the number of degree-days is taken to be zero. [NB The same definition applies to **number of cooling degree-days**, for periods when energy (mainly electricity) is used to cool buildings in summer heat.]

For example, for an average ambient temperature over a half hour period of 10.7 °C and a base temperature of 15.5 °C, the number of degree-days is (15.5–10.7) × (1/48) = 0.1 °C. The number for the whole day is then the sum of the degree-days for each half-hour period. If temperature readings of that frequency are not available, then the average ambient temperature for the day can be used.

Rather than taking data on energy consumption for every day, a plot of monthly consumption against the number of degree-days in that month is often used. An example from a house in the UK is shown in Fig. 11.4. The scatter is much reduced from that of daily data, since the variations from changes in occupancy at weekends, thermal mass effects, and heat gain fluctuations are smoothed out over a period of a month. The intercept and slope of the best-fit

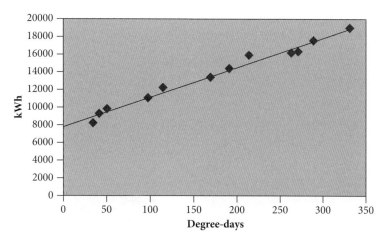

Fig. 11.4 Degree-days per month and the performance line for a year.

straight line, called the **performance line**, gives the non-space heating requirement and the heat loss per degree-day, respectively.

EXAMPLE 11.2

The average number of degree-days in the UK for base temperatures of 18.5, 15.5, and 10 °C are 3422, 2463, and 1060, respectively. The house whose data are shown in Fig. 11.4 has a base temperature of 15.5 °C^{-1} and a floor area of 200 m^2. The heat loss per degree-day is 33.3 kWh d^{-1}. Find the heating loss per degree per square metre.

The heating loss corresponds to an average heat loss of 1.39 kW. For a base temperature of 15.5 °C, the average temperature difference between the outside and the base temperature is 2463/365 = 6.75 °C. The heat loss per degree is therefore 206 W °C^{-1}, and the heat loss per degree per square metre 1.03 W °C^{-1}m^{-2}.

The monthly performance data for a house can be used to estimate the base temperature when it is not known, by calculating the number of degree-days with the base temperature as a variable and then looking for the value of the base temperature that gives the best-fitting straight line to the monthly consumption values.

11.1.6 The Transition to Passive Buildings

Most housing stock around the world is not energy efficient and will still be around in 2050. Moreover, climate change is affecting the energy demand for heating and cooling buildings. Fig. 11.5 shows the population-weighted number of degree-days in Europe for heating and cooling. The number of heating degree-days (mainly in northern Europe) during 1981–2017 dropped by 6%, while there was an increase of 33% in the number of cooling degree-days (mainly in southern Europe) over the same period.

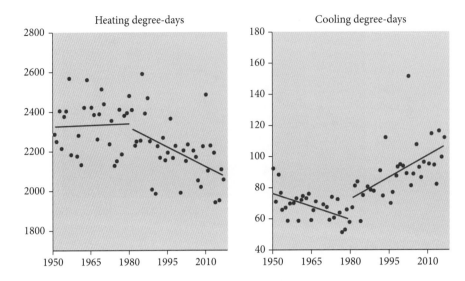

Fig. 11.5 Trend in heating and cooling degree-days in Europe (HCE) [*Source*: GEOSTAT 2011 grid dataset provided by Statistical Office of the European Union (Eurostat)].

The demand for electricity for cooling is expected to rise further in the coming decades, especially during summer heatwaves. This further emphasizes the need for renewables to displace fossil-fuel as the main source of electricity, for more new houses to be built to the Passivhaus standard (Box 11.2), and for older buildings to be retrofitted to higher energy standards (see Section 11.1.7).

Box 11.2 The Passive House (Passivhaus) design

A building can achieve the largest savings in energy by minimizing energy loss and maximizing passive energy inputs. When the loss is small enough, passive sources like the occupants, each emitting about 80 W, appliances, and sunlight can meet most of the heating requirements.

The building must be very well insulated, with U-values for the opaque parts of the building envelope less than 0.15 W m^{-2} K^{-1} and, for the windows, less than 0.8 W m^{-2} K^{-1}. Thermal bridging must also be minimized. Air infiltration typically accounts for a significant heat loss, so the building must be air-tight and fresh air brought in to the house through *mechanical ventilation, with heat recovery* (MVHR). The much better insulation of a Passive House is shown in a thermal image of one, between two conventionally constructed adjacent houses, in Fig. 11.6. In MVHR, the warm air leaving the house goes through a heat exchanger that transfers 75% of its warmth to the incoming cool fresh air. Using daylight can reduce the need for artificial lighting and can aid the solar gain.

Other passive design concepts can be incorporated, such as evaporative cooling and solar-heated desiccant dehumidification. This latter method uses a material that

Fig. 11.6 Thermal image of a Passive House (light areas are warmer than dark areas).

Table 11.2 Average energy consumption values for housing settlements in Germany

Existing Houses Germany	Low-Energy Houses Niedernhausen	Passive Houses Wiesbaden	Passive Houses Hannover	Passive Houses Stuggart
112 kWh m^{-2}y^{-1}	65 kWh m^{-2}y^{-1}	13.4 kWh m^{-2}y^{-1}	12.8 kWh m^{-2}y^{-1}	12.8 kWh m^{-2}y^{-1}

removes water and is regenerated by solar heat (see Chapter 5 Section 5.1.1), which can yield over 50% saving in energy. In hot and humid areas, when combined with evaporative cooling, it provides an alternative to refrigeration air-conditioning systems. Zero energy use is possible through incorporating sufficient renewable sources of energy such as solar thermal collectors and PV panels.

The performance of three passive house settlements in Germany are compared in Table 11.2 with the average heating consumption of existing houses and with one group of 'low-energy' houses in Germany. It can be seen that all the passive house settlements' heating consumption are lower than the 'Passivhaus' standard of 15 kWh m^{-2} y^{-1}, and that the energy savings are considerable.

11.1.7 Retrofitting Existing Buildings

Retrofitting a house first requires monitoring its current energy use and noting the requirements of the occupants; thermal imaging can be used to see what is required in the building envelope. The energy losses can then be reduced, low-carbon technology added, and any modifications to the house made as appropriate. For example, for a semi-detached house with solid brick walls, a slate roof, and two bedrooms, the following measures could be taken. The insulation of the building envelope could be improved significantly by adding insulating panels, triple glazing, insulated external doors, and ensuring that thermal bridging is minimized. Air changes could be reduced by sealing gaps, and ventilation supplied by MVHR, and efficient appliances, LED lights, solar hot water, and PV panels could be installed.

For a larger office building, a green roof and PV panels could be installed on the roof, daylight could be increased by adding double glazed windows, air tightness improved, natural and cross ventilation utilized, phase change materials used to reduce overheating in rooms on the top floor, and heating provided by a biomass-fed CHP plant. District heating could also be considered if cost-effective. Solar heating collectors could be installed for water heating.

Trials have shown that it is possible to achieve reductions of 50–75% in detached houses, and to cut space heating by 80–90% in apartment blocks and thereby achieve the Passivhaus standard for new buildings. In developing countries, such as China, improvements to the building envelope have reduced the energy used for cooling by as much as a half and for heating by two-thirds. For commercial buildings, upgrading heating, ventilating, and air conditioning (HVAC) equipment can reduce energy consumption by up to 50%, and more could be achieved by improving the building envelope, particularly when it is mainly glass and there is a large solar gain. Using efficient lighting can also achieve significant savings.

The experience in the UK, though, where much of the building stock is old, has been that there is often a significant discrepancy between expected and realized improvements in retrofit projects, and a 50% reduction is more common. Some of this is due to a thermal comfort 'take back' effect where occupants use more heating after the retrofit. Thermal bridges and gaps in the insulation also reduce the performance. Another factor is that several of the skills required are new to the construction industry and the occupants are often unfamiliar with the new technologies, so installation and use are not optimal. Better building regulations would help—as would setting minimum standards and clear labelling of a building's energy use, as has proved effective in Europe for appliances.

In hot regions, passive energy savings can reduce demand considerably. Vegetation can make the environment cooler through the evaporation of water from its leaves, shading, and through channelling the wind. Avoiding direct sunlight into rooms and having light-coloured external surfaces reduces the solar gain. In 2021, a 98% reflective white paint was reported by scientists at Purdue University, which could help reduce the need for air-conditioning. It is also important for comfort to reduce the air humidity and have air flow, as both help the human body to perspire.

How the challenge of reducing energy demand at relatively low cost has been met in a hot climate is illustrated in the case study carried out at Shinawatra University in Thailand (see Case Study 11.1).

Case Study 11.1 Energy-efficient building at Shinawatra University, Thailand

---

Thailand is a hot and humid country which has undergone rapid urbanization in the last few decades, and most people now live in high-density conurbations. Air conditioning has become the standard way to keep cool, unlike in the traditional Thai wooden house, which has natural ventilation and shady areas to escape the Sun. The situation is exacerbated by the fact that urban buildings are typically constructed with reinforced concrete, brick, plaster, and concrete roof tiles, which act as heat sinks. Air conditioning and other energy-consuming devices for modern lifestyles have led to a massive rise in energy consumption in recent years. Given the general lack of skilled workers in the construction industry and the need for high-quality heat-insulating materials to be imported, a new approach to construction is needed that makes best use of the natural resources and significantly reduces energy demand.

A novel solution to the problem has been developed at Shinawatra University in Bangkok for an energy-efficient building that incorporates a number of different functions. The challenge was to create an attractive, multifunctional environment that significantly reduced the energy usage for cooling purposes. The design minimizes the amount of hard-top surface on the outside of the building and uses natural vegetation and water resources to create a comfortable microclimate in three separate zones: a lecture theatre, a laboratory, and a canteen. The specific requirements of each zone (lighting, temperature, humidity, and air flow) are satisfied in such a way as to minimize the overall energy consumption of the building. Pond cooling is used to reduce the cost of air conditioning in the lecture theatre and the computer centre in the laboratory, the dining area is cooled by rerouting cool air streams, the heat generated in the canteen is vented through the upper openings of the building using the stratification effect, the large thermal mass of concrete is exploited to provide thermal stability, and glazing material is chosen that allows the penetration of the visible part of the spectrum but not the infrared. The net result is that the lighting and air-conditioning systems use only about one-sixth of the electrical energy of typical buildings. The capital cost is somewhat larger than that for a typical building, but this is soon paid off by savings in running costs. Thus, the concept makes economic sense, reduces the carbon footprint, and significantly improves the quality of life of the people who use the building (see Fig. 11.7). It provides a model for other countries with similar climates.

Fig. 11.7 Shinawatra University in Thailand: the Main Hall building (*Source*: Wikimedia Commons (Vincentvalentine289)).

11.1.8 Energy Storage in Buildings

The large-scale deployment of electrically driven heat pumps will increase the demand for low-carbon electricity. Photovoltaics are increasingly providing such power during the day, so it is desirable to shift the evening load to daytime by using energy storage. Li-ion batteries could be used, or flow batteries, provided their cost is competitive. However, a much cheaper alternative may be to use a thermal energy store fed by a heat pump. Water is already used in domestic hot water tanks and in large ($\sim$75,000 m^3) volumes for inter-seasonal storage. Latent heat and thermochemical stores, although with higher energy densities, are not so well developed. We can estimate the amount of thermal storage needed by looking at the heat loss estimates in Table 11.2 (see Example 11.3).

Alternatively, the whole building can be used as an energy store. An Italian study on energy storage in well-insulated buildings found that the inside temperature of a four-storey building fell from 24 °C to just 19.5 °C in five days when outside it was 7 °C, compared with under 12 hours before renovation.

EXAMPLE 11.3

A low-carbon built house in the UK has a space heating requirement of 100 W K^{-1}. Estimate the size of hot-water tank required for the heating demand in winter to be met by heat pumps heating the hot-water tank (60 °C) during periods of low demand.

The UK has an average 6.75-degree difference between inside and outside for a base temperature of 15.5 °C. Assuming 70% occupancy, the average space heating requirement S is given by $S = 100 \times 6.75 \times 0.7 = 0.47$ kW. Space heating takes about 60% and hot water about 25% of the total energy demand of a house, which makes the total heating demand about 0.67 kW. Winter load will be about twice as high as the averag.e, so the daily heat requirement R would be about 32 kWh. Assuming an ambient temperature of 20 °C, a V m^3 tank of hot water at 60 °C would provide $E = 4.18 \times 10^6 \times (60 - 20) \times V$ joules of thermal storage. Putting $E = 32 \times 3.6 \times 10^6$ joules gives $V = 0.69$ m^3. The tank could be charged in 3 hours using a 6 kW air source heat pump with a conservative COP of 2 during periods when the demand was low.

11.1.9 Low-carbon Hydrogen as a Source of Heat

Provided it is made with low CO_2 emissions, **hydrogen** can be used to provide low-carbon heat, as well as being used in fuel cells to provide electricity. An interesting project is underway in the UK to see if it feasible to decarbonize domestic heating by changing over from natural gas to hydrogen. The details are described in Case Study 11.2. There are various investigations in Europe on the use of gas grids to transport hydrogen, either pure or mixed with natural gas, as a way of reducing emissions from heating both in homes and also in industry, where hydrogen is seen as a means of helping to decarbonize.

Case Study 11.2 H21 hydrogen project

By 2050, the UK is committed to reducing its carbon emissions to zero. Approximately 30% of the final energy consumption in the UK is used for domestic heating, which is mainly provided by natural gas (predominantly methane) from the North Sea gas fields, and distributed throughout the UK by a gas grid. Natural gas emits around 180 gm kWh^{-1} CO_2eq, and decarbonizing the domestic heating sector would contribute significantly to meeting the UK's 2050 and Paris Agreement commitments.

Burning hydrogen produces water vapour, not CO_2, and a recent feasibility study known as the **H21 Leeds City Gate Project**, has confirmed that:

i.   the methane can be converted to hydrogen using standard **steam-methane reforming technology***;

ii. the existing gas network of polyethylene pipes can be used as a hydrogen network;

iii. nearby salt caverns could be used for hydrogen storage, for several months;

iv. the system has the capacity to satisfy the entire heat demand for Leeds (population 475,000).

[*High-temperature steam (700 °C–1,000 °C) is reacted with methane under pressure to produce hydrogen and carbon monoxide:

$$CH_4 + H_2O \ (+ \ heat) \rightarrow CO + 3H_2$$

The carbon monoxide is then converted into carbon dioxide and hydrogen in the 'water-shift reaction':

$$CO + H_2O \rightarrow CO_2 + H_2$$

The carbon dioxide is separated from the hydrogen and would be pumped into depleted rock formations under the North Sea.]

The project has now been expanded, and called the H21 Northern Cities project, to consider supplying a number of cities in the North of England. If successful, the rest of the country could be converted by 2050. Appliances (cookers and boilers) would need to be changed, but the convenience of using the existing gas network and the ability to store large amounts of energy make the overall cost of decarbonizing this sector attractive. If adopted, the network could be converted incrementally with little disruption to customers. The cost would be spread over all customers and amortized over 40 years, and as a result would have little impact on the cost of heating. In addition, the availability of low-cost hydrogen could help the development of hydrogen-fuelled vehicles.

11.1.10 Summary of Actions Needed to Reduce the CO_2 Emissions and Energy Use of Buildings

A raft of strong actions is needed to significantly reduce CO_2 emissions and energy use in buildings to combat global warming. These include:

- **switching to renewable sources** of electricity and heating and cooling via **heat pumps** (air- or ground-source, see Chapter 5 Section 5.2), **solar thermal panels**, and **hydrogen**;

- **retrofitting** existing buildings to high standards and insisting that new homes have **near zero energy** consumption by improving **insulation** and **glazing**;

- **changing behaviour**, e.g. only heating and lighting rooms being occupied, turning down the thermostat and wearing warmer clothing, not using a tumble dryer;

- **using passive design** concepts to reduce energy demand; particularly, additional cooling from **climate change**;

- **ensuring that building regulations** are in line with targets to reduce CO_2 emissions and energy use of buildings;

- **installing smart controls** and **meters**, and installing **more efficient domestic appliances**.

Government policies that overcome the barriers to implementing these measures need to be introduced, to reverse the possibility that global building energy demand doubles by 2050 and ensure a stable or even declining energy demand relative to our current usage. Furthermore, the energy that is used must be decarbonized.

Shifting energy use away from peak times would reduce capacity requirements and enable more PV to be used. The cost of energy is relatively cheap compared with other living costs in industrialized societies, so charging more in the evening may not be effective. Subsidized storage—batteries and/or hot water tanks—could be cost-effective.

11.2 Energy Demand in Industry

In 2016, the industry sector accounted for around 116 EJ (29%) of final energy use and 13.5 $GtCO_2$ (~39%) of energy-related emissions. The production of cement, iron and steel, chemicals and petrochemicals continue to be responsible for the bulk of direct (generated on site) and process CO_2 emissions, as can be seen in Fig. 11.8. (The indirect emissions of about 5 $GtCO_2$ are those arising from the supply of electricity, fuels and materials on site.)

In recent years there has been significant growth in the economies of China and India and other countries in the developing world, and the demand for cement and steel and other materials has risen very sharply. As a result, the emissions could double by 2050, unless significant reductions are made. Furthermore, there has been a significant shift of industrial production to the Far East. To keep on track to reduce global warming to less than 1.5 °C by 2050, **industrial emissions need to reduce by at least 90%**.

Technical options for decarbonizing industry are much less well developed than those for buildings or transport. What makes it difficult to reduce emissions is that there are many

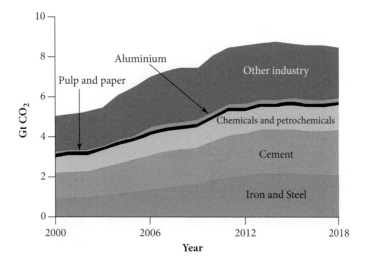

Fig. 11.8 Industry direct (including process) CO_2 emissions (IEA 2019).

different processes in industry, many of which are energy-intensive, and many involve heat generation by fossil fuel combustion. Moreover, the investment timescales in industry are typically 20–40 years, so working practices are more entrenched. It is important that new plant avoids lock-in to higher carbon-intensive processes.

Industry is also more affected by international competition and is reluctant to reduce its competitiveness. Increased costs through tighter controls on emissions can lead to **carbon leakage**: the transfer of production to another country, where production costs are lower and limits on carbon emissions are less stringent, which can result in an increase in overall emissions.

Mitigation can come about by switching to low-carbon sources, which reduces the CO_2 per kWh (the **carbon intensity**), and by increasing the **energy intensity** (the amount of energy used per unit of industrial output) through improvements in **energy efficiency**; for example, in cross-cutting technologies, like motors. Decarbonizing the electricity supply reduces the indirect emissions. This is particularly important in the aluminium industry, where over 80% of emissions are indirect due to its huge electricity consumption, which accounts for 3.5% of global electricity production.

Energy savings can be made by improvements in heat recovery, through increased insulation in furnaces and steam systems, and by reducing losses in driven devices. Also important is improving the efficiency by which materials are used, through '**reduce, reuse, and recycle**', e.g. by reusing components, making objects lighter, substituting other low-C materials, and repairing and extending the life of objects by making individual components replaceable. The IPCC estimated in its 5th Assessment Report (AR5) that the **energy intensity** in industry could be improved by 25% by using the best available technology, and possibly by another 20% through innovation. Hence, improving the **carbon intensity** of manufactured goods is essential to reach the <1.5 °C goal.

Motors and steam generation account for around 40% of the final energy consumed by industry, and savings of about 15% are achievable. A considerable amount of heat is dissipated in electricity generation, and this could be utilized, e.g. by combined heat and power (CHP).

In some countries CHP is widely used, e.g. in Finland, where 90% of urban housing is linked to district heating systems, and CHP can be applied to industrial processes as well. Globally, however, the use of CHP is low, because the heat source usually needs to be close to where the heat is needed, and the capital cost of CHP plant is higher than for conventional plant.

The options for fuel switching are biofuels, solar thermal, carbon capture, power-to-gas, and low-carbon electricity:

Biofuels can provide both low-temperature and high-temperature heat, but their availability and cost limit their mitigation potential. Furthermore, there are concerns over biofuels about land use and competition with food production. However, the use of agricultural residues for co-firing would avoid these issues.

Solar thermal can provide process heat but requires good levels of solar irradiance, and market penetration is small.

Carbon capture and storage costs are high and its uptake has been very slow. There may be liability issues arising from the possibility of leakage, and there has been public concern over inland underground storage sites. Furthermore, only a limited number of suitable storage sites have been identified in some important regions. It could though be valuable for certain industrial processes, such as cement production. And CCS, if combined with biofuels (BECCS), can produce negative emissions.

The increasing availability of competitively priced renewable electricity from **PV** and **wind** make it possible to use **power-to-gas** and **electrothermal processes** as alternatives to fossil-fuel heat.

For any particular industry, the amount of greenhouse gas emissions per annum, G, can be expressed as the product:

$$G = \frac{G}{E} \times \frac{E}{M} \times \frac{M}{P} \times P \tag{11.5}$$

where G/E is the carbon intensity of the energy source, E/M is the industry's energy consumption per tonne of material produced (the energy intensity), M/P is the amount of material per product (the material intensity), and P is the number of products. Hence, reducing carbon, energy, and material intensity, and the number of new products by improving energy efficiency and implementing life-cycle changes, can be very effective (see Example 11.4).

EXAMPLE 11.4

CCS and renewables are employed to reduce emissions in the steel (25%), cement (27%), and chemical (14%) industries by 70%, with other emissions (34%) reduced by 50%. Energy intensity and materials intensity are improved by 20% throughout industry. What reduction in the use of steel products would be required to produce a total emissions reduction of 80%?

The fraction of emissions left in the cement, chemical, and other industries is $(0.41 \times 0.3 + 0.34 \times 0.5) \times 0.8 \times 0.8 \times 1 = 18.8\%$. The emissions from a fraction x of the steel production must be below 1.2%, so $0.30 \times 0.3 \times 0.8 \times 0.8\, x = 0.012$, which yields $x = 0.14$. Steel production would therefore have to drop by 14% in order to reach a total emissions reduction of 80%.

We now consider the power-to-gas, electrothermal, and CCS processes in more detail.

11.2.1 **Power-to-gas**

Power-to-gas technologies exploit the surplus energy from low-carbon sources (notably wind and solar) to produce gaseous fuels such as **hydrogen, methane**, and **ammonia**, that can be stored and used for transportation or for generating heat (see Case Studies 11.3 and 11.4).

Hydrogen can be made by electrolysis: a well-established technology that uses electricity to split water into its basic elements of hydrogen and oxygen (see Box 10.2). When produced with renewable generated electricity it is called **green hydrogen**, and globally many green hydrogen projects are underway. It is increasingly seen as a cost-effective way to provide clean energy for things that cannot easily be electrified or run using batteries, such as heavy manufacturing and long-distance transport, and for long-term energy storage. There is also the infrastructure for its transport by pipelines, ships, trucks, or as ammonia, and the transition away from fossil fuels to green hydrogen will create many jobs and can improve local air quality. Currently, the largest industrial scale electrolysers use alkaline electrolytes and capacities of around 50 MW are planned. In Germany, there are proposals for the large-scale generation of hydrogen from offshore wind in order to overcome the lack of grid capacity, but it requires an electrolytic process that can operate with an intermittent source (see Case Study 11.3). Solid-oxide electrolysis cells (fuel cells operating in reverse—see Chapter 11 Section 11.4.4) with much greater production rates at high efficiency that operate at higher temperatures are under development.

Hydrogen is currently nearly all produced from fossil fuels; mainly by the steam-reforming of methane (see Case Study 11.2); when typically the CO_2 is not captured it is called **grey hydrogen**; when it is, **blue hydrogen**, which is then low-carbon as only 80–90% of the CO_2 is captured. But the price of green hydrogen is falling fast; for example, the Norwegian electrolyzer company Nel announced in 2021 plans to build a 2 GW plant that would produce green hydrogen at $1.5 per kg by 2025, about the same price as grey hydrogen, assuming renewable electricity at $20 per MWh. The hydrogen can be used in fuel cells to generate electricity or as a gaseous fuel for heating or in internal combustion engines (ICEs).

Case Study 11.3 Wind power-to-hydrogen

Alkaline electrolysers have efficiencies in the range 50–80%, but typically need a fairly *continuous* supply of electricity, so intermittent sources like wind and solar can present a problem. The challenge is to produce an electrolyser that can operate with a fluctuating energy input, and one of the advantages of using polymer electrolyte membranes (PEM) is their ability to operate at both low and high current densities.

A number of pilot projects around the world are developing technologies that overcome the intermittency problem. One of these is the proposed **Gigastack project**, a joint venture between ITM Power, Ørsted, Phillips 66 and Element Energy, with funding from the UK Government. Phase 1, completed in 2019, involved the design of a low-cost stack of 5 MW PEM electrolysers, and showed that the electrolysers could be operated successfully using electricity supplied from the world's largest offshore wind farm, Hornsea Two, in the North Sea. The consortium argues that using the economies of scale and high-volume production will enable it to become competitive with existing electrolyser technology, and thus produce green hydrogen on a scale that will transform the heating and transport needs of the future.

Most steel manufacturers in Europe are proposing to use hydrogen as a route to decarbonization instead of CCS; for instance, Thyssen Krupp has started switching to hydrogen in their blast furnaces. Hydrogen is also used in the chemical industry, and the marked increase in global electrolyser capacity indicates great growth in the deployment of hydrogen.

There are also plans supported by the German government for a dedicated offshore wind farm to power a green hydrogen plant in north west Germany to produce carbon-neutral aviation fuel. In the pilot project the oxygen, generated along with the hydrogen in a 30 MW alkaline electrolyser, will be used in a cement plant, which could enable easier capture of the emitted CO_2 (see Chapter 3 Section 3.7), with the waste heat from the electrolysis process sold to a district heating system. The location of the green hydrogen plant is close to salt caverns where hydrogen could be stored and then transported by pipeline for use elsewhere. Also under investigation is the electrolysis of seawater on disused oil platforms powered by offshore wind farms, with the hydrogen transported to shore via existing gas pipelines. This could be more economic than building underwater electric cables for hydrogen production on land. Special electrodes that resist corrosion by the chlorine ions in the water are being developed that would avoid the need to desalinate the seawater first.

Methane can be formed by reacting hydrogen with carbon dioxide at elevated temperatures (300–400 °C) and pressures, in the presence of a nickel catalyst (Sabatier process). The reaction is:

$$CO_2 + 4H_2 \rightarrow CH_4 + 2H_2O \quad \Delta H = 165\,\text{kJ mol}^{-1} \tag{11.6}$$

When methane burns (i.e. combusts with oxygen), the same amount of carbon dioxide is released as was used in its production, so the methane produced in this way provides a low-carbon energy store and can be used to substitute for natural gas to generate heat. Methane can be converted by several processes to olefins, which are the basic chemicals in the production of many plastics.

A proposed power-to-methane plant in Finland plans to produce natural gas from hydrogen and carbon dioxide from an existing power station. The commercial viability of the project is determined by the cost of the electricity used in the electrolysis process.

Case Study 11.4 Power-to-ammonia

A process under development is the use of renewably produced hydrogen in the Haber process to make ammonia. Ammonia is a valuable and expensive product, and an efficient low-carbon method of manufacture could give a near-term profitable business. About 160 Mt NH_3 are produced a year, and its price (2016) was around $550 tonne^{-1}. In addition, it would reduce CO_2 emissions by about 450 Mt a year and be easy to store and transport. Ammonia is an essential component of fertilizers: it is combined first with about 200 Mt of CO_2 a year to produce urea (see Exercise 11.20), which is then processed further to make fertilizers.

The Haber process takes nitrogen from the air and combines it with hydrogen at high pressure (~200 bar) and temperature (400–450 °C) in the presence of an iron-based catalyst to produce ammonia in an exothermic reaction:

$$N_2 + 3H_2 \rightleftharpoons 2NH_3 \quad \Delta H = -92 \text{ kJ mol}^{-1}$$

The gaseous ammonia is cooled to a liquid state, and the nitrogen and hydrogen that have not reacted (~85%) are recycled. By Le Chatelier's principle, to shift the reaction to the right would require as low a temperature as possible, since the reaction is exothermic, but this has to be tempered by having a sufficiently high reaction rate, which favours a high temperature, to produce reasonable quantities of ammonia per hour. High pressure will also favour ammonia production, as the reaction which produces fewer molecules will be favoured, as well as improve the rate. Around 200 bar is a compromise between the cost of high-pressure pipes and reactor vessels and the rate of reaction.

An attractive method under investigation is to make ammonia directly from steam and nitrogen by electrolysis, as this would avoid the need for high pressure and could be low-carbon. The reaction is:

$$N_2 + 3H_2O \rightarrow 2NH_3 + \frac{3}{2}O_2$$

Ammonia can be used in an internal combustion engine, gas turbine, or fuel cell to provide energy. On combustion, it produces N_2 and H_2O:

$$4NH_3 + 3O_2 \rightarrow 2N_2 + 6H_2O(g) \quad \Delta H = -1267 \text{ kJ mol}^{-1}$$

The heat of combustion which includes the energy from condensing water (44 kJ mol^{-1}) is 380 kJ mol^{-1} for ammonia or 22.5 MJ kg^{-1}, which is approximately half of that of petrol or diesel. It can be burnt cleanly with little NO_x pollutants, and cars have already been powered by ammonia. Ammonia is also being considered as the fuel to decarbonize long distance shipping.

As a fuel for fuel cells, ammonia is liquid at 10 bar and room temperature and has 11.5 MJ litre^{-1}, compared with 8.5 MJ litre^{-1} for liquid hydrogen at −253 °C (20 K) or

4.5 MJ litre^{-1} for compressed H_2 at 690 bar and 15 °C, and a higher energy density than metal hydrides. It is, however, toxic at high concentrations but detectable by people at small concentrations. The infrastructure for transporting ammonia exists already, unlike for hydrogen, so it is a good candidate for fuel cells. It can also be decomposed to give hydrogen by using light metal imides at modest temperatures; in that case liquid ammonia could be an effective store of hydrogen, if ammonia synthesis were cheap enough.

Together with renewable production, ammonia could be a valuable fuel, energy vector, and energy store.

11.2.2 Electrification of Heat

Industry requires low (<100 °C), medium (100–400 °C) and high temperature (400–2000 °C) heat, which accounts for about 10% of all greenhouse gas emissions. This heat can be supplied by electric furnaces (radiant and convective heating), and by heat pumps for low to medium temperatures. Induction heating, microwave, laser, electron beam, and plasma technologies can also be used and are efficient, since they can provide localized heating. Electric arc furnaces are already used in steel-from-scrap production, and electric arc heated ionized gas, plasma heating, can provide temperatures greater than 2000 °C. Most of these would need to be adapted for specific industrial processes.

An EU study has looked at the implications of decarbonizing the energy-intensive industry in Europe entirely by electrification. It estimated that electricity demand would increase by about 50%, with use in industry rising by near 150%. The cost of basic materials like steel and cement might rise by 20–100% assuming a carbon price of €100 tonne CO_2. While this would make decarbonization difficult, the impact on overall prices is generally very small: a doubling of cement costs resulting in less than a 1% increase in the cost of a residential building.

With renewable electricity prices falling such that they could be the most economic source of energy, the energy intensive industries could, if electrified, become flexible consumers of excess solar and wind production.

11.2.3 Carbon Capture in Industrial Processes

Some of the largest localized sources of CO_2 emissions are industrial sites. For example, a large steel plant could contain five blast furnaces each producing 3.5 $MtCO_2$ per year—a total of 17.5 $MtCO_2$ compared with about 7.6 $MtCO_2$ from a 1 GW coal fired power plant—so industrial sites could provide economies of scale for CCS. The cost of using CCS depends on the CO_2 concentration and on the ease of integration. Traditional ammonia production, for example, yields an almost pure stream of CO_2 that only needs compressing. However, most industrial applications produce low concentration streams, for which chemical or pressure swing absorption, and membrane or cryogenic separation can produce 85–90% removal. To achieve close to 100% removal, pure oxygen can be used instead of air in the combustion, so that the flue gases are not diluted with nitrogen, or the separation occurs as part of the process (in-process), as in the gasification of biomass (see Chapter 3 Section 3.7 for further details on CCS).

Fig. 11.9 An industrial plant, showing multiple sources of emissions (zhaojiankang/istock).

The integration of CCS will generally be complicated, as there are often several sources (see Fig. 11.9) of CO_2 with different concentrations on the same site; e.g. a steel works with 10 smoke stacks. The capture process requires energy, e.g. for regeneration of the absorbent and for pressurizing. Aiming for close to 100% capture by using in-process or oxygen combustion would require substantial changes to existing plants. There are a few pilot projects underway but these are mainly designed to capture 90% of emissions. Transport of the CO_2 seems to be quite straightforward and is a small part of the overall cost of CCS. Globally, several storage sites have been identified (e.g. in the North Sea), but not all are near to hand and their long-term security is still under review.

Retrofitting CCS in industrial processes will probably result in only 65–85% of emissions being captured at most, going by estimates in the steel and cement industries, and the cost has been estimated at around €100 per tonne CO_2 avoided. If industrial plants are redesigned and rebuilt, then costs could be lower and the percentage captured higher. Implementing carbon capture and utilization (CCU), where the CO_2, instead of being buried, is used to make products, would offset some of the costs.

Globally, there are numerous initiatives underway to capture the CO_2 generated in industrial processes, including:

Japan (Mitsubishi), using CO_2 from an oil refinery and hydrogen from electrolysis to produce methanol, as a substitute for gasoline and as a feedstock for the chemical engineering industry;

Canada (Svante), using nano-materials to capture CO_2 from cement-making plants and other industries.

Norway (Northern Lights project) providing storage of CO_2 from industrial sites in Norway and Europe in a reservoir 2.6 km under the seabed in the North Sea, with the first phase expected by mid-2024;

UK (Net-zero Teesside) aiming to decarbonize a cluster of industrial emitters, which include producers of blue hydrogen (H_2 from steam reforming natural gas), in the NE of England by 2030.

But these are all are relatively small scale and much development and investment will be needed to achieve a significant reduction by CCS in industrial emissions—possibly 5% by 2050.

We now look at mitigation options more closely in the iron and steel, cement, and chemical industries, which accounted in 2018 for 66% of direct (including process) emissions from industry.

11.2.4 Steel Industry

According to the World Steel Association, steel production accounts for about 7-9% of CO_2 emissions from fossil fuels, about a fifth of all industrial emissions, each tonne of steel resulting in 1.8 tonnes of CO_2. Over 1300 million tonnes of steel are currently produced per annum, and the IEA predicts that this will rise to 3000 million tonnes by 2050.

The reduction of iron ore to iron using coke (a carbon-rich solid made from coal) in blast furnaces is responsible for about 80% of the CO_2 emissions in steel production. Increased recovery of heat and gases in the blast furnaces would improve their efficiency, and reducing energy use in final processing would also help. Steel can also be produced by recycling steel through an electric arc furnace where the material is melted by passing a very high current through it. Over 55% of steel comes from blast furnaces and about 40% from electric arc furnaces, but the availability of scrap steel is limited.

Alternatives to making steel with coke are to use hydrogen, electrolysis, or biochar. Reduction is much faster with hydrogen than with coke, but heat must be added. Electrolysis of iron ore to iron is possible and is known as electrowinning: the ore can be suspended in an acid or alkaline solution which produces a sponge-like iron that has to be melted for alloying; plasma or inductive ovens or gasified biomass can be used. The use of hydrogen could be utilized as a way of storing energy for load levelling in the electricity grid.

However, the timescale to develop alternative technologies on a large enough scale to compete with existing steel plants is likely to be well over a decade. In the meantime, some manufacturers will be looking for ways of modifying the steel-making process to reduce the CO_2 emissions. For example, Tata Steel claims that it could reduce its CO_2 emissions by 80%, by simplifying the preparation of raw materials and using carbon capture and storage. Such measures may well drive up costs, and there is a need for an international agreement on creating a greener global steel industry. But already in 2020, Germany's Thyssenkrupp announced plans to use green hydrogen to reduce its iron ore, and Japan's Nippon Steel promised carbon neutrality by 2050.

11.2.5 Cement Industry

The current global production of cement is about 3300 million tonnes a year, and is estimated to reach nearly 4500 million tonnes a year by 2050. Globally, cement production produces about 27% of all CO_2 direct emissions from industry (see Fig. 11.8), of which 60% comes from the calcination of limestone to lime ($CaCO_3$ to CaO), and 40% from the burning of fossil fuels for heat. Cement is made from a mixture of limestone, clay, and other materials, which is first heated to dry and calcinate the limestone, and then fed into a kiln and heated up to around 1500 °C where the lime reacts and produces calcium silicates and other compounds. The result is called clinker, which after grinding and mixing with some gypsum makes cement.

To avoid process emissions, alternatives to limestone-based cement are being investigated. A novel example is the idea of **plastic roads**, using consumer plastic waste. The Dutch cities of Zwolle and Giethorn have built bicycle paths using modular plastic elements. Also, India has developed a composite mixture of bitumen and plastic for road construction; by 2019, India had built 33,700 km of plastic roads, requiring 1 million plastic bags per km.

Another example is to make much more use of natural materials such as timber, as in the following Case Study 11.5.

Case Study 11.5 World's tallest timber building (Brumunddal, Norway)

Brumunddal is a small town north of Oslo which has a new tourist attraction: an 18-storey structure built using engineered wood (which includes cross-laminated timber, CLT), making it the world's tallest timber building (Fig. 11.10). Wood is a carbon store, and 40% of Norway is covered by forest, and using this natural resource to construct timber buildings and then re-planting with saplings to absorb more carbon dioxide in the atmosphere seems an obvious winner.

Fig. 11.10 Mjøstårnet in Brumunddal, Norway, the world's tallest wooden building (photo courtesy of Voll Arkitekter AS and VizWorks).

Encouraged by government policy to become carbon neutral and the global movement towards greener economies, Norway aims to become a world leader in the manufacture of zero energy buildings. France has decreed that by 2022 at least 50% of the structure of all publicly-funded buildings should be made from natural materials, and other countries are moving in the same direction. Properly manufactured and treated, buildings made from such materials can meet the required safety standards, particularly fire regulations. [Wooden beams can be treated with fire-resistant paints or covered by fire-resistant dry walling; also, steel beams soften with heat and are normally encased by concrete to prevent direct exposure to flames and for extra strength.]

A world in which people live in eco-friendly buildings made from timber and other plant-based materials, is becoming a serious alternative to the traditional energy-intensive brick, steel and concrete structures with their high embodied emissions that dominate the planet today.

11.2.6 Chemical Industry

Unlike the steel and cement industries, the chemical industry is very diverse with many different processes used. There are, however, a few groups of chemicals that are the primary feedstock for many final products. These are shown in Fig. 11.11, which also indicates the relative amounts of fossil fuel-based raw materials and energy that are used. The inorganic group labelled 'other' contains carbon black, soda ash, chlorine, and sodium hydroxide. An important olefin is ethylene. Around 0.85 $GtCO_2eq$ is stored in plastics and fibres each year. While plastics have many benefits, particularly in medical applications, their disposal has created enormous pollution, which is a particular challenge.

In the petrochemical industry an important process is **steam cracking**, in which hydrocarbons are heated in the presence of steam and often a catalyst to convert them to lighter hydrocarbons. Ammonia is an essential raw material for fertilizers, and over 75% is produced with hydrogen from the steam reforming of natural gas, with the rest using coal, oil, or biomass.

The efficiency of many processes could be increased by improving heat and energy recovery, more effective catalysts, and new membrane separation methods. Olefin production takes the most energy, and the global average energy intensity is ~23 GJ per tonne compared with the best obtained of 16 GJ per tonne. Biomass-based production can also improve carbon intensity and help in reducing the use of coal. But the challenge is to make green production processes cost-competitive and here a carbon price would be a great help.

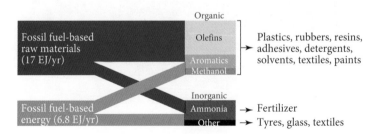

Fig. 11.11 Primary processes in the chemical industry (1 Gtoe ≈ 42 EJ) (GICC2012).

11.2.7 Summary of Industrial Emissions and Demand Reduction

Industry will need significant investment on a global scale for substantial emissions savings to be achieved and to avoid lock-in to more carbon-intensive technologies. Collaboration within and across industries will help attain best practice everywhere. The multitude of processes makes the task very challenging.

Significant deployment of CCS has been advocated but currently is expensive. More widespread electrification will be easier as low-carbon electricity becomes very cheap and plentiful. Renewable power-to-gas, in particular the production of low-carbon hydrogen from electrolysis of water using electricity from renewables, could help significantly.

Energy savings can also be made by improvements in heat recovery, through increased insulation, and by reducing losses. Considerable effort needs to be directed to **reduce, re-use, and recycle materials**. The search for **alternatives to concrete**, which contributes so much to emissions and generally cannot be recycled, is a major priority.

11.3 Energy Demand in Transport

Transport emissions have been growing significantly in the last two decades, rising from 5.8 $GtCO_2$ in 2000 to 8.2 $GtCO_2$ in 2018. In 2016, the transport sector accounted for around 116 EJ (~29%) of final energy use and 7.9 $GtCO_2$ (~22%) of energy-related emissions. Over the next few decades, the number of light duty vehicles (LDVs) is expected to double from its current total of about 1 billion, with two-thirds of the growth in countries outside the OECD as their GDP increases. Limiting global warming to 1.5 °C will require a reduction in transport emissions to about zero by 2050.

About 95% of the energy used in the transport sector comes from oil, and accounts for about 60% of the total global oil consumption. How this energy was divided among the different modes of transport in 2012 is shown in Fig. 11.12.

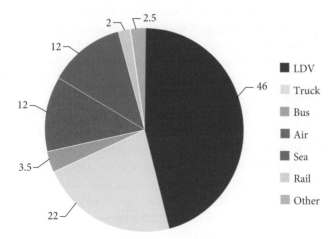

Fig. 11.12 Percentages of total transportation energy in 2012 (EIA).

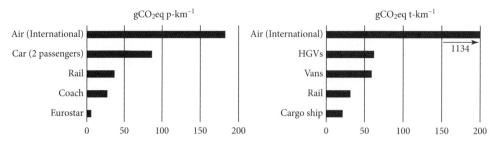

Fig. 11.13 GHG emissions (CO_2eq) per passenger-km and per tonne-km. For air transport, the contributions from CO_2 and from the effects of high altitude are approximately equal (BEIS Defra Conversion Factors 2020).

The largest share is due to LDVs with 46%, next are trucks with 22%, then boats and planes with 12% each, and buses, trains, and other modes, such as pipelines, with 8%. The typical carbon emissions per passenger-km and per tonne-km for freight are significantly different, as Fig. 11.13 shows.

Mitigation measures can be introduced through low-carbon technology (such as EVs and Airbus's planned H_2 powered jets), and policy and behavioural changes. Examples of the latter are: avoiding journeys, e.g. by urban design, local manufacturing, internet shopping, the use of video conferencing, and increasing the use of public transport and of walking and cycling. More compact cities with cheaper public transport will have the additional benefit of improving productivity and access to facilities. An example of this is in the city of Freiburg in Germany (see Case Study 11.6).

Case Study 11.6 Transportation in Freiburg (Germany)

In the 1970s, Freiburg was clogged-up with cars and decided to become a 'Green City'. When other German cities were getting rid of their trams in favour of cars, Freiburg decided to go in the opposite direction. It has become a model for how dedicated city planning can promote substantial reductions in both energy use, carbon emissions, and pollution from transport, and improve the quality of life in the process.

Now 70% of the population of 224,000 live within 500m of a tram stop. The trams are cheap, extensive, and provide a service every 7.5 minutes in rush hours. A decision to cut the price of a public transport pass by a third more than doubled the number of people using public transport.

Preference is given to walking, cycling, and public transport, rather than cars, and neighbourhood centres have been built to provide shops within easy walking distance. Bicycle lanes are clearly marked out on roads, and cyclists have priority at intersections. Freiburg also has over 400 km of cycle paths and 9000 parking spaces for bikes. The speed limit on most streets is 20 mph, and just walking pace on some.

In the development of the Vauban residential district, large parts were made car free, with parking only on the edge of the area. And the improved access to public transport, the introduction of car sharing, and the encouragement of bicycling has reduced car ownership there considerably.

Policy changes could discourage cars by introducing congestion charges, and encourage the use for freight of high-speed rail and shipping rather than trucks and aircraft. Efficiency improvements and an increased number of occupants can reduce the energy intensity (kWh per passenger-km or kWh per tonne-km), but most trucks already have very efficient (~45%) diesel engines and are often streamlined. So, to obtain deeper cuts in emissions will require lowering the carbon intensity of the fuels by using, for example, electricity, hydrogen from low-carbon sources, ammonia, bio-methane, biofuels, or (though not currently cost-competitive) low-carbon synthetic fuels.

Obstacles to the adoption of these measures include high initial capital costs, long lifetime of existing stock, and existing consumer preferences; for example, owning a car rather than car sharing. One counterproductive result of energy (and hence cost) savings is the *rebound effect* (see Chapter 12 Section 12.6.1) whereby money saved is spent on other energy consuming activities. In practice, efficiency gains are not entirely reversed by the rebound effect, but are typically reduced by between 5% and 40%. In the transport sector, the immediate gains will be achieved by improving the efficiency of petrol and diesel ICEs; see Example 11.5 and Exercise 11.23.

EXAMPLE 11.5

A 1600 kg car has a drag coefficient of 0.4, a frontal area of 2.5 m^2, and a rolling resistance 1% of its weight. Estimate (a) the energy in kWh required travelling 100 km at 67 mph, (b) the litres of petrol consumed, (c) the kilograms of CO_2 emitted.

(a) From eqn (2.41), drag force = $0.5 \times 0.4 \times 1.2 \times 30^2 \times 2.5 = 540$ N, rolling resistance = 157 N, and energy = force × distance = $6.97 \ 10^7$ J = 19.4 kWh.

(b) Assume 20% efficiency for a petrol ICE, so that the energy required = $3.49 \ 10^8$ J = 8.0 kg = 11 litres.

(c) Assume petrol acts like octane so emissions are 24.7 kg CO_2.

Improvements in the last couple of decades have tended to be used to increase the size, weight, and power of vehicles, rather than their fuel economy; e.g. a 10% increase in weight reduces fuel economy by ~7%. Regulations that ensure that cars have good fuel economy, and hence low emissions per km, would counter this trend. Ultimately, the introduction of electric vehicles coupled with a decarbonized electricity supply will be needed or, if cheap hydrogen made from low-carbon energy is available, or suitable biomass can be produced sustainably, vehicles powered by fuel cells or biofuels.

We now first look at electric vehicles.

11.3.1 Electric Vehicles

For electric vehicles, the electric motor can provide a good torque from rest (unlike an internal combustion engine, which has a minimum operational rpm), and can also provide braking by operating the motor as a generator: regenerative braking. One type of motor that is increasingly being used is a DC brushless motor in which a constant magnetic field is generated in

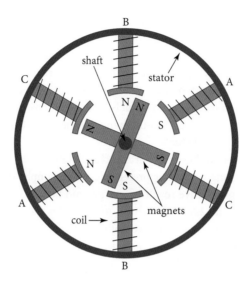

Fig. 11.14 DC brushless motor.

turn by pairs of poles AA, BB, and CC on a stator (see Fig. 11.14). Two permanent magnets are attached to the rotor. When the current through the pairs of poles AA and BB is as shown, the forces between the poles cause the shaft to rotate. Sensors detect the angular position of the rotor magnets and switch the current to the next pair of stator coils so as to maintain the torque. Only two pairs of stator coils are on at any one time. The efficiency of these motors is good, typically 80–90%, and their power density is high. Moreover, their speed is easy to control by simply changing the voltage applied to the coils (see Box 11.3).

Box 11.3 Brushless DC motor

The rotating permanent magnets on the rotor cause a changing field through the stator coils. Fig. 11.15 shows one coil leaving a B field from one of the permanent magnets on the rotor with a relative speed v. We will make the approximation that the field is uniform across a pole tip and then zero; i.e. the flux change $\Delta\phi = -BLv\,dt$. Hence the back emf is $\varepsilon = BLv$ (see eqn (10.2)). The force F experienced by the conductor carrying the current I is given by $F = BIL$. By Newton's third law there is an equal and opposite force on the rotor magnet.

The power P_R produced by the rotor is $P_R = Fv$, while the electrical power P_S to drive a current I across a potential ε is $P_S = \varepsilon I$; i.e.

$$P_R = Fv = BILv \quad P_S = \varepsilon I = BILv \tag{11.7}$$

Therefore, the electrical power supplied to overcome the back emf equals the mechanical power generated.

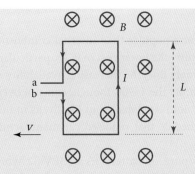

Fig. 11.15 Coil leaving a B field with relative speed v.

In general, for a rotor in a DC brushless motor, since the relative speed of a coil and magnet is proportional to the angular velocity of the rotor ω, the back emf is of the form $\varepsilon = k\omega$, where k is a constant. For a coil with resistance R, the current I flowing through the coil is related to the voltage V_P across it by the equation:

$$V_P - k\omega = IR \tag{11.8}$$

Since the current flows through two identical coils, the electrical power $P_S = 2\varepsilon I$. Equating the electrical and mechanical powers then gives the torque τ produced by the rotor:

$$P_R = \tau\omega = P_S = 2\varepsilon I = 2k\omega I \quad \text{so } \tau = 2kI$$

In an electric car with a brushless DC motor, the speed is altered by simply changing the voltage applied to the stator coils. If the voltage is reduced below the back emf voltage, then by eqn (11.8) the sign of the current I changes and the torque produced is in the opposite direction, i.e. braking occurs. The motor is now acting as a generator and the charge produced can be stored in a battery or super-capacitor. This regenerative braking improves the efficiency of the car.

EXAMPLE 11.6

A brushless DC motor is driven at 1500 rpm and the back emf is then 40 V. Each coil has a resistance of 0.6 Ω. Find (a) the coil current and (b) the efficiency of the motor when the motor develops 2 Nm of torque, and (c) the minimum value of DC voltage required to drive the motor at 3000 rpm and 2 Nm.

(a) $k = \varepsilon/\omega = 40/157.1 = 0.255$ N m A^{-1}, as 1500 rpm corresponds to $\omega = 157.1$ rad s^{-1}
$I = \tau/2k = 2/(2 \times 0.255) = 3.92$ A

(b) The mechanical power equals $2\varepsilon I$; the resistive loss equals $I^2 2R$, so the efficiency $2\varepsilon I/(2\varepsilon I + I^2 2R) = \varepsilon(\varepsilon + IR) = 40/42.35 = 94.5\%$.

(c) $V_P = k\omega + IR = (3000/1500)40 + 3.92 \times 0.6 = 82.35$ V. Since the DC supply is across two coils at any one time, $V_D = 2V_P = 164.7$ V

11.3.2 Electric Cars

While electric cars were more popular than internal combustion engine cars (ICEs) in the first decade of the 20th century, the lower cost of ICEs, particularly following the introduction by Henry Ford of the assembly line, the increasing availability of oil, and their much longer range made ICEs dominant. By the 1990s, though, concern over global emissions of CO_2 led to renewed interest in EVs not just globally but locally, as in the California ZEV (zero emission vehicle) regulation schedule that required car manufacturers to increase their market share of ZEVs.

Besides a relatively few pure or battery EVs (BEVs), hybrid electric vehicles (HEVs), which combine an internal combustion engine and a fuel tank with an electric motor and a battery, were produced and these gave reductions in CO_2 emissions of up to 35% compared with similar conventional cars. Plug-in hybrid electric vehicles (PHEVs) were also introduced to provide a useful bridging technology to pure long range EVs. They could enable a significant fraction of many people's daily car journeys to be made just using the battery supply of electricity, with longer journeys using the ICE to generate electricity. In Europe, 50% of trips are estimated <10 km and 80% <25 km, while in the US, ~60% are <50 km and 85% <100 km. The plug-in electric light vehicles BEVs and PHEVs are referred to as PEVs.

Pure or battery electric vehicles give the greatest scope for emissions savings, though the amount depends on the decarbonization of the electricity grid (see example 11.7). Although by 2010 electric cars with a range of 100 miles were available, e.g. the Nissan Leaf, there was still the 'range anxiety' of running out of charge on the occasional long journey, since there were few charging stations around, and, in particular, the price of BEVs was still high. But by 2020, typical ranges were over 200 miles and BEVs were starting to be more competitive.

EXAMPLE 11.7

Estimate the global reduction in carbon emissions if cars were electrified and charged by an electricity supply with a carbon intensity of 0.5 kg CO_2 kWh^{-1}. EVs require 15 kWh through the wheels per 100 km, and petrol cars 9 litres per 100 km. Globally there are ~1 billion cars, each travelling on average 15,000 km per year. Take the overall electrical efficiency of charging and of the motor as 80%.

Assume petrol (density 0.73 kg per litre) acts like octane, i.e. 352 kg CO_2 per 114 kg of octane, equivalent to 2.25 kg per litre and 20.3 kg per 100 km. Global car emissions produce 3.05 Gt CO_2. EVs emit $\left(15\times\dfrac{0.5}{0.8}\right)$ kg CO_2 per 100 km, so global emissions would then be 1.41 Gt CO_2, a saving of 1.64 Gt CO_2 per year.

11.3.3 Future of Electric Vehicles

While the sale of electric vehicles (PEVs) was still only 4.2% of total car sales in 2020, their growth is significant (globally 43% in 2020). It has been driven by the continuing improvement in batteries, both in their cost and their capacity, and by the increasing pressure to drastically reduce CO_2 emissions. Banning ICEs is easier for governments than imposing a carbon tax to phase out fossil fuels, and Norway is stopping the sale of new ICEs after 2025. Fourteen other countries have pledged to follow suit over the next two decades with the UK by 2030, and California by 2035. And China also plans to halt production of conventional ICE cars by 2035.

In order for EVs to reach parity with ICEs, lithium-ion battery packs need to get down to around \$100 kWh^{-1}, which should happen by 2023 (see Chapter 10 Section 10.12.1). But already electric cars are typically more affordable than equivalent ICEs, because of the significant lifetime saving in fuel and maintenance costs from the high efficiency and simplicity of electric motors; e.g. Volkswagen's ID.3 has a range up to 340 miles at a price of ~£30 k. Moreover, the performance of electric cars is good.

Besides helping to reach the targets for reducing climate change, switching to electric cars cuts out the harmful particulates and nitrous oxides emitted by engines, in particular those powered by diesel. These contribute significantly to pollution in cities around the world, which causes health problems such as asthma. Electric cars emit fewer particulates from their brake pads, as they mainly brake by using their motors as generators. The contribution from tyre wear is similar for all cars; but if cars can be made lighter in future, and acceleration and deceleration reduced by self-driving (autonomous) vehicles, such as Amazon's Zoox EV, then this will improve.

Regulations over the mining of cobalt, which is used in many car batteries, are needed to avoid human rights abuses. But there are lithium-ion battery chemistries without cobalt, such as lithium iron phosphate, and battery demand should be able to be met sustainably, particularly when recycling becomes the norm.

The major car manufacturers are now investing massive amounts of money into producing electric cars. In 2020, the range of new electric cars was typically 200 miles or more, helping to overcome 'range anxiety'. However, countries will need to install a huge number of fast charging points in cities (Fig. 11.16) and the countryside, to replace the massive infrastructure of fossil fuel stations. Improvements in batteries are also expected to reduce charging times. Batteries typically (2020) take around 30 minutes with a fast charger to be 80% full. But in

Fig. 11.16 An electric car charging station (*Source*: Wikimedia Commons under creative commons license 2.0: https://creativecommons.org/licenses/by-sa/2.0/deed.en).

2021, the first thousand Li-ion batteries with a charging time of 5 minutes were manufactured. In these batteries, developed by StoreDot in Israel and made by Eve Energy in China, the graphite is replaced by germanium-based nanoparticles, though the plan is to use silicon to reduce costs. Also, overhead charging wires are a possibility for trucks.

All the large auto makers are now producing electrics cars, with General Motors in the US committing in 2021 to all electric LDVs by 2035. Hybrid and plug-in hybrid cars are only an interim solution (too complex and therefore too expensive). EVs had a 4.2% share of the market in 2020 and fast expansion is expected. Bank of America Merrill Lynch predicted that EV sales will achieve 12%, 34%, and 90% of the market by 2025, 2030, and 2050, respectively. Bloomberg has a similar projection with BEVs equal to ICEs by 2035, with PHEVs by then only a few per cent of sales.

Electric buses, motor cycles, mopeds, bicycles and scooters are also taking off around the world, making green transport a reality. The transition to heavy goods vehicles, ships and aeroplanes, is more problematic: while there are now some electrified trucks, urban waste vehicles, short-haul ships, and some small electric aeroplanes are being trialled, for heavy goods vehicles and long-distance shipping, ammonia (see case study 11.4), fuel cells, and bio-fuels could make a significant difference.

11.3.4 Biofuels and Fuel Cells

Biofuels supply approximately 3% of transport's energy demand, while the penetration of fuel-cell powered vehicles is still very small. The use of biofuels and of hydrogen for fuel cells has been affected by the lack of availability of cheap sustainably produced (low-carbon) supplies. As discussed in Chapter 4 Section 4.5, the production of first-generation biofuels from biomass can lead to significant emission of CO_2 from the change in land use and can also compete with food production. They would also require a considerable land area (see Example 11.8).

EXAMPLE 11.8

A sugarcane plantation produces 4 tonnes of ethanol per hectare per year. (a) What area of plantation would be required to displace 4×10^{10} litres of petrol per year (France's annual consumption) with bioethanol? (b) Compare this area to that of France (6.45×10^5 km^2). (Density of petrol is 0.73 kg litre^{-1}.)

(a) Petrol has a LHV of ~43.5 MJ kg^{-1}, so there are ~32 MJ per litre. The energy content of 4×10^{10} litres of petrol is therefore 1.28×10^{12} MJ. An area A ha of sugarcane would produce $4000A$ kg of ethanol per year. Ethanol has a LHV of ~27 MJ kg^{-1}. Because a higher compression ratio is possible, the thermal efficiency of an ICE optimized for ethanol is ~40%, compared with ~25% for one optimized for petrol. So the area required is given by

$$4000 \times 27 \times A \times 1.6 = 1.28 \times 10^{12}, \text{ so } A = 7.41 \times 10^6 \text{ ha} = 7.41 \times 10^4 \text{ km}^2$$

(b) i.e. a square 345×345 km (215×215 miles), which is 11.5% of the area of France.

Biofuels

Biofuels can be used in internal combustion engines, but often modifications to the engines are required. Only small changes are required for methane to be used in Otto-cycle spark engines, but larger alterations are needed in the case of diesel engines. The carbon intensity of methane is lower than that of petrol and diesel, and produces up to 25% reduction in emissions (gCO_2 km^{-1}). Biomethane from anaerobic digestors can be used. Biofuels could help in reducing emissions from aviation, and in 2021 Boeing announced that they would deliver planes that could run on 100% biofuel by 2030.

Bioethanol has been used for many years in Brazil, where the large land area and suitable climate enables sugarcane to be grown sustainably. But elsewhere, first generation biofuel production is limited, and the need is for second generation or advanced biofuels. These are biofuels from lingo-cellulosic, CAM plants, or from algae that grow on marginal land (or water) that could provide a large volume, but are still under development (see Chapter 4 Section 4.5). However, biofuels in internal combustion engines still produce harmful particulates and nitrous oxides.

Fuel cells

A **fuel cell** is an electrochemical device that can be used to generate electricity cleanly. Unlike a battery, a fuel cell is fed continuously by chemicals. The chemical feed consists of hydrogen and oxygen, and the fuel cell provides a means of combining them to make water, i.e. the opposite of **electrolysis**. There are two electrodes (the anode and cathode) where the chemical reactions take place, and a catalyst speeds up the reactions at the electrodes. There is virtually no pollution, and the only by-product is water. Also, since a fuel cell is not a heat engine operating in a closed cycle, the Carnot limit to its efficiency does not apply.

The fuel cell was invented in 1839 by **William Grove**, but interest in it soon fell away when cheap fossil fuels became widely available. It was not until the 1920s and 1930s that significant progress was made in their development. In the early 1960s, NASA decided to use fuel cells to provide electricity for the Gemini and Apollo space capsules, which led to further improvements. With their carbon-free energy and high efficiencies, interest in fuel cells is high. Fuel cell development has centred on the materials used for the electrodes, the catalysts, and the choice of electrolyte. Some devices use liquid electrolytes (alkalis, molten carbonate, and phosphoric acid), while others use solid electrolytes (proton exchange membrane (PEM) and solid oxide). We concentrate here on PEM cells, whose main features are shown in Fig. 11.17, since they illustrate the physical principles involved (see Box 11.4). Hydrogen fuel cells are well developed and suitable for use in heavy vehicles and trains, where they are starting to have an impact; the space to store compressed hydrogen cylinders is available, and their high-energy density by weight and speed of refuelling are attractive (see Chapter 10 Section 10.16 and Exercise 10.26). They are also quiet and reliable.

The output voltage of one cell is around 1 volt and the efficiency of conversion to electricity of the energy of combustion (the enthalpy) of hydrogen with oxygen to form water is about 60%, with the rest released as heat. While the oxygen input is from the air, the hydrogen has to be first generated and stored. Hydrogen is mainly produced commercially by steam reforming, in which a hydrocarbon is reacted with water to generate H_2 and CO_2. For this process to

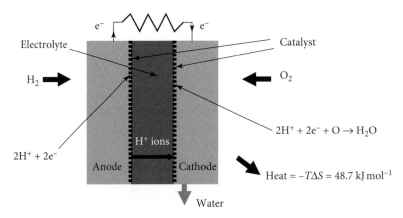

Fig. 11.17 PEM fuel cell (schematic). The electrolyte is a thin plastic membrane and the anode and cathode are porous. Inputs are H_2 and O_2, outputs are water and heat.

Box 11.4 Proton exchange membrane (PEM) fuel cell

Hydrogen is introduced on one side of the cell and flows through a porous anode, where it dissociates into hydrogen ions and electrons on contact with a catalyst (see Fig. 11.17). The porous electrodes are often made of a porous carbon-impregnated cloth or paper 100–300 μm thick. The membrane acts as the electrolyte, since it is permeable to hydrogen ions but is impermeable to electrons. The electrons flow through the external circuit, after which they combine on the catalyst with oxygen and the hydrogen ions, which have passed through the plastic electrolytic membrane, to form water. The overall reaction is:

$$H_2 + \frac{1}{2}O_2 \rightarrow H_2O, \quad \Delta H = -285.8 \text{ kJ mol}^{-1}$$

where ΔH is the change in enthalpy. The reaction occurs at constant pressure, so $-\Delta H$ gives the energy released when hydrogen and oxygen combine plus the work done when the volume of the gases changes (as explained in Chapter 2 Derivation 2.1). The entropy of the gases decreases in this process, since the number of moles is reduced. In a reversible process the entropy of the entire system, reactants and surroundings, remains constant. As a result, an amount of heat equal to $-T\Delta S$, where ΔS is the change in the specific entropy of the gases, is transferred to the surroundings. The amount of energy available as electrical energy is minus the change in the Gibbs free energy $-\Delta G$, where:

$$\Delta G = \Delta H - T\Delta S = (-285.8 + 48.7) = -237.1 \text{ kJ mol}^{-1}$$

In this reversible (ideal) process the efficiency of conversion is $\Delta G/\Delta H = 83\%$. The voltage V generated is given by equating ΔG to the total charge ΔQ that flows, multiplied by the voltage V, i.e.:

$$\Delta G = \Delta QV = nN_A eV$$

where n is the number of electrons released per mole, N_A is the number of molecules in 1 mole (Avogadro constant), and e is the charge on an electron. The product $-N_A e \equiv F$ (known as the **Faraday constant**) is equal to the amount of charge in one mole of electrons and has the value 9.65×10^4 coulombs per mole. Since two electrons are released per molecule of hydrogen, $n = 2$, the voltage generated is:

$$V = 237.1 \times 10^3 / (2 \times 9.65 \times 10^4) = 1.23 \text{ volts}$$

This is the open circuit voltage of a PEM cell, but with current flow through an external circuit, the voltage drops to speed up the reactions and to overcome the internal resistance. Eventually, the current is limited by the formation of water blocking reaction sites on the catalyst.

be clean the carbon dioxide must be captured and stored. Alternatively, hydrogen can be produced cleanly by electrolysis using renewably generated electricity (see Chapter 10 Box 10.2) and this production is increasing rapidly as decarbonization efforts quicken.

11.3.5 **Summary of Transport Emissions Reduction**

To obtain significant emission reductions in **road** and **rail** transport, one study concluded that a multi-pronged strategy, combining electric drive, advanced biofuels, and fuel cells, would be needed. Light-duty vehicles (LDVs) and rail would be mostly EVs, while heavy-duty (trucks) would use advanced biofuels, hydrogen, and electric power. Global investments in biofuels are on the decline though, and the rapid improvement in batteries makes electrification of LDVs with them look most likely, with fuel cells having a share for heavy transport.

The emission of CO_2 from combustion powered vehicles and ships can be reduced by imposing speed restrictions, increasing taxation on less fuel-efficient vehicles, and reducing the weight of cars.

Behavioural changes could also provide additional savings, encouraged by urban re-design and policies to promote public (low-carbon) transport, cycling, electric vehicles, and walking; and by avoiding journeys by using video conferencing and internet shopping.

For **planes**, which are the next largest user of energy (see Fig. 11.12), a move to fuels derived from biomass or using power-to-gas/fuel could reduce emissions. Ammonia, biofuels, and fuel cells could also be used for **shipping**, while sails and on-board PV generation could provide some power. Electrification for long range shipping, though, would require a significant improvement in battery storage.

Economies of scale and the 'learning effect' will hasten the transition to EVs and better batteries. The electrification of LDVs will bring about significant emission reductions.

SUMMARY

- Buildings, industry, and transport account for ~89% of global final energy demand, and electricity use accounts for ~25%, and to limit global warming to 1.5 °C will require reducing their carbon emissions to almost zero by 2050.

- Actions to reduce and shift demand are as important as those to decarbonize the supply of both electricity and heat.

- Decarbonizing heat in industry will be a particular challenge and will require an increased use of heat pumps, biofuels, and the development of renewable power-to-gas; in particular, hydrogen.

- Improvements in the insulation of buildings and the efficiency of machines and processes, and good urban design, will help reduce demand significantly, as will reusing and recycling materials and sharing resources.

- Carbon capture may best contribute through reducing industrial process emissions, though CCS with hydrogen production may be an important source of decarbonized heat.

- Electric vehicles can provide significant reductions in carbon emissions, as will shifting transport from road and air to rail and sea.

- These actions will require a very significant increase in the generation of low-carbon electricity and heat, the development of heat and electrical storage, and the implementation of effective regulations and policies.

FURTHER READING

BloombergNEF, Electric vehicle outlook 2021.

Brown, T. et al. (2012). *Reducing CO2 emissions from heavy industry: a review of technologies and considerations for policy makers.* Grantham Institute for Climate Change, Imperial College, London (GICC2012).

Cho, J. et al., *White light-emitting diodes,* Laser & Photonics Reviews (2017) (LED2017).

Fulton, F. and Miller, M. (2015). *Strategies for transitioning to low-carbon emission trucks in the United States.* Institute of Transportation Studies, University of California at Davis.

IEA Tracking Industry 2019 (IEA2019).

Lechtenböhmer, S. et al. *Decarbonising the energy intensive basic materials industry through electrification,* SDEWES (2015).

REN21 Renewables 2019 Global Status Report, Energy Efficiency.

World Resources Institute, Global greenhouse gas emissions in 2016 by sector.

nicola.qeng-ho.org/housemodel/interactive.php Home heat-loss calculator (HHLC).

www.carboncommentary.com/Goodall, C. Useful source of information about low-carbon issues.

www.ecotippingpoints.org/our-stories/indepth/germany-freiburg-sustainability-transporta-tion-energy-green-economy.html Freiberg: a Green City.

www.eea.europa.eu/data-and-maps/indicators/heating-degree-days-2/assessment (HCE)

www.eia.gov/outlooks/ieo/pdf/transportation.pdf (EIA).

www.ipcc.ch/report/ar5/ IPCC's Fifth Assessment Report, Working Group III contributions on buildings, industry, and transport (IPCC-AR5).

www.h21.green/projects/h21-leeds-city-gate/ (H21Leeds).

www.siemens-energy.com/global/en/offerings/renewable-energy/hydrogen-solutions.html Information on green hydrogen.

EXERCISES

11.1 A meeting room requires an illumination of 9000 lm and is used 20% of the time. How much electricity is saved per year in kWh if fluorescent lights rated at 90 lm W^{-1} are replaced by white LEDs rated at 300 lm W^{-1}.

11.2 Calculate the U-value (W m^{-2} K^{-1}) of a thermal block wall of thickness 100 mm and thermal conductivity 0.132 W m^{-1} K^{-1}, assuming that the thermal resistance of the air layers on the inner and outer surfaces are 0.12 m^2 K W^{-1} and 0.06 m^2 K W^{-1}, respectively.

11.3 Calculate the U-value for the glass component (i.e. neglecting the frame) of a triple-glazed window with dimensions $d_g/2 = d_a = 8$ mm, $k_g = 1.5$ W m^{-1} K^{-1}, and $k_a = 2.1 \times 10^{-2}$ W m^{-1} K^{-1}, where d_g and d_a are thicknesses of the glass and of the airgaps, and the resistivity of the inner and outer surface air layers is 0.16 W^{-1} m^2 K.

11.4 An older domestic property loses 25% of its heat through the roof and 10% through the windows. Estimate the percentage decrease in the energy used for space heating by (a) replacing the existing boiler (75% efficient) by a condensing boiler (95% efficient), (b) decreasing the U-value of the roof insulation by 60% and the windows by 70% (by installing double-glazing), (c) reducing the average temperature difference between inside and outside from 12 °C to 10 °C (by turning down the air thermostat).

11.5 In Table 11.1 the house has two floors separated in height by 2.5 m, and a wall, floor, window, and door area of 117.9, 98.6, 18.8, and 3.7 m^2 respectively. Use the values given to verify the percentage energy losses for a 'recent building'.

11.6 The heat loss through air movement is given by $\dot{Q} = V \rho c_P (T_i - T_o)$, where V is the air volume exchanged per hour, ρc_p is the heat capacity of air (0.018 Btu $°F^{-1}$ ft^{-3}), and T_i and T_o are the air temperatures inside and outside the house. Calculate the heat loss for a house with 2000 square feet of floor space and 8-foot ceilings for $T_i = 65$ °F and $T_o = 10°$F. Assume the number of air changes per hour is 0.5.

11.7 Calculate the total number of degree-days over a whole year for (a) Long Island and (b) San Francisco, using the data in the following tables. The base temperature is 60 °F.

Long Island	Jan	Feb	Mar	Apr	May	Jun	Jul	Aug	Sep	Oct	Nov	Dec
High °F	37	40	50	61	71	80	85	84	76	65	54	42
Low °F	25	26	34	44	54	64	68	67	60	48	41	30

San Francisco	Jan	Feb	Mar	Apr	May	Jun	Jul	Aug	Oct	Sep	Nov	Dec
High °F	55	58	60	64	66	70	71	72	74	70	62	56
Low °F	41	45	45	47	48	52	54	55	55	51	47	42

11.8 A house in Pittsburgh designed for an outdoor temperature of 4 °F (the temperature is lower than this only ~1% of the time) has a heating load of 36 MJ h^{-1} for an inside temperature of 65 °F. Calculate the heating required (in kWh) for a year with 5500 degree-days (for a nominal temperature of 65 °F).

11.9 In a location with an average number of degree-days of 2000 for a base temperature of 16 °C, a house with a floor area of 100 m^2 has a heat loss per degree-day of 20 kWh d^{-1}. Calculate the heat loss per degree per square metre of floor area.

11.10 Assess the relative merits of the materials given in the following table from the point of view of heat storage in building construction.

	Melting point °C	Latent heat of fusion kJ kg^{-1}
Potassium fluoride tetrahydrate	18.5	231
Zinc nitrate hexahydrate	36.4	147
Butyl stearate	19	140
1-dodecanol	26	200

Source: Feldman et al. (1991). *Solar Energy Materials* 22, 231–42.

11.11 Describe the mitigation options that are available for buildings, and discuss those that are most suitable for where you live.

11.12 In 2012 the total space heating requirement for the ~26 million homes in the UK was 330 TWh. Find the average heat loss per degree of each home. Take the average number of degree-days as 2463, and the occupancy as 0.7.

11.13 What are the difficulties with retrofitting old houses to a good standard, and why is retrofitting important?

11.14 What techniques can be used to reduce the need for air-conditioning in a building?

11.15 In a certain region there are 10 million houses and each house has a 1 m^3 hot water tank. Each tank can be charged by a 5 kW air source heat pump with a coefficient of performance (COP) of 3 for 4 hrs in the middle of the day. (a) Estimate the amount of heat stored and the amount of electricity used? (b) How might this operation help in managing a grid with a high penetration of renewables?

11.16 Explain the advantages and disadvantages of heating buildings using renewably generated hydrogen.

11.17 Explain why it is particularly difficult to decarbonize industry.

11.18 What options are available to lower the carbon and energy intensity of manufactured goods?

11.19 Discuss the advantages for industry of generating hydrogen from wind power.

11.20 Urea is produced from the reaction of ammonia and carbon dioxide. The urea is formed in a 2-step reaction:

$$2NH_3 + CO_2 \rightleftarrows NH_2COONH_4 \text{ (ammonium carbamate)}$$

$$NH_2COONH_4 \rightleftarrows H_2O + NH_2CONH_2 \text{ (urea)}$$

For 160 Mt of ammonia, what is the tonnage of carbon dioxide required, and of urea produced?

11.21 Explain what power-to-gas means and how methane and ammonia might be produced in this process, and the advantages of doing so.

11.22 Explain why cement manufacture produces so much CO_2 emissions, and how these might be reduced.

11.23 Discuss the implications of the electrification of heat in industry in Europe for renewable energy production.

11.24 How many kg of carbon dioxide per MWh of heat must be captured and stored to decarbonize heat in the combustion of (a) hydrogen from steam reforming methane (b) methane. What are the implications for CCS?

(LHV: H_2 120 MJ kg^{-1}; CH_4 50 MJ kg^{-1})

11.25 Discuss the advantages of using wood as a construction material.

11.26 Describe how behavioural changes can reduce the emissions from transport.

11.27 Estimate the global savings in emissions from cars if all were electrified and charged from a supply with a carbon intensity of 0.05 kg CO_2 kWh^{-1}.

11.28 Estimate the difference in the number of kilometres per litre of petrol when driving a car at 108 km h^{-1} and 86.4 km h^{-1}.

11.29 Explain the principles of operation of a brushless DC motor and how the motor can be used in regenerative braking.

11.30 In the UK, the average electricity consumption per household is around 3700 kWh per year. Estimate the size of battery an EV would need to supply sufficient energy to cover a round-trip journey of 40 km, and also provide for the electricity needed by a home at night-time. Estimate the extra cost of this electricity for the home due to the cost of the battery and comment.

11.31 Discuss to what extent city planning can promote significant reductions in carbon emissions.

11.32 Discuss whether electrification of vehicles will occur to a significant extent by 2035.

11.33 Where might fuel cells have the biggest impact on decarbonizing transport?

 For further information and resources visit the online resources
www.oup.com/he/andrews_jelley4e

12 Energy and Society: Making the Transition from Fossil Fuels to Renewables

✔ List of Topics

- ☐ Impact of more CO_2
- ☐ Policies supporting renewables
- ☐ Carbon pricing
- ☐ Carbon abatement policies
- ☐ CO_2 emissions and the Kaya identity
- ☐ Lowering energy demand

- ☐ Decarbonizing energy supply
- ☐ Economics of global decarbonization
- ☐ Adaptation and geoengineering
- ☐ Current global situation
- ☐ Actions required

→ Introduction

The amount **of energy consumed per capita** by any country is closely related to its **standard of living** (see Fig.1.7 in Chapter 1). As developing countries have become more industrialized, and their populations have grown, their demand for energy and standard of living have increased. However, most of the world's energy is still predominantly produced from fossil fuels, and unless this situation changes very significantly over the next few decades, the projected increase of carbon dioxide in the atmosphere means that the world will experience dangerous climate change due to further global warming.

In this chapter we review:

- the impact of energy production on the environment and global warming, and the different measures to cap the cumulative emissions of carbon dioxide to limit climate change;

- the policies to stimulate the growth of low-carbon energy sources—notably renewable energy—after the oil crises of the 1970s;

- the **Kyoto Protocol** in 1997—the first international effort to tackle global warming and climate change—and why it failed to slow global emissions;

- the **Paris Agreement** in 2015, which introduced national targets, but so far is insufficient to bring CO_2 emissions to net zero by 2050 and hold warming to less than 1.5°C (the limit required to keep climate change in check).

We outline the actions that must be taken to limit climate change, in particular the importance of reducing the overall energy demand, in order to hasten the essential decarbonization of the energy supply. We explain why decarbonizing transport and heat are as important as decarbonizing electricity and summarize how this might be achieved by electrification and by power-to-gas technologies.

Kicking our addiction to fossil fuels and eliminating their subsidies presents enormous social and economic challenges. The global fossil fuel industry is vast and employs a huge number of people who will need to be employed elsewhere. We describe the interrelation between the standard of living of a country, its energy requirement, and the carbon emissions associated with the energy supply, via the **Kaya identity**.

Fortunately, wind and solar power are now widely at grid parity, and the technology for them to replace fossil fuels as the world's predominant source of energy is now cost-effective. However, in order to avoid the damaging climate change that current projections indicate, we need to speed up the transition from fossil fuel to renewables, and we identify some of the key actions and proposals, such as the **Green New Deal**.

12.1 The Likely Impact of Continued Fossil Fuel Emissions

The concentration of carbon dioxide in the Earth's atmosphere has risen sharply over the last 50 years. As explained in Chapter 1 Section 1.3, **water vapour** and **carbon dioxide** are the two main greenhouse gases that trap the infrared radiation emitted by the Earth and thereby raise the temperature of the Earth's surface. The current level (2019) of CO_2 is 410 ppmv, which compares with 315 ppmv in 1955 and 280 ppmv in the pre-industrial era before ~1750.

Over the last 50 years the concentration of CO_2 has increased by about 85 ppmv and the global mean temperature has risen by about 0.7°C; the two effects are strongly correlated with **anthropogenic** causes (human-induced) during this period. This warming is extremely fast on a geological time scale. Although some of the observed temperature changes (see Chapter 1 Fig. 1.10) are due to variations in solar irradiance and to natural phenomena such as volcanic activity, the rise over the period 1970–2019 can only be explained by also taking the anthro-pogenic impact on CO_2 concentration into account. The problem cannot be ignored, because carbon dioxide remains in the atmosphere for hundreds of years, and what we do about it over the next few decades will largely determine the quality of life for generations to come.

Furthermore, **global warming is accelerating**. Between 2015–2019, the average tempera-ture rose about 0.2°C, compared to the average for the five-year period 2011–2015. Global temperatures are now 1.1°C higher than in the pre-industrial period. (Over land, the cor-responding temperatures were 0.3°C and 1.7°C warmer.) Also, sea levels have risen more quickly, and there is a continuing decline in the extent of sea-ice in both the Arctic and Antarctic, in the amount of ice in glaciers, and in the Greenland and Antarctic ice sheets. (Land ice melt and thermal expansion dominate the sea-level rise).

Global warming is causing significant **climate change**, which is damaging to societies and the global economy. Severe heat waves have affected all continents, causing unprecedented wildfires in North America, Europe, Australia, and elsewhere (see Fig. 12.1), with the damage just from wildfires in California in 2018 estimated at $150 billion. Extreme rainfall events and associated floods were more frequent, notably in W. Europe and China in 2021. Tropical cyclones also caused significant damage, with the cost of **Hurricane Harvey** in 2017 alone estimated at >US$125 billion. Africa, in particular, was badly affected by droughts. The poor-est have been worst affected, and climate change has partly been responsible for flows of

Fig. 12.1 Bushfire in Queensland, Australia, 2019 (Getty Images/iStockphoto).

refugees that are destabilizing politics worldwide. The probability of many of these extreme weather events has been raised significantly by the increase in global warming.

An analysis of **ocean temperatures** has shown an increase of ~0.7°C in surface temperatures, which is consistent with the predictions of those climate models that include the anthropogenic increase in CO_2. Marine life has been affected: the population of walruses has dropped sharply off the coast of Alaska, affecting the livelihood of the Eskimos, and many species of plankton (an important part of the marine food chain) are moving to cooler waters. Ocean warming and acidification are killing large amounts of coral worldwide, and there has been a large loss of coral reef in the Caribbean off Puerto Rico, the Virgin Islands, and the Great Barrier Reef off Australia. **Biodiversity** has also been affected badly, with a million species under threat of extinction.

The burning of fossil fuels, forests, and crop wastes is causing massive pollution that is very damaging to health and affecting millions in cities. These pollutants can form **aerosols**—tiny liquid and solid particles suspended in the air—that can disperse worldwide. Their overall effect on the climate is to slow global warming by shading the Earth. However, it is expected that the increase in the level of aerosols is unlikely to match the rise in CO_2 because of emission controls, so the effects of CO_2 on temperature will be exacerbated. Aerosols also have a considerably shorter lifetime in the atmosphere than CO_2.

12.1.1 What Would a World 3 Degrees Warmer Be Like?

In their fifth assessment (AR5) in 2014, the IPCC found that by the end of the twenty-first century, if we carried on as we are (the so-called a *business-as-usual scenario*) the temperature rise by 2100 relative to the average from year 1861–1880 would be likely to be about **4°C** (IPCC-AR5).

To avoid this frightening outcome, the **Paris COP 21** meeting of nations in 2015 agreed that emissions must be reduced to limit global warming to **2°C and if possible 1.5°C**, and pledged to make reductions (COP21). However, the unconditional nationally determined contributions (NDCs) so far pledged would only likely limit global warming to about **3.2°C** by 2100 (UN2019), and would still lead to wider-ranging and more destructive impacts, and risk the possibility of catastrophic climate change.

In this scenario significant combustion of fossil fuels would continue. As a result, the pollution and heat in many cities would become unbearable, with equatorial regions most badly affected. Wildfires and logging would cause extensive deforestation by 2100, and coral reefs would also have vanished. Global warming would be exacerbated by there no longer being any ice in the Arctic in the summer, and there would be extreme and widespread flooding in many coastal cities across the globe, including many of our largest cities like London and New York. Heat waves and droughts would become intolerable, causing mass migrations, triggering widespread refugee problems and political instability. Water availability would be badly affected by melting glaciers and by shrinking rivers and water tables, and some developing countries would experience chronic food shortages. Food production would have been hurt by shifting climate zones, higher temperatures, and extreme weather, with yields dropping by about 3-7% per degree rise, depending on the type of crop (wheat, rice, maize, and soybean) and region. Tropical diseases would also be more widespread and there would be mass extinctions of species.

Even more concerning is that the warming could have passed a threshold beyond which the changes become irreversible and extremely dangerous to life on Earth.

12.1.2 Could the Global Climate Reach a Tipping Point?

When a small change in a driver of climate change (e.g. global mean temperature) causes a disproportionate response in the climate system, then the climate is said to be at a **tipping point**. If the event is self-reinforcing and causes global temperatures to continue rising even as emissions are reduced, then it could lead to a **Hothouse Earth** (HE2018). In this case, the warming is predicted to eventually stabilize at around 4–5°C above pre-industrial temperatures with sea levels 10–60 m higher than today—conditions that would undermine the ability to feed a large global population.

An example of such a tipping point, estimated to be above 5°C, would be if the global warming released **methane** frozen under the Siberian permafrost at a significant rate. Methane is roughly 25 times as potent a greenhouse gas as CO_2, so the release of methane would speed up global warming, and this in turn would cause further methane release. This positive feedback effect could lead to a dangerously fast rise in global temperatures that could continue even with no fossil fuel emissions. Another climate driver is the clearance of the **Amazon rain forest** for farming. This is expected to exhibit a tipping point when warming is around 3–5°C, when a large area could convert to dry Savannah and release a vast amount of CO_2. There could be a cascade of these and other tipping points as global warming gathers pace, increasing the risk of a Hothouse Earth. This possibility strongly strengthens the case for trying to limit global warming to 2°C at most, to at least minimize this risk.

12.2 Economic Policies to Promote Renewable Energy

The 1970s saw a number of huge hikes in oil prices by OPEC, which caused Western governments to develop renewable energy sources and nuclear power, and to look for other sources of fossil fuels, e.g. North Sea oil exploration. The initial aim was to reduce dependency on OPEC controlled oil and provide energy security. Later, it became seen as a means of combating the threat of global warming and climate change, by reducing carbon emissions. In the early years, renewable energy generators (apart from hydropower) were more expensive than fossil fuel plants, and required subsidies in order to compete. A range of economic initiatives were proposed to encourage the development of renewable energy and reduce the dependence on fossil fuels. We now describe these policies and international efforts to curb emissions, before discussing how we can decarbonize our energy supply and reduce demand, in order to hasten the transition to net-zero emissions.

12.2.1 Feed-in Tariffs (FITs)

A **feed-in tariff** is a guaranteed price that a producer of renewable electricity will receive over a long period, and is set at a level such that the producer makes a reasonable profit. The long-term period of the tariff reduces the risk for investors and the extra cost of production over that for fossil fuel generation is usually shared by all the consumers of electricity in the province or country. Germany (with its 'Energiewende' policy), Denmark, Spain and the United States pioneered the creation of markets for renewables, which led to advances in technology and economies of scale. The manufacture and installation of renewables is now dominated by China. The level of a FIT can be based on the levelized cost and will therefore be different for different technologies. The value may also depend on the size of the plant, to encourage both centralized and distributed generators. It can also be revised to take into account decreasing costs arising from the learning effect. FITs have been used in many countries and have been successful in promoting renewable energy. Also, the set-up costs of the scheme are relatively low.

12.2.2 Tradeable Green Certificates (TGCs)

A **Tradeable Green Certificate** is a device that requires suppliers of electricity to show that they have obtained a specified fraction of renewable energy. Typically, there is one TGC per MWh of generated electricity (though this can vary to encourage less developed technologies) and suppliers can either buy them directly from a renewable generating company or on the open market. The extra source of revenue enables the renewable energy companies to compete with fossil fuel companies, and the cost to the suppliers is passed on to the consumers. Market forces determine the price of a TGC and investment goes to the most competitive companies. Furthermore, the increase in production as the fraction of renewable energy rises helps to reduce electricity prices as the technologies improve through the learning effect (see Chapter 1 Section1.5.8). However, in practice, the price volatility of TGCs has deterred investors, and FITs have achieved a larger deployment of renewable electricity generators at a lower cost than have TGC schemes.

12.2.3 Renewable Energy Auctions

By around 2015, the cost of electricity generated from wind and solar farms in several parts of the world was starting to approach that from fossil fuels. Further falls in cost in the following five years have meant that wind and solar power are now at **grid parity** in many countries, with some producing electricity cheaper than that from existing fossil fuel plants. Auctions are now becoming an increasingly popular way of promoting renewable energy deployment. In these auctions, generator companies submit a bid with a price per kilowatt-hour (kWh) for a certain amount of electricity. The competitive nature of these auctions brings out the real price of generation, which is often difficult for a regulator to determine. Moreover, the revenue for the successful company is guaranteed for a period, which makes it attractive for investors. These auctions have led to significant cost reductions over the last few years.

12.2.4 Contracts for Difference (CfD)

Another mechanism, which is used in the UK to aid investment in renewable energy, is to ensure that generators receive an agreed price per kWh for a fixed period, e.g. 15 years. If the market price is lower that this 'strike 'price then the generator receives the difference, but if higher, the generator rather than the customer pays. This way the customer is protected from high prices while the generator is assured of a steady income, which encourages investment. Generator companies bid for CfDs in an auction, with the lowest bids based on the strike price—the amount per MWh the generator wants to be paid—ranked highest. The scheme is funded by a levy on all electricity suppliers.

12.3 How to Put a Price on Carbon Emissions

Besides the need to promote renewables, it was recognized by the 1990s that emissions of carbon dioxide must be curbed in order to limit global warming and some economists began to promote the idea of imposing a price on carbon. They argued that without a price on carbon there would be little incentive to reduce emissions and that the principle of polluter pays should apply. This means that the party responsible for the emissions should pay for the damage done to the environment and to human health. The effective price would be that which causes the emissions to fall fast enough.

Two different systems have been proposed to put a price on carbon: by governments imposing a levy on carbon emissions (i.e. a carbon tax) or by a quota system, known as cap-and-trade (or emissions trading).

A **carbon tax** gives a company an incentive to reduce its carbon emissions wherever the cost of reduction is less than the tax. However, if the tax is too low then there is no reduction, and if it is too high then costs can seriously affect productivity and prices. A carbon tax can also affect the less well-off adversely, but this can be offset by directing the tax revenue to those most disadvantaged.

The alternative **cap-and-trade** system sets a limit on emissions and distributes permits that allow firms to emit amounts that total up to this limit. These permits can be obtained either through auctions, initial allocations, or trading with other firms, since some firms can reduce emissions more easily than others. The price is determined by demand, rising when an economy grows and falling in a recession.

Initially, permits were often given out free, called **grandfathering**. This is politically more acceptable, since firms then only have to pay for extra permits, but governments miss out on initial revenues that could promote green energy. A better system of allocation of permits is one that ensures scarcity and a high carbon price, so reductions really do occur. To get over the volatility of the price of permits caused by the demand varying, a floor or ceiling price can be introduced. On the other hand, taxes generate immediate revenue for governments, and can provide an incentive to reduce independent of demand. The level of a tax can be adjusted to try and give the reduction in emissions required. Trading appeals to industry as there is the possibility of profits. Caps on emissions also tend to be easier to agree than taxes, but a long-term trading period is needed to reduce market risk. Setting up the scheme can prove difficult, since it requires records to be kept and emissions monitored.

For those activities for which no low-carbon solution currently exists (such as flying long distances that cannot be avoided), **offsetting** is a transitional solution, whereby the carbon emissions are compensated for by financing a reduction in emissions elsewhere. This is best done by paying into an independently certified scheme that captures carbon dioxide, such as through enabling effective and sustainable reforestation, or displaces it through renewable generation. Offsetting has been criticized for allowing continued fossil fuel use when a low-carbon substitute is possible, and in particular when the offsetting scheme is not additional; i.e. the carbon reductions would have taken place in any case. Offsetting more, say double, than what is emitted in certified schemes should provide some certainty of reduction while emissions are still significant.

Governments can also impose **regulations** that reduce carbon emissions, such as:

- raising **building standards**, e.g. increasing the minimum amount of insulation in buildings, and thereby reducing the use of fossil fuels for heating
- placing **fuel efficiency** (mpg or km per litre) requirements on vehicles.

The usual market forces provide an incentive to meet the standard as cost-effectively as possible, but there is no market incentive to innovate and go beyond the standard. Hence, regulation alone is unlikely to find the most cost-effective way of reducing carbon emissions, but together with price mechanisms it can be very useful. It can also encourage the deployment, and research into reducing the cost, of efficient technologies.

12.4 **Carbon Abatement Policies**

By the 1980s, the world was starting to wake up to the dangers of continued emissions of CO_2 from fossil fuel combustion. One of the first initiatives to curb carbon dioxide emissions was the introduction of a carbon tax, which is simpler to set up than an emissions trading scheme (see Section 12.3). The tax (collected by the government) is levied on all fuels that emit CO_2, and the level is based on the carbon intensity of the fuel, so coal would have a higher tax per kWh than gas. The scheme was introduced in Finland, the Netherlands, Norway, Sweden, and Denmark in the 1990s, and provided an inducement for everyone to reduce emissions, and was applied to transport, domestic consumers, and industrial consumers. It has been successful in the Nordic countries where low carbon energy is available. However, it potentially damages the competitiveness of those countries with high emissions per unit of output, which was one reason why the United States favoured emissions trading rather than a carbon tax, and a carbon trading scheme was adopted in the negotiations set up by the United Nations (UN) in the 1990s.

12.4.1 **Why the Kyoto Protocol Was Ineffective**

Scientific concern over global warming grew in the 1980s, and in 1988 the **Intergovernmental Panel on Climate Change (IPCC)** was founded by the World Meteorological Organisation (WMO) and the UN. The first United Nations Earth Summit was held in Rio de Janeiro, Brazil, in 1992, where the United Nations Framework Convention on Climate Change (UNFCCC) was adopted. A series of annual meetings, called the **Conference of Parties (COP)**, was also set up to support the climate change negotiations. The Framework stated a goal of stabilizing the concentration of greenhouse gases at a level that would prevent dangerous climate change, though the level was not specified. It acknowledged that the priority for developing countries was economic and social development and that developed countries should take the lead in reducing emissions. This led to the **Kyoto Protocol** at COP3 in 1997. Under this protocol, an emissions trading scheme was set up among industrialized countries, who agreed to reduce their emissions of greenhouse gases by 5.2% for the period 2008–2012 compared with those of 1990. **China, India, and other developing countries were not required to reduce emissions**, as it was recognized that the bulk of the increased levels of CO_2 in the atmosphere had been produced by the developed nations. It was also agreed that the rich developed countries should help fund the efforts of poorer countries to adapt to climate change.

Offsetting was also allowed, whereby countries received credit for any CO_2 sinks that they developed, such as creating a forest, or for work on carbon abatement in developing countries where costs tend to be lowest. A criticism of this clean development mechanism (CDM) was that a significant number of the projects were 'non-additional'. Another important criticism was over manufactured items and the assigning of carbon emissions to the country producing the goods rather than to the country importing them. This has resulted in several developed countries having lower emissions than they are actually responsible for!

The Kyoto protocol came into force in February 2005. Over 160 countries ratified the agreement, with the notable exception of the United States, which was concerned that the reduction in emissions would be damaging to its economy, and felt that the protocol should have included targets for developing countries as well as developed countries. While the small reductions agreed by industrialized countries under the Kyoto Protocol were largely met, the agreement failed to slow the growth in global emissions, which were dominated by those arising from the huge expansion in the economies of developing nations, all of which were exempted, in particular China. In 2000, China emitted about 3.35 Gt of CO_2, compared with 6.0 Gt by the United States, but by 2010 China had increased its emissions to 8.5 Gt, whereas those of the United States had decreased to 5.7 Gt. A scheme for emissions trading was also set up by the European Union in 2005. However, an initial over-allocation of credits resulted in a surplus, and as a result a low carbon price and little impact on emissions. The overall impact of the Kyoto Protocol was disappointing, with global CO_2 emissions rising from 25.2 Gt y^{-1} in 2000 to 35.0 Gt y^{-1} by 2013.

It was hoped that the UNFCCC meeting in Copenhagen in 2009 would produce an agreement to follow on the Kyoto Protocol which ended in 2012, thereby establishing a further international and more effective commitment to reduce emissions. However, the **Copenhagen Accord** failed to produce any agreement on emissions targets, and only managed to resolve that, on the basis of the scientific evidence for global warming, the world should strive to keep the temperature rise since pre-industrial times to below 2°C—a target chosen to avoid the worst impact of climate change that had been first suggested in 1990. It was also the subject of a misinformation campaign by **climate change deniers**. After the Copenhagen meeting there was little further progress, in part because of the global economic crisis in 2008. No significant progress was made until **COP21 in Paris in 2015**, resulting in the much-hailed **Paris Agreement**. *But even this failed to halt the growth in emissions.*

12.4.2 **Why Progress on Curbing CO_2 Emissions Has Been so Slow**

Although the science was clear by 1990 that continued emissions would damage the environment, all countries wanted economic growth which was rooted in fossil fuel technologies. This led to protracted negotiations following the Rio meeting and only a small reduction in emissions for developed and none for developing countries in the Kyoto Protocol. Moreover, there were many debates following the Kyoto meeting as to the scientific basis for attributing global warming to human caused emissions, with the **oil industry lobbying hard to deny the link**.

With more frequent extreme weather events and the accumulation of scientific evidence, in particular the IPPC 4th assessment in 2007, more people began to realize the dangers, but governments were still reluctant to act.

Wind and solar power were still more expensive than fossil fuels, and energy was vital for the expansion of developing nations' economies and raising their standard of living. Moreover, these countries felt that they were not responsible for the bulk of emissions, and that the developed nations, often much less vulnerable to climate change impacts, should bear the brunt of the cuts. But these rich countries did not want to damage their economic competitiveness and growth by investing in reducing emissions since clean alternatives were expensive. Climate change still seemed to most governments to be a distant threat that did not warrant an immediate response. Balancing the responsibility for the present generation against that for future generations has been a particular challenge, and the desire for economic growth has been at the expense of the environment.

Attempts to decarbonize have also been hindered by governments trying to help their citizens by **subsidizing fossil fuels** (and fossil fuel companies), even though the G20 countries, which account for 85% of the world's wealth (gross GDP), have made repeated pledges that these subsidies should be phased out. Many are consumer subsidies that reduce prices through government price controls on fuels; these are popular policies as they can make, for example, car travel or cooking cheaper. But often only the well-off can afford the fuels, and these subsidies divert government funds from helping the poor and from spending on other priorities such as education and health. Such subsidies also make it harder for low-carbon technologies to compete.

12.4.3 **The Paris Agreement in 2015 and the Revised Target of 1.5°C**

By COP21 in Paris in December 2015, attitudes to decarbonization were changing. This time, to avoid negotiations faltering yet again on the vexed question of a fair allocation of the reductions in emissions between countries, it was proposed that **each country should determine its own contribution**. The argument was that it was in their interest as well as others to do so: decarbonizing energy was increasingly more economic and would create new jobs, efficient transportation, and more habitable cities with less air pollution (see Fig. 12.2), as well as reducing the risk of dangerous climate change.

This was the basis of the Paris Agreement in which nations reaffirmed the **goal of keeping global warming below 2°C, while also moving towards 1.5°C**, and agreed two emission goals: a peaking of emissions as soon as possible, and a carbon neutrality, where carbon emissions are balanced by carbon sinks, in the latter half of this century. The agreement came into force in November 2016. With the US re-joining in 2021, all the countries with high emissions, in particular China, the US, and India, are now committed to the agreement.

Fig. 12.2 The air pollution in Delhi before and during the COVID-19 shutdown in 2020 (Photo by SAJJAD HUSSAIN/AFP via Getty Images).

However, the intended nationally determined contributions (NDCs) proposed so far—even if all implemented—have been estimated to only limit global warming to about **3.2°C by 2100** (UN2019). A further report in 2018 from the IPCC concluded that the effects of 2°C warming would be significantly worse than 1.5°C with, for instance, the frequency of extreme heat waves, and the loss of plants, vertebrates, insects, marine fisheries, and crop yields, all two to three times greater, and significantly worsen the risks of floods, extreme heat, droughts and poverty for hundreds of millions of people. The report also put a limit on the emissions of carbon dioxide compatible with keeping within 1.5°C warming.

As explained in Chapter 2 Section 2.3.4, the temperature rise in global warming is given to a good approximation by the **cumulative emissions** of carbon dioxide. The IPCC 1.5°C paper in 2018, after allowing for some non-CO_2 effects, estimated a **limit of 580 gigatonne on further emissions of CO_2 after 2017 to restrict global warming to 1.5°C**, with a 50% probability. There would then be about a 90% chance of lying within the range 1.2 to 1.9°C. The spread arises due to uncertainties in the response of the carbon cycle and the climate to increases in the concentration of greenhouse gases, and in the effects of non-CO_2 greenhouse gases and aerosols.

In particular, **clouds** play an important role in affecting the radiation balance: they cause a significant cooling, relative to no cloud cover, by reflecting sunlight that is partly offset by warming from the absorption of terrestrial infrared radiation. The changes in cloud properties that accompany global warming decrease the overall cooling from clouds, but the uncertainty is large and is a significant one in determining the climate sensitivity, and hence in the carbon budget available to limit a certain amount of global warming.

The global warming was already at 1.1°C in 2017 and, if we carry on as we are (emitting about 37 gigatonne of CO_2 a year from fossil fuels alone), we will probably exceed 1.5°C by 2035, only a short time away. We need to reach net-zero emissions by around 2050, and for this to happen, countries must ratchet up their contributions, and emissions must fall significantly over the next decade.

12.5 Estimating Future CO_2 Emissions: The Kaya Identity

The total emissions of CO_2 each year depends on the world's annual energy demand, and on the amount of CO_2 emitted per unit of energy produced, called the **carbon intensity,** with **coal-fired plants** the worst polluters at about one kilogram of CO_2 per kWh. Projections for the total energy demand depend in particular on how the global economy grows in the next few decades. The world's wealth (i.e. before the COVID-19 pandemic), as measured by the gross domestic product (GDP) of all countries, is increasing at around 3% per year, mostly in the developing world, where it reflects an increase in population and a rise in the standard of living that is correlated with the energy consumption per capita in these countries (see the **Human Development Index**, Chapter 1 Section1.2): as their GDP grows, their energy requirement also increases. The amount of energy required per unit increase in GDP (its **energy intensity**) reduces with time, through a shift from industrial to service economies, such as seen in China and India, and improvements in the efficiency of processes, and is currently decreasing worldwide at an average rate of 1.8%. The net result is that energy consumption in 2050 would be expected to be about 40% higher than that in 2015.

The dependence of global emissions on energy consumption and carbon intensity is summarized concisely in the **Kaya identity**:

$$\text{CO}_2\text{ emissions} = \text{population} \times \frac{\text{GDP}}{\text{population}} \times \frac{\text{energy}}{\text{GDP}} \times \frac{\text{CO}_2\text{ emissions}}{\text{energy}}$$

$$= \text{population} \times \text{affluence} \times \text{energy intensity} \times \text{carbon intensity}$$

Limiting population, which is helped by the increasing urbanization in the world, family planning programmes, and providing education and empowering women, will lower the rise in emissions and the pressure on resources. The global birth rate is falling and is now close to two per family, but there are currently more young than old people and the UN estimates that the population will increase from 7.6 billion in 2019 to about 9.7 billion by 2050, peaking around the end of this century at nearly 11 billion (though the effects of the COVID-19 pandemic have not been accounted for). Attempts to control population through limiting family sizes raises ethical and social concerns. Also, the personal wealth or **affluence** of citizens is increasing as the standard of living improves throughout the world. Attention has therefore focused on **energy intensity** (energy/GDP) and **carbon intensity** (CO$_2$ emissions/energy). Energy intensity is improved by energy efficiency and energy demand reduction, and carbon intensity by energy supply switching to lower-carbon sources of energy (e.g. coal to gas or renewables).

The commitments made by countries in Paris in 2015 to reduce global warming if all were enacted, together with current and planned policies, would mean that about 70%, rather than 80%, of the global energy supply would *still* be from fossil fuels in 2050, with coal use being much reduced. But this change in fossil fuel consumption, when taking account of the estimated rise in energy demand, would leave carbon dioxide emissions much the same as now. The cumulative amount of CO$_2$ emitted in the interval 2017–2050 would be more than 1200 gigatonne, and have likely taken global warming close to 2°C.

What is required is to reduce net emissions to zero by about 2050, and to achieve this we need first to identify the main sources of carbon emissions.

12.5.1 Current Global Energy-related CO$_2$ Emissions

In 2016 the primary energy consumption of fossil fuels was close to 471 EJ, contributing 82 % of the primary energy supply of 576 EJ. The carbon intensity of oil, coal, and gas are ~0.06–0.11 kgCO$_2$ per MJ, with an average of ~0.075 GtCO$_2$ per EJ of heat. For electricity consumption, the average carbon intensity is ~0.6 kgCO$_2$ per kWh, which is equivalent to ~0.16 GtCO$_2$ per EJ of electricity. The energy-related CO$_2$ emissions come from the combustion of fossil fuels, so we can estimate the global emissions as 471 × 0.075 ≈ 35 GtCO$_2$, close to the actual emissions of ~36 GtCO$_2$ in 2016.

In Chapter 11 we looked at how the building, industry, and transport sectors accounted for about 28%, 39%, and 22%, respectively, of the energy-related CO$_2$ emissions in 2016, corresponding to 30%, 29%, and 29%, respectively of the final energy demand. The shares of emissions and of final energy demand only changed slightly over the period 2010–2016 with the demand in transport ending up more than a per cent and that in buildings and industry each down about one per cent. The percentage of the world's primary energy demand supplied by fossil fuels in 2010 and 2016 was also within 0.5%, while fossil fuel's share of electricity generation fell slightly from 67.5% to 65.7%, reflecting that consumption of fossil fuels as well as renewable energy grew during this period (see Chapter 1 Fig. 1.6).

We can now draw up a table of how the emissions are approximately distributed across the various sectors, and also determine the final energy demand in 2016. The electricity consumption was about 22,000 TWh or 80 EJ, with approximately half in the building sector. The final energy demand is the primary consumption less that lost in conversion of heat to electricity and in other processes. The result is shown in Table 12.1, with estimates of the demand and emissions to nearest 5 EJ, and 0.5 $GtCO_2$, respectively. The final energy demand was ~400 EJ, about three quarters of the primary energy consumption of 576 EJ. Of the ~175 EJ of energy not used, most was rejected in the generation of electricity.

Biomass used in buildings includes traditional biomass in the developing world for heating and cooking which amounted to about 7% of final energy. In the final energy demand about 10% is non-energy (or embedded energy) like plastics, asphalt and lubricants. The industry, buildings, and transport sectors account for around 90% of the demand and emissions, while other activities, which include agriculture, account for the other 10%. The total final energy consumption (TFEC), which is the final energy used and not stored, was about 360 EJ, or 100,000 TWh, in 2016. We can see that the decarbonization of heat is just as important as that of electricity in the building and industry sectors, since their associated emissions are both

Table 12.1 Approximate global energy-related emissions and final energy demand in 2016

Sector	$GtCO_2$	EJ	% CO_2	% Final
Industry (heat)	5.5	75	37	30
Industry (electricity)	5.5	35		
Industry (biomass)		10		
Industry (process)	2.5			
Transport (heat)	8.5	110	23	29
Transport (biofuel)		5		
Buildings (heat)	4.0	50	29	30
Buildings (electricity)	6.5	40		
Buildings (biomass)		30		
Other demand (heat)	3.0	40	11	11
Other demand (electricity)	1	5		
Totals	36.5	400	100	100

significant. Also, transport is primarily powered by fossil fuels. These three sectors dominate the final energy demand.

In all sectors the future demand for energy can be very significantly reduced by making efficiency savings, either through **improved conversion efficiency**, which refers to improvements in technology (e.g. LEDs and electric cars rather than incandescent bulbs or petrol cars), or **savings in providing services,** which refers to reducing the demand in passive devices, for instance the demand for heat to make a building comfortably warm, or through **behavioural changes**, for example: taking public transport rather than driving a car.

We now describe methods for lowering the energy demand, which will reduce energy intensity, before discussing decarbonization, which will improve carbon intensity. Improvements in the energy intensity can be used as a proxy for those brought about by efficiency savings, but will not account for such changes in GDP caused by, for instance, an economy becoming more reliant on services and commerce rather than on heavy industry.

12.6 **Lowering Energy Demand**

There is great scope for decreasing energy usage in buildings, transport, and industry, as discussed in Chapter 11. The provision of energy to perform a service like maintaining a building at a comfortable temperature involves a number of steps. First, the primary supply of energy (a fuel or a renewable energy source) has to be converted into more useful forms. Fuel refineries, thermal power stations, electric motors, heat pumps, gas burners, compressors, and internal combustion engines are all examples of **energy conversion devices**. A chain of these devices, such as a fuel refinery, thermal power station, and electric motor, provide the end-use energy, which is mainly in the form of motion, heat, or light. Then this energy can be used to run machines, like cars or washing machines, or to heat furnaces or ovens, or in general **energy consuming units.**

Considerable savings can be made in the amounts of energy these energy consuming units use, for instance in the number of gallons of petrol per kilometre for a car, or in the number of megajoules per square metre of floor space per year for keeping a building warm. These are **passive systems** since energy is not converted but used, and generally ends up as low-grade heat or as embodied potential or chemical energy. A study (UCamb2011) on the practical upper limits of reducing energy demand in transport, industry, and buildings concluded that around 73% of global energy could be saved through practical design changes that reduced losses, and mechanical and thermal inertia in the energy consuming units. The largest savings of **83% were in buildings, 63% in transport, and 62% in industry**.

There is also considerable scope for improving the efficiency of **conversion devices**; often, by ensuring that they are not over-designed in capacity and work away from their optimal efficiency point. Many are fossil fuel powered, and in these there are also significant losses occurring in heat transfer and in combustion (Chapter 2 Section 2.6). The University of Cambridge study (UCamb2011) estimated that improvements could reduce losses in conversion devices by 56%.

Such significant conversion improvements may well be more easily made by changing from combustion engines and hot gases to electric motors and heat pumps, powered by electricity from wind and solar farms, to provide motion and heat. Electric motors and heat pumps can have efficiencies some three times higher than those of internal combustion engines and direct combustion heating, respectively. The conversion of wind and solar radiation to electricity by wind turbines and commercial silicon solar cells to electricity is also very efficient: they are now typically about 85% and 75% of their theoretical maximum, the Betz and Shockley-Queisser limits (see Chapter 7 Section 7.5 and Chapter 8 Derivation 8.1, respectively. (Their carbon intensity is also markedly lower.)

12.6.1 Economic Considerations

Combining the possible efficiency improvements in conversion devices with the savings that could be made in passive systems would result in a total reduction in demand of about 85% (UCamb2011). While this is a practical upper limit, it shows the enormous potential gains from demand reduction. Economic considerations will limit some measures (see Section 12.8) but there are many that not only can save energy but can be cost-effective; for instance, improving home insulation or switching to renewables. But there can also be hidden costs to improving efficiency, e.g. the cost of redecoration after installing cavity wall insulation, and the disruption to production while installing new technology.

Lack of capital in the developing world can also hinder making improvements to existing systems. Another effect that can reduce savings, probably more in the developed than developing world, is what is called the **rebound effect**. This refers to a situation where the lower costs of energy services arising through efficiency improvements release spending money that is used for additional activities or objects whose associated emissions (e_a) offset those saved through the improved efficiency (e_s); when $e_a > e_s$ is referred to as the '**Jevons paradox**'.

For example, improving the efficiency of a ship's engines would result in a reduction in emissions provided the total mileage remained the same. But the emissions would rise if the reduction in the cost of transport opened up new markets that increased the ship's total mileage significantly. However, it would appear in practice that efficiency gains are not entirely reversed by the rebound effect, and, while difficult to assess, are probably typically reduced by between 5% to 40%. Therefore, there is the potential for efficiency and energy savings to help considerably in meeting an increasing global demand for energy, or for the same energy supply to provide more services.

12.6.2 Energy Intensity Improvement

The International Renewable Energy Agency has estimated that the **synergy that electrification with renewables brings of greater efficiency** in several areas, notably in transport (electric vehicles) and heating (heat pumps), together with the actions described above for reducing the energy demand, could help achieve a greater fall in global energy intensity of around 2.8% per year (as well as a fall in carbon intensity) than the current 1.8% per year (NB if the demand were reduced by 85% in the 30-year period 2020–2050, this would correspond to a 6% annual reduction in energy demand). This would keep the world's energy demand in the industrial, building, and transport sectors, roughly constant at about 100,000 TWh a year.

However, even if the commitments made by countries in the Paris meeting in 2015 were achieved, **most of our energy in 2050 would still be from fossil fuels**, so it is imperative to decarbonize our energy supply quickly, as we now discuss.

12.7 Decarbonizing the Global Energy Supply

Several organizations, from fossil fuel companies like BP or Shell to the International Energy Agency (IEA) or the International Renewable Energy Agency (IRENA), provide projections, called *scenarios*, of the growth in the total amount of anthropogenic greenhouse gas emissions under certain assumptions affecting these emissions. These can include the population, economic activity, lifestyle, land use, technology, and climate policy over the period of the projection. A baseline or **business-as-usual (BAU) scenario** is one in which no additional efforts are made to limit emissions, and whereby following the Paris Agreement global warming would be around 3.2°C by 2100.

In one of the International Energy Agency (IEA) scenarios, the **sustainability development scenario (SDS)**, which aims to address limiting global warming, curtailing air pollution, and giving universal access to energy, three of the UN sustainability goals, energy efficiency (44%) and renewables (36%) account for 80% of emissions reduction compared with BAU by 2040. These changes would mean an estimated 1.6 billion fewer premature deaths from pollution, and over a billion more people with access to electricity and clean cooking.

Unfortunately, in this IEA scenario emissions would only be halved by 2040, the world still reliant on fossil fuels, in particular natural gas, and it would need substantial negative emissions technology after 2050 in order to limit warming to close to 1.5°C. Even in their **faster transitions scenario (FTS)**, emissions would not reach net-zero until ~2060, and it assumes that CCS would be used extensively by industry to capture emissions to limit warming to 1.5°C. However, reliance on CCS and negative emissions technologies is very risky since they have not been deployed at all at scale, and it also delays the transition away from fossil fuels. And it is why they are only used to a relatively small extent to account for 10% of emissions reduction in the scenario discussed in this section. So, what could the world do to **achieve net-zero emissions by 2050 and limit warming to about 1.5°C?**

Essentially, there are various different combinations of low-carbon technologies and demand limitations that would give the same outcome. The prescription below is based on an **IRENA (2018) scenario** with the TFEC equal to 100,000 TWh. The contribution from biomass remains much the same as now but with modern rather than traditional sources. Biomass is attractive as it can be stored and is therefore a dispatchable source that can be used to help balance supply and demand. However, as we saw in Chapter 4, there are concerns over the sustainability of biomass, and the large increase in renewable supply required comes from wind and solar PV.

As discussed in the relevant technology chapters, the growth of wind and solar power has been such that their total installed capacity could produce 60,000 TWh per year by 2050

Table 12.2 Accessible potentials by 2050 of low-carbon sources of electricity

Source	Potential (TWh y^{-1})	Source	Potential (TWh y^{-1})
CCS[*]	10,000	Solar PV	40,000
Hydro	7000	Solar CSP[***]	5000
Marine[**]	1000	Biomass	15,000
Geothermal	1800	Nuclear	5000
Wind	20,000	**Total**	**104,800**

[*]industrial and air capture; [**]tidal and wave; [***]includes solar thermal

with massive investment. Together with the production from biomass, hydropower, solar CSP, geothermal power, and marine power, the total output from renewable supplies could be ~90,000 TWh per year by 2050. Nuclear power by then could generate about 5000 TWh per year and fossil fuel combustion another 10,000 TWh, with the emissions captured either directly or by air capture to give a total of ~105,000 TWh per year from low-carbon supplies. The accessible potentials by 2050 of all the low-carbon technologies are given in Table 12.2.

In 2015, the fraction of energy from renewables was about 18%, mostly from bioenergy and hydropower. By 2037 this will need to be close to 60%, and the fraction from fossil fuels down to about 35%, to be on track for zero emissions by 2050, by which time around 85% of the energy demand should be from renewables, 5% from nuclear, with carbon capture removing the remaining emissions from fossil fuel use. Lowering the carbon intensity can be achieved by fossil fuel switching to a fuel with lower emissions, for instance coal to gas or to wind or solar, or using CCS and nuclear power.

Figure 12.3 shows how the contribution of renewables to the global final energy consumption might look in 2037 for the world to be on course. By 2037, fossil fuel consumption is limited to oil for transport and gas for heating and some power, with coal no longer used. Under this scenario, emissions would be about 9 GtCO$_2$ y^{-1}. If this target is not reached until 2060,

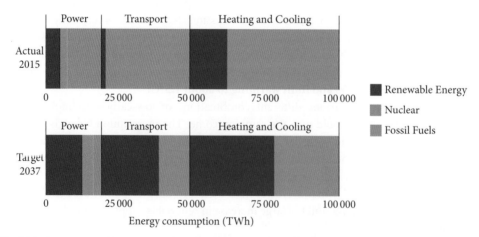

Fig. 12.3 The increase by 2037 in renewable energy to be on target to limit global warming to 1.5 °C by 2050. Power—for machines, appliances, and lighting—was supplied in 2015 by about 80% of the electricity produced; the remaining 20% was used on heating and cooling.

with zero emissions by 2090, then the warming will be limited to about 2°C, but the consequences of climate change would be significantly worse.

The growth in electric vehicles, in heat pumps, and in power-to-gas doubles the electricity demand to 50,000 TWh a year by 2037. The higher efficiency of electric motors keeps the transport demand about the same. The use of traditional biomass in the developing world decreases significantly, with solar panels providing access to electricity for cooking and lighting, but modern biomass for heating and transport has increased its share, and biomass (traditional and modern) provides some 15,000 TWh per year, mostly as heat, much the same as in 2015.

Nuclear and hydropower produce some more electricity, 4000 and 6000 TWh y^{-1}, respectively, but by far the largest increase is in wind to 10,000 TWh y^{-1} and solar PV to 20,000 TWh y^{-1}, which all together will supply some 80% of the target electricity consumption of 50,000 TWh in 2037. The deployment of wind and solar farms will be helped by the rapidly falling cost of battery storage, which will help manage the variability in their supply. The remaining electricity generation could come principally from other renewables and gas power plants. This would need an aggressive expansion of renewable generation, which, if maintained, would enable some 75,000 TWh y^{-1} of electricity generation by 2050 and enable there to be net-zero emissions by then.

Besides the build-up of technology to decarbonize heat, **carbon capture and storage** particularly for some industrial processes will be required, as it will for some gas-fired generators used for balancing electricity grids. The technology is developed but must be promoted, possibly through some combination of subsidies, regulation, and a carbon price, as it is not currently at all widely used (see Chapter 3 Section 3.7). But it is not just capture at the production site, the scale needed will also require air capture, as a very real danger exists that CO_2 emissions will not drop fast enough to restrict warming to 1.5°C.

12.7.1 Nature-based Solutions and Negative Emissions Technologies

Nature-based Solutions (NbS) can provide both mitigation and adaption by absorbing and reducing emissions of CO_2, and by giving protection against extreme weather. These actions can also improve biodiversity and wellbeing. Of an estimated 10 $GtCO_2$ y^{-1} mitigation from NbS (all at a cost of <$100 tCO_2^{-1}, with a third < $10 tCO_2^{-1}), protecting and managing lands could each contribute 4 $GtCO_2$ y^{-1}, while restoring forests 2 $GtCO_2$ y^{-1}. These actions could reduce global warming by 0.3°C by 2075, but only if combined with rapid decarbonization, otherwise rising temperatures will cause ecosystems to be net emitters. After 2050 some of the NbS saturate.

Planting more trees may appear one of the easiest and cheapest ways of removing carbon dioxide from the atmosphere, and it has been estimated that globally 900 million hectares (Mha) are available, but the first priority is to halt the huge annual loss of trees. Already about 10 million hectares of forest are being cleared each year for soy, palm oil and other crops, and also for cattle grazing. This loss causes about a tenth of the annual global CO_2 emissions and a significant loss of biodiversity. Moreover, the area of trees required to sequester a significant amount of CO_2 is huge— about a quarter of the area of the United States (~250 Mha) for

200 GtCO$_2$ when the trees are mature. After this time the trees would need to be replaced, with the timber used in buildings, for example. It has been suggested that the forestry residues could be burnt to produce energy (heat or electricity) and the emitted CO$_2$ captured and stored. This **bioenergy carbon capture (BECC)** is controversial, and care would have to be taken to ensure that any change in land-use resulted in a net decrease in emissions. Moreover, it is undeveloped and could be in competition with other demands for arable land and fresh water.

Climate change is also increasing the risk of forest wildfires. It is possible, though, to capture CO$_2$ directly from the air using chemical absorbers, called **direct air-capture (DAC)**, which is a much more compact and certain process than using biomass, but it is currently expensive. Origen Power is looking to reduce costs by combining carbon capture with the production of lime, which has a commercial value. Another method, under development by Carbon Engineering, uses potassium hydroxide (KOH), which forms potassium carbonate (K$_2$CO$_3$) on contact with carbon dioxide. Lime (Ca(OH)$_2$) is used to regenerate the potassium hydroxide and form calcium carbonate (CaCO$_3$), which is heated to release the carbon dioxide for compression and storage—in this latter process the lime is regenerated (see Fig. 12.4). They estimate that the cost of capturing the CO$_2$ could be as low as \$100 per tonne. The reactions involved are:

$$CO_2 + 2KOH \rightarrow K_2CO_3 + H_2O; \quad K_2CO_3 + Ca(OH)_2 \rightarrow 2KOH + CaCO_3$$

$$CaCO_3 \rightarrow CaO + CO_2; \quad CaO + H_2O \rightarrow Ca(OH)_2$$

Fig. 12.4 Design of Carbon Engineering air contactor that would collectively capture about a million tons of CO$_2$ per year, equivalent to the annual emissions of 250,000 cars (R. de Richter et al. Progress in Energy and Combustion Science 60 (2017) 68–96).

Another DAC method, already in operation, is that of Climeworks, where the carbon dioxide is absorbed from the air by a selective filter and is released when the filter is heated to 80–100°C. The pure CO_2 is mixed with water and pumped underground where it reacts with hot basaltic rocks and turns into stone over a few years. The heated water is piped back to the surface where it is used to heat the selective filters. Another company, Global Thermostat, which uses an amine-base sorbent to capture the CO_2 from the air in their pilot plants, is working with the oil giant ExxonMobil to help scale up their technology.

Rather than sequester the CO_2, the carbon dioxide can be transformed into a fuel or compound by reacting it with hydrogen. Recently a group at Oxford University reported a high yield of aviation jet fuel using inexpensive iron-based catalysts; using green hydrogen, the fuel would be carbon neutral. Another method being developed by the Karlsruhe Institute of Technology in Germany is to capture CO_2 from the atmosphere and convert it into carbon black, for tyre production.

As the emitted CO_2 is dispersed throughout the atmosphere, air-capture plants could be located anywhere, and be close to cheap low-carbon sources of energy, as in very windy regions or locations with high irradiance. Other ways under investigation are: enhancing the weathering of magnesium silicate rocks such as olivine by mining it and spreading it finely in mainly humid regions. The weathering transforms CO_2 into bicarbonate and rain washes this into rivers that remove it to the oceans, where it eventually precipitates as carbonate; burying biochar (though the amount of land that can be used for biomass and biochar production is uncertain); fertilizing parts of the oceans with nutrients to increase CO_2 uptake– in particular adding iron, but undesirable side effects are a concern; and increasing the alkalinity of the oceans to increase their CO_2 uptake and offset ocean acidification. The challenge with all these methods is both scale and cost, and there can be risks to the environment. Moreover, we must avoid the **moral hazard** of relaxing mitigation efforts, as the failure of negative emission technologies (NETs) to deliver would affect those most vulnerable to climate change the most.

12.8 **Economics of Global Decarbonization**

Countries need to generate energy cleanly, change technologies where necessary, and use this energy to provide all the services required. The cost of this transition will depend on the costs of generation and of switching technologies, particularly in industry, and on the amount of energy needed. Estimates of the current levelized cost of generation in the USA are given in Fig.12.5.

The **marginal cost** is the cost of fuel plus the maintenance and operations (M&O) costs. It is dominated by that for fuel for conventional plants, but by the M&O cost for renewable and nuclear plants, as there are no fuel costs for renewables and the cost of nuclear fuel is relatively small. It is defined as the cost of increasing the supply of electricity by one MWh from an existing generator. The high end of the LCOE from coal plants incorporates the cost of carbon capture at 90%, but without the costs of transportation and storage of the CO_2.

Offshore wind costs have been falling sharply in recent years and look like they will be competitive with conventional generation by the mid-2020s. Without storage, wind and solar

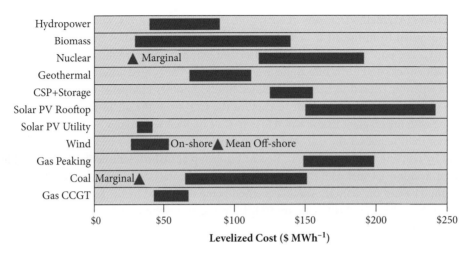

Fig. 12.5 USA levelized cost of energy (unsubsidized) *Source*: Lazard (2019); IRENA for hydropower and biomass LCOE. (CCGT: combined cycle gas turbine; CSP: concentrated solar power; marginal: marginal cost of generation).

PV lack the dispatchability of conventional generators, but their lower costs compared to gas peaking plants mean that, with batteries, they are increasingly cost competitive for meeting short duration spikes in demand, as the cost of batteries (mainly Li-ion) fall. We can see that onshore wind and utility scale solar PV, which became cost-competitive with new conventional power plants in many countries several years ago, are now becoming competitive with the marginal cost of such plants, and they could provide the world's total energy demand.

The total final energy demand will be affected by population, economic growth, policies, and lifestyle factors. The transition will also cause job losses in the fossil fuel industries. But there will be many new jobs in the renewable industries. This social cost and upheaval will need to be addressed through financial support and retraining. The idea of a '**just transition**'—that this cost should be borne by governments rather than the individual fossil-fuel workers—was recognized in the Paris Agreement.

The renewable energy companies worldwide will require the investment of many billions of dollars (some of which could come from divesting from fossil fuel companies). However, the total cost of continuing our dependence on fossil fuels would be many times more than that required for renewables, due to the effect on health and the damage to the environment from climate change. These externalities need to be fully appreciated and costed by imposing a carbon price. In addition to a price on carbon emissions, the large subsidies for fossil fuels need to be removed.

12.8.1 **Carbon Price of Mitigation Policies**

In most countries there are mitigation policies that subsidize low-carbon or energy efficiency technologies or restrict high carbon ones, e.g. grants for EVs or home insulation, or fuel efficiency standards. As these actions are reducing the amount of carbon dioxide emissions, their costs can be expressed in terms of pounds or dollars per tonne of carbon dioxide saved. For example, consider a new hydroelectric power plant generating electricity at 3p kWh^{-1}, compared with building a coal-fired plant providing electricity at 7p kWh^{-1} and with emissions of 1 $kgCO_2$ kWh^{-1}. Then the saving, sometimes called a negative cost, would be £40 per $tonneCO_2$. Likewise, insulating a building at a cost of £500 could save 25,000 kWh of heat over the lifetime of the building, derived from burning gas that emits 0.25 $kgCO_2$ kWh^{-1} at a cost of 4p kWh^{-1}. This would translate, neglecting discounting, to a cost of minus £80 per $tonneCO_2$. On the other hand, a new onshore wind farm generating at 5p kWh^{-1} compared to an existing coal-fired power plant where the capital cost has been repaid and the price of electricity only reflects that of coal and is 3p kWh^{-1}, the cost would be £20 per tonne CO_2. Any cost savings though can cause behavioural changes that offset some of the gains—the rebound effect. For instance, some of the savings on heating from improving insulation can be reduced by keeping homes warmer.

Lazard has estimated that policies aimed at encouraging wind and solar PV could be particularly cost-effective at reducing CO_2 emissions with \$36–\$41 per ton vs. coal and \$23–\$32 per ton vs. gas CCGT (neglecting any integration costs) (Lazard2019). These **abatement costs** are what have been called the short-term, or static, costs of reducing CO_2, as they just cover the lifetime of the project. They can also be called the **cost neutral carbon price** since they are the price of CO_2 per tonne that, when multiplied by the lifetime savings in CO_2 in tonne, equals the cost of the project. But there are also long-term, or dynamic, costs, as the effects of the project can outlive its lifetime. In particular, the reduction in climate change damage caused by the saving in CO_2 emissions is long-term and is referred to as the **social cost of carbon;** it is expressed in terms of the present value of the damage per tonne of CO_2. This is particularly difficult to quantify since it involves the economic effects of climate damage, the cost of adaptation to the changing climate, the value placed on damages to future generations, i.e. the appropriate discount rate to apply, and because the damage per tonne of CO_2 will increase as cumulative emissions rise. Also, there are health benefits from the improvement in air-quality from less CO_2 emissions that are not included in the social cost, and these can be substantial and galvanize action as they are immediate.

The Obama administration in the USA calculated the social cost of carbon as \$46 per ton of CO_2 in 2017. The power generating technologies that have a static abatement cost less than that value have been estimated recently (Gillingham 2018) to be onshore wind, natural gas combined cycle, utility scale photovoltaic, and new natural gas with 90 per cent carbon capture and storage (though it was noted that the CCS cost estimate may be optimistic). Putting a price on carbon equal to the social cost of carbon would be economically the best course of action, as it would promote actions that mitigated the damages from carbon emissions most cheaply. However, this can be difficult politically due to strong vested interests in fossil fuels, and there are also considerable uncertainties over the true social cost.

A recent study (Ricke2018) put the global social cost of carbon lying much higher between about \$200 to \$800 and most likely close to \$400 per $tonCO_2$. Further global warming places developing countries at a disadvantage since they are generally hotter than developed countries, so higher temperatures affect them more, particularly from poor agricultural yields and flooding. The study found that India would be most affected, with a social cost of \$90 per ton CO_2. And though the percentage effect on GDP is less in the USA than in India as it is cooler, its high GDP (it has the world's highest) made the USA next at \$50 per ton CO_2, which argues that it would be in the USA's long-term interest to curb emissions.

But is not just the long-term social or health costs that should be considered. There can be long-term economic benefits from a new technology that in the short term is expensive, but with the learning and economies of scale from mass production can prove very competitive. A notable example is the government and state support for the manufacture of solar panels in Germany, California, and China in the 2000s that led to massive cost reductions, and to solar PV now being competitive with fossil fuel generation in many parts of the world. On the other hand, in the short term, fuel switching from coal to natural gas may be attractive, but it is an investment that locks in fossil fuel infrastructure, reduces the expansion and potential cost reduction of renewables, and neglects the social cost of carbon. Subsidizing EV charging stations may appear costly but encourages EV sales that in turn boosts battery demand. Subsequent battery price reductions from mass production make solar and wind farms with battery storage more competitive with conventional dispatchable generation, which promotes renewables. Hence, a subsidy that is costly in the short term can lead to very significant long-term, or dynamic, cost savings.

A wide range of carbon prices have been estimated for keeping global warming to within 1.5°C, with a median carbon price of $105 per tCO_2 in 2020 rising to $145 per tCO_2 in 2030. Research suggests that a carbon price needs to be complemented with stringent regulatory policies such as minimum performance standards and building codes to be most cost effective. And to avoid 'carbon leakage', i.e. when carbon-intensive manufacturing moves to a country with lower emissions standards, the EU proposed in 2020 imposing a **carbon border tax** on imports, unless major emitting countries agreed to a carbon pricing floor.

12.8.2 Investment Required for the Transition

Under a BAU scenario, the International Renewable Energy Agency (IRENA2019) has estimated that investments in the global energy sector will total 95 trillion by 2050. They estimate that making the transition to low-carbon technologies would require an additional USD $15 trillion, equivalent to an average of 2% of global GDP per year over the period. Investment would shift away from fossil fuels towards renewables, and related infrastructure, and energy efficiency. The rapid fall in the cost of renewables and in electrification solutions have helped to lower the transition cost, but the additional investment would be needed early on, particularly in the next decade.

Energy sector subsidies totalled at least USD $605 billion in 2015 and are forecast to increase to USD $850 billion annually under BAU. But in the transition to low-carbon, they would decline to USD $470 billion per year by 2050 away from fossil fuels to renewables and technologies mainly in the transport and industry sectors. The additional energy sector costs for the transition are estimated to be some three to seven times less than the total cost of the environmental and health damages and fossil fuel subsidies under BAU.

The transition promises health and GDP benefits and the creation of more jobs in renewables, energy efficiency, and flexibility than those lost in fossil fuel industries. These are not likely to be geographically aligned nor occur at the same time, and some other jobs could decline. Policies will be needed to address these problems to gain wide support for the transition, that is to make it a just transition. Furthermore, funds to alleviate the impacts in the less wealthy countries must be available to ensure investment in low-carbon technology.

12.9 **Adaptation and Geo-engineering**

However, whether the world will manage to achieve this goal without emitting a dangerous amount of CO_2 is far from clear. As we have already incurred adverse climate change and may well exceed our emission targets, we will have to work on **adaptation** as well as on mitigation, and should also carry out research on geo-engineering as a means of reducing global warming.

12.9.1 **Adaptation**

There is already a need to adapt due to the emissions we have already made with global temperatures on average higher, flooding more severe, and wildfires more frequent for a start. Adapting to such climate impacts also becomes more expensive the longer it is put off since their severity increases with more emissions. Cutting emissions has been seen rightly as essential and adaptation often viewed as a distraction and sometimes as a capitulation to inevitable climate change. But with more extreme weather, building resilience to the effects of climate change is now realized as vital.

Some adaptations can worsen global warming, for instance, air-conditioning using electricity from coal-fired power plants, but others can have several benefits, such as planting coastal mangroves that buffer floods, mitigate by absorbing carbon dioxide, and increase biodiversity (see Fig.12.6). Mangroves are an example of a **blue carbon**, rather than a **green carbon**, store, which are those in coastal or marine, rather than in land, ecosystems. Coastal communities are most at risk of flooding, but high upfront costs in adaptation, which can

Fig. 12.6 The mangrove coast of Card Sound, Florida, surrounding South Biscayne Bay.

involve a managed retreat to higher ground or building flood defences, can be offset by long-term economic, environmental, and safety improvements; but can also make for difficult choices as to what and who should be protected.

Other examples of adaptation are switching to drought-resistant crops, creating early warning systems for extreme weather, planning for power outages, and changing to buildings designed to stay cool during the hotter summers that are expected from climate change. Financing these projects can be difficult in poorer countries, and working out the best strategy with limited funds presents a huge challenge.

12.9.2 Geoengineering

Geoengineering can be defined as an action that tries to reduce the effects on the global climate caused by increased levels of CO_2 by ways other than by removing the CO_2, e.g. by increasing the albedo of the planet. One method would be to inject **sulphate aerosols** into the upper atmosphere, which would reflect some of the sunlight and thereby cool the Earth down. Alternatively, the **reflectivity of land surfaces** could be increased by making areas white, or **space reflectors** could be deployed. However, such actions are likely to cause significant changes in precipitation and temperature in some regions and would not reduce the acidification of the oceans. But besides the need to reduce costs, many geo-engineering schemes affect large regions, and there are governance issues that would have to be resolved.

Air-capture of CO_2 is often included as an example of geoengineering, but it has been discussed separately in Section 12.7.1 as an example of a negative emissions technology, since it tackles the root cause of global warming and not just its symptoms. For this reason, these technologies are generally to be favoured.

12.10 Current Global Situation

Huge investment in renewables is occurring in the world's major economies of China, European Union, USA, and India, with China expected to account for 40% of the increase in renewables in the five-year period 2019–2024. There has also been considerable growth in the USA, despite the negative attitude of the Trump Administration, through the initiatives of States, cities and businesses. Wind and solar farms are the fastest growing sector, and are putting a strain on transmission grids. Many grids are being upgraded or expanded; for instance, in 2018 four new interconnectors across Europe received funding to aid in the integration of more renewable power.

The deployment of renewables has been slow in Russia, where there is an abundance of gas and oil. However, some growth has occurred in Japan, which has reduced its reliance on nuclear power following the Fukushima accident, but is still quite dependent on coal. Investment in Latin America where good hydropower and biomass resources exist, is increasing fast, but Brazil and, in particular, Venezuela, have considerable oil resources which may slow their transition to renewable generation. In the developing countries of the world, investment in renewables is growing rapidly, and distributed generation with solar panels is helping millions gain access to electricity (see Fig. 12.7).

Fig. 12.7 India, Rajasthan, Jodhpur, rural houses with solar panels (Tibor Bognar/Alamy Stock Photo).

What is happening in the countries with the largest emissions is what matters most, and the CO_2 emissions by country in 2018 were (in order): China (28%), United States (15%), India (7%), Russian Federation (5%), Japan (3%), Germany (2%), Iran (2%), South Korea (2%), Saudi Arabia (2%), Indonesia (2%), Canada (2%), Mexico (1%), South Africa (1%), Brazil (1%), Turkey (1%), Australia (1%), UK (1%), Poland (1%), Italy (1%), France (1%), and the rest of the world 21%. **China, the US, and India account for 50% of the global emissions**. The IEA reported that in 2019 emissions flattened after two years of growth, helped by the expansion in advanced economies of wind and solar PV, fuel switching from coal to gas and an increase in nuclear power. The average carbon intensity of electricity generation in these countries fell by almost 6.4% in 2019 to 0.34 $kgCO_2$ per kWh. The US saw the largest fall in emissions of any country, equivalent to a drop of 2.7%, while emissions in the EU (including the UK) dropped by 5.7%. Japan's also fell 4.9% as it restarted nuclear reactors following the Fukushima accident. Emissions outside the advanced economies grew mainly in Asia, where coal demand continued to expand to provide over 50% of energy use and accounted for around 10 $GtCO_2$ (~30%) of global emissions. Even so, simply flattening CO_2 emissions is not enough.

12.11 Actions Required to Contain Temperature Rise

The world must switch from fossil fuels to low-carbon sources of energy, in particular renewables, by 2050. This will be an enormous challenge that will be helped by reducing energy demand as much as possible, while still increasing access to energy across the world. The global standard of living (as measured by the average GDP per capita) has improved by a factor of 2.8 in the last 50 years through the increase in energy supply. However, that energy has predominantly come from the combustion of fossil fuels and has led to global warming and climate change that has put the world's population and environment in great peril.

The global **COVID-19 pandemic** that swept the world in 2020 has caused a devasting loss of life and will affect the world's economy profoundly. It has reduced globalization and hopefully countries will look to renewables, as the world recovers, to provide jobs and the energy security they require. But the sharp fall in economic activity will make the transition more difficult to achieve as funds will be needed to address the plight of the present generation. Nonetheless, climate change is an existential threat, so we must speed up decarbonizing our energy supply and reducing fossil fuels as well as consumption as quickly as we can.

12.11.1 **Reducing Fossil Fuels**

One of the major immediate challenges is to cut the use of coal, the burning of which causes a third of the world's CO_2 emissions. The capacity of coal-fired power plants almost doubled during the period 2000–2019 to 2000 GW, largely because the Chinese economy grew so quickly. China has now half the world's total capacity, with the USA and India accounting for another quarter. Coal use also expanded quickly in India, but in both China and India growth is slowing as renewables become cheaper, and due to concern over air pollution and climate change. However, new coal-fired plants associated with China's '**Belt and Road Initiative**' are a concern. Coal demand in the USA and Europe is declining (though not quickly in Germany, partly because of its phasing-out of nuclear power), and emissions from coal burning look like they have almost peaked. However, globally they are not falling fast enough to meet the 1.5°C target: **there should be no new coal plants built and existing ones should be phased out more quickly**. Oil demand must also fall; and this is beginning to happen with the switch from petrol and diesel cars and vans (which account for about 10% of global CO_2 emissions) to electric vehicles; as must the demand for gas, which would be helped by halting fracking. And the pace of the change in technology to provide heat cleanly through power-to-gas and heat pumps in industry and in buildings must accelerate greatly; the increasing development of water electrolysis using electricity from renewables for hydrogen production is encouraging.

The speed of the transition to renewables has been greatly hindered by **climate change deniers** in positions of influence, notably in the US. This denial that climate change is caused by the emissions of carbon dioxide has been promoted by vested interests in the status quo, in particular by the fossil fuel industry. Continued fossil fuel extraction is already being justified by relying on carbon capture for a significant fraction of global CO_2 emissions. But this would be very unwise, since the technology for carbon capture is not established at that scale, nor would it generally be as cheap as using more renewables; it may well be only cost effective for selected processes and remove only 10% of global emissions. For example, in aviation United Airlines announced in 2020 that it would achieve carbon-neutrality by 2050 through investing in direct air-capture technology.

Many billions of dollars are tied up in companies that own fossil fuel deposits that should be left underground, and in 2020 the major oil companies BP and Shell announced net-zero targets by 2050. These '**stranded assets**' will become worthless, as will many fossil fuel power plants. Not only that, many jobs that are dependent on the combustion of fossil fuels will be lost, and one way to address this problem might be to use some of the revenues from a carbon price to support people made unemployed by the transition to a low-carbon economy. Currently a **price on carbon emissions** is put via several schemes, either through carbon taxes or through emission trading schemes. In the European Union scheme, political pressure has led to too low a price (per tonne of CO_2 emitted) for it to put a brake on fossil fuels; as a result, the UK introduced a minimum carbon price that is helping cause the early closure of its coal-fired plants (by 2025). In the US, some states (and the federal government under President Biden) are taking the initiative, with schemes already in place in the Northeast and in California. China plans to introduce a large trading scheme that should help in lowering its emissions, and in its shift to renewables.

In addition to a price on carbon emissions, the **large subsidies for fossil fuels** need to be removed. The global annual health costs alone caused by burning fossil fuels has been estimated as at least six times these subsidies, and shows just how important is speeding up the transition to renewables. Climate related litigation puts pressure on carbon-emitting companies and states, and globally there has been a significant increase in such litigation. Although lawsuits have argued, so far unsuccessfully, that fossil fuel companies should pay for climate change related damages, in December 2019, the supreme court in the Netherlands ruled that the state has a duty of care to protect its citizens from climate change under its human rights obligations. Many companies now acknowledge the threat of climate change and have pledged to reach net-zero by 2050. But it is important that plans are realistic, open to scrutiny, and have interim milestones that are not reliant on unproven technologies. Also having investment linked to such targets would incentivize action.

12.11.2 Reducing Consumption

Finding ways to reduce consumption and the demand for energy is essential, since they decrease the rate required for decarbonizing the power supply. Since 1970 the global population has doubled and annual carbon dioxide emissions have increased by 250%; the world's wealth (GDP) has also grown by a factor of 4.5. While the average standard of living has increased through this use of fossil fuels, it has led to a massive depletion and deterioration of the world's resources. In the period 1970–2014, populations of vertebrates declined by 60% on average, with around one million species of plants and animals now at risk of extinction, and vast areas of forests lost. Also, severe pollution in the oceans and the atmosphere, and a precipitous fall in the number of insects resulting from intensive farming (in particular pesticides) and global warming, threatens a catastrophic loss in biodiversity.

Built-in obsolescence, with many products thrown away and not reused, has created massive waste (notably plastics). It is vital that we move away from consumerism to a more sustainable lifestyle, encouraging a circular economy with recycling and reuse. Emissions from land clearances must cease, and significant reforestation must occur. And limiting population, which is helped in particular by providing education and empowering women, will lower emissions and the pressure on resources.

12.11.3 **Practical Actions to Promote Decarbonization**

States, cities and individuals can contribute enormously through promoting the use of **renewables**, and are already doing so; e.g. in California and New York State, San Diego, Jaipur, Hamburg, Toronto, Bangalore, and in many other places. Electrification of city transport (buses and trains) and of cars and bicycles will speed the transition, as will companies and cities that commit to reducing their carbon footprint. And changing modes of transporting goods, with road and air to rail and shipping, would also help. By 2021, over 800 cities, 100 regions, and 1500 companies had pledged net-zero emissions.

Individuals can opt to receive their electricity and gas from renewable energy companies, or generate their own renewable energy through rooftop solar PV and solar thermal panels. They can also reduce their energy consumption and take public transport, or hire electric vehicles. For those activities for which no low-carbon solution currently exists, such as flying long distances, and which cannot be avoided, **offsetting** the carbon emissions is a transitional solution. It is not a substitute, though, for making all possible reductions in personal emissions. **Community involvement** can lead to much greater acceptance of wind and solar farms; and movements such as the 'Sunrise Movement' can inspire people to take action on climate change, as has Greta Thunberg and her advocacy of student strikes.

Energy policy must urgently facilitate a huge and rapid transition to renewables from fossil fuels, an upheaval akin to going to war. A very ambitious proposal in the US is for a massive investment in decarbonizing the energy supply within a decade, with the jobs created used to tackle inequality. It has been called the '**Green New Deal**' from its similarities to the economic stimulus of Roosevelt's 'New Deal' in the 1930s. In December 2019, the European Union announced a plan with similar objectives, which is similarly called the '**Green Deal**', with the aim of no net emissions by 2050, and where economic growth is decoupled from resource use and the environment is protected. Any such plan needs to be a long-term commitment to produce results and to give confidence to investors. Pay-back may be slow and may be in conflict with private companies' short-term profits for shareholders, in which case strong regulation or state involvement is required. The economist Marianna Mazzucato argues for a '**Mission Economy**' where capitalism is more stakeholder rather than shareholder orientated, and where governments actively shape markets to promote the transition.

Non-CO_2 emissions, e.g. methane from livestock, can be reduced by using vegetable or cultured protein. **Eating less meat** would reduce the land area required for food and enable ecosystems to be restored, and help in capturing carbon dioxide through reforestation. And in agriculture, conservation tillage, in which seeds are planted without replMoving the land, can reduce CO_2 emissions significantly. **Labelling food and consumables** with their carbon footprint (though it can be difficult to take account of wide variations in production methods) would help make people aware that some products have high associated emissions, and would help in 'greening' consumption.

The consequences of climate change are already being felt around the world; for example, in some coastal areas where the risk of flooding has markedly increased and in some tropical regions where the risk of extreme heat waves is much higher. Investment in **adaptation** must occur in these regions, but only emphasizes the importance of **mitigation**. Schemes designed to reduce global warming, such as through artificially increasing atmospheric aerosols, are also being considered, but much more research is needed to assess whether the benefits of **geoengineering** outweigh the risks from unintended extreme weather, and they must not distract from curbing CO_2 emissions.

Individuals must discuss with family, friends, and colleagues the importance of supporting renewables, reducing fossil fuels, and the need for urgent action; and citizen's climate assemblies can help develop effective policies.. The increase in extreme weather events, such as wildfires and severe flooding, is making people realize that much more concerted action is needed. By the beginning of 2021, over 100 countries—responsible for about two-thirds of greenhouse gas emissions—had announced plans to cut their emissions to "net zero" by mid-century. The countries included the UK, Japan, South Korea, the EU, China (by 2060); and the US intends to under the new President Joe Biden, who announced in 2021 an interim target of a 50% reduction in GHG emissions from 2005 levels by 2030. These commitments will help reduce the cost of capital for green investments.

We already have the technologies to solve the problem of global warming: power can be decarbonized by renewables, and buildings, transport and industry by electrification and using green hydrogen, but we need to inject more impetus for their deployment. Significant progress by 2030 is crucial. Renewables are now affordable as the costs of electricity from wind and solar farms have dropped dramatically (halving in some places within the last few years), and their deployment is being helped by the rapidly falling cost of battery storage; also, electrification brings big energy savings. A huge investment—which will also generate many new jobs—will be required, but nothing like the cost of inaction. And divesting from fossil fuels to renewables is fast becoming the most economic, as well as sustainable, choice.

As the world recovers from the COVID-19 pandemic, its energy supply must become low carbon to ensure that the world can have power that does not damage our environment, or cause climate change that would profoundly affect our lives.

 SUMMARY

- Global CO_2 emissions from burning fossil fuels threatens dangerous climate change, and the damage will be considerably mitigated if warming can be limited to 1.5°C above pre-industrial levels

- Climate change has already led to more frequent extreme weather events and loss of biodiversity

- Subsidies for renewables, in particular feed-in-tariffs, led to increased production and massive cost reductions through the learning effect and economies of scale, such that wind and solar PV power are now at or below grid-parity in many parts of the world

- Global attempts at carbon abatement have been ineffective, with global carbon dioxide emissions rising by 40% in the period 2000–2013. Nations recognized in the Paris Agreement in 2015 that warming must be held to 1.5°C above pre-industrial levels

- But national contributions to cutting CO_2 emissions are currently insufficient to reach net zero by 2050. It will require a massive world-wide effort and investment, with very significant reductions in energy demand coupled with decarbonizing both electricity and heat

- Progress on decarbonizing power is encouraging but the pace of the transition to low carbon technologies must accelerate sharply in the industry, transport, and building sectors

- Imposing a carbon price together with tougher regulations—plus removing the large fossil fuel subsidies—will help the transition to low-carbon technologies

- Governments need to give impetus to the transition, but states, cities, communities, and individuals can all make significant contributions

 FURTHER READING

Affordable and clean energy: *www.undp.org/content/undp/en/home/sustainable-development-goals/*.

Bach, L.T. et al. CO_2 removal with enhanced weathering and Ocean Alkalinity enhancement, Front. Clim.(2019), *doi.org/10.3389/fclim.2019.00007*.

Figueres, C. and Rivett-Carnac, T. *The Future We Choose* (Manilla Press 2020).

Gillingham, K. and Stock, J.H. The cost of reducing greenhouse gas emissions, Journal of Economic Perspectives 32 (2018) 53; *pubs.aeaweb.org/doi/pdf/10.1257/jep.32.4.53* (Gillingham 2018).

Goodall, C. *What we need to do now* (Profile Books 2020).

Haszeldine, R. S. et al. Negative emissions technologies and carbon capture and storage to achieve the Paris Agreement commitments. *Phil. Trans. R. Soc. A* 376 (2018): *dx.doi.org/10.1098/rsta.2016.0447*.

Hove, A. Current direction for renewable energy in China, Oxford Institute for Energy Studies (2020); *www.oxfordenergy.org/wpcms/wp-content/uploads/2020/06/Current-direction-for-renewable-energy-in-China.pdf*.

IEA World Energy Outlook 2019; *www.iea.org/reports/world-energy-outlook-2019*.

IRENA, *Global Energy Transformation: A roadmap to 2050* (International Renewable Energy Agency): *www.irena.org/publications* (IRENA2018 and IRENA2019).

Jelley, N. *Renewable Energy: A Very Short Introduction* (Oxford University Press 2020).

Mazzucato, M. *Mission Economy – A Moonshot Guide to Changing Capitalism* (Allen Lane, 2021).

REN21 Renewables 2019 Global Status Report; *www.ren21.net/gsr-2019/*

Ricke, K. et al. Country-level social cost of carbon, Nature Climate Change (2018) *pinguet.free.fr/ricke918.pdf* (Ricke 2018).

Special report on Global Warming of 1.5°C, Intergovernmental Panel on Climate Change, 2018: *research.un.org/en/climate-change/reports*.

Standard of living, the human development index: *hdr.undp.org/en/content/human-development-index-hdi*.

Steffen, W. et al. Trajectories of the Earth System in the Anthropocene (2018): *www.pnas.org/* (HE2018).

Twidell, J. and Weir, T. (2015). *Renewable energy resources*, 3rd ed. Taylor & Francis, London.

UN emissions gap report 2019: *www.unenvironment.org/resources/emissions-gap-report-2019* (UN2019).

https://ourworldindata.org/ Very useful data on Energy and Environment.

https://ukcop26.org/ 26th UN Climate Change Conference, COP26, Glasgow, 2021.

https://unfccc.int/climate-action/race-to-zero-campaign Global campaign to encourage net-zero carbon emissions by 2050 in support of the Paris Agreement.

https://www.ipcc.ch/assessment-report/ar6/ IPCC Sixth Assessment Report (2021/2022). Unequivocal that human influence has warmed the atmosphere, ocean and land.

pubs.acs.org/doi/abs/10.1021/es102641n Reducing energy demand: what are the practical limits? Cullen, J.R. et al. (UofCamb2011).

www.c2es.org/international/negotiations/cop21-paris/summary (COP21).

www.ipcc.ch/report/ar5/ IPCC's Fifth Assessment Report (2013–2014) (IPPC-AR5).

www.lazard.com Levelized cost of energy version 13.0 (Lazard2019).

EXERCISES

12.1 Discuss the evidence that global warming is accelerating.

12.2 Explain three possible global catastrophes if nothing is done to stop global warming.

12.3* Show that a concentration of 405 ppmv corresponds to ~3250 Gt of CO_2 in the atmosphere.

12.4 Give two examples of effects that could amplify the rise in global temperatures arising from an increased level of CO_2 in the atmosphere, and two that could be 'tipping points'.

12.5 What economic policies have been adopted to boost the development of renewable energy and discourage the use of fossil fuels. To what extent have these been successful?

12.6 Why was the Kyoto Protocol ineffective?

12.7 Are there some positive outcomes for climate change coming out of the impact of the COVID-19 pandemic?

12.8 Describe the Kaya identity and explain why the carbon intensity and energy intensity are focused upon.

12.9 Why is electrification so important for improving energy intensity in many sectors?

12.10 Is it necessary for a government to adopt a BANANA (Build Absolutely Nothing Anywhere Near Anyone) policy to counter the NIMBY (Not In My Back Yard) attitude that a project can engender?

12.11 Discuss the relative merits of regulation, carbon emissions trading, carbon taxes, and feed-in tariffs as ways of reducing carbon emissions.

12.12 Show that a reduction of about 1 Gt y^{-1} of carbon could be achieved by (a) improving fuel efficiency from 30 mpg to 60 mpg for an annual mileage of 10,000 miles and assuming there are 2×10^9 cars globally, (b) substituting about 1400 GWe of coal-based generating capacity with gas-based power (assume a plant efficiency of 50% and that a plant of 33% efficiency produces 1 kg of CO_2 per kWh from coal and 0.5 kg of CO_2 per kWh from gas), (c) placing wind turbines over about 40 million hectares, and (d) planting about 400 million hectares of land with cellulose-based crops.

12.13 Explain why energy savings are crucial in combating global warming, and why there can be difficulties in reducing carbon emissions through conservation measures.

12.14 The IEA predicted that 75% of global energy will still come from fossil fuels in 2030. Would a policy of requiring energy providers to sequester a small but rising fraction of any related emissions be an effective mitigation strategy? Would increasing the captured fraction by 1% for every 10 Gt of carbon emitted keep the temperature rise below 2°C?

12.15 Successive energy conversions result in 100 units of input fuel energy delivering 10 units of energy output. Explain why end-use efficiency improvements are particularly important.

12.16 To what extent should emerging economies be expected to comply with limits on environmentally harmful gases compared with richer economies?

12.17 What are the main actions that should be taken to reduce carbon emissions in the building, industry, and transport sectors?

12.18 How does one balance the responsibility for future generations with the (energy) needs of the present?

12.19 In many countries there is a strong seasonal demand for heat for building. Why does this make electrifying heating using heat pumps a challenge? Would using renewable electricity to generate hydrogen be a better strategy?

12.20 Does the expected decrease through the learning effect of renewable energy costs make power-to-gas a better mitigation strategy than CCS?

12.21 Estimate what global warming would be limited to if zero CO_2 emissions was only reached by 2075? Investigate whether then using negative emission technology could lower warming to 1.5°C.

12.22 Write a *critical* account of the prospects of geo-engineering as a solution to global warming.

12.23 Comment critically on the following statements:

(a) The risks associated with burning fossil fuels are much greater than those associated with nuclear power.
(b) It is essential to develop carbon capture technology on a large scale to tackle global warming.
(c) Money should be spent on mitigating the effects of global warming, e.g. by building dykes or relocating people away from low-lying lands, rather than on preventing global warming.

(d) The development of low-cost energy storage is the key challenge in providing sustainable low-carbon energy.

12.24 Discuss the implications of the LCOE values given in Fig. 12.5 for energy policy in the USA.

12.25 Discuss the importance of establishing a global carbon price to tackle climate change.

12.26 Is the Green New Deal an effective policy to enable a fast transition away from fossil fuels to renewables?

12.27 What mitigation strategy would you advise your government to adopt? Give reasons.

12.28 Assess the potential for hydrogen from water electrolysis to help combat global warming.

For further information and resources visit the online resources
www.oup.com/he/andrews_jelley4e

Numerical answers

Chapter 1

1.2 73 m

1.4 2×10^4 N

1.9 ~10 billion

1.13 Greater by (a) 3.5 (b) 10.7

1.14 (a) 2 °C (b) 1 °C

1.16 (a) 0.8% (b) 6.6%

1.19 ~0.65 $k kW^{-1}

1.21 (a) 6.28 c kWh^{-1} (b) 7.74 c kWh^{-1}

1.22 (a) 6.35 c kWh^{-1} (b) 7.81 c kWh^{-1}

1.24 3.08 c kWh^{-1}-> 2.59 c kWh^{-1} (b) 2.41 c kWh^{-1}
      (c) 2.46 c kWh^{-1} (d) 2.96 c kWh^{-1}

1.27 (a) 18 y (b) 28.6 y

1.31 (a) 68.7 MJ (b) 72.4 MJ (c) 72.1 MJ

Chapter 2

2.1 5×10^3 s

2.3 > 5.8 m for < 10% variation

2.4 0.59 m

2.5 $Re \approx 21000$

2.6 63.4 °C and 36.6 °C

2.8 74.6 W cf 1.03 W

2.9 ≈ 14.9 W

2.10 374 W cf 272 W

2.14 289 K

2.24 13.8%

2.25 1.37 J

2.27 371.5 K

2.29 1379 kJ mol^{-1}; 1247 kJ mol^{-1}

2.30 11370 kJ mol^{-1}

2.33 ≈ 0.44 m s^{-1}

2.36 10^4

2.40 6.67

2.43 $\approx 1.07 \times 10^6$ N

2.44 ≈ 221 m s^{-1}

Chapter 3

3.2 (a) 81.8 y (b) 38.7 y (c) 38.7 y

3.8 ≈ 15 kJ

3.9 861 kJ

3.11 (a) 0.56 (b) 778 kJ kg^{-1} (c) 342 kJ kg^{-1}
      (d) 1015 kJ kg^{-1}

3.12 (a) 20 kJ kg^{-1} (b) 3702 kJ kg^{-1} (c) 1882 kJ kg^{-1}
      (d) 0.50

3.13 854 K; 47%

Chapter 4

4.1 1.2×10^5 ha

4.7 ~600 litres

4.9 17.5 MtCo$_2$ y^{-1}

4.11 ~2.2×10^7 t

4.14 110 Mt of CO_2

4.15 ~2×10^4 sq km

4.16 0.28

4.19 (a) 5.3 Mt (b) 10.5 Mt

4.20 (a) ~5×10^5 ha (b) 2.15 Mt C (c) 2.3×10^7 sq km

4.21 0.75 kg

4.22 2.3 kgCO$_2$

4.24 (a) 38300 sq km (b) 44 Mt CO_2

4.26 19 million ha

Chapter 5

5.1 11.2 m

5.2 11560

5.3 597 K

5.6 3.0

5.7 (a) 35% (b) 57%

5.8 1700; 62%

5.10 ± 1.4°; 450; 142.7 MW

5.11 39.8%; 45.2%

5.12 28600 tonne; ~34%

5.15 2.2

5.16 0.43 MW

5.17 941 m

5.18 11 W

5.19 0.75 km; 4 bar

5.22 ~34 y

5.23 ~36 m; ~100 MW

Chapter 6

6.2 9.8 MW

6.3 62.8 MW

6.4 ½; 0.92

6.7 0.86

6.11 43.5

6.19 36500 TWh

6.21 (a) 44% increase (b) 44% decrease

6.25 20%; 10 m

6.26 1.6 MW

6.28 5.2 GW

6.30 463 m s^{-1}

6.32 12.5 m s^{-1}

6.37 10 m s^{-1}

6.39 25.6 MW

Chapter 7

7.1 9.85 MW

7.4 43 tonne-weight

7.9 3.57 m, 9.9°; 1.79 m, 2.6°; 1.19 m, 0.1°;
     0.89 m, −1.2°

7.12 3.6×10^8

7.13 97 − 131 Pa

7.14 1.68 MW

7.15 (a) 72 c kWh^{-1}; 21 c kWh^{-1} (b) 1372 c kWh^{-1}
      (c) 53 c kWh^{-1}; 16 c kWh^{-1}

7.16 (a) 0.028 MW ha^{-1} (b) 0.036 MW ha^{-1}

7.26 2.04 kS kW^{-1}

7.29 CF ~0.34

Chapter 8

8.1 42%

8.2 (a) 300 W m^{-2} 520 W m^{-2}

8.4 5777 K

8.12 (a) 84.6 mW; 3.61 Ω (b) 83.0 mW

8.13 ~4.7 m^2

8.14 13.65 V; 14.07 V; 14.32 V; 14.48 V; 14.62 V

8.15 ~250 ha

8.16 (a) 1.25 eV; 0.69 eV (b) 38.2 mW (c) 38.2%

8.18 50.4%

8.20 26.7% cf 18.0%

8.22 1.66 MW; 30.8%

8.25 1250 euro

8.26 (a) 4.3 c kWh^{-1} (b) (i) 5.5 c kWh^{-1} (ii) 6.8 c kWh^{-1}

Chapter 9

9.1 ~200 MeV

9.2 0.25 t

9.3 ~100 kt

9.4 48 kt; 98 y

9.7 (b) 0.77 m

9.9 ~32 h

9.11 65 s

9.12 1.6%

9.13 6.1%

9.14 (a) 1.7×10^4 TBq

9.15 (a) 6.8 × 10^7 m^{-2} s^{-1} (b) 400 µSv h^{-1}

9.16 27 y

9.17 (a) ~12 m^3 (b) ~320 m^3

9.28 2400 kg

9.30 1.45 × 10^{20}

9.31 12.5 keV; 1.9 × 10^6

9.32 8 × 10^7 K

9.33 ~2500 km

9.34 3.5 × 10^{21}

9.35 1.2 s

Chapter 10

10.1 64

10.4 11.3 A

10.6 (a) 3.3% (b) 0.25%

10.8 ~41%

10.9 (a) 2% (b) 0.01%

10.10 33

10.19 [CO$_2$] = 0.142; [H$_2$] = 0.142, [CO] = 0.058,
       [H$_2$O] = 0.058 mol litre^{-1}

10.20 2.00, 1.98, 1.92, 1.88, 1.80, 1.72 V

10.21 720 Wh

10.27 (a) 1685 MJ (b) 0.7 m^3; 380 kg

10.28 347 MW

10.29 100 km^2; 1.0 km^2

10.31 8.81 MJ; 2.45 kWh

10.32 2.33 × 10^3 MWh; 7.5 h

10.35 2.74 MJ

Chapter 11

11.1 122.6 kWh

11.2 1.05 W m^{-2} K^{-1}

11.3 1.07 W m^{-2} K^{-1}

11.4 (a) 21% (b) 22% (c) 17%

11.6 7920 Btu

11.7 (a) 3620 (b) 1517

11.8 21640 kWh

11.9 1.52 W $^\circ$C^{-1}m^{-2}

11.12 307 W $^\circ$C^{-1}

11.15 (a) 2.16×10^{15} J; 0.2 TWh

11.20 207.1 Mt CO_2; 282.4 Mt urea

11.24 (a) 165 kg MWh^{-1} (b) 198 kg MWh^{-1}

11.27 2.9 Gt CO_2

11.28 3.5 km

11.30 16 kWh; ~3.5 c kWh^{-1}

Chapter 12

12.21 ~1.9 $^\circ$C

Combined list of symbols and acronyms

A	albedo, area, or revenue	$\mathscr{L}^*$	lift force per unit length
a	wave amplitude or induction factor (wind)	L	characteristic length or latent heat
B	magnetic field	m	mass
$b(A)$	binding energy per nucleon	Nu	Nusselt number
C	capacitance, cost, concentration, or conductance	n	number, n-type, or neutron
		n	number density or refractive index
C_D	drag coefficient	P	power
C_L	lift coefficient	Pr	Prandtl number
C_p	power coefficient (wind)	$\mathscr{P}$	power per unit area
C_p	heat capacity at constant pressure	p	p-type
C_v	heat capacity at constant volume	p	pressure
c	specific heat, wave speed, wing chord, or speed of light	Q	heat, volume flow rate, quality factor, or reaction quotient
D	diameter	$\dot{Q}$	heat flow rate
$\mathscr{D}$	drag force	q	total speed
d	distance	R	gas constant, R-value, degree of reaction, resistance, wind turbine blade radius, or rate
E	energy		
E_g	band gap energy		
e^-	electron	Re	Reynolds number
e	electron charge	r	pressure ratio
F	flux or Faraday constant	S	entropy, shape factor, or solar flux
f	thermal utilisation factor	s	specific entropy or wing span
G	Gibbs energy, gravitational constant, or irradiance	T	temperature or thrust
		t	time
g	acceleration due to gravity	U	internal energy, thermal conductance or speed
H	enthalpy		
h	specific enthalpy, heat transfer coefficient, head, tidal range, Planck constant, or water depth	u, v, w	velocity components
		u	specific energy or speed
		V	volume, potential energy, voltage, or value
I	current, intensity, or moment of inertia		
I_C	photocell current	V_B	standard battery potential
I_F	forward current	V_g	band gap voltage
I_L	light-induced current	V_{OC}	open circuit voltage
I_S	saturation current	V_T	thermal motion voltage ($\equiv kT/|e|$)
I_{SC}	short circuit current	v	specific volume or velocity
J	current density	Wh	Watt hour
K	equilibrium constant or hydraulic conductivity	W	work, width, or total plasma energy
		Wp	Watt-peak
k	thermal conductivity, wave number, or neutron multiplication constant	w	wind turbine blade width
		Y	crop yield
$\mathscr{L}$	lift force	X	exergy
		x	steam quality or coordinate

x, y, z	coordinates	μ	coefficient of viscosity, permeability, or chemical potential
z_0	surface roughness parameter		
α	alpha particle	ν	kinematic viscosity or average number of fission neutrons
α	angle of attack		
α_s	wind shear coefficient	ρ	density, Larmor radius, reflectance, or resistivity
β	beta particle		
β	angle or radiative loss coefficient	σ	Stefan–Boltzmann constant or cross-section
γ	gamma-ray		
γ	ratio of specific heats	Σ_x	reaction x macroscopic cross section
Γ	circulation or width of excited state	τ	torque or mean lifetime
ε	emissivity, fast fission factor, back emf, dielectric constant, or efficiency	ϕ	angle, magnetic flux, or porosity
		Φ	neutron flux or magnetic flux linkage
η	efficiency or number of neutrons produced per neutron absorbed		
		Ω	solid angle or angular velocity
θ	angle	ω	angular velocity
κ	thermal diffusivity	ω_c	cyclotron frequency
Γ	circulation	ψ	streamline
λ	wavelength or tip speed ratio	^{o}	standard state symbol or degree

---

AD	anaerobic digestion	CRF	capital recovery factor
AM	air mass	CSP	concentrated solar power
ASHP	air-source heat pump	DAC	direct air capture
BAU	business-as-usual	DNI	direct normal insolation
BECC	bioenergy carbon capture	DSM	demand side management
BEV	battery electric vehicle	DSSC	dye sensitized solar cell
BIPV	building integrated photovoltaics	EfW	energy from waste
BOS	balance-of-system	EGS	enhanced geothermal system
C3	3-carbon plant-based molecule	EIA	Energy Information Administration or environmental impact assessment
C4	4-carbon plant-based molecule		
CAM	crassulacean acid metabolism (plant)	EOR	enhanced oil recovery
CAES	compressed air energy storage	EPBT	energy payback time
CAGR	compound annual growth rate	ERL	effective radiating level
CBA	cost benefit analysis	ERoEI	energy invested ratio
CCGT	combined cycle gas turbine plant	EV	electric vehicle
CCS	carbon capture and storage	FAME	fatty acid methyl esters (biodiesel)
CCU	carbon capture and utilization	FAO	Food and Agricultural Organisation of the UN
CDM	clean development mechanism		
CF	capacity factor	FER	fossil-fuel energy replacement ratio
CFCs	chlorofluorocarbons	FF	fill factor
CfD	contracts for difference	FIT	feed-in tariff
CHP	combined heat and power	FT	Fischer-Tropsch process
COP	Conference of the Parties or coefficient of performance	GDP	gross domestic product
		GE	genetic engineering
CO_2eq	carbon dioxide equivalent	GHG	greenhouse gas
CPC	compound parabolic concentrator	GSHP	ground-source heat pump
CPV	concentrated photovoltaics	GWe	gigawatts of electrical power

HAWT	horizontal axis wind turbine	O&M	operations and maintenance
HDI	human development index	OLR	outgoing long wavelength radiation
HEV	hybrid electric vehicle	OPEC	Organisation of Petroleum Exporting Countries
HHV	higher heating value		
HPP	hydropower plant	OPV	organic photovoltaic cell
HTF	heat transfer fluid	OSW	organic solid waste
HVAC	heating ventilation and air conditioning or high voltage alternating current	OTEC	ocean thermal energy conversion
		PCM	phase change material
HVDC	high voltage direct current	PEM	proton exchange membrane
ICE	internal combustion engine	PEV	pure electric vehicle
ICS	improved cooking stove	PHEV	plug-in hybrid electric vehicle
IEA	International Energy Agency	PPA	power purchase agreement
INDC	intended nationally determined contribution	PSH	pumped storage hydropower
		PSP	pumped storage plant
IPPC	Intergovernmental Panel on Climate Change	PV	photovoltaics
		PWR	pressurized water reactor
IRENA	International Renewable Energy Agency	R&D	research and development
ITER	International thermonuclear experimental reactor	RCBA	risk cost benefit analysis
		ROR	run-of-river
JET	Joint European Torus	R/P	reserve-to-production ratio
LCA	life-cycle analysis	SMR	small modular reactor
LCOE	levelized cost of energy	STEG	solar thermoelectric generation
LDV	light duty vehicle	TCRE	transient climate response to emissions
LED	light emitting diode	TEG	thermoelectric generator
LHV	lower heating value	TES	thermal energy store
LOCA	loss-of-cooling accident	TGC	tradable green certificate
LWR	light water reactor	TPV	thermo-photovoltaic cell
M&O	maintenance and operations	UNFCC	United Nations Framework Convention on Climate Change
MSW	municipal solid waste		
MVHR	mechanical ventilation with heat recovery	VAWT	vertical axis wind turbine
		WEC	World Energy Council
NDC	nationally determined contribution	WHO	World Health Organisation

Index